Intelligent Autonomous Control of Spacecraft with Multiple Constraints

Qinglei Hu · Xiaodong Shao · Lei Guo

Intelligent Autonomous Control of Spacecraft with Multiple Constraints

Springer

Qinglei Hu
School of Automation Science
and Electrical Engineering
Beihang University
Beijing, China

Xiaodong Shao
School of Aeronautic Science
and Engineering
Beihang University
Beijing, China

Lei Guo
School of Automation Science
and Electrical Engineering
Beihang University
Beijing, China

ISBN 978-981-99-0680-2 ISBN 978-981-99-0681-9 (eBook)
https://doi.org/10.1007/978-981-99-0681-9

This Springer imprint is published by the registered company Springer Nature Singapore Pte Ltd.
The registered company address is: 152 Beach Road, #21-01/04 Gateway East, Singapore 189721, Singapore

This work is dedicated to our families, who with love and encouragement, have always supported us in our academic research.

Preface

Aerospace technology is the strategic cornerstone of national security and modern national defense construction, and its comprehensive political, military and economic benefits are increasingly prominent. With the rapid development of space technology and continuous deepening of human space exploration, the world's space launch activities are showing an increasing trend year by year. Especially some emerging space missions, such as on-orbit servicing, spacecraft formation flying, deep space exploration, etc., have received considerable attention and investment from major space powers in recent years. For example, in April 2021, the Northrop Grumman Corporation and the company's wholly-owned subsidiary, SpaceLogistics LLC, successfully accomplished the docking of the "Mission Extension Vehicle-2" (MEV-2) to the "Intelsat 10-02" (IS-10-02) commercial communication satellite and the follow-up life-extension services; in December 2021, the National Aeronautics and Space Administration (NASA) launched the $10 billion James Webb Space Telescope (JWST) for deep space exploration; the European Space Agency (ESA) and NASA jointly initiated the space gravitational wave detection project "Laser Interferometer Space Antenna" (LISA). Compared with the traditional space missions like Earth observation, these emerging space missions impose higher autonomy, safety, and precision requirements for spacecraft attitude and orbital motions.

A key trait of spacecraft is the ability to autonomously plan attitude and orbital motion trajectories and track the trajectories afterwards in order to accomplish the given spaceflight missions, by using the Attitude and Orbit Control System (AOCS). The AOCS, as one of the most important subsystems onboard the spacecraft, is mainly in charge of measuring, estimating, and controlling the spacecraft attitude and orbital position using sensors and actuators for autonomous position-attitude determination and control as well as pre-programmed maneuvers. Figuratively speaking, the AOCS is the "brain" of spacecraft. In space applications, the spacecraft is usually required to autonomously and safely perform precision attitude and orbital motion in a complex space environment, while complying with multiple complex constraints. In this book, three kinds of constraints are considered: (1) state constraints; (2) physical constraints; and (3) performance constraints. State constraints mainly include sensor pointing constraints, path constraints, and linear/angular velocity constraints.

The pointing constraints can be further classified into forbidden pointing constraints and mandatory pointing constraints. The former arises from the fact that the optical sensitive payloads (e.g., infrared telescopes) should be kept away from direct exposure to the bright objects (e.g., the Sun) to avoid functional damage, while the latter is to keep the antenna pointed towards the ground stations or neighboring spacecraft for a stable communication link. Pointing constraints exist for many spacecraft, such as JWST, Hubble Telescope, Infrared Space Observatory (ISO), etc. Path constraints arise from safety concerns. More specifically, when spacecraft executing an orbital motion, it should move to the goal location along a safe path, in order to avoid collision with any obstacles, like the space debris or large structural components of other spacecraft (e.g., antenna and solar array). Linear/angular velocity constraints are caused by the limited range of velocity sensors or specific mission requirements. The second type of constraints—physical constraints—refer to actuator magnitude saturation caused by the limited output capability of actuators. The third type of constraints mainly include execution time and energy consumption constraints as well as control performance constraints. The execution time constraint is caused by "mission time-window", while the energy consumption constraint is due to the limited working medium carried by the spacecraft. In addition, space missions usually impose stringent transient and steady-state performance requirements (e.g., time-window, overshoot, and control accuracy) for the spacecraft AOCS. This is the so-called control performance constraints. The aforementioned constraints pose significant challenges for the AOCS design, and ensuring their satisfaction is crucial both for spacecraft safety and mission success.

The past decades have witnessed remarkable progress in the field spacecraft constrained motion planning and control. But nonetheless, most of the existing solutions have a limited constraint-handling capability and, moreover, are highly dependent on model information. However, the attitude and position dynamics of spacecraft usually tend to be uncertain and perturbed due to, for example, fuel consumption, payload motion, appendage deployment, and environmental disturbances (e.g., solar radiation pressure, aerodynamic forces, and magnetic disturbances). On the other hand, the actuators have a risk of occurring faults during their long-term operations, which may degrade the control performance and cause constraint violation or, even worse, lead to mission failure. These adverse factors necessitate the design of safe, reliable, and robust controllers with constraint-handling capability for the spacecraft AOCS. Thus far, it still remains an open problem to deal with multiple types of constraints in the presence of parameter uncertainties, multi-source disturbances, and actuator faults. Therefore, a further study is needed to develop advanced attitude and orbit control strategies with enhanced safety, reliability, robustness, and constraint-handling capability for emerging space missions. In recent years, the new generation of Artificial Intelligence (AI) technology has been vigorously developed and provides a promising way to enhance the autonomy and intelligence of spacecraft AOCS. It can endow the spacecraft AOCS with higher computational efficiency as well as stronger complex problem solving and constraint-handling capabilities.

In particular, with the development of onboard micro processors, the AI technology has attracted extensive attention in the aerospace industry and academia.

This book gathers a collection of the authors' recent research results that can reflect the theoretical and technological advances in the field of intelligent autonomous control of spacecraft under multiple constraints. The main contents of this book are as follows. For the spacecraft reorientation problem with pointing and angular velocity constraints, we report a date-driven adaptive control scheme, two adaptive robust Fault-Tolerant Control (FTC) schemes, and a Reinforcement Learning (RL)-based approximate optimal FTC scheme. By using artificial potential functions in conjunction with the adaptive and learning control techniques, the proposed methods enable the spacecraft to achieve the constrained pointing reorientation, despite the presence of inertia uncertainties and/or actuator faults. For the fault-tolerant attitude stabilization problem of spacecraft equipped with Control Moment Gyroscopes (CMGs), a neural network (NN)-based intelligent Fault Detection and Diagnosis (FDD) strategy is proposed, and by using the disturbance and fault estimation results, an adaptive fault-tolerant controller is designed to achieve active FTC. Later, an RL-based dynamic control allocation scheme is proposed for spacecraft attitude maneuver, with the aim of solving the problems of CMG singularity avoidance and energy saving. Considering the relative position tracking control problem for spacecraft formation flying, we develop an RL-based optimal event-triggered control algorithm using a single-critic NN, which not only achieves energy-optimal tracking control, but also greatly reduces the execution frequency of the control commands. We also consider the six-Degrees-of-Freedom (6-DOF) pose tracking control problem for spacecraft Rendezvous and Proximity Operations (RPOs) with spatial motion constraints (including the sensor field-of-view constraint, approaching corridor constraint, and relative linear/angular velocity constraints). Two Immersion and Invariance (I&I) adaptive pose tracking control schemes, a composite learning pose tracking control scheme, and a dual-quaternion-based intelligent pose control scheme are successively proposed to enable the pursuer spacecraft to safely and autonomously approach the desired position (near the target spacecraft) with a specific attitude, while complying with spatial motion constraints. Stability analysis and simulation results are provided to demonstrate their effectiveness and performance.

This book explores the intelligent autonomous multi-constraint control problems for different spacecraft applications, with the aim of providing a almost self-contained presentation of dynamics modeling, controller design and analysis, as well as simulation results. This book introduces the authors' up-to-date research works that can reflect the latest theoretical and technological advances in spacecraft intelligent autonomous control with multiple constraints. It provides a theoretical guideline for the AOCS design of modern spacecraft, and can serve as a reference guide for follow-up engineering developments in emerging missions like on-orbit servicing and deep space exploration.

Beijing, China
January 2023

Qinglei Hu
Xiaodong Shao
Lei Guo

Acknowledgements

The results in this book would not have been possible without the efforts and support of our colleagues and students. In particular, we are indebted to Prof. Yang Shi, Prof. Youmin Zhang, Prof. Maruthi R. Akella, Prof. Zheng H. Zhu, and Dr. Bowen Yi for their constant support and professional inspirations. We also acknowledge all laboratory members at Beihang University, especially the Masters & Ph.D. students Yuan Tian, Yuandong Li, Yongxia Shi, Haoyang Yang, Chao Duan, Haoran Li, Biru Chi, and Han Wu, for their contributions and efforts dedicated to this book. In addition, the authors would also like to thank the editors in Springer Nature for their great help to accomplish the publication of this book.

We greatly acknowledge the support of our research by the National Key R&D Program of China under Grants 2021YFC2202600 and 2021YFC2202603, the National Natural Science Foundation of China under Grants 62227812, 61960206011 and 62103027, the China Postdoctoral Science Foundation under Grant 2021M690300, and the Stable Supporting Fund of the Science and Technology on Space Intelligent Control Laboratory under Grant HTKJ2021KL502014. This book provides a almost self-contained presentation of our recent work in intelligent autonomous control of spacecraft with multiple constraints. Partial contents of this book are adapted from a number of our recent publications. We acknowledge IEEE, Elsevier Masson SAS, and Chinese Society of Aeronautics and Astronautics for granting us the permission to reuse materials from our publications copyrighted by these publishers in this book.

Contents

Acronyms

ACS	Attitude Control System
ADP	Approximate/Adaptive Dynamic Programming
ADRC	Active Disturbance Rejection Control
AI	Artificial Intelligence
AOCS	Attitude and Orbit Control System
APF	Artificial Potential Function
ARCSS	Autonomous Rendezvous and Capture Sensor System
ARPs	Airy-Rodrigues Parameters
ATPF	Appointed-Time Performance Function
BGF	Bounded Gain Forgetting
BIBO	Bounded-Input Bounded-Output
BLF	Barrier Lyapunov Function
CE	Certainty-Equivalence
CL	Concurrent Learning
CMG	Control Moment Gyro
COBE	Cosmic Background Explorer
COC	Constrained Optimal Control
CoM	Center of Mass
DE	Differential Evolution
DLR	German Aerospace Center
DOB	Disturbance Observer
DOBC	Disturbance Observer-Based Control
DOF	Degree-Of-Freedom
DQN	Deep-Q Network
DREM	Dynamic Regressor Extension and Mixing
DRL	Deep Reinforcement Learning
ECI	Earth-Centered Inertial
EM-VSD	Electric Motor-Variable Speed Drive
ESA	European Space Agency
ESO	Extended State Observer
ETC	Event-Triggered Control

FDD	Fault Detection and Diagnosis
FDI	Fault Detection and Isolation
FE	Finite Excitation
FOV	Field-of-View
FTC	Fault-Tolerant Control
FZ	Forbidden Zone
GA	Genetic Algorithm
GNC	Navigation, Guidance, and Control
HGA	High-Gain Antenna
HIL	Hardware-In-the-Loop
HJB	Hamilton–Jacobi–Bellman
HJI	Hamilton-Jacobi-Isaacs
I&I	Immersion and Invariance
iBLF	Integral Barrier Lyapunov Function
IDVD	Inverse Dynamics in the Virtual Domain
IE	Interval Excitation
IRAS	Infrared Astronomical Satellite
ISO	Infrared Space Observatory
ISS	International Space Station
JWST	James Webb Space Telescope
LiDAR	Light Detection and Ranging
LISA	Laser Interferometer Space Antenna
LMI	Linear Matrix Inequality
LOS	Line-of-Sight
LQR	Linear Quadratic Regulator
LRE	Linear Regression Equation
LTI	Linear Time-Invariant
LTV	Linear Time-Varying
LVLH	Local-Vertical Local-Horizontal
MPC	Model Predictive Control
MRAC	Model Reference Adaptive Control
MRPs	Modified Rodrigues Parameters
MW	Momentum Wheel
MZ	Mandatory Zone
NASA	National Aeronautics and Space Administration
NDO	Neural Network Disturbance Observer
NN	Neural Network
non-CE	Non-Certainty-Equivalence
PD	Proportional-Derivative
PDE	Partial Differential Equation
PE	Persistent Excitation
PID	Proportional-Integration-Derivative
PPC	Prescribed Performance Control
PPO	Proximal Policy Optimization
PSO	Particle Swarm Optimization

PWPF	Pulse-Width Pulse-Frequency
QCQP	Quadratically Constrained Quadratic Programming
RBF	Radial Basis Function
RL	Reinforcement Learning
RPOs	Rendezvous and Poximity Operations
RRT	Rapidly Exploring Random Tree
RW	Reaction Wheel
SAC	Simple Adaptive Control
SAMPEX	Solar, Anomalous, and Magnetospheric Particle Explorer
SCP	Sequential Convex Programming
SFF	Spacecraft Formation Flying
SGCMG	Single-Gimbal Control Moment Gyro
SMC	Sliding Mode Control
SOCP	Second Order Cone Programming
SVM	Support Vector Machine
TD	Tracking Differentiator
VSC	Variable Structure Control
w.r.t.	with respect to
XTE	X-ray Timing Explorer
ZOH	Zero-Order Hold

Chapter 1
Introduction

Aerospace technology is the strategic cornerstone of national security and modern national defense construction, and it is also an important indicator to measure a country's comprehensive strength. With the rapid development of space technology and its comprehensive benefits in politics, military and economy, many countries have increased their investment in this field, and issued a series of development strategies. With the rapid development of space technology and the increasing frequency of human space exploration, the world's space launch activities are showing an increasing trend year by year, and space missions are constantly developing towards diversification and unmanned autonomy. Emerging space missions, such as on-orbit servicing, formation flying, deep space exploration, etc., have received continuous attention and investment from major space powers in recent years. All these missions require the spacecraft to have high-precision and high-stability position and attitude control capabilities. At the end of 2021, the National Aeronautics and Space Administration (NASA) successfully launched the James Webb Space Telescope (JWST) (see Fig. 1.1), which costs tens of billions. The attitude control accuracy of its coarse-level platform is required to be better than $6.5''$, and the pointing stability is $1''/0.1\,\mathrm{s}$, while the precise-level imaging control accuracy is as high as $0.0073''$ [1]; Space gravitational wave detection projects, such as the Laser Interferometer Space Antenna (LISA) jointly carried out by the European Space Agency (ESA) and NASA, and "TianQin" and "Taiji" Projects of China, use three spacecraft to form an equilateral triangle configuration formation in space (see Fig. 1.2), and invert the source of gravitational waves in the middle and low frequency bands by measuring the distance change between the spacecraft. Among them, the LISA project requires the spacecraft attitude control accuracy to be better than $1 \times 10^{-7}\,\mathrm{rad} \cdot \mathrm{Hz}^{-\frac{1}{2}}$. Thus far, how to ensure that the spacecraft safely, autonomously, and precisely operates in orbit in the harsh space environment is a key issue in aerospace engineering.

A key trait of a spacecraft is the ability to autonomously plan its own motion trajectory and track the trajectory afterwards in order to accomplish specified flight tasks via jointly applying motion planning and control algorithms. The Attitude and

Q. Hu et al., *Intelligent Autonomous Control of Spacecraft with Multiple Constraints*,
https://doi.org/10.1007/978-981-99-0681-9_1

Fig. 1.1 Schematic of the JWST

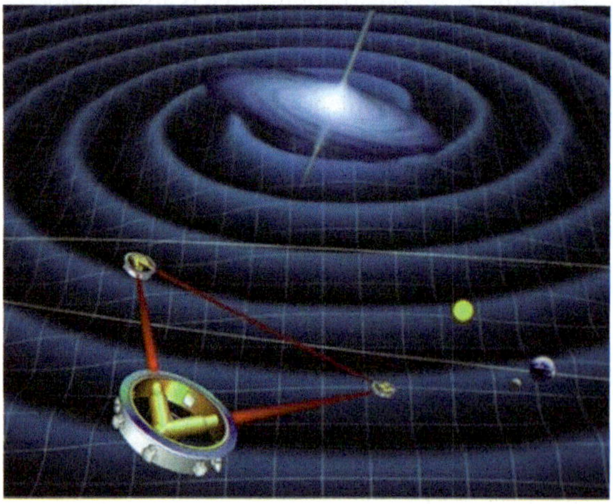
Fig. 1.2 Schematic of satellite formation for space gravitational wave detection

Orbit Control System (AOCS) is one of the most important subsystems on-board the spacecraft, whose main tasks are measuring, estimating, and controlling of spacecraft rotational and translational motions. It is particularly critical for the successful implementation of the assigned space missions, such as initial sun acquisition, staring imaging, rendezvous and docking, formation flying, etc. The AOCS of spacecraft is mainly composed of the following four parts [2]: measurement unit, planning unit, control unit and actuator, as shown in Fig. 1.3. The measurement unit uses spaceborne measurement sensors, such as star sensors, rate gyroscopes, Light Detection and Ranging (LiDAR), etc., to monitor the motion information (i.e., attitude, position, linear/angle velocity) of the spacecraft in real time and feed it back to the planning and control unit. The planning unit generates motion trajectories that satisfies various constraints, according to the requirements of the space missions. After that, the control unit outputs control signals for the actuators (e.g., reaction wheels, thrusters,

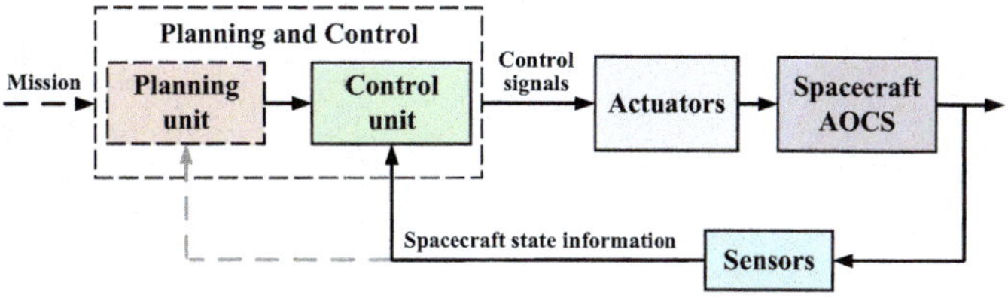

Fig. 1.3 Structure of the spacecraft attitude and orbit control system

and control momentum gyros) to regulate the spacecraft to track the desired trajectories. The spacecraft AOCS with high safety, strong autonomy, and high-precision is the key to ensure the stable and reliable operations of spacecraft in complex space environments and the safe implementation of specific space missions.

In space applications, the spacecraft is usually required to operate in a complex environment without any collision with obstacles, while complying with complex motion and physical constraints, such as actuator input saturation, sensor pointing constraints, approaching path constraints, linear/angular velocity constraints, which can be categorized into three types. The first type is state constraints, including pointing constraints, path constraints, and (relative) linear/angular velocity constraints. The pointing constraints can be further classified into forbidden pointing constraints and mandatory pointing constraints [3, 4]. A typical mission scenario with pointing constraints is shown in Fig. 1.4. The forbidden pointing constraints arise from the fact that some spaceborne sensitive payloads (e.g., infrared telescopes, laser interferometers, etc.) should always be kept away from direct exposure to the Sun vector or other bright objects [5], to avoid functional damage. For example, the Infrared Astronomical Satellite (IRAS) requires that the angle between the pointing of its infrared telescope and the Sun vector is not less than 60° to avoid direct light, so that it can work in the low temperature environment required for infrared observation. Similar pointing constraints also exist for the JWST, Hubble Telescope, Infrared Space Observatory (ISO), and Cassini Deep Space Explorer [6]. The scenario of mandatory pointing constraints is that the spacecraft undertaking tasks, such as relay communication, deep space exploration, formation flight, etc., needs to keep their high-gain antenna (HGA) pointed towards ground stations or neighboring spacecraft to maintain a stable communication link [7]. Path constraints mean that the spacecraft should move to the desired position along a safe collision-avoidance path. A typical scenario is that when the servicing spacecraft performs rendezvous and proximity operations with a tumbling target, it should approach the desired docking point along a certain approaching corridor (in general, a cone-shaped zone around the docking axis), while avoiding collision with any obstacles, for safety concerns [8, 9]. Linear/angular velocity constraints stem from specific mission requirements or the limited range of velocity sensors, such as rate gyroscopes, star sensors, LiDAR, etc. This requires the linear/angular velocity of the spacecraft to be maintained within a specific range [3, 10]. The second type of constraints is physical constraints, that is, actuator output magnitude constraints, due to the limited output capability of actuators (such as

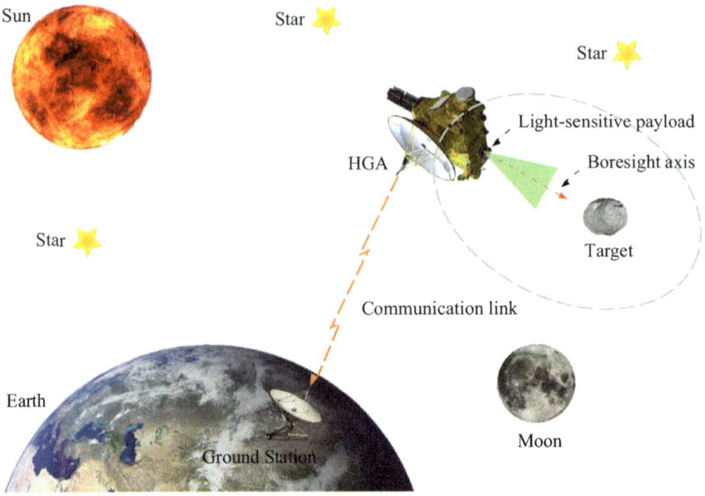

Fig. 1.4 A typical mission scenario with pointing constraints

reaction wheels, control moment gyros, thrusters, etc.) [11]. The third type of constraints is mission performance constraints, including time and energy consumption constraints and control performance constraints. The time constraint is caused by the short "mission time window" [12], while the energy consumption constraint is due to the limited working medium carried by the spacecraft [13]. On the other hand, the control performance constraints arise from the specific mission requirements (e.g., time window, overshoot, and control accuracy), which usually pose stringent requirements on the control algorithm design. These constraints should be taken into account in the design of AOCS to ensure the safe translational and rotational maneuvers of the spacecraft.

Although considerable progress has been made in the field of constrained motion planning and control, most of the existing methods have limited constraint-handling capability, and moreover, are susceptible to model uncertainties, multi-source disturbances, actuator faults, and some other adverse affects, which may degrade the control performance and cause safety issues or, even worse, lead to mission failure. This necessitates the design of safe, reliable, and high-performance three-degrees-of-freedom (3-DOF) attitude/position controllers or 6-DOF pose (i.e., concurrent position-attitude) controllers with strong constraint-handling capability. While a significant challenge arises when some of those issues are treated simultaneously. Despite recent advances, a further study is still needed to develop advanced control schemes with enhanced autonomy, safety, robustness, constraint-handling capability for space applications. With the development of on-board micro processors and efficient algorithms, the artificial intelligence (AI) technology represented by machine learning has made great progress in recent years. When compared with the traditional technologies, AI technology has obvious advantages in terms of efficiency, complex problem solving, and constraint handling. It plays an increasingly important role in the high-tech field, especially in the aerospace field. This book aims to gather a collection of our recent research works that can reflect the theoretical and technological advances in safe autonomous control of spacecraft under multiple constraints, and provide theoretical tools for the design of AOCS of modern spacecraft. Before proceeding with the introduction of the main contents of this book, we shall review the state of art spacecraft motion planning and control methods along with possible future trends.

1.1 Review of Spacecraft Motion Planning

To meet the increasing demands for current and near-future space missions such as deep space exploration, on-orbit servicing, etc., the spacecraft should be equipped with more advanced payloads. Therefore, the design of motion planning algorithms should not only consider the mission objectives, but also take into account various state and physical constraints introduced by payloads, actuators, and other aspects. For example, the ISO satellite designed a maneuver strategy that can protect payloads from the impact of earth thermal radiation source [14]; Cassini satellite required an attitude avoidance mechanism to protect the onboard sensor from solar irradiation [15]; the solar arrays should be pointed to the Sun for power supply; The X-ray Timing Explorer (XTE) satellite required that its maneuver speed should not exceed the upper limit of gyro measurement. Therefore, multi-constraint motion planning is an urgent key technology for spacecraft attitude and position control. The existing multi-constraint motion planning algorithms can be generally divided into six categories: geometric method, artificial potential function (APF) method, discretized method, randomized sampling method, optimization-based method, and intelligent method. These methods can handle different types of constraints, and have different optimization capability and computational efficiency. Table 1.1 briefly summarizes their

Table 1.1 Constraint-handling capability comparison of different planning methods

Methods	Pointing constraint	Angular velocity constraint	Actuator saturation
Geometric method	Single constraint	✓	✓
APF-based method	Multiple constraints	✓	✗
Discretized method	Multiple constraints	✗	✗
Randomized method	Multiple constraints	✗	✗
Optimization-based method	Multiple constraints	✓	✓
Intelligence-based method	Multiple constraints	✓	✓

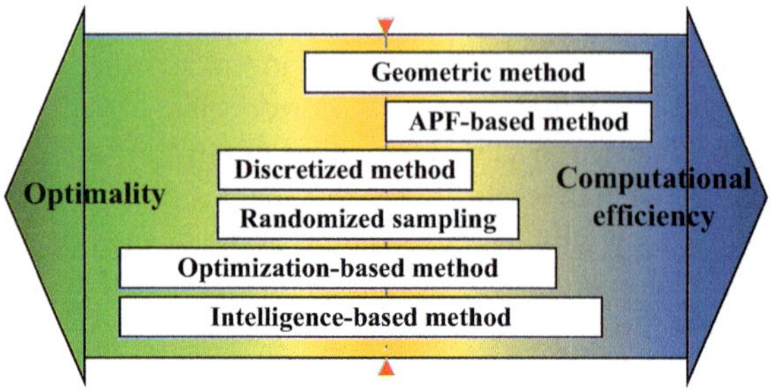

Fig. 1.5 Optimality and computational efficiency comparison of different planning methods

capabilities of handling attitude pointing constraints, angular velocity constraints, and actuator saturation. Figure 1.5 compares the optimality and computational efficiency of these methods.

1.1.1 Geometric Method

By establishing the geometric relationship between the sensor's boresight axis and the specific pointing directions, the geometric methods can effectively deal with attitude pointing constraints. Generally speaking, the geometric method first designs an optimal control strategy that can accurately drive the spacecraft from the initial attitude to the desired attitude in an unconstrained environment. Then, it finds an intermediary attitude so that the unconstrained control strategy can be directly applied to a two-segment path without violating constraints. The geometric method is simple to design and can obtain analytical planning strategy, which has significant advantages in computational efficiency.

For the solar, anomalous, and magnetospheric particle explorer (SAMPEX), Frakes et al. [16] explored a method to avoid the velocity direction of stars, which reduces the influence of space debris on the exploration precision. Considering the avoidance of bright celestial bodies and the communication to ground, Hablani [7] proposed two alternative attitude maneuver schemes for minimum rotation path so that the sensor boresight could bypass the forbidden zones; moreover, a pointing-mandatory strategy is further designed for spacecraft to keep the communication to ground. In particular, for an under-actuated spacecraft, non-standard Euler axis/angle parameters were used in [17, 18] to describe the vertical geometric relationship between the actuated and under-actuated axes. Two two-step tangent attitude maneuver strategies were proposed to avoid a certain pointing constrained zone. Ayoubi et al. [5] proposed an optimal geometric planning strategy for the avoidance of Sun.

Compared with the traditional geometric methods, differential geometric methods simplify the planning problem with help of Lie algebra, vector field, and some other tools. As shown in Fig. 1.6, complex motions in Euclidean space can be significantly simplified by using the special Euclidean group (SE(3)) and dual unit sphere (S_D^2). In addition, the geometric discretization method can preserve the invariance and sym-

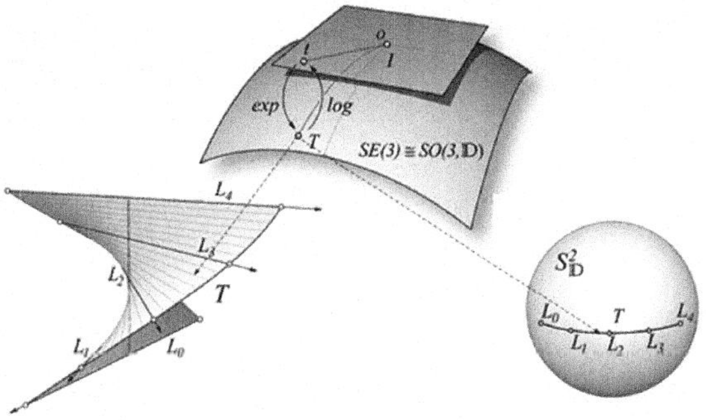

Fig. 1.6 Line motion in Euclidean space, SE(3), and S_D^2

metry of continuous time system, which can improve the accuracy and convergence of nonlinear programming algorithm. Based on the differential geometry method, Sorensen [14] analytically derived key parameters of maneuver trajectory for the ISO satellite, which avoids the effects arising from the Sun and Earth thermal radiation on observation missions. Spindler [19] proposed a geometric planning strategy for arbitrary forbidden pointing constraints.

Biggs et al. [20] proposed a semi-analytic optimal planning strategy under the special orthogonal group (SO(3)). By optimizing weight factors and control parameters, it can restrict the system trajectory and control amplitude. Henninger et al. [21] converted the optimal planning problem into a two-point boundary value problem and derived the analytical solution based on Lax pair. Combining the APF method and sliding mode control, Geng et al. [22] proposed an attitude control law based on the dual unit groups, and obtained an analytical form of system trajectory in the sliding stage. Then, by adjusting the initial states and control parameters, the curvature of system trajectories can be tuned to avoid obstacles.

Based on the above argument, the geometric planning method can obtain analytical or semi-analytical optimal maneuver paths, thus having advantages in computational efficiency. Moreover, it can remain the group structure and physical properties and, therefore, is conducive to numerical calculation. However, the geometric method has a limited ability to deal with multiple constraints, and can only consider relatively single optimization index.

1.1.2 Artificial Potential Function Method

The APF method is a very effective tool to deal with kinematic constraints. Its basic idea is shown in Fig. 1.7. Consider a target moving in an artificial potential field, where the obstacles has high potentials and the desired position has the lowest potential. Intuitively, if the control force is designed according to the negative gradient direction of the potential field, the target can be driven to the desired goal, while avoiding the constrained regions. The APF-based controller does not need to plan the complete

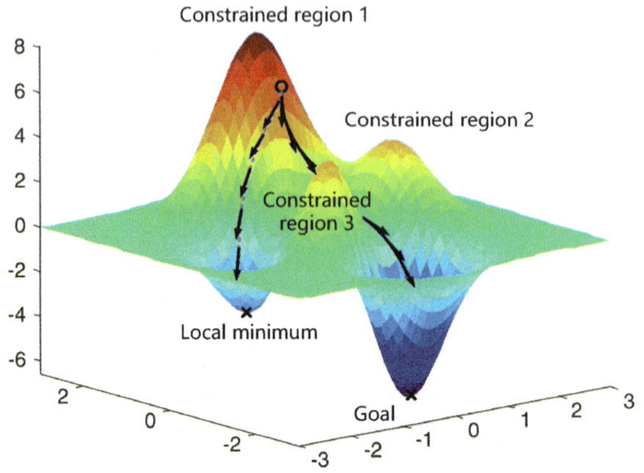

Fig. 1.7 Illustration of artificial potential fields

system trajectory in advance and has a good real-time performance. Some scholars constructed a variety of APFs to form a specific potential field to achieve obstacle-avoidance. The commonly used APFs are exponential function (including Gaussian function), quadratic function, logarithmic function, navigation function, etc.

Mclnnes [23] used an unambiguous Gaussian function to establish high potential regions for bright celestial direction in space, and proposed a bounded attitude control strategy. Wisniewskia and Kulczycki [24] established the attitude dynamics of Hamiltonian form using the quaternion representation, and designed an attitude constrained control law using the energy shaping method. Considering multiple attitude forbidden and mandatory zones, Lee et al. [4, 25] proposed a series of APF-based methods, which not only achieve smoothness and approximate global convergence of the control laws, but also ensure the feasibility and stability of multi-constraint problem solving. In addition, the APF methods can deal with angular velocity constraints. Aiming at the rest-to-rest attitude maneuver problem of spacecraft, Shen et al. [26] designed a logarithmic potential function using the sliding mode variables, such that the system has a high potential when the angular velocity approaches to the barrier boundary. Hu et al. [27] and Shao et al. [3] considered, respectively, the spacecraft reduced-attitude system and uncertain rigid-body attitude dynamics, and designed logarithmic potential functions to deal with angular velocity constraints. The APF methods are also used by the authors' group to solve the problems of attitude maneuvers and close-range proximity operations of spacecraft. For the rendezvous and proximity operations, Dong et al. [28] used the dual quaternion to design a 6-DOF pose control law. By constructing a new potential function, the proposed method ensures that the satisfaction of both the field-of-view (FOV) and approaching corridor constraint during the proximity operations. Considering multiple attitude-forbidden constraints, Hu et al. [29] presented a logarithmic potential function to avoid the unwinding phenomenon inherent in the quaternion representation. Hu et al. [30, 31] designed a navigation function for spacecraft constrained attitude maneuvers, and

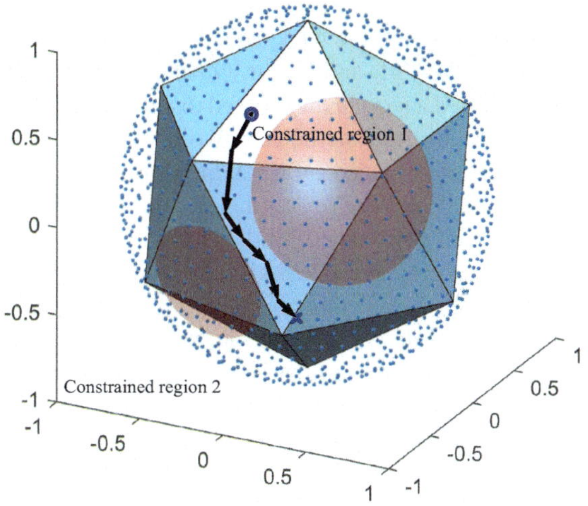

Fig. 1.8 Discretization of the attitude sphere and constraints

proved that its minimum point infinitely approaches the system equilibrium point, without the need for transforming the attitude constraints into convex ones [30, 31].

From the above analysis, it can be seen that the APF methods can effectively deal with multiple attitude/path constraints and does not depend on the global information. Thus, this kind of methods can achieve dynamic planning. In addition, they can be directly used in the design of analytic control law, and has advantages in terms of real-time computation. However, a well-known drawback of the APF methods is the local minimum problem, which may cause the system trajectory residing at a local minimum point, thus failing to achieve the given control objective.

1.1.3 Discretized Method

Compared with the analytic features shared by the aforementioned methods, a series of numerical optimization methods are mainly introduced in the following. These methods have stronger trajectory optimization abilities and provide new ideas for solving multi-objective multi-constraint planning problems. Firstly, the discretized method is introduced, which uses the polyhedron (such as cube and icosahedron) to discretize the attitude unit sphere uniformly, converts the pointing vector into integers, and establishes the topological structure for the pointing vector. Then, the constrained set is transformed into the pixel set on the unit sphere by path planning algorithm. Finally, a cost function in terms of the path length is defined, and the path searching algorithms are used to obtain the near-optimal path, as depicted in Fig. 1.8. After obtaining the desired path, a series of attitude describing the rotation sequence is established, and the desired trajectory can be further realized through a feedback controller.

At present, the existing methods for the discretization of attitude unit sphere mainly include cosmic background explorer (COBE) cubic algorithm, icosahedron algorithm [32–34], and so on. On the basis of COBE cubic algorithm, Tegmark [32] proposed a icosahedron-based discretization method. Compared with the COBE cubic algorithm, this method has two advantages: (1) square pixels in the COBE cubic algorithm are replaced by hexagonal pixels, thus requiring fewer pixels for space discretization; (2) each pixel has six similarly equidistant neighbors, which makes the generated trajectory smoother. Considering the sensitivity of optical instruments carried by Bevo-2 or ARMADILLO cubic satellite to strong light, Kjellgerg [33, 34] proposed a path planning method to handle pointing and control constraints, using the icosahedral discretization algorithm. In this method, continuous points were discretized into regular hexagon pixels connected to icosahedron in the attitude unit sphere, and A^* graph-searching algorithm was used to search the feasible paths. However, this method can only consider a single sensitive instrument and lacks the ability of attitude trajectory planning for multiple sensitive payloads. To solve this problem, Tanygin [35, 36] proposed the definition of ARPs (Airy-Rodrigues Parameters) and a new minimum deformation projection. Attitude constraints are represented and discretiezed by the new parameters, and the optimal path planning is achieved by using the improved A^* algorithm.

Discretized method has a clear physical meaning and has been widely used in the field of robot motion planning. The discretized method can effectively deal with attitude pointing constraints and actuator saturation, while achieving the shortest maneuver path in the probabilistic sense. However, it also has obvious disadvantages. On the one hand, it depends on the structure of off-line solution; on the other hand, compared with the analytical methods (i.e., geometric methods and APF methods), the discretized method requires a larger amount of calculation. As such, it is difficult to realize dynamic planning, thus limiting its practical application.

1.1.4 Randomized Planning Method

The planning method based on space discretization in Sect. 1.1.3 is more suitable for path planning problems in low dimensional space. In such problems, it often has a good completeness, but requires a complete environment model; moreover, in high-dimensional spaces, it is easy to cause the "curse of dimensionality". To overcome these disadvantages, the randomized planning method was proposed in [37], which replaces completeness with probabilistic completeness to improve search efficiency and is suitable for the planning problems in high-dimensional spaces. The key idea behind this method is to randomly generate state points in a feasible space, and then find a group of feasible connections between these state points to get a planning trajectory [38], as shown in Fig. 1.9. Feron et al. [39] proposed a randomized attitude planning method using the Rapidly Exploring Random Tree (RRT), and established a local guidance law based on the Lyapunov function, which achieves spacecraft large-angle attitude path planning under various types of attitude constraints. Yershova et

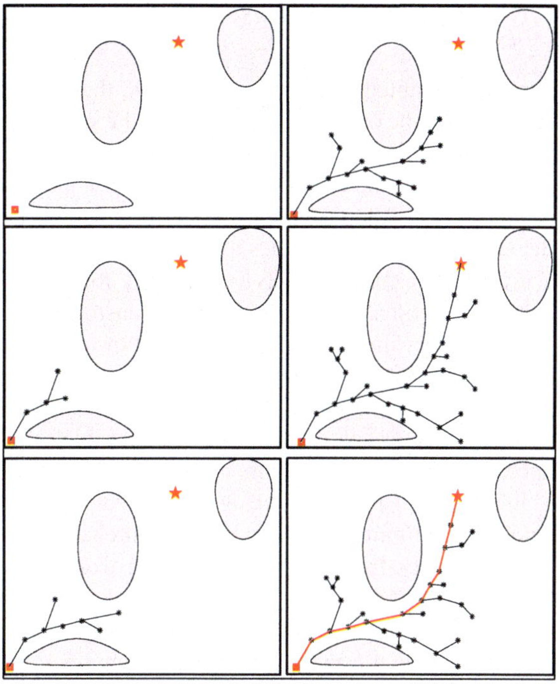

Fig. 1.9 Illustration of the randomized planning method

al. [40] proposed a deterministic sampling algorithm in the rotation matrix space. This resolution-complete sampling method provides a new direction for the development of randomized planning algorithms. It is noted that using the deterministic sampling method to replace the completely random sampling method in the traditional RRT algorithm can reduce the time used by the RRT algorithm to a certain extent.

The advantage of randomized planning method is that it can deal with various types of motion constraints with high computational efficiency, with the aid of the probability-based method. However, due to the mechanism of random probability, this kind of methods can only plan the spacecraft to a neighborhood of the desired state in the sense of probability. The state deviation converges exponentially to zero as the number of vertices in the state graph increases. Similar to the discretized method in Sect. 1.1.3, the randomized planning algorithm is also difficult to deal with multiple different types of constraints. Moreover, most of these methods only focus on planning a feasible safe path, while ignoring the dynamics constraint and angular velocity constraints, so it is difficult to achieve optimality.

1.1.5 Optimization-Based Method

Due to the limited fuel and time window, the spacecraft attitude and position controller should be optimized according to the specific performance index, while complying with various constraints. The numerical optimization method is widely used to achieve time and/or energy optimal motion planning of spacecraft. The core idea of this method is to take the given performance index as an objective function, take the kinematics and dynamics as equality constraints, and characterize all kinds of state and physical constraints as inequality constraints. Within this setting, the multi-constraint optimal control problem is transformed into the numerical optimization problem under multiple constraints, as shown in Fig. 1.10.

Optimization-based methods have been widely used for spacecraft motion guidance and control under multiple constraints. Zhao et al. [41] studied the problem of optimal trajectory planning for aerospace vehicles with nonconvex collision avoidance constraints. This problem is solved in the framework of sequential convex programming (SCP) by introducing a specified performance index. Boyarko et al. [42] studied the minimum-time and minimum-energy optimal trajectory planning problem for spacecraft rendezvous with a passive tumbling target, with explicit consideration of both the translational and rotational dynamics, and a direct collocation method was proposed based on the Gauss pseudo-spectral approach. Leomanni et al. [43] investigated the trajectory planning problem for spacecraft autonomous rendezvous and docking with a space tumbling target. The involved optimization problem, which is nonconvex with nonlinear constraints, was effectively tackled by solving a finite number of linear programs. To solve the optimal control problem for spacecraft

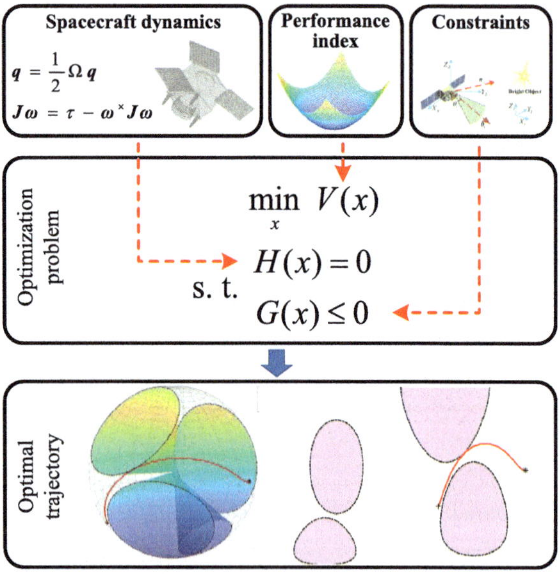

Fig. 1.10 Illustration of the numerical optimization method

attitude maneuvers, in [44, 45] the optimal control problem was transformed into a two-point boundary value problem using the indirect method. The discretized method of Lie group variation integral is helpful to improve discretization accuracy as well as solving accuracy and efficiency. As a relatively mature optimization problem solver, convex optimization technology has been extensively studied and employed in spacecraft motion planning [46]. Aiming at the constrained attitude reorientation tasks of typical satellites like Planck space observation satellite, Kim and Mesbahi [47] transformed the attitude constraints into constraints in the form of convex linear matrix inequality (Linear Matrix Inequality, LMI), and proved that the transformed problem is equivalent to the original problem. Kim et al. [48] solved the convex optimal planning problems for spacecraft attitude maneuvers, and performed constraint convexity for more complex constraints including soft constraints, dynamic constraints, and mixed constraints. Sun and Dai [49] proposed an iterative algorithm using the quadratically constrained quadratic programming (QCQP) technique. On the basis of [47], the attitude forbidden and mandatory constraints were transformed into quaternion inequality constraints, and angular velocity constraints and actuator saturation were also considered. Moreover, the semi-definite relaxation technique was used to transform the problem into a semi-definite programming problem with rank constraint, for which a rank-minimization iterative algorithm was proposed and convex optimization was used to obtain the optimal attitude maneuver trajectory. Tam and Lightsey [50] used the mixed integer programming approach to solve the spacecraft attitude planning problem under multiple constraints.

The aforementioned numerical optimization methods are to obtain the optimal motion trajectory under specific conditions by describing an optimal problem and solving its numerical solution. However, solving this kind of problem is time-consuming and needs high-level hardware support, especially when the constraints are nonconvex, nonlinear and coupled. On the other hand, such methods are model-dependent, and their solutions often depend on accurate model information.

1.1.6 Artificial Intelligence-Based Method

As discussed in Sect. 1.1.5, the numerical optimization methods have disadvantages of low solving efficiency and high model dependence. It is difficult to effectively solve complex problems, such as the "grey/black box problem" whose model is partially unknown, and the non-convex game problems. It is worth noting that recent advances in artificial intelligence (AI) technology provide theoretical, methodological and technical supports for spacecraft intelligent autonomous planning. Meta-heuristic-based intelligent optimization methods have been extensively studied in spacecraft attitude maneuver, for example, genetic algorithm (GA) and its extension [51, 52], particle swarm optimization (PSO) [53–55], differential evolution (DE) [56], etc. Benefiting from the characteristics and mechanism of bionics, these intelligent algorithms can alleviate the local minimum problem caused by non-convexity.

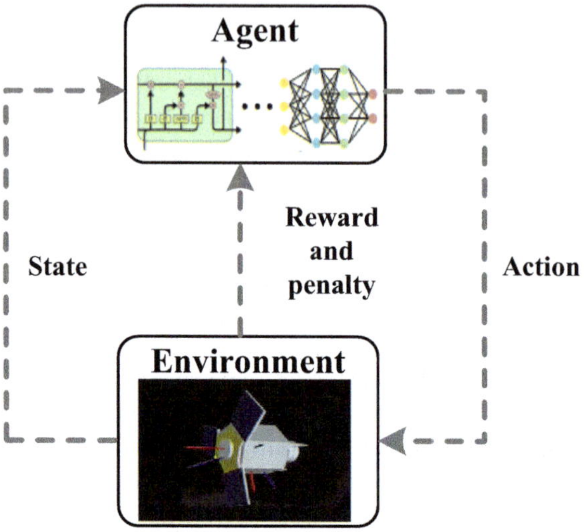

Fig. 1.11 Illustration of reinforcement learning method

In recent years, the new generation of AI technology represented by reinforcement learning (RL) has attracted considerable attention and has been extensively applied to spacecraft systems. The basic idea of RL is shown in Fig. 1.11. By constructing the environmental feedback ("reward" or "penalty"), the agent is driven to learn and adjust the current strategy to obtain the control policy that can maximize the total reward. By visualizing the desired motion states as "reward" and characterizing the constraint-violating motion states as "penalty", the RL technical provides an effective way to deal with motion constraints. As a branch of AI, deep reinforcement learning (DRL) also serves as an efficient technology for motion planning under multiple constraints. Oestreich et al. [57] developed a spacecraft docking policy and implemented it as a feedback control law. A reward function was designed to avoid collisions and minimize the control and error costs, and then the agent was trained to maximize the total reward. Hovell and Ulrich [58] introduced a guidance strategy for spacecraft proximity operations, which leveraged DRL to derive a velocity commands. It is verified through experiments that the learned guidance law can achieve spacecraft docking to a rotating targets while avoiding some obstacles. Qu et al. [59] studied the relative position tracking problem of the autonomous spacecraft rendezvous under the requirement of collision avoidance. Deep deterministic policy gradient (DDPG) algorithm was employed in conjunction with meta-learning to train a control scheme for this mission. In [60], a set of attitude stabilization strategy were designed for non-cooperative target capture based on the deep-Q network (DQN) algorithm. This method uses historical data to train the networks and does not depend on the target's mass, thus having relatively strong intelligence and adaptability. Based on the proximal policy optimization (PPO) algorithm, an attitude maneuver strategy was proposed in [61], which significantly improves performance without requiring the inertia information. Elkin et al. [62] proposed a new RL framework for spacecraft attitude maneuvers, in which a high-fidelity digital simulation environment was used for network training. However, most of these methods focus on the attitude maneuvers without constraints.

Dong et al. [63] designed an online learning algorithm for spacecraft constrained attitude maneuvers in the framework of adaptive dynamic programming (ADP). By combining the historical data with real-time measurement data, the Hamilton–Jacobi–Bellman (HJB) equation corresponding to the optimal problem is solved online, and the attitude maneuver strategy is updated in real-time, thus achieving attitude reorientation under forbidden pointing and angular velocity constraints. Yang et al. [64] further considered actuator misalignment, and improved the online learning strategy to achieve high-precision attitude maneuvers under forbidden pointing constraints. The proportional-differential (PD) control law was designed to obtain an initial planning strategy, which is updated in real-time through online reward feedback. This method achieves attitude avoidance with low computational costs and online performance optimization. For spacecraft rendezvous and proximity operations, Hu et al. [65] proposed an ADP-based control scheme using the dual-quaternion representation to obtain an approximate optimal solution. Both FOV and path constraints are dealt with by designing a special cost function.

The AI technique like RL provides a new way to solve the multi-constraint spacecraft motion planning problems. In the aforementioned works, constraints are dealt with by constructing environmental rewards. How to design an efficient and reasonable reward mechanism according to the practical mission demands is still an open problem. In addition, fast online learning using small sample data is preferable for on-orbit applications. Thus, the design of efficient learning algorithms is a key technology in future works.

1.2 Review of Spacecraft Attitude and Position Control

Attitude and position control is one of the key technologies to ensure the stable flight of spacecraft. In actual engineering, proportional-integral-derivative control (PID) has been widely used in the design of spacecraft AOCS, due to its simple structure, intuitive parameter adjustment, and easy implementation. However, traditional PID control cannot effectively deal with parameter uncertainties, multi-source disturbances, and actuator faults, while failing to satisfy some underlying state and physical constraints. Therefore, it is difficult to meet the development requirements of emerging space missions for the AOCS with high autonomy, high reliability, and high precision. In recent years, many scholars at home and abroad have carried out theoretical and application research on the problems of spacecraft attitude and position control, and have proposed a series of advanced control methods for safe autonomous attitude and position maneuvers of spacecraft. The following will introduce the recent progress of 3-DOF attitude/position control and 6-DOF pose control methods, with particular attention to adaptive control, anti-disturbance control, fault-tolerant control, state-constrained control, and intelligent control.

1.2.1 Adaptive Control of Spacecraft

The mass and inertia uncertainties of the spacecraft caused, for example, by fuel consumption, payload variation, and appendage deployment, may give rise to performance degradation, and even lead to system instability. Adaptive control can effectively deal with parameter uncertainties, through online estimating the unknown parameters. The traditional adaptive control methods are based on the certainty-equivalence (CE) principle, the basic idea of which is as follows [66]. First, the feedback controller is designed under the assumption that the uncertain parameters are known. Then, an estimated value is used to replace the uncertain parameters in the control law, and the corresponding adaptive law is derived to update the estimated values online, while ensuring the stability of the closed-loop system and the convergence of output errors. Due to the simple and intuitive ideas, the CE-based adaptive control has been widely applied in spacecraft attitude and position control [67–69]. Under the CE framework, the adaptive law is mainly designed to precisely cancel the additional "disturbance" term introduced by the parameter uncertainties. However, when the parameter estimates fail to converge to their true values, this fragile cancellation operation tends to degrade the transient performance and robustness of the closed-loop system. But, in practice, parameter convergence can be achieved only if the reference trajectory satisfies the persistent excitation (PE) conditions.

In order to improve the performance of adaptive control, some scholars have deviated from the CE principle and proposed a non-certainty-equivalent (non-CE) adaptive control method based on Immersion and Invariance (I&I) technology. The I&I methodology was first proposed by Astolfi and Ortega in [70] based on the concept of differential geometry. The I&I adaptive control method mainly constructs a differential manifold by adding an additional function item in the parameter estimation, and designs the control law and the adaptive law to make the constructed manifold invariant and attractive. It deals with parameter uncertainties in a robust way. It should be pointed out that the I&I adaptive control needs to solve a specific partial differential equation (PDE) to form the differential manifold, but for multi-input nonlinear systems such as spacecraft attitude dynamics, such a PDE usually has no analytical solution, that is, the so-called "integrability obstacle". This obstacle greatly limits the applicability of the traditional I&I adaptive control methods in spacecraft attitude and position control. To overcome this problem, Seo and Akella [71] proposed a regressor filtering based I&I adaptive control method for spacecraft attitude tracking. By introducing the state and regressor filters, an augmented filtering system is constructed and then the analytical solution of the PDE is directly gave using the filtered states and regressor matrix. In [9] and [72], I&I adaptive attitude and pose controllers were developed, respectively. Karagiannis et al. [73] proposed the dynamic scaling technology, which provides an alternative idea for overcoming the integrability obstacle. More specifically, the integral variables in the original regressor matrix are replaced with the corresponding filtered ones to make the regressor matrix integrable; after that, an approximate solution of the ODE is given, and the dynamic scaling factor is designed to compensate for the approximate error.

Yang et al. [74] applied this method to the spacecraft attitude tracking control, and improved the dynamic gains, thus effectively avoiding the high gain issue. Wen et al. [75] further improved the dynamic scaling factor, which removes the requirements of exact knowledge on the minimum eigenvalue of the inertia matrix that required in [74]. Xia and Yue [76] proposed a dynamic scaling based I&I adaptive control method for anti-unwinding attitude stabilization of spacecraft, and introduced a saturated scaling factor so that the controller does not require any dynamic gain. Shao et al. [3, 10] proposed dynamically scaled I&I adaptive control schemes, which address the constrained attitude reorientation and pose tracking maneuvers of spacecraft, despite the presence of parameter uncertainties.

Note that both the CE and non-CE I&I adaptive control methods can achieve parameter convergence only if the regression matrix satisfies the restrictive PE condition [77], which is rarely met in practical applications. In recent years, many scholars at home and abroad have carried out research on relaxing the dependence of parameter convergence on the PE conditions. Chowdhary and Johnson [78] proposed a concurrent learning (CL)-based adaptive control scheme under the framework of model reference adaptive control (MRAC). By simultaneously using rich historical data and current measurement data to design a parameter learning law, which can ensure parameter convergence when the system states only satisfy the strictly weaker finite excitation (FE) condition than PE. Zhao and Duan [79] proposed a saturated finite-time CL adaptive control scheme for 6-DOF attitude and position tracking of combined spacecraft. Later, [80] further extended the results in [79] to the attitude tracking problem of combined spacecraft, with explicit consideration of external disturbances. Since the CL algorithm requires data selection and storage procedures, it is thus necessary to select an appropriate recording scheme, in order to make the stored data contain as many linearly independent elements as possible. Of course, this inevitably increases the algorithm complexity. Moreover, the traditional CL-based adaptive control methods need unmeasurable state derivatives (e.g., angular/linear acceleration) to construct parameter estimation errors. Although online estimation of state derivatives can be performed using fixed-point smoothers, sliding mode differentiators, etc., they are sensitive to measurement noise. To avoid the use of unmeasurable state derivatives, Cho et al. [81] proposed a composite adaptive control method that can effectively avoid the usage of state derivative information and achieve parameter convergence under the FE condition; moreover, it adopts the integral of filtered regressor for data storage, thus avoiding online data selection. Based on a similar idea to [81], Pan and Yu [82] proposed a composite learning control scheme and successfully applied it to the control of manipulators. Aiming at the problems of spacecraft attitude tracking and multi-constraint attitude reorientation, Dong et al. [83] and Shao et al. [3, 84] proposed several adaptive and learning control methods based on the idea of composite learning, which simultaneously achieve high-precision attitude tracking/reorientation and online parameter identification.

1.2.2 Anti-Disturbance Control of Spacecraft

In practice, the spacecraft dynamics usually tend to be disturbed. The multi-source disturbances may affect the accuracy and stability of the spacecraft AOCS, which necessitates the design of controllers with strong robustness against disturbances. The existing spacecraft anti-disturbance control methods can be classified into three categories: (1) robust control methods; (2) active disturbance rejection control (ADRC) methods; (3) disturbance observer-based control (DOBC) methods. Robust control emerged in the late 1970s, initially to solve the control problems in the aerospace engineering. Until the famous DGKF method was published in 1988 [85], the relevant theory became mature. Liu et al. [86] proposed a non-fragile robust $\mathcal{H}_\infty$ control method to solve the problems of attitude stabilization and vibration suppression of flexible spacecraft in the presence of modeling uncertainties, control perturbations, external disturbances, and control input limitations. Some scholars proposed a hybrid $\mathcal{H}_2/\mathcal{H}_\infty$ robust attitude control method by combining the $\mathcal{H}_2$ and $\mathcal{H}_\infty$ robust control methods [87, 88]. Aiming at the problem of spacecraft attitude tracking subject to external disturbances, Luo et al. [89] proposed a $\mathcal{H}_\infty$ inverse optimal controller to achieve the $\mathcal{H}_\infty$ optimal to disturbances. Wang and Li [90] proposed a robust optimal control strategy to address the problem of spacecraft attitude stabilization in the presence of installation deviations and external disturbances. Due to its strong robustness to uncertainties and disturbances, the variable structure control (VSC) is also the most widely used robust anti-disturbance control method. Pukdeboon and Kumam [91] proposed a robust optimal sliding mode control scheme for spacecraft 6-DOF pose tracking maneuvers in the presence of external disturbances. Hu [92] proposed a sliding mode control (SMC) scheme for large-angle attitude maneuvers of flexible spacecraft, which effectively suppresses the effects of external disturbances and flexible vibrations. In the past decade, many terminal SMC schemes have been proposed to achieve finite-time convergence of output errors [93, 94]. Aiming at the attitude control problem of spacecraft subject to external disturbances and input saturation, Wallsgrove and Akella [95] proposed a saturated smooth VSC scheme. By constructing time-varying filtered variables and introducing a sharpness function, asymptotic attitude convergence is achieved despite the presence of external disturbances. Following the line of [95], Hu et al. [96, 97] proposed several smooth VSC methods, which achieve asymptotic anti-unwinding attitude stabilization, while ensuring the satisfaction of angular velocity constraints and input saturation. The central idea behind the robust control is "maintaining the status quo", that is, considering the worst disturbance situation that the system may face, design a structurally fixed robust controller to achieve suppression of bounded disturbances. It is well-known that the robust control has a strong conservatism. In contrast, the ADRC and disturbance observer-based methods can effectively reduce the conservatism by estimating and compensating disturbances.

Since the ADRC technique was proposed by Han in 1998 [98], this method has been widely studied by the control community and has been successfully applied in aerospace and industrial fields. ADRC inherits the error feedback nature of PID control, and introduces the disturbance estimation and compensation technologies to suppress multi-source disturbances. ADRC is generally composed of tracking differentiator (TD), extended state observer (ESO), nonlinear state error feedback control law, and disturbance compensation. Xia et al. [99] proposed an ESO-based SMC scheme for the spacecraft attitude tracking problem. By designing the ESO to estimate and compensate the lumped disturbances, the conservatism of the traditional SMC was greatly relaxed. Although the aforementioned nonlinear ADRC methods have advantages of good disturbance rejection capability, its structure is relatively complex and there are many parameters to be adjusted, making the parameter tuning a painful task. To overcome this problem, Gao [100] proposed a linear ADRC method and gave a bandwidth-based observer gain tuning scheme. By simplifying the multi-parameter tuning problem to a single parameter tuning problem, it greatly simplifies the complexity of controller design and theoretical analysis. Bai et al. [101] studied the attitude tracking problem of spacecraft subject to external disturbances. A linear ESO was designed to estimate and compensate for external disturbances, and an adaptive controller was then derived to achieve fast and high-precision convergence of attitude and angular velocity tracking errors. One caveat here is that, due to the limitation of observer gains, measurement noise may cause inaccurate disturbance estimation and control performance degradation accordingly.

The DOBC method was originated in the 1980s. At first, Ohishi et al. [102] proposed a frequency-domain disturbance observer to control the DC servo motor. The basic idea of the DOBC method is to design a disturbance observer to offset the influence of disturbances on the control performance, and to achieve specific control requirements by combining with other control methods. Due to its simple structure and convenient parameter setting, the DOBC method has attracted considerable attention from the control community. Although the frequency domain disturbance observer has been successfully applied to various systems, its analysis and design are based on linearized models and linear system theory, which limits its application range. Chen et al. [103] proposed for the first time a nonlinear disturbance observer using the time domain method, and applied it to the control of the manipulators to achieve accurate friction compensation. Since then, nonlinear DOBC theory has been developed rapidly. For the spacecraft AOCS, the structure of the DOBC method is shown in Fig. 1.12. For the spacecraft attitude stabilization, Sun and Zheng [104] proposed a nonlinear disturbance observer (DOB) to estimate and feed-forward compensate the lumped disturbances (including the measurement noise, parameter perturbations, and external disturbances), and designed a saturation robust controller based on an anti-windup compensator. Later, Sun et al. [105] further developed a DOB-based relative pose controller for spacecraft rendezvous and proximity operations. Zhang et al. [106] designed an adaptive sliding mode controller to achieve finite time attitude tracking, by combining the integral-type DOB and the terminal SMC technology. Zhu et al. [107] developed an adaptive DOB and a flexible vibration observer to address the problems of active vibration suppression and anti-disturbance attitude control of flexible spacecraft. Wu et al. [108, 109] designed a finite-time DOB and an iterative learning DOB to compensate internal and external disturbances, which achieve high-performance attitude control of flexible spacecraft.

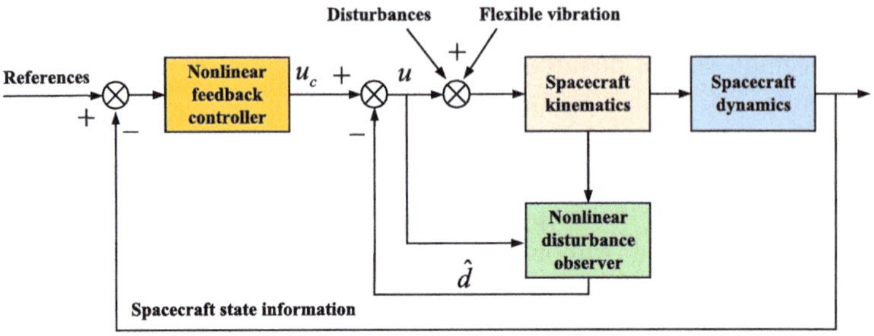

Fig. 1.12 Structure of DOBC

It is worth noting that the spacecraft AOCS is often affected by multi-source disturbances with different representation forms, such as norm-bounded variables, harmonic variables, step variables, non-Gaussian/Gaussian random variables, and rate-of-change bounded variables, etc. Most of the traditional DOBC methods can only compensate for a single class of disturbances, and intrinsic characteristics of different types of disturbances are not fully excavated. Thus, it is hard to achieve fine disturbance compensation and rejection. On the basis of DOBC, Guo and Chen [110] first proposed a composite layered anti-disturbance control framework, where a double-loop structure of "inner loop" + "outer loop" is adopted to design the controller. This method has been successfully applied in spacecraft systems, robotic systems, and some other systems. Zhu et al. [111] combined the ADRC and DOBC methods to form an enhanced composite anti-disturbance control method, established an external model to describe the disturbances caused by flexible vibrations, and regarded the other disturbances (e.g., parameter uncertainties and environmental disturbances) as a differential bounded equivalent disturbance. Then, the DOBC is used to finely estimate and compensate the flexible vibrations, while the ADRC is used to compensate other disturbances. The hybrid construct has a stronger anti-disturbance ability than a single DOBC or ADRC method. Recently, Yu et al. [112] proposed the concept of disturbance estimability and disturbance compensability for the first time, and design an enhanced anti-disturbance attitude controller for the flexible spacecraft, which achieves fine disturbance compensation and high-precision attitude tracking.

1.2.3 Fault-Tolerant Control of Spacecraft

All space agencies in the world pay consideration attention to the reliability and operational safety of spacecraft during its design, production and on-orbit operation. However, due to the manufacturing level and cost limitation, as well as the impact of harsh space environments such as high/low temperature, strong radiation, and electromagnetic interference, spacecraft failure incidents still happen from time to time,

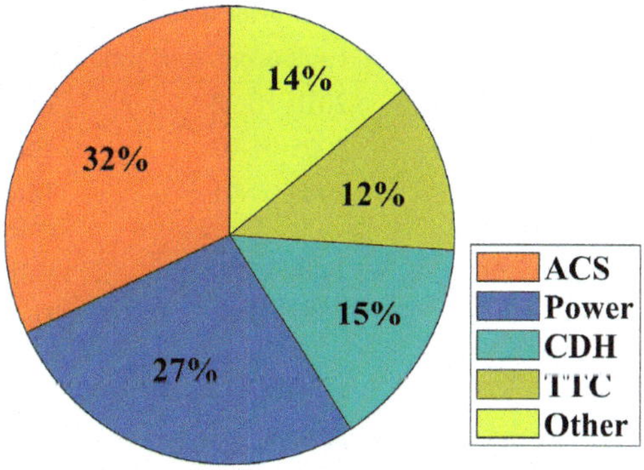

Fig. 1.13 Subsystem faults

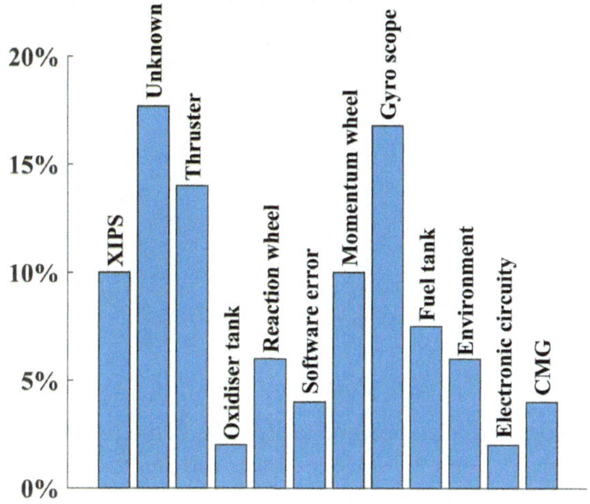

Fig. 1.14 Component fault of the AOCS

such as solar panel damage and gyroscope failure, instruction system abnormality, etc. Faults inevitably bring great hidden dangers to the spacecraft safety. According to statistics [113], among the 156 failures of on-orbit spacecraft from 1980 to 2005, the failures of the AOCS and power system account for 59% of the total faulty cases, as shown in Fig. 1.13. Further analyzing the impact of failures, it is found that nearly 65% of the failures are non-fatal and only lead to the degradation of space missions, while nearly 40% of the failures are fatal, leading to the complete failure of space missions and causing huge economic losses.

Some anomalies are extremely easy to occur in the AOCS, due to its complex structure and frequent work. According to [113], among the 156 failure events that occurred from 1980 to 2005, AOCS failures accounted for 32% of the total failures, as shown in Fig. 1.13. The component fault proportion of the spacecraft AOCS is depicted in Fig. 1.14, from which it can be observed that nearly 50% of AOCS failures were caused by actuator anomalies. During the long-term operation of the spacecraft, the force and torque actuators need to work frequently to drive the spacecraft to

Table 1.2 Typical cases of thruster faults

Spacecraft	Fault occurrence time	Fault analysis	Fault impact
Eutelsat W3B	2010-10-28	Thruster fuel leakage	Loss of control
JCSat-1B	2005-01-02	Thruster abnormal	Mission interruption
Galaxy 8I	2000-09-01	Three XIPS thrusters failed	Lifetime reduction
Nozomi	1998-12-20	Valve stuck at open	Mission interruption
Iridium 27	1997-09-14	Thruster faults	Loss of control

Table 1.3 Typical cases of flywheel faults

Spacecraft	Fault occurrence time	Fault analysis	Fault impact
Kepler	2013-05-14	Two RWs failed	Mission degradation
TOPEX	2005-10-09	MW on the pitch axis failed	Total failure of satellite
FUSE	2001-12-01	One RW failed	Mission time reduction
EchoStar V	2001-07-01	One of three MWs failed	Lifetime reduction
Radarsat-1	1999-09-15	Excessive friction of MWs	Performance degradation
GPS BII-07	1996-05-21	One RW failed	Loss of control

accomplish the specified tasks. This is the reason why they have a high failure rate. Tables 1.2 and 1.3 analyze typical thruster and flywheel fault cases. In Table 1.3, RW and MW denote, respectively, the reaction wheel and the momentum wheel. It can be seen that the actuator faults will lead to the mission degradation, and even worse the complete failure of the spacecraft or the disintegration of the spacecraft, causing huge economic losses and catastrophic consequences. Therefore, it is particularly important to enable the spacecraft to have autonomous fault-handling capabilities.

In order to meet the increasing safety, reliability and maintainability requirements of spacecraft, it is urgent to develop fault-tolerant control (FTC) technology such that the spacecraft AOCS can accommodate actuator faults. FTC originated from the high reliability requirements of aerospace engineering. In the 1980s, the U.S. Air Force proposed the concept of "self-repairing flight control system" to ensure that flight vehicles can still land safely in case of faults. According to [114, 115], the existing spacecraft FTC methods can be divided into two categories: passive FTC and active FTC. The passive FTC is essentially a robust control method. In the design process, it is necessary to fully consider all potential fault types, and treat them together as system uncertainties. This method does not require online fault information and control reconstruction, and moreover, it can handle multiple types of faults at the same time. Thus, it has the advantages of simple structure and strong practicability, and has received extensive attention in the field of spacecraft control.

Existing passive FTC methods for the design of spacecraft AOCS mainly focus on "adaptive control + X" methods (such as adaptive control + SMC, adaptive control + robust control, etc.). Cai et al. [116] proposed an indirect adaptive robust FTC method for spacecraft attitude tracking, which not only can accommodate thruster faults under limited thrusts, but also has strong robustness against parameter uncertainties and external disturbances. Shen et al. [117] proposed a finite-time FTC scheme for spacecraft attitude tracking. Xia and Zou proposed an adaptive saturated FTC method for spacecraft rendezvous and docking [118]. Dong et al. [119] and Hu et al. [120] developed two finite-time FTC problem of 6-DOF pose tracking of spacecraft using time-varying sliding mode, which can ensure that the relative position and attitude errors converge to zero at an user-defined time. Based on the sequential Lyapunov analysis, Xiao et al. [121] considered the attitude tracking problem of rigid-flexible coupled spacecraft and designed an adaptive fault-tolerant controller, which achieve tracking error convergence with guaranteed performance bounds, despite the presence of actuator faults, measurement noise, parameter uncertainties, and external disturbances, etc. In recent years, Shao et al. [122, 123] proposed the "adaptive control + prescribed performance control" FTC method for spacecraft attitude tracking, which ensure that the attitude tracking errors evolve strictly within the prescribed performance envelops, under the parameter uncertainties, external disturbances, and actuator faults. Although the passive FTC can effectively handle a large class of actuator faults and is robust to parameter uncertainties and multi-source disturbances, the controller is too conservative to recover the ideal control performance.

Different from the passive FTC, the active one mainly introduces a fault detection and diagnosis (FDD) mechanism for online detection and diagnosis of actuator faults, and then uses the diagnostic information from the FDD module to reconstruct the controller. The framework of the active FTC method is shown in Fig. 1.15. This method can make full use of the physical and analytical redundancy of the system and actively react to actuator faults, making it less conservative. An active FTC system generally contain two parts: fault diagnosis and fault-tolerant control, where fault diagnosis can be further divided into: (1) fault detection–to detect whether the system has a fault and the time when the fault occurs; (2) fault isolation–to know which part of the system occurred faults, fault type and specific location; (3) fault identification–to identify the size of fault based on the system measurement information. At present, a large number of scholars have studied the actuator fault diagnosis of the spacecraft AOCS, and the existing results can be divided into three categories: model-based methods, data-based methods, and knowledge-based methods. The book [124] provides a comprehensive review for the existing spacecraft fault diagnosis algorithms. Fonod et al. [125] presented a robust fault detection and isolation (FDI) strategy for spacecraft rendezvous, which can achieve fast and accurate diagnosis and isolation of thruster faults. Although the recent advances in fault diagnosis, most existing methods encounter various problems, such as high false alarm rate, difficult isolation, low identification accuracy, etc., in the presence of multi-source disturbances. How to improve the accuracy fault diagnosis under multi-source disturbances needs to be further studied.

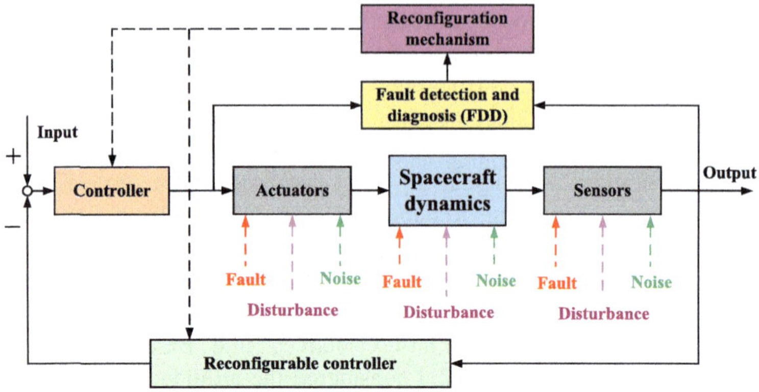

Fig. 1.15 Structure of the disturbance observer based control

At this point, we shall review the research progress of fault identification and fault-tolerant control methods. For spacecraft attitude tracking, Shen et al. [126] proposed a fault detection scheme that can avoid false alarms, and constructed an exponentially convergent fault identifier to estimate the total actuator faults. On this basis, an adaptive sliding mode fault-tolerant controller was further designed, which can effectively compensate the effect of actuator faults under input saturation. Aiming at the fault diagnosis problem of spacecraft single-gimbal control moment gyros (CMGs), Li et al. [127] proposed a neural network-based disturbance observer to learn the periodic disturbances and therefore decouple the disturbances and faults; moreover, a fault diagnosis scheme was presented based on the neural network and adaptive estimator, which achieves the isolation and estimation of the CMG faults. In recent years, some scholars have also viewed actuator faults and disturbances to "lumped disturbances", and proposed a series of anti-disturbance fault-tolerant control methods based on the ADRC or DOBC methods [128–130]. By designing an iterative learning disturbance observer, Hu et al. [130] proposed an anti-disturbance FTC method, and verified its practical effectiveness via hardware-in-the-loop experiments. Gui [131] studied the observer-based spacecraft attitude FTC problem, and designed a continuous sliding mode fault-tolerant controller. The sequential Lyapunov method was used to predict the convergence bounds of the steady-state tracking errors. The control allocation-based FTC methods have also been widely used in the design of active FTC systems. Shen et al. [132] proposed a fault-tolerant control allocation method for spacecraft attitude tracking, by using the fault identification results in conjunction with the SMC and non-robust control allocation techniques. This method can make full use of the remaining active actuators to deal with actuator faults. Later, Shen et al. [133] further proposed a robust fault-tolerant control allocation scheme. Li et al. [134] proposed an attitude FTC method in the closed-loop control allocation framework. Hu et al. [135] developed a closed-loop robust fault-tolerant control allocation method, by transforming the control allocation problem to a robust minimum variance problem, for which an analytical allocation solution is provided.

1.2.4 State-Constrained Control of Spacecraft

Due to safety concerns or performance requirements, almost all on-orbit missions require the spacecraft operate under certain state constraints, such as safe path constraints, sensor pointing constraints, and linear/angular velocity constraints, etc. The violation of constraints may degrade the mission quality or even cause catastrophic accidents. The state-constrained attitude and/or position control problem of spacecraft has attracted considerable attention from the aerospace and control community. During the past decade, various methods for state constraints have been reported in the literature, such as reference governors [136, 137], model predictive control [25, 138–140], artificial potential function (APF) [3, 10], prescribed performance control (PPC) [9, 122, 123], etc. MPC has inherent constraint-handling ability and thus is a useful tool for motion planning. Lee et al. [25] proposed a nonlinear MPC control law for constrained precision landing mission, which generated a fuel-optimal trajectory while keeping geometrical constraints (i.e., line of sight, glide-slope and thrust direction constraints). Li et al. [140] investigated the application of MPC strategy for controlling a chaser spacecraft to dock with a tumbling target, wherein several engineering constraints such as control input saturation, collision avoidance, velocity constraint, and dock-enabling condition are considered simultaneously. Weiss [139] presented a strategy and case studies of spacecraft relative motion guidance and control based on the application of linear quadratic MPC. Obstacle avoidance is considered in the rendezvous phase, while line-of-sight constraint, bandwidth constraint and exhaust plume direction constraint are addressed during the docking phase.

Besides, barrier Lyapunov function (BLF) has been extensively employed to deal with state constraints for systems. In fact, the basic idea of BLF is similar to the APF. Tee et al. [141] originally developed a logarithmic symmetric/asymmetric BLF for strict feedback nonlinear systems. Later, BLFs have been extended to various applications. The recent advances of the APF-based methods have been summarized in Sect. 1.1.2. In this subsection, we will concentrate on the PPC method, since some parts of this book involve it. Note that most of the existing solutions to the state-constrained problem focus on analyzing the satisfaction of constant or time-varying state (error) constraints, while failing to guarantee a prescribed transient and steady-state specifications for the output errors (e.g., attitude and position tracking errors). In fact, guaranteeing the prescribed performance metrics specified a priori by the designer is of great importance for mission success [122], due to the direct relationship of stabilization/tracking performance with mission specific requirements. Fortunately, the PPC, originally developed in [142], provides an effective way to guarantee prescribed transient and steady-state performance and has been extensively applied to the controller design for spacecraft AOCS with stringent performance requirements (e.g., see [122, 123, 143, 144]). The core idea of this method is to pre-specify a performance envelope for the state or error of the controlled system, which describes the transient and steady-state performance like convergence speed, maximum overshoot, and steady-state error, as shown in Fig. 1.16. Then, by using the error transformation technique, the original performance constrained problem is

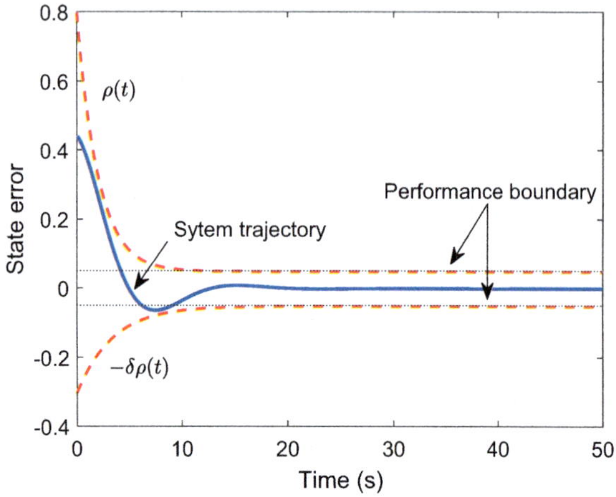

Fig. 1.16 Transient and steady-state performance constraints

transformed into an equivalent non-constrained one, for which many mature control methods, such as SMC, DOBC, adaptive control, etc., can be directly used to design the controller according to the actual control requirements.

Wei et al. [145] reviewed the research advances and future trends of spacecraft PPC methods. Considering the anti-disturbance and fault-tolerance capabilities of the PPC methods, we divides the existing PPC methods into two categories: PPC + estimation/observation methods and model-free PPC methods, where the first type of methods mainly include "PPC + adaptive control", "PPC + DOBC", "PPC + neural network approximation", etc. Shao et al. [9] considered the spacecraft rendezvous and proximity operations with spatial motion constraints (including approaching corridor constraints and sensor's field-of-view constraints), and proposed an I&I adaptive PPC scheme. By tactfully transforming the spatial motion constraints and specific mission requirements into the transient and steady-state performance constraints of the pose tracking errors, the designed controller can enable the spacecraft to accomplish the rendezvous and proximity operations with a tumbling target, while satisfying the preassigned performance requirements and complying with the motion constraints. In [122, 123], two adaptive fault-tolerant PPC schemes were presented, which ensure prescribed performance attitude tracking of spacecraft in the presence of parameter uncertainties, external disturbances, actuator faults, and input saturation. It should be pointed out that [122, 123] introduce the BLFs in conjunction with the PPC methods to ensure the satisfaction of prescribed performance specifications. Liu et al. [143] proposed an adaptive fault-tolerant PPC approach for spacecraft attitude tracking, which guarantees the prescribed performance of both attitude and angular velocity errors, despite the presence of external disturbances and actuator faults. Huang and Duan [146] developed an ESO-based anti-windup fault-tolerant PPC scheme for attitude tracking of combined spacecraft.

Although the aforementioned PPC methods have strong robustness as well as anti-disturbance and fault-tolerance capabilities, most of them require online estimation or observation, which inevitably result in high algorithm complexity and large calculation burden, making them difficult to be implemented in orbit. To improve the practicability of the PPC approaches, some scholars have explored the low-complexity model-free PPC technique in recent years [147]. Zhou et al. [148] proposed a coordinate-free robust PPC method for the spacecraft attitude tracking problem. This method has a simple controller structure and does not require system parameter information. Aiming at the problems of spacecraft attitude stabilization, tracking and combined spacecraft attitude takeover control, several low-complexity model-free PPC methods have been proposed in [144, 149]. These methods do not require prior estimation or online identification of model parameters and disturbances, and can guarantee the output errors evolve strictly within the prescribed performance envelops, regardless of parameter uncertainties and external disturbances. In particular, an appointed-time performance function was introduced in [143, 144] based on the terminal sliding mode, whereby a model-free two-layer PPC method was proposed for spacecraft attitude tracking. Hu et al. [150] further designed a model-free PPC method for flexible spacecraft attitude tracking, which ensures the prescribed performance of both the attitude and angular velocity errors.

Most of the existing PPC methods can only deal with transient and steady-state performance constraints, but it is difficult to take into account various motion and physical constraints. Yong et al. [151] introduced an auxiliary system using the positive system theory to relax the performance envelopes when input saturation occurs, which overcomes the PPC design problem under input saturation. But nonetheless, how to design an effective PPC algorithm under the coexistence of multiple motion and physical constraints is an urgent problem to be solved.

1.2.5 Intelligent Control of Spacecraft

With the rapid development of artificial intelligence technology, it has been widely applied in multi-constraint spacecraft control. References [152, 153] provided a comprehensive review for the application of intelligent technology in solving the problems of fault diagnosis, uncertainty compensation, and performance optimization in space missions. An intelligent PD controller for uncertain flexible spacecraft was proposed in [154], where a radial basis function (RBF) neural network (NN) was introduced to compensate for unknown perturbation terms. Li et al. [155] proposed an adaptive NN-based feedback controller for the attitude cooperation problem of distributed spacecraft. Chen et al. [156] also used the RBF NN to build a closed-loop adaptive fault-tolerant controller, which achieves high-precision attitude tracking in the presence of unknown dead-zones and disturbances. Liu et al. [157] proposed a Q-learning-based optimal control method for the attitude tracking of combined spacecraft using the off-line data learning. In [158], a DNN was used to replace the traditional PID control method to design the attitude tracking controller, whilst a genetic algorithm was used to select the initial weights of the network, so as to improve the deployment efficiency of the intelligent algorithm. Zhang et al. [159]

designed a PID-Guide TD3 learning algorithm to train the spacecraft attitude controller. By using the PID controller as a guided training, the off-line learning efficiency of the traditional TD3 algorithm has been greatly improved.

Spacecraft optimal control is a benchmark problem in the aerospace engineering, and has aroused extensive research attention. From a theoretical viewpoint, to achieve optimal control, one needs to solve the HJB equation subject to a user-defined cost function. However, due to the high nonlinearity and strong coupling of spacecraft AOCS, analytically solving optimal control problems is a challenging task. RL, commonly called adaptive dynamic programming (ADP) in the control system community, provides a promising way for solving the optimal control problem. By employing the NN, the ADP algorithm can derive a near-optimal control policy by iteratively approximating the cost function and optimal control law online. In this respect, Vamvoudakis and Lewis [160] proposed a policy iteration algorithm for learning the continuous-time optimal control solution of nonlinear systems. Both convergence of the learning algorithm and stability of the closed-loop system have been guaranteed. By appropriately selecting value functions for the nominal system, Liu et al. [161] derived an RL-based robust control algorithm for a class of uncertain nonlinear systems subject to input constraints. Recently, the RL-based optimal control solutions have been extended to the spacecraft applications. By employing an estimator-based critic-only ADP, Dong et al. [162] solved the tracking control problem with guaranteed prescribed performance subject to uncertain system parameters. Hu et al. [65] proposed an RL-based six degree-of-freedom control scheme for the proximity operations of spacecraft, where both the field-of-view constraint and approaching path constraint are addressed. Yang et al. [64] proposed an online RL-based control scheme to achieve optimal spacecraft attitude reorientation under pointing constraints, and verified its effectiveness via a hardware-in-the-loop (HIL) experimental platform.

1.3 Contents of the Book

Having reviewed the state of art spacecraft motion planning and control technologies, we provide now an outline of the contents of this book.

Chapter 2 introduces notations, coordinate frames, spacecraft dynamics, and mathematical preliminaries, which are the foundation for the subsequent chapters.

Chapter 3 addresses spacecraft attitude control problem subject to attitude and angular velocity constraints in the presence of inertia uncertainties. The basic framework of the developed control scheme is built upon I&I adaptive control methodology, which helps remove the restrictive realizability condition that does not hold in the Lyapunov sense when angular velocity constraints are taken into account. Two judiciously constructed potential functions are employed to handle double-level state constraints, and then an I&I adaptive control law is proposed to ensure asymptotic convergence of attitude error and angular velocity. In addition, to further relax the dependence of parameter convergence on PE condition, the I&I adaptive law is extended to a data-driven counterpart through adding a learning term that is acquired by adopting the regressor filtering in conjunction with the dynamic regressor extension and mixing (DREM) procedure.

Chapter 4 investigates the spacecraft fault-tolerant attitude control problem under attitude and angular velocity constraints, and we propose two FTC laws for constrained spacecraft attitude reorientation. Firstly, an adaptive FTC algorithm with saturated virtual control is derived in the framework of backstepping. Both unwinding phenomenon and forbidden/mandatory attitude constraints are addressed with a judiciously developed potential function. Then, an integral BLF is employed to further handle angular velocity constraint, wherein the uniform strong controllability assumption is established with sufficient conditions and feasibility analysis. Secondly, a learning-based approximate optimal FTC scheme is proposed for constrained spacecraft attitude reorientation. A specially designed cost function is developed, which can accommodate actuator faults and deal with attitude and angular velocity constraints. Then, by using reinforcement learning technology, a single-critic NN is developed to online approximate the cost function, wherein CL methodology is adopted to relax the PE condition.

Chapter 5 is dedicated to intelligent FDD and FTC of spacecraft. An NN-based fault diagnosis scheme is proposed to address the problem of fault isolation and estimation for the Single-Gimbal Control Moment Gyroscopes (SGCMGs) of spacecraft in a periodic orbit. A disturbance observer is developed for active anti-disturbance, based on which, the orbital periodic disturbance can be decoupled with fault by resorting to the fitting and memory ability of NN. In addition, the fault diagnosis scheme is established based on information fusion idea, wherein data of spacecraft attitude and gimbals position are combined to implement fault isolation and estimation. Finally, an adaptive sliding mode controller incorporating the disturbance and fault estimation results is designed to achieve active fault-tolerant control.

In Chap. 6, a dynamic control allocation scheme based on reinforcement learning is proposed to solve the problem of singularity avoidance and energy saving in spacecraft attitude maneuver. Firstly, the null space is used to decouple the outer loop control from the inner loop control allocation, and the control allocation equation without torque deviation is constructed. Then the control moment gyro and attitude dynamics are modeled as an augmented system using the control linearization assumption, and the control allocation is transformed into a dynamic problem. The cost function is constructed and transformed into the Bellman equation. As it is difficult to obtain the analytical solution of the partial differential equation, an integral reinforcement learning algorithm based on off policy strategy is designed to estimate the parameters. The algorithm does not require system model and the adjustability of disturbances.

Chapter 7 focuses on the optimal tracking control for the leader-follower spacecraft formation flying system. In order to solve the Hamilton-Jacobi-Bellman (HJB) equation, a single-critic NN is developed to approximate the optimal cost function. Moreover, by combining the parameter projection rule and gradient descent algorithm, a semi-global adaptive update law is derived to tune the critic NN. In doing so, a continuous near optimal tracking controller is presented. Subsequently, an input-state-dependent event-triggered mechanism is designed to ensure that the near optimal tracking controller is implemented only when specific events occur, which significantly reduces the execution frequency of the control command. Remarkably, benefiting from the construction of an input-based triggering error, the conventional assumption on the Lipschitz continuity of the controller is tactfully removed, thus erasing the computable demand on the unknown Lipschitz constants. Rigorous analysis on the system stability and Zeno-free behavior are provided successively.

In Chap. 8, an adaptive prescribed performance pose tracking control scheme is presented for spacecraft rendezvous and proximity operations (RPOs) with a freely tumbling target, under parameter uncertainties as well as motion and performance constraints. By visualizing the motion constraints and the prescribed performance metrics as the pose tracking error bounds, the original constrained tracking error dynamics is transformed into an equivalent "state-constrained" one. Then, a non-CE adaptive controller is designed using the barrier function in conjunction with backstepping control, which is capable of guaranteeing the transformed errors remain within the specified ranges, despite the presence of parameter uncertainties. As a consequence, the overall control scheme can accomplishes spacecraft RPO whilst complying with the underlying motion and performance constraints. Besides, the underlying singularity problem in the attitude extraction algorithm is avoided through properly choosing the performance bounds for the position tracking errors.

Chapter 9 studies the 6-DOF pose control issue of spacecraft RPOs subject to kinematic and dynamic constraints, as well as parameter uncertainties. A class of APFs that are free of local minima are firstly constructed to deal with kinematic and dynamic constraints imposed by certain safety and physical requirements. Secondly, using the dynamic scaling technique, an I&I adaptive pose controller is tactfully designed. This controller is shown to be able to circumvent the realizability condition that is required for most existing adaptive control approaches, but nonetheless may not hold under dynamic constraints. Thirdly, through Lyapunov's direct method, the asymptotic stability of the closed-loop system is analyzed. The proposed method enables the pursuer to arrive at the desired anchoring point with a specified pointing, whilst satisfying both kinematic and dynamic constraints. Moreover, it depends upon the I&I adaptive design philosophy and, therefore, introduces an attracting manifold, whereby the deterministic case of closed-loop performance (no effect of parameter uncertainties) can be asymptotically recovered.

In Chap. 10, a composite learning pose tracking control strategy is proposed based on the CL and DREM techniques, which can synchronously enhance parameter convergence and tracking performance under a strictly weak IE condition. Firstly, a CE-based adaptive control law is given, and the filtered system dynamics is established to avoid the use of unmeasured state derivatives when constructing parameter estimation errors. Then, a traditional composite adaptive law is derived, and on this basis, the composite learning law is further designed. Lyapunov stability analysis shows that if the regressor matrix satisfies the IE condition, the proposed composite learning control scheme can ensure both the tracking errors and parameter estimation errors converge to zero by making use of the stored historical information. Moreover, benefiting from the DREM procedure and some special designs, the parameter estimation error dynamics are independent of each other, and parameter convergence rate does not depend on the signal excitation strength, which make the gain selection simpler and clearer. Finally, simulation results verify the effectiveness of the problem control scheme.

In Chap. 11, an RL-based pose control scheme is proposed for spacecraft RPOs under spatial motion constraints. As a stepping stone, the dual-quaternion formalism is employed to characterize the 6-DOF spacecraft relative motion dynamics and motion constraints. Then, an RL-based control scheme is developed under the dual-quaternion algebraic framework to approximate the optimal control solution subject to a cost function and a Hamilton-Jacobi-Bellman equation. In addition, a specially constructed barrier function is embedded in the reward function to deal with spatial motion constraints. Lyapunov stability analysis shows the ultimate boundedness of the state errors and network weight estimation errors. Besides, it is also shown that a PD-like controller under dual-quaternion formulation can be employed as the initial control policy to trigger the online learning process. Its boundedness is proved by a special Lyapunov strictification method. Finally, simulation results of prototypical spacecraft missions with RPOs are provided to illustrate the effectiveness of the proposed method.

In Appendix concludes this book by a review of our major findings, along with our ideas about the future of the field.

References

1. Xu G, Wu J, Gou Z, Zhang B (2017) High accuracy high stability and high agility pointing technology of spacecraft. Spacecraft Engineering 26(1): 91–99
2. Li Y, Huang H (2019) Current trends of spacecraft intelligent autonomous control. Aerospace Control and Application 45(4): 7–18
3. Shao X, Hu Q, Shi Y, Yi B (2022) Data-driven immersion and invariance adaptive attitude control for rigid bodies with double-level state constraints. IEEE Transactions on Control Systems Technology 30(2): 779–794
4. Lee U, Mesbahi M (2014) Feedback control for spacecraft reorientation under attitude constraints via convex potentials. IEEE Transactions on Aerospace and Electronic Systems 50(4): 2578–2592
5. Ayoubi MA, Hsin J (2020) Sun-avoidance slew planning with keep-out cone and actuator constraints. Journal of Spacecraft and Rockets 57(6): 1175–1185
6. Fabinsky B (2006) A survey of ground operations tools developed to plan and validate the pointing of space telescopes and the design for wise. In: Proceedings of SPIE - The International Society for Optical Engineering, Orlando, FL, United states, pp 383–395
7. Hablani HB (1999) Attitude commands avoiding bright objects and maintaining communication with ground station. Journal of Guidance, Control, and Dynamics 22(6): 759–767
8. Hu Q, Dong H, Zhang Y, Ma G (2015) Tracking control of spacecraft formation flying with collision avoidance. Aerospace Science and Technology 42: 353–364
9. Shao X, Hu Q, Shi Y (2021) Adaptive pose control for spacecraft proximity operations with prescribed performance under spatial motion constraints. IEEE Transactions on Control Systems Technology 29(4): 1405–1419

10. Shao X, Hu Q (2021) Immersion and invariance adaptive pose control for spacecraft proximity operations under kinematic and dynamic constraints. IEEE Transactions on Aerospace and Electronic Systems 57(4): 2183–2200
11. Akella MR, Valdivia A, Kotamraju GR (2005) Velocity-free attitude controllers subject to actuator magnitude and rate saturations. Journal of Guidance, Control, and Dynamics 28(4): 659–666
12. Wang X, Wu G, Xing L, Pedrycz W (2020) Agile earth observation satellite scheduling over 20 years: Formulations, methods, and future directions. IEEE Systems Journal 15(3): 3881–3892
13. Marsh H, Karpenko M, Gong Q (2016) Energy constrained shortest-time maneuvers for reaction wheel satellites. In: AIAA/AAS astrodynamics specialist conference, Long Beach, CA, United states, pp 5579–5598
14. Sorensen A (1993) Iso attitude maneuver strategies. NASA STI/Recon Technical Report A 95: 975–987
15. Singh G, Macala G, Wong E, Rasmussen R, Singh G, Macala G, Wong E, Rasmussen R (1997) A constraint monitor algorithm for the cassini spacecraft. In: Guidance, Navigation, and Control Conference, New Orleans, LA, United states, pp 272–282
16. Frakes JP, Henretty DA, Flatley TW, Markley F, San JK, Lightsey E (1992) Sampex science pointing with velocity avoidance. In: In: Spaceflight mechanics 1992; Proceedings of the 2nd AAS (AIAA Meeting, Colorado Springs, CO, Feb. 24-26, 1992. Pt. 2 (A93-48426 20-12), Univelt, Inc., AAS PAPER 92-182
17. de Angelis EL, Giulietti F, Avanzini G (2015) Single-axis pointing of underactuated spacecraft in the presence of path constraints. Journal of Guidance, Control, and Dynamics 38(1): 143–147
18. Duan C, Hu Q, Zhang Y, Wu H (2020) Constrained single-axis path planning of underactuated spacecraft. Aerospace Science and Technology 107: 106345
19. Spindler K (1998) New methods in on-board attitude control (aas 98-308). Spaceflight Dynamics 1998, Volume 100 Part 1, Advances in Astronautical Sciences 100: 111
20. Biggs JD, Colley L (2016) Geometric attitude motion planning for spacecraft with pointing and actuator constraints. Journal of Guidance, Control, and Dynamics 39(7): 1672–1677
21. Henninger HC, Biggs JD (2018) Optimal under-actuated kinematic motion planning on the epsilon-group. Automatica 90: 185–195
22. Geng Y, Biggs JD, Li C (2021) Pose regulation via the dual unitary group: An application to spacecraft rendezvous. IEEE Transactions on Aerospace and Electronic Systems 57(6): 3734–3748
23. Mclnnes CR (1994) Large angle slew maneuvers with autonomous sun vector avoidance. Journal of Guidance, Control, and Dynamics 17(4): 875–877
24. Wisniewski R, Kulczycki P (2005) Slew maneuver control for spacecraft equipped with star camera and reaction wheels. Control Engineering Practice 13(3): 349–356
25. Lee U, Mesbahi M (2017) Constrained autonomous precision landing via dual quaternions and model predictive control. Journal of Guidance, Control, and Dynamics 40(2): 292–308
26. Shen Q, Yue C, Goh CH, Wu B, Wang D (2018) Rigid-body attitude stabilization with attitude and angular rate constraints. Automatica 90: 157–163
27. Hu Q, Chi B, Akella MR (2019) Reduced attitude control for boresight alignment with dynamic pointing constraints. IEEE/ASME Transactions on Mechatronics 24(6): 2942–2952
28. Dong H, Hu Q, Liu Y, Akella MR (2019) Adaptive pose tracking control for spacecraft proximity operations under motion constraints. Journal of Guidance, Control, and Dynamics 42(10): 2258–2271
29. Hu Q, Chi B, Akella MR (2019) Anti-unwinding attitude control of spacecraft with forbidden pointing constraints. Journal of Guidance, Control, and Dynamics 42(4): 822–835
30. Hu Q, Liu Y, Dong H, Zhang Y (2020) Saturated attitude control for rigid spacecraft under attitude constraints. Journal of Guidance, Control, and Dynamics 43(4): 790–805
31. Hu Q, Liu Y, Zhang Y (2021) Velocity-free saturated control for spacecraft proximity operations with guaranteed safety. IEEE Transactions on Systems, Man, and Cybernetics: Systems 52(4): 2501–2513

32. Tegmark M (1996) An icosahedron-based method for pixelizing the celestial sphere. The Astrophysical Journal 470(2): L81
33. Kjellberg HC, Lightsey EG (2013) Discretized constrained attitude pathfinding and control for satellites. Journal of Guidance, Control, and Dynamics 36(5): 1301–1309
34. Kjellberg HC, Lightsey EG (2016) Discretized quaternion constrained attitude pathfinding. Journal of Guidance, Control, and Dynamics 39(3): 713–718
35. Tanygin S (2012) Attitude parameterizations as higher-dimensional map projections. Journal of Guidance, Control, and Dynamics 35(1): 13–24
36. Tanygin S (2015) Fast three-axis constrained attitude pathfinding and visualization using minimum distortion parameterizations. Journal of Guidance, Control, and Dynamics 38(12): 2324–2336
37. Kavraki LE, Svestka P, Latombe JC, Overmars MH (1996) Probabilistic roadmaps for path planning in high-dimensional configuration spaces. IEEE Transactions on Robotics and Automation 12(4): 566–580
38. Karaman S, Frazzoli E (2011) Sampling-based algorithms for optimal motion planning. The International Journal of Robotics Research 30(7): 846–894
39. Feron E, Dahleh M, Frazzoli E, Kornfeld R (2012) A randomized attitude slew planning algorithm for autonomous spacecraft. In: AIAA Guidance, Navigation, and Control Conference and Exhibit, p 4155
40. Yershova A, LaValle SM (2004) Deterministic sampling methods for spheres and so(3). In: IEEE International Conference on Robotics and Automation, 2004. Proceedings. ICRA'04. 2004, IEEE, vol 4, pp 3974–3980
41. Zhao Z, Shang H, Wei B (2022) Tackling nonconvex collision avoidance constraints for optimal trajectory planning using saturation functions. Journal of Guidance, Control, and Dynamics 45(6): 1002–1016
42. Boyarko G, Yakimenko O, Romano M (2011) Optimal rendezvous trajectories of a controlled spacecraft and a tumbling object. Journal of Guidance, Control, and Dynamics 34(4): 1239–1252
43. Leomanni M, Quartullo R, Bianchini G, Garulli A, Giannitrapani A (2022) Variable-horizon guidance for autonomous rendezvous and docking to a tumbling target. Journal of Guidance, Control, and Dynamics 45(5): 846–858
44. Lee DY, Gupta R, Kalabić UV, Di Cairano S, Bloch AM, Cutler JW, Kolmanovsky IV (2017) Geometric mechanics based nonlinear model predictive spacecraft attitude control with reaction wheels. Journal of Guidance, Control, and Dynamics 40(2): 309–319
45. Gupta R, Kalabic UV, Di Cairano S, Bloch AM, Kolmanovsky IV (2015) Constrained spacecraft attitude control on so(3) using fast nonlinear model predictive control. In: Proceedings of the American Control Conference, Chicago, IL, United states, pp 2980–2986
46. Liu X, Lu P, Pan B (2017) Survey of convex optimization for aerospace applications. Astrodynamics 1(1): 23–40
47. Kim Y, Mesbahi M (2004) Quadratically constrained attitude control via semidefinite programming. IEEE Transactions on Automatic Control 49(5): 731–735
48. Kim Y, Mesbahi M, Singh G, Hadaegh FY (2010) On the convex parameterization of constrained spacecraft reorientation. IEEE Transactions on Aerospace and Electronic Systems 46(3): 1097–1109
49. Sun C, Dai R (2015) Spacecraft attitude control under constrained zones via quadratically constrained quadratic programming. In: AIAA Guidance, Navigation, and Control Conference, Kissimmee, FL, United states, pp 2010–2026
50. Tam M, Lightsey EG (2016) Constrained spacecraft reorientation using mixed integer convex programming. Acta Astronautica 127: 31–40
51. Kornfeld R (2003) On-board autonomous attitude maneuver planning for planetary spacecraft using genetic algorithms. In: AIAA Guidance, Navigation, and Control Conference and Exhibit, Austin, TX, United states, p 5784
52. Wu C, Han X, An W, Gong J, Xu N (2022) Application of the improved grey wolf algorithm in spacecraft maneuvering path planning. International Journal of Aerospace Engineering p 8857584

53. Spiller D, Ansalone L, Curti F (2016) Particle swarm optimization for time-optimal spacecraft reorientation with keep-out cones. Journal of Guidance, Control, and Dynamics 39(2): 312–325
54. Spiller D, Melton RG, Curti F (2018) Inverse dynamics particle swarm optimization applied to constrained minimum-time maneuvers using reaction wheels. Aerospace Science and Technology 75: 1–12
55. Wu C, Xu R, Zhu S, Cui P (2017) Time-optimal spacecraft attitude maneuver path planning under boundary and pointing constraints. Acta Astronautica 137: 128–137
56. Melton RG (2018) Differential evolution/particle swarm optimizer for constrained slew maneuvers. Acta Astronautica 148: 246–259
57. Oestreich CE, Linares R, Gondhalekar R (2021) Autonomous six-degree-of-freedom spacecraft docking with rotating targets via reinforcement learning. Journal of Aerospace Information Systems 18(7): 417–428
58. Hovell K, Ulrich S (2021) Deep reinforcement learning for spacecraft proximity operations guidance. Journal of Spacecraft and Rockets 58(2): 254–264
59. Qu Q, Liu K, Wang W, Lü J (2022) Spacecraft proximity maneuvering and rendezvous with collision avoidance based on reinforcement learning. IEEE Transactions on Aerospace and Electronic Systems
60. Ma Z, Wang Y, Yang Y, Wang Z, Tang L, Ackland S (2018) Reinforcement learning-based satellite attitude stabilization method for non-cooperative target capturing. Sensors 18(12): 4331
61. Vedant JT (2019) Reinforcement learning for spacecraft attitude control. In: 70th International Astronautical Congress, Washington, DC, United states
62. Elkins JG, Sood R, Rumpf C (2022) Bridging reinforcement learning and online learning for spacecraft attitude control. Journal of Aerospace Information Systems 19(1): 62–69
63. Dong H, Zhao X, Yang H (2020) Reinforcement learning-based approximate optimal control for attitude reorientation under state constraints. IEEE Transactions on Control Systems Technology 29(4): 1664–1673
64. Yang H, Hu Q, Dong H, Zhao X (2021) ADP-based spacecraft attitude control under actuator misalignment and pointing constraints. IEEE Transactions on Industrial Electronics 69(9): 9342–9352
65. Hu Q, Yang H, Dong H, Zhao X (2021) Learning-based 6-dof control for autonomous proximity operations under motion constraints. IEEE Transactions on Aerospace and Electronic Systems 57(6): 4097–4109
66. Ioannou PA, Sun J (2012) Robust adaptive control. Courier Corporation
67. Egeland O, Godhavn JM (1994) Passivity-based adaptive attitude control of a rigid spacecraft. IEEE Transactions on Automatic Control 39(4): 842–846
68. Thakur D, Srikant S, Akella MR (2015) Adaptive attitude-tracking control of spacecraft with uncertain time-varying inertia parameters. Journal of Guidance, Control, and Dynamics 38(1): 41–52
69. Singla P, Subbarao K, Junkins JL (2006) Adaptive output feedback control for spacecraft rendezvous and docking under measurement uncertainty. Journal of Guidance, Control, and Dynamics 29(4): 892–902
70. Astolfi A, Ortega R (2003) Immersion and invariance: A new tool for stabilization and adaptive control of nonlinear systems. IEEE Transactions on Automatic control 48(4): 590–606
71. Seo D, Akella MR (2008) High-performance spacecraft adaptive attitude-tracking control through attracting-manifold design. Journal of Guidance, Control, and Dynamics 31(4): 884–891
72. Lee KW, Singh SN (2019) Immersion-and invariance-based adaptive control of asteroid-orbiting and-hovering spacecraft. The Journal of the Astronautical Sciences 66(4): 537–553
73. Karagiannis D, Sassano M, Astolfi A (2009) Dynamic scaling and observer design with application to adaptive control. Automatica 45(12): 2883–2889
74. Yang S, Akella MR, Mazenc F (2017) Dynamically scaled immersion and invariance adaptive control for euler–lagrange mechanical systems. Journal of Guidance, Control, and Dynamics 40(11): 2844–2856

75. Wen H, Yue X, Yuan J (2018) Dynamic scaling–based noncertainty-equivalent adaptive spacecraft attitude tracking control. Journal of Aerospace Engineering 31(2): 04017098
76. Xia D, Yue X (2022) Anti-unwinding immersion and invariance adaptive attitude control of rigid spacecraft with inertia uncertainties. Journal of Aerospace Engineering 35(2): 04021137
77. Boyd S, Sastry SS (1986) Necessary and sufficient conditions for parameter convergence in adaptive control. Automatica 22(6): 629–639
78. Chowdhary G, Johnson E (2010) Concurrent learning for convergence in adaptive control without persistency of excitation. In: Proceedings of 49th IEEE Conference on Decision and Control (CDC), Atlanta, GA, USA, pp 3674–3679
79. Zhao Q, Duan G (2020) Finite-time concurrent learning adaptive control for spacecraft with inertia parameter identification. Journal of Guidance, Control, and Dynamics 43(3): 574–584
80. Zhao Q, Duan G (2021) Concurrent learning adaptive finite-time control for spacecraft with inertia parameter identification under external disturbance. IEEE Transactions on Aerospace and Electronic Systems 57(6): 3691–3704
81. Cho N, Shin HS, Kim Y, Tsourdos A (2017) Composite model reference adaptive control with parameter convergence under finite excitation. IEEE Transactions on Automatic Control 63(3): 811–818
82. Pan Y, Yu H (2018) Composite learning robot control with guaranteed parameter convergence. Automatica 89: 398–406
83. Dong H, Hu Q, Akella MR, Yang H (2019) Composite adaptive attitude-tracking control with parameter convergence under finite excitation. IEEE Transactions on Control Systems Technology 28(6): 2657–2664
84. Shao X, Hu Q, Li D, Shi Y, Yi B (2022, https://doi.org/10.1109/TAES.2022.3194846) Composite adaptive control for anti-unwinding attitude maneuvers: An exponential stability result without persistent excitation. IEEE Transactions on Aerospace and Electronic Systems
85. Doyle J, Glover K, Khargonekar P, Francis B (1988) State-space solutions to standard H_2 and H_∞ control problems. In: 1988 American Control Conference, Atlanta, GA, USA, pp 1691–1696
86. Liu C, Shi K, Sun Z (2019) Robust H_∞ controller design for attitude stabilization of flexible spacecraft with input constraints. Advances in Space Research 63(5): 1498–1522
87. Chen BS, Wu CS, Jan YW (2000) Adaptive fuzzy mixed H_2/H_∞ attitude control of spacecraft. IEEE Transactions on Aerospace and Electronic Systems 36(4): 1343–1359
88. Liu C, Ye D, Shi K, Sun Z (2017) Robust high-precision attitude control for flexible spacecraft with improved mixed H_2/H_∞ control strategy under poles assignment constraint. Acta Astronautica 136: 166–175
89. Luo W, Chu YC, Ling KV (2005) H_∞ inverse optimal attitude-tracking control of rigid spacecraft. Journal of Guidance, Control, and Dynamics 28(3): 481–494
90. Wang Z, Li Y (2020) Rigid spacecraft robust optimal attitude stabilization under actuator misalignments. Aerospace Science and Technology 105: 105990
91. Pukdeboon C, Kumam P (2015) Robust optimal sliding mode control for spacecraft position and attitude maneuvers. Aerospace Science and Technology 43: 329–342
92. Hu Q (2008) Sliding mode maneuvering control and active vibration damping of three-axis stabilized flexible spacecraft with actuator dynamics. Nonlinear Dynamics 52(3): 227–248
93. Lu K, Xia Y (2013) Adaptive attitude tracking control for rigid spacecraft with finite-time convergence. Automatica 49(12): 3591–3599
94. Guo Y, Huang B, Song Sm, Li Aj, Wang Cq (2019) Robust saturated finite-time attitude control for spacecraft using integral sliding mode. Journal of Guidance, Control, and Dynamics 42(2): 440–446
95. Wallsgrove RJ, Akella MR (2005) Globally stabilizing saturated attitude control in the presence of bounded unknown disturbances. Journal of Guidance, Control, and Dynamics 28(5): 957–963
96. Hu Q, Li L, Friswell MI (2015) Spacecraft anti-unwinding attitude control with actuator nonlinearities and velocity limit. Journal of Guidance, Control, and Dynamics 38(10): 2042–2050

97. Hu Q, Tan X (2017) Unified attitude control for spacecraft under velocity and control constraints. Aerospace Science and Technology 67: 257–264
98. Han JQ (1998) Auto disturbance rejection controller and its applications. Control and Decision 13(1): 19–23
99. Xia Y, Zhu Z, Fu M, Wang S (2010) Attitude tracking of rigid spacecraft with bounded disturbances. IEEE Transactions on Industrial Electronics 58(2): 647–659
100. Gao Z (2003) Scaling and bandwidth-parameterization based controller tuning. In: Proceedings of the American Control Conference, Denver, CO, United states, pp 4989–4996
101. Bai Y, Biggs JD, Zazzera FB, Cui N (2018) Adaptive attitude tracking with active uncertainty rejection. Journal of Guidance, Control, and Dynamics 41(2): 550–558
102. Ohishi K, Nakao M, Ohnishi K, Miyachi K (1987) Microprocessor-controlled dc motor for load-insensitive position servo system. IEEE Transactions on Industrial Electronics (1): 44–49
103. Chen WH, Ballance DJ, Gawthrop PJ, O'Reilly J (2000) A nonlinear disturbance observer for robotic manipulators. IEEE Transactions on Industrial Electronics 47(4): 932–938
104. Sun L, Zheng Z (2017) Disturbance-observer-based robust backstepping attitude stabilization of spacecraft under input saturation and measurement uncertainty. IEEE Transactions on Industrial Electronics 64(10): 7994–8002
105. Sun L, Huo w, Jiao Z (2018) Disturbance-observer-based robust relative pose control for spacecraft rendezvous and proximity operations under input saturation. IEEE Transactions on Aerospace and Electronic Systems 54(4): 1605–1617
106. Zhang J, Zhao W, Shen G, Xia Y (2020) Disturbance observer-based adaptive finite-time attitude tracking control for rigid spacecraft. IEEE Transactions on Systems, Man, and Cybernetics: Systems 51(11): 6606–6613
107. Zhu W, Zong Q, Tian B, Liu W (2022) Disturbance observer-based active vibration suppression and attitude control for flexible spacecraft. IEEE Transactions on Systems, Man, and Cybernetics: Systems 52(2): 893–901
108. Yan R, Wu Z (2019) Super-twisting disturbance observer-based finite-time attitude stabilization of flexible spacecraft subject to complex disturbances. Journal of Vibration and Control 25(5): 1008–1018
109. He T, Wu Z (2021) Iterative learning disturbance observer based attitude stabilization of flexible spacecraft subject to complex disturbances and measurement noises. IEEE/CAA Journal of Automatica Sinica 8(9): 1576–1587
110. Guo L, Chen WH (2005) Disturbance attenuation and rejection for systems with nonlinearity via dobc approach. International Journal of Robust and Nonlinear Control: IFAC-Affiliated Journal 15(3): 109–125
111. Zhu Y, Guo L, Qiao J, Li W (2019) An enhanced anti-disturbance attitude control law for flexible spacecrafts subject to multiple disturbances. Control Engineering Practice 84: 274–283
112. Yu X, Zhu Y, Qiao J, Guo L (2021) Antidisturbance controllability analysis and enhanced antidisturbance controller design with application to flexible spacecraft. IEEE Transactions on Aerospace and Electronic Systems 57(5): 3393–3404
113. Tafazoli M (2009) A study of on-orbit spacecraft failures. Acta Astronautica 64(2-3): 195–205
114. Zhang Y, Jiang J (2008) Bibliographical review on reconfigurable fault-tolerant control systems. Annual Reviews in Control 32(2): 229–252
115. Yin S, Xiao B, Ding SX, Zhou D (2016) A review on recent development of spacecraft attitude fault tolerant control system. IEEE Transactions on Industrial Electronics 63(5): 3311–3320
116. Cai W, Liao X, Song Y (2008) Indirect robust adaptive fault-tolerant control for attitude tracking of spacecraft. Journal of Guidance, Control, and Dynamics 31(5): 1456–1463
117. Shen Q, Wang D, Zhu S, Poh K (2015) Finite-time fault-tolerant attitude stabilization for spacecraft with actuator saturation. IEEE Transactions on Aerospace and Electronic Systems 51(3): 2390–2405
118. Xia K, Zou Y (2019) Adaptive saturated fault-tolerant control for spacecraft rendezvous with redundancy thrusters. IEEE Transactions on Control Systems Technology 29(2): 502–513

119. Dong H, Hu Q, Friswell MI, Ma G (2016) Dual-quaternion-based fault-tolerant control for spacecraft tracking with finite-time convergence. IEEE Transactions on Control Systems Technology 25(4): 1231–1242
120. Hu Q, Shao X, Chen WH (2017) Robust fault-tolerant tracking control for spacecraft proximity operations using time-varying sliding mode. IEEE Transactions on Aerospace and Electronic Systems 54(1): 2–17
121. Xiao Y, de Ruiter A, Ye D, Sun Z (2021) Adaptive fault-tolerant attitude tracking control for flexible spacecraft with guaranteed performance bounds. IEEE Transactions on Aerospace and Electronic Systems 58(3): 1922–1940
122. Hu Q, Shao X, Guo L (2017) Adaptive fault-tolerant attitude tracking control of spacecraft with prescribed performance. IEEE/ASME Transactions on Mechatronics 23(1): 331–341
123. Shao X, Hu Q, Shi Y, Jiang B (2018) Fault-tolerant prescribed performance attitude tracking control for spacecraft under input saturation. IEEE Transactions on Control Systems Technology 28(2): 574–582
124. Hu Q, Xiao B, Li B, Zhang Y (2021) Fault-Tolerant Attitude Control of Spacecraft. Elsevier
125. Fonod R, Henry D, Charbonnel C, Bornschlegl E, Losa D, Bennani S (2015) Robust fdi for fault-tolerant thrust allocation with application to spacecraft rendezvous. Control Engineering Practice 42: 12–27
126. Shen Q, Yue C, Goh CH, Wang D (2018) Active fault-tolerant control system design for spacecraft attitude maneuvers with actuator saturation and faults. IEEE Transactions on Industrial Electronics 66(5): 3763–3772
127. Li Y, Hu Q, Shao X (2022) Neural network-based fault diagnosis for spacecraft with single-gimbal control moment gyros. Chinese Journal of Aeronautics 35(7): 261–273
128. Li B, Hu Q, Yu Y, Ma G (2017) Observer-based fault-tolerant attitude control for rigid spacecraft. IEEE Transactions on Aerospace and Electronic Systems 53(5): 2572–2582
129. Ran D, Chen X, de Ruiter A, Xiao B (2018) Adaptive extended-state observer-based fault tolerant attitude control for spacecraft with reaction wheels. Acta Astronautica 145: 501–514
130. Hu Q, Zhang X, Niu G (2019) Observer-based fault tolerant control and experimental verification for rigid spacecraft. Aerospace Science and Technology 92: 373–386
131. Gui H (2021) Observer-based fault-tolerant spacecraft attitude tracking using sequential lyapunov analyses. IEEE Transactions on Automatic Control 66(12): 6108–6114
132. Shen Q, Wang D, Zhu S, Poh EK (2015) Inertia-free fault-tolerant spacecraft attitude tracking using control allocation. Automatica 62: 114–121
133. Shen Q, Wang D, Zhu S, Poh EK (2016) Robust control allocation for spacecraft attitude tracking under actuator faults. IEEE Transactions on Control Systems Technology 25(3): 1068–1075
134. Li B, Hu Q, Ma G, Yang Y (2018) Fault-tolerant attitude stabilization incorporating closed-loop control allocation under actuator failure. IEEE Transactions on Aerospace and Electronic Systems 55(4): 1989–2000
135. Hu Q, Li B, Xiao B, Zhang Y (2021) Closed-loop based control allocation for spacecraft attitude stabilization with actuator faults. In: Control Allocation for Spacecraft Under Actuator Faults, Springer, pp 185–217
136. Nicotra MM, Liao-McPherson D, Burlion L, Kolmanovsky IV (2019) Spacecraft attitude control with nonconvex constraints: an explicit reference governor approach. IEEE Transactions on Automatic Control 65(8): 3677–3684
137. Dang Q, Liu K, Wei J (2022) Explicit reference governor based spacecraft attitude reorientation control with constraints and disturbances. Acta Astronautica 190: 455–464
138. Guiggiani A, Kolmanovsky I, Patrinos P, Bemporad A (2015) Fixed-point constrained model predictive control of spacecraft attitude. In: Proceedings of the American Control Conference, Chicago, IL, United states, pp 2317–2322
139. Weiss A, Baldwin M, Erwin RS, Kolmanovsky I (2015) Model predictive control for spacecraft rendezvous and docking: Strategies for handling constraints and case studies. IEEE Transactions on Control Systems Technology 23(4): 1638–1647

140. Li Q, Yuan J, Zhang B, Gao C (2017) Model predictive control for autonomous rendezvous and docking with a tumbling target. Aerospace Science and Technology 69: 700–711
141. Tee KP, Ge SS, Tay EH (2009) Barrier lyapunov functions for the control of output-constrained nonlinear systems. Automatica 45(4): 918–927
142. Bechlioulis CP, Rovithakis GA (2008) Robust adaptive control of feedback linearizable mimo nonlinear systems with prescribed performance. IEEE Transactions on Automatic Control 53(9): 2090–2099
143. Liu M, Shao X, Ma G (2019) Appointed-time fault-tolerant attitude tracking control of spacecraft with double-level guaranteed performance bounds. Aerospace Science and Technology 92: 337–346
144. Yin Z, Suleman A, Luo J, Wei C (2019) Appointed-time prescribed performance attitude tracking control via double performance functions. Aerospace Science and Technology 93: 105337
145. Wei C, Chen Q, Liu J, Yin Z, Luo J (2021) An overview of prescribed performance control and its application to spacecraft attitude system. Proceedings of the Institution of Mechanical Engineers, Part I: Journal of Systems and Control Engineering 235(4): 435–447
146. Huang X, Duan G (2020) Fault-tolerant attitude tracking control of combined spacecraft with reaction wheels under prescribed performance. ISA Transactions 98: 161–172
147. Bechlioulis CP, Rovithakis GA (2014) A low-complexity global approximation-free control scheme with prescribed performance for unknown pure feedback systems. Automatica 50(4): 1217–1226
148. Zhou ZG, Zhang YA, Shi XN, Zhou D (2017) Robust attitude tracking for rigid spacecraft with prescribed transient performance. International Journal of Control 90(11): 2471–2479
149. Luo J, Yin Z, Wei C, Yuan J (2018) Low-complexity prescribed performance control for spacecraft attitude stabilization and tracking. Aerospace Science and Technology 74: 173–183
150. Hu Y, Geng Y, Wu B, Wang D (2020) Model-free prescribed performance control for spacecraft attitude tracking. IEEE Transactions on Control Systems Technology 29(1): 165–179
151. Yong K, Chen M, Shi Y, Wu Q (2020) Flexible performance-based robust control for a class of nonlinear systems with input saturation. Automatica 122: 109268
152. Liu F (2018) Application of artificial intelligence in spacecraft. Flight Control Detection 1(1): 16–25
153. Shirobokov M, Trofimov S, Ovchinnikov M (2021) Survey of machine learning techniques in spacecraft control design. Acta Astronautica 186: 87–97
154. Hu Q, Xiao B (2012) Intelligent proportional-derivative control for flexible spacecraft attitude stabilization with unknown input saturation. Aerospace Science and Technology 23(1): 63–74
155. Li D, Ma G, Li C, He W, Mei J, Ge SS (2018) Distributed attitude coordinated control of multiple spacecraft with attitude constraints. IEEE Transactions on Aerospace and Electronic Systems 54(5): 2233–2245
156. Chen M, Tao G (2015) Adaptive fault-tolerant control of uncertain nonlinear large-scale systems with unknown dead zone. IEEE Transactions on Cybernetics 46(8): 1851–1862
157. Liu Y, Ma G, Lyu Y, Wang P (2022) Neural network-based reinforcement learning control for combined spacecraft attitude tracking maneuvers. Neurocomputing 484: 67–78
158. Cheng CH, Shu SL (2010) Application of ga-based neural network for attitude control of a satellite. Aerospace Science and Technology 14(4): 241–249
159. Zhang Z, Li X, An J, Man W, Zhang G (2020) Model-free attitude control of spacecraft based on PID-guide TD3 algorithm. International Journal of Aerospace Engineering 2020
160. Vamvoudakis KG, Lewis FL (2010) Online actor–critic algorithm to solve the continuous-time infinite horizon optimal control problem. Automatica 46(5): 878–888
161. Liu D, Yang X, Wang D, Wei Q (2015) Reinforcement-learning-based robust controller design for continuous-time uncertain nonlinear systems subject to input constraints. IEEE Transactions on Cybernetics 45(7): 1372–1385
162. Dong H, Zhao X, Luo B (2022) Optimal tracking control for uncertain nonlinear systems with prescribed performance via critic-only adp. IEEE Transactions on Systems, Man, and Cybernetics: Systems 52(1): 561–573

Chapter 2
Dynamics Modeling and Mathematical Preliminaries

2.1 Introduction

As the basis of the following chapters, this chapter will establish the spacecraft (relative) translational and rotational dynamics, and introduce Lyapunov stability theory involved in the attitude and position control design and stability analysis. The spacecraft translational and rotational motions are normally described in specific reference frames to facilitate the dynamics modeling and to define the desired motions. First, several reference frames necessary for describing the spacecraft translational and rotational motions are defined. Then, we establish the (relative) translational and rotational dynamics of spacecraft and the 6-DOF integrated one. In particular, considering the rendezvous and proximity operations (RPOs), the close-range relative motions of two spacecraft (e.g., a pursuer spacecraft and a tumbling target) are divided into two synchronous maneuvers: relative position tracking and line-of-sight (LOS) pointing adjustment. For the former, to simplify the design of the position controller, based on the transport theorem [1], also known as the acceleration synthesis theorem, we deduce a modified relative translational dynamics under the target's body-fixed frame, instead of the classical fully nonlinear Clohessy-Wiltshire (CW) equations; while for the latter, according to the LOS requirements of the visual sensor of onboard the pursuer, we introduce an LOS frame and extract the desired attitude for the relative attitude tracking maneuvers. On this basis, a new 6-DOF integrated relative translational and rotational dynamics of spacecraft are further established. Finally, some mathematical preliminaries on which the works of the following chapters depends is given.

Q. Hu et al., *Intelligent Autonomous Control of Spacecraft with Multiple Constraints*,
https://doi.org/10.1007/978-981-99-0681-9_2

2.2 Notations

Throughout the book, we denote $\mathbb{R}$ as the set of all real numbers, $\mathbb{R}^n$ as the n-dimensional Euclidean space, and $\mathbb{R}^{m\times n}$ as the vector space of $m \times n$ real matrices. $\boldsymbol{I}_n \in \mathbb{R}^{n\times n}$ is the $n \times n$ identity matrix, $|\cdot|$ is the absolute value, while $\|\cdot\|$, $\|\cdot\|_\infty$, and $\|\cdot\|_F$ denote the 2-norm, ∞-norm, and Frobenius-norm, respectively. For a square matrix $\boldsymbol{A}$, $\lambda_{\max}(\boldsymbol{A})$ and $\lambda_{\min}(\boldsymbol{A})$ denote its maximum and minimum eigenvalues, respectively. The notation $(\cdot)^\top$ is the transpose, $(\cdot)^\dagger$ is the pseudo-inverse of a matrix, adj$(\cdot)$ denotes the adjoint matrix, whereas det$(\cdot)$ and rank$(\cdot)$ represent the determinant and rank of a matrix, respectively. The cross-product operator $\boldsymbol{S}(\cdot) : \mathbb{R}^3 \to \mathbb{R}^{3\times 3}$ is defined such that $\boldsymbol{S}(\boldsymbol{x})\boldsymbol{y} = \boldsymbol{x} \times \boldsymbol{y}$, $\forall \boldsymbol{x}, \boldsymbol{y} \in \mathbb{R}^3$, that is, for any vector $\boldsymbol{x} = [x_1, x_2, x_3]^\top \in \mathbb{R}^3$, we have

$$\boldsymbol{S}(\boldsymbol{x}) = \begin{bmatrix} 0 & -x_3 & x_2 \\ x_3 & 0 & -x_1 \\ -x_2 & x_1 & 0 \end{bmatrix}, \tag{2.1}$$

$SO(3) = \{\boldsymbol{R} \in \mathbb{R}^{3\times 3} \mid \boldsymbol{R}^\top \boldsymbol{R} = \boldsymbol{I}_3,\ \ \det(\boldsymbol{R}) = 1\}$ is the special orthogonal group. The unit 3-sphere, a 3-D manifold embedded in $\mathbb{R}^3$, is defined as $\mathbb{S}^3 = \{\boldsymbol{q} \in \mathbb{R}^4 \mid \|\boldsymbol{q}\| = 1\}$. We define the vector spaces $\mathcal{L}_2 = \{\boldsymbol{u}(t) : \mathbb{R} \mapsto \mathbb{R}^n | (\int_0^\infty \boldsymbol{u}^\top(t)\boldsymbol{u}(t)dt)^{\frac{1}{2}} < \infty\}$ and $\mathcal{L}_\infty = \{\boldsymbol{u}(t) : \mathbb{R} \mapsto \mathbb{R}^n | \sup_{t\geq 0} \|\boldsymbol{u}(t)\| < \infty\}$. We denote by $\odot$ the multiplication operator of quaternions or dual quaternions. $\boldsymbol{R}_{\mathcal{AB}}$ is the rotation matrix from the frame $\mathcal{B}$ to the frame $\mathcal{A}$, and $\boldsymbol{x}^{\mathcal{A}}$ refers to the vector $\boldsymbol{x}$ expressed in the frame $\mathcal{A}$. In addition, the standard sign function sgn(x) is determined by

$$\operatorname{sgn}(x) = \begin{cases} -1, & \text{if } x < 0 \\ 0, & \text{if } x = 0\,. \\ +1, & \text{if } x > 0 \end{cases} \tag{2.2}$$

2.3 Coordinate Frames

The coordinate frames involved in describing the translational and rotational motions of an Earth orbiting spacecraft are quite typical and commonly used in the literature even if minor exceptions may exist for different missions. In the following, we introduce the earth-centered inertial frame where the laws of dynamics are naturally written, the orbit reference frame, a reference frame with axes defined by the desired attitude trajectory, the body-fixed frames of the deputy and chief spacecraft (only needed for space missions like RPOs or leader-follower formation flying), as shown in Fig. 2.1.

- **Earth-centered inertial (ECI) frame** $\mathcal{I}$. This is a quasi-inertia coordinate system, and can be assumed as inertia for most applications of Earth orbiting spacecraft.

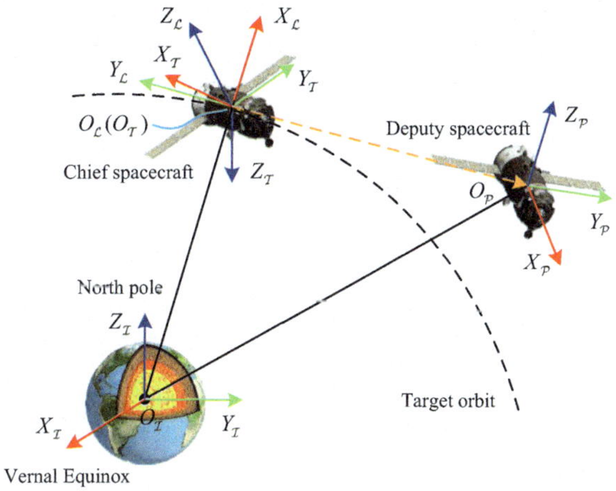

Fig. 2.1 Coordinate systems

It should be noted that the ECI frame is not well-suited as an inertial frame, if the effect of the other celestial bodies cannot be ignored, for example, interplanetary trajectories. The J2000 ECI frame is a specific ECI frame, which is defined on the 1st January, 2000:

- the origin $O_{\mathcal{I}}$ is located at the center of the Earth;
- $\boldsymbol{X}_{\mathcal{I}}$ axis lies in the Earth's equatorial plane along the intersection with the ecliptic plane towards the Vernal Equinox;
- $\boldsymbol{Z}_{\mathcal{I}}$ axis is perpendicular to the equatorial plane and points towards the North Pole, and $\boldsymbol{Y}_{\mathcal{I}}$ axis forms a right-handed frame.

- **Orbit reference frame** $\mathcal{L}$, also known as the Local-Vertical-Local-Horizontal (LVLH) frame. It is useful in describing the relative translational motion of two spacecraft. The triad axes depend on the target's position and velocity.
  - the origin $O_{\mathcal{L}}$ is located at the center of mass (CoM) of the target;
  - $\boldsymbol{X}_{\mathcal{L}}$ axis points radially outward from the Earth's center;
  - $\boldsymbol{Z}_{\mathcal{L}}$ axis is in the direction of the angular momentum of the orbit, and $\boldsymbol{Y}_{\mathcal{L}}$ completes the right-handed coordinate system.
- **The body-fixed frame of the deputy spacecraft** $\mathcal{P}$. The origin $O_{\mathcal{P}}$ is attached to the CoM of deputy spacecraft, and the triad axes are coincide with its principal axes of inertia.
- **The body-fixed frame of the target spacecraft** $\mathcal{T}$, with the origin $O_{\mathcal{T}}$ located at the target's CoM, and the triad axes are coincide with the principal axes of inertia of the target.

With a little abuse of terminologies, in the RPOs mission (resp. formation flying mission), the chief spacecraft is called the target (resp. leader), while the deputy spacecraft is called the pursuer (resp. follower). When considering the RPOs, without loss of generality, we assume that both the docking port (or capture mechanism) and the onboard visual sensors of the pursuer are installed in the $+\boldsymbol{X}_{\mathcal{P}}$ axis of the frame $\mathcal{P}$, and that the docking axis of the target is aligned with $-\boldsymbol{X}_{\mathcal{T}}$ of the frame $\mathcal{T}$.

2.4 Mathematical Models of Spacecraft Dynamics

2.4.1 Spacecraft Attitude Dynamics

We use quaternion for spacecraft attitude representation and denote it as $\boldsymbol{q} = [\boldsymbol{q}_v^\top, q_4]^\top \in \mathbb{R}^4$, where $\boldsymbol{q}_v^\top = [q_1, q_2, q_3]^\top$ and q_4 are the vector and scalar parts, respectively. For any quaternion $\boldsymbol{q}$, $\boldsymbol{q}_x$, $\boldsymbol{q}_y$, and $\boldsymbol{q}_z$, some basic definitions and operations are listed below [2–4].

Conjugate:

$$\boldsymbol{q}^* = [-\boldsymbol{q}_v^\top, q_4]^\top. \tag{2.3}$$

Norm:

$$\|\boldsymbol{q}\| = \sqrt{\boldsymbol{q}_v^\top \boldsymbol{q}_v + q_4^2}. \tag{2.4}$$

Inverse:

$$\boldsymbol{q}^{-1} = \boldsymbol{q}^*/\|\boldsymbol{q}\|. \tag{2.5}$$

Multiplication:

$$\boldsymbol{q}_x \otimes \boldsymbol{q}_y = \begin{bmatrix} q_{x4}\boldsymbol{q}_{yv} + q_{y4}\boldsymbol{q}_{xv} + \boldsymbol{S}(\boldsymbol{q}_{xv})\boldsymbol{q}_{yv} \\ q_{x4}q_{y4} - \boldsymbol{q}_{xv}^\top \boldsymbol{q}_{yv} \end{bmatrix}. \tag{2.6}$$

Algebraic properties:

$$\boldsymbol{q}_x^\top(\boldsymbol{q}_y \otimes \boldsymbol{q}_z) = \boldsymbol{q}_z^\top(\boldsymbol{q}_y^* \otimes \boldsymbol{q}_x) = \boldsymbol{q}_y^\top(\boldsymbol{q}_x \otimes \boldsymbol{q}_z^*). \tag{2.7}$$

For a quaternion satisfying $\|\boldsymbol{q}\| = 1$, it is called a unit quaternion. We denote $\mathbb{Q} = \{\boldsymbol{q} \in \mathbb{R}^4 \mid \boldsymbol{q}_v^\top \boldsymbol{q}_v + q_4^2 = 1\}$ as the set of unit quaternions. Unit quaternion are transformed from Euler-axis/angle and can be written as $\boldsymbol{q} = [\boldsymbol{q}_v^\top, q_4]^\top \in \mathbb{Q}$, where $\boldsymbol{q}_v$ and q_4 are defined as

$$\boldsymbol{q}_v = \boldsymbol{n}\sin(\frac{\theta}{2}),\ q_4 = \cos(\frac{\theta}{2}), \tag{2.8}$$

where $\boldsymbol{n}$ and θ denote the Euler axis and angle, respectively. Define $\boldsymbol{q}_{\mathcal{AB}}$ as the quaternion from frame $\mathcal{A}$ to $\mathcal{B}$, then for any $\boldsymbol{x} \in \mathbb{R}^3$, its expression $\boldsymbol{x}^{\mathcal{A}}$ in $\mathcal{A}$ and expression $\boldsymbol{x}^{\mathcal{B}}$ in $\mathcal{B}$ satisfy:

$$\vec{\boldsymbol{x}}^{\mathcal{A}} = \boldsymbol{q}_{\mathcal{AB}}^* \otimes \vec{\boldsymbol{x}}^{\mathcal{B}} \otimes \boldsymbol{q}_{\mathcal{AB}}, \tag{2.9}$$

where $\overset{\rightarrow\mathcal{A}}{\boldsymbol{x}} = [(\boldsymbol{x}^{\mathcal{A}})^\top, 0]^\top, \overset{\rightarrow\mathcal{B}}{\boldsymbol{x}} = [(\boldsymbol{x}^{\mathcal{B}})^\top, 0]^\top$. In addition, unit quaternion $\boldsymbol{q}_{\mathcal{AC}}$ from frame $\mathcal{A}$ to C can be obtained by $\boldsymbol{q}_{\mathcal{AB}}$ and $\boldsymbol{q}_{\mathcal{BC}}$ (unit quaternion from frame $\mathcal{B}$ to C), that is

$$\boldsymbol{q}_{\mathcal{AC}} = \boldsymbol{q}^*_{\mathcal{CB}} \otimes \boldsymbol{q}_{\mathcal{AB}}. \tag{2.10}$$

As a matter of fact, the conjugate of unit quaternion represents its inverse transformation, satisfying

$$\boldsymbol{q}^*_{\mathcal{AB}} \otimes \boldsymbol{q}_{\mathcal{AB}} = \boldsymbol{q}_I, \tag{2.11}$$

where $\boldsymbol{q}_I = [0, 0, 0, 1]^\top$. In addition, the corresponding rotation matrix for a given unit quaternion $\boldsymbol{q}_{\mathcal{AB}} = [(\boldsymbol{q}_{\mathcal{AB}v})^\top, q_{\mathcal{AB}4}]^\top \in \mathbb{Q}$ is [5]

$$\boldsymbol{R}_{\mathcal{AB}} = (q^2_{\mathcal{AB}4} - \boldsymbol{q}^\top_{\mathcal{AB}v}\boldsymbol{q}_{\mathcal{AB}v})\boldsymbol{I}_3 + 2\boldsymbol{q}_{\mathcal{AB}v}\boldsymbol{q}^\top_{\mathcal{AB}v} - 2q_{\mathcal{AB}4}\boldsymbol{S}(\boldsymbol{q}_{\mathcal{AB}v}). \tag{2.12}$$

Based on unit quaternion representation, the spacecraft attitude kinematics and dynamics are described as

$$\dot{\boldsymbol{q}} = \frac{1}{2}\boldsymbol{Q}(\boldsymbol{q})\boldsymbol{\omega}, \tag{2.13}$$

$$\boldsymbol{J}\dot{\boldsymbol{\omega}} = -\boldsymbol{S}(\boldsymbol{\omega})\boldsymbol{J}\boldsymbol{\omega} + \boldsymbol{\tau}_c + \boldsymbol{\tau}_d, \tag{2.14}$$

where $\boldsymbol{q} = [\boldsymbol{q}_v^\top, q_4]^\top \in \mathbb{Q}$ is the unit quaternion describing the orientation of the body-fixed frame $\mathcal{P}$ with respect to the inertial frame $\mathcal{I}$, $\boldsymbol{\omega} \in \mathbb{R}^3$ is spacecraft angular velocity in $\mathcal{P}$, $\boldsymbol{J} \in \mathbb{R}^{3\times 3}$ is spacecraft inertia matrix resolved in $\mathcal{P}$, $\boldsymbol{\tau}_c \in \mathbb{R}^3$ is the control torque exerting on the spacecraft expressed in $\mathcal{P}$, and $\boldsymbol{\tau}_d \in \mathbb{R}^3$ is the disturbance torque caused by gravity gradient torque, solar radiation pressure torque, etc. In addition, $\boldsymbol{Q}(\cdot)$ is defined as

$$\boldsymbol{Q}(\boldsymbol{q}) = \begin{bmatrix} q_4\boldsymbol{I}_3 + \boldsymbol{S}(\boldsymbol{q}_v) \\ -\boldsymbol{q}_v^\top \end{bmatrix}, \quad \forall \boldsymbol{q} = [\boldsymbol{q}_v^\top, q_4]^\top \in \mathbb{Q}. \tag{2.15}$$

Considering the attitude tracking problem, we denote $\boldsymbol{q}_d$ and $\boldsymbol{\omega}_d$ as the desired attitude and angular velocity, respectively. Then, attitude error $\boldsymbol{q}_e = [\boldsymbol{q}_{ev}^\top, q_{e4}]^\top \in \mathbb{Q}$ is computed as

$$\boldsymbol{q}_e = \boldsymbol{q}_d^{-1} \otimes \boldsymbol{q}. \tag{2.16}$$

Based on (2.12), angular velocity error $\boldsymbol{\omega}_e \in \mathbb{R}^3$ is

$$\boldsymbol{\omega}_e = \boldsymbol{\omega} - \boldsymbol{R}\boldsymbol{\omega}_d \tag{2.17}$$

where the rotation matrix $\boldsymbol{R} \in SO(3)$ from the reference frame to the spacecraft body-fixed frame $\mathcal{P}$ is given by

$$\boldsymbol{R} = (q^2_{e4} - \boldsymbol{q}^\top_{ev}\boldsymbol{q}_{ev})\boldsymbol{I}_3 + 2\boldsymbol{q}_{ev}\boldsymbol{q}^\top_{ev} - 2q_{e4}\boldsymbol{S}(\boldsymbol{q}_{ev}). \tag{2.18}$$

Therefore, attitude error kinematics and dynamics are

$$\dot{\boldsymbol{q}}_e = \frac{1}{2}\boldsymbol{Q}(\boldsymbol{q}_e)\boldsymbol{\omega}_e, \tag{2.19}$$

$$\boldsymbol{J}\dot{\boldsymbol{\omega}}_e = -\boldsymbol{S}(\boldsymbol{\omega})\boldsymbol{J}\boldsymbol{\omega} + \boldsymbol{J}(\boldsymbol{S}(\boldsymbol{\omega}_e)\boldsymbol{R}\boldsymbol{\omega}_d - \boldsymbol{R}\dot{\boldsymbol{\omega}}_d) + \boldsymbol{\tau}_c + \boldsymbol{\tau}_d. \tag{2.20}$$

It should be noted that for the attitude stabilization or maneuver, $\boldsymbol{\omega}_d = \boldsymbol{0}$ and $\boldsymbol{\omega}_e = \boldsymbol{\omega}$, and the dynamic equation (2.20) reduces to

$$\boldsymbol{J}\dot{\boldsymbol{\omega}}_e = -\boldsymbol{S}(\boldsymbol{\omega})\boldsymbol{J}\boldsymbol{\omega} + \boldsymbol{\tau}_c + \boldsymbol{\tau}_d. \tag{2.21}$$

2.4.2 Spacecraft Relative Position Dynamics

Ignoring the space perturbation forces, the orbit dynamics of the deputy (e.g., pursuer or follower) and chief (e.g., target or leader) spacecraft in the inertial frame $\mathcal{I}$ can be described as [6]:

$$\ddot{\boldsymbol{r}}_p^{\mathcal{I}} = -\mu\frac{\boldsymbol{r}_p^{\mathcal{I}}}{r_p^3} + \frac{\boldsymbol{f}^{\mathcal{I}}}{m_p}, \tag{2.22}$$

$$\ddot{\boldsymbol{r}}_t^{\mathcal{I}} = -\mu\frac{\boldsymbol{r}_t^{\mathcal{I}}}{r_t^3}, \tag{2.23}$$

where $\mu = 398600.4418\,\mathrm{kg}^3/\mathrm{m}^2$ is the gravitational constant of earth, $\boldsymbol{r}_p^{\mathcal{I}} \in \mathbb{R}^3$ and $\boldsymbol{r}_t^{\mathcal{I}} \in \mathbb{R}^3$ denote, respectively, the position vectors of the deputy and chief spacecraft, expressed in $\mathcal{I}$, $r_p = \|\boldsymbol{r}_p^{\mathcal{I}}\|$, $r_t = \|\boldsymbol{r}_t^{\mathcal{I}}\|$, m_p is the mass of the deputy spacecraft, and $\boldsymbol{f}^{\mathcal{I}} \in \mathbb{R}^3$ is the control force exerting on the deputy spacecraft in $\mathcal{I}$.

Define $\boldsymbol{r}^{\mathcal{I}} = \boldsymbol{r}_p^{\mathcal{I}} - \boldsymbol{r}_t^{\mathcal{I}}$ as the relative position of the deputy spacecraft with respect to the chief spacecraft, as shown in Fig. 2.2. Based on (2.20) and (2.21), we can obtain the relative position dynamics in $\mathcal{I}$:

$$\ddot{\boldsymbol{r}}^{\mathcal{I}} = -\mu\left(\frac{\boldsymbol{r}_p^{\mathcal{I}}}{r_p^3} - \frac{\boldsymbol{r}_t^{\mathcal{I}}}{r_t^3}\right) + \frac{\boldsymbol{f}^{\mathcal{I}}}{m_p}. \tag{2.24}$$

Considering that the measurement of relative position is convenient in $\mathcal{L}$, (2.22) is transformed in $\mathcal{L}$ for traditional dynamics modeling. By resorting to the transport theorem [1], the acceleration $\ddot{\boldsymbol{r}}^{\mathcal{I}}$ in $\mathcal{I}$ satisfies:

$$\ddot{\boldsymbol{r}}^{\mathcal{I}} = \frac{d^2\boldsymbol{r}}{dt^2} + 2\boldsymbol{\omega}_l \times \frac{d\boldsymbol{r}}{dt} + \frac{d\boldsymbol{\omega}_l}{dt} \times \boldsymbol{r} + \boldsymbol{\omega}_l \times (\boldsymbol{\omega}_l \times \boldsymbol{r}) \tag{2.25}$$

Fig. 2.2 Illustration of relative position motion of two spacecraft

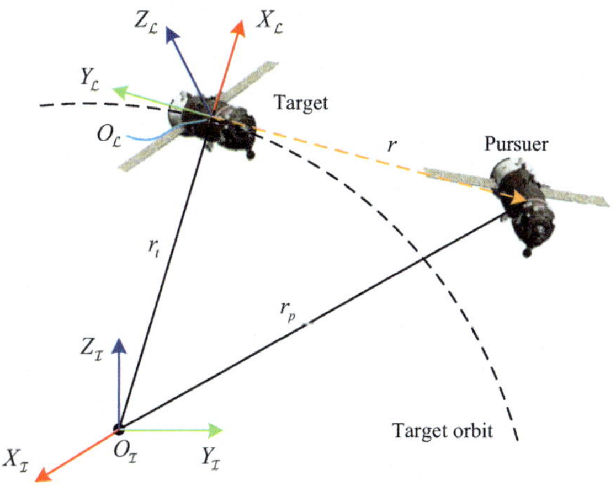

where $\boldsymbol{r} \in \mathbb{R}^3$ is the relative position $\boldsymbol{r}^{\mathcal{I}}$ described in $\mathcal{L}$, $\boldsymbol{\omega}_l = [0, 0, \dot{\upsilon}_o]^\top$ is the orbital angular velocity, υ_o is true anomaly of the chief orbit. Based on (2.22) and (2.23),the Euler-Lagrangian relative position dynamics in $\mathcal{L}$ is [7]

$$\boldsymbol{M}_p^* \ddot{\boldsymbol{r}} + \boldsymbol{C}_p^*(\dot{\upsilon}_o)\dot{\boldsymbol{r}} + \boldsymbol{D}_p^*(\dot{\upsilon}_o, \ddot{\upsilon}_o, r_p)\boldsymbol{r} + \boldsymbol{n}_p^*(r_p, r_t) = \boldsymbol{f}^{\mathcal{L}} \tag{2.26}$$

where $\boldsymbol{M}_p^* = m_p \boldsymbol{I}_3$, $\boldsymbol{C}_p^*(\dot{\upsilon}_o) = m_p \boldsymbol{S}(\boldsymbol{n}_1)$ is the Coriolis-like matrix, $\boldsymbol{n}_1 = [0, 0, 2\dot{\upsilon}_o]^\top$, $\boldsymbol{D}_p^*(\dot{\upsilon}_o, \ddot{\upsilon}_o, r_p)\boldsymbol{r} = m_p \boldsymbol{a}(\dot{\upsilon}_o, \ddot{\upsilon}_o, r_p)\boldsymbol{r}$ can be regarded as a potential force, $\boldsymbol{n}_p^*(r_p, r_t) = m_p \boldsymbol{n}_2(r_p, r_t)$ is a nonlinear term, $\boldsymbol{a}(\dot{\upsilon}_o, \ddot{\upsilon}_o, r_p)$ and $\boldsymbol{n}_2(r_p, r_t)$ are given by

$$\boldsymbol{a}(\dot{\upsilon}_o, \ddot{\upsilon}_o, r_p) = \frac{\mu}{r_p^3}\boldsymbol{I}_3 + \begin{bmatrix} -\dot{\upsilon}_o^2 & -\ddot{\upsilon}_o & 0 \\ \ddot{\upsilon}_o & -\dot{\upsilon}_o^2 & 0 \\ 0 & 0 & 0 \end{bmatrix}, \quad \boldsymbol{n}_2(r_p, r_t) = \mu \left[\frac{r_t}{r_p^3} - \frac{1}{r_t^2}, 0, 0 \right]^\top \tag{2.27}$$

where $r_t = a_o(1 - e_o^2)/(1 + e_o \cos \upsilon_o)$, a_o and e_o are semi-major axis and orbital eccentricity respectively, r_t and υ_o are given by [7]

$$\begin{cases} \dot{r}_t = -\dfrac{r_t \ddot{\upsilon}_o}{2\dot{\upsilon}_o}, \quad \dot{\upsilon}_o = \dfrac{n_o(1 + e_o \cos \upsilon_o)^2}{(1 - e_o^2)^{\frac{3}{2}}} \\ \ddot{\upsilon}_o = \dfrac{-2n_o^2 e_o (1 + e_o \cos \upsilon_o)^3 \sin \theta}{(1 - e_o^2)^3} \end{cases} \tag{2.28}$$

where $n_o = \sqrt{\mu / a_o^3}$ is the average orbital angular velocity, $r_p = \|\boldsymbol{r} + \boldsymbol{r}_t\|$ with $\boldsymbol{r}_t = [r_t, 0, 0]^\top$, and $\boldsymbol{f}^{\mathcal{L}}$ is the description of $\boldsymbol{f}$ in $\mathcal{L}$.

2.4.3 Spacecraft Relative Position-Attitude Coupled Dynamics

In this subsection, a relative position-attitude coupled dynamics is established for spacecraft RPOs with a tumbling target. For spacecraft RPOs, the pursuer needs to move towards the desired position while adjusting its attitude such that the boresight of its onboard vision sensor points toward the target for real-time and accurate relative state measurements. This, in fact, involves two synchronously occurring maneuvers: relative position tracking and boresight pointing adjustment. For the former, a new relative translational dynamics is established to facilitate its problem formulation and solving, while, for the latter, an LOS frame is introduced to align the boresight of the pursuer's onboard vision sensor toward the target and the desired attitude is extracted, whereby a new relative rotational dynamics is established. After that, by integrating the two models, a new integrated 6-DOF relative position-attitude dynamics is formulated.

Assumption 2.1 Since the engagement times are much shorter than the target's orbital period, it is reasonable to neglect the target orbital perturbations, and suppose that the pursuer is immune to external disturbances.

2.4.3.1 Modified Relative Position Dynamics

For the relative position tracking maneuvers, the traditional relative position dynamics is established in the LVLH frame $\mathcal{L}$, like the fully nonlinear CW equation (2.26). However, since such a model involves complex coordinate transformations and algebraic operations, it inevitably increases the difficulty of position controller design, especially when some motion constraints are considered. To solve this problem, a new relative position dynamics model is derived under the target body-fixed frame $\mathcal{T}$ by using the transport theorem [1], which can effectively simplify the design and analysis of the position controller.

Similar to (2.25), the acceleration $\ddot{\boldsymbol{r}}^I$ satisfies

$$\ddot{\boldsymbol{r}}^I = \frac{d^2\boldsymbol{\rho}}{dt^2} + 2\boldsymbol{\omega}_t \times \frac{d\boldsymbol{\rho}}{dt} + \frac{d\boldsymbol{\omega}_t}{dt} \times \boldsymbol{\rho} + \boldsymbol{\omega}_t \times (\boldsymbol{\omega}_t \times \boldsymbol{\rho}), \tag{2.29}$$

where $\boldsymbol{\rho} \in \mathbb{R}^3$ is the vector $\boldsymbol{r}^I$ expressed in the frame $\mathcal{T}$. Similar to the deduction of (2.26), it follows from (2.3) that

$$\boldsymbol{M}_p\ddot{\boldsymbol{\rho}} + \boldsymbol{C}_p(\boldsymbol{\omega}_t)\dot{\boldsymbol{\rho}} + \boldsymbol{D}_p(\boldsymbol{\omega}_t, \dot{\boldsymbol{\omega}}_t, \rho_s)\boldsymbol{\rho} + \boldsymbol{n}_p(\boldsymbol{\rho}_s, \boldsymbol{\rho}_t) = \boldsymbol{f}^{\mathcal{T}}, \tag{2.30}$$

where $\boldsymbol{M}_p = m_p\boldsymbol{I}_3$ is the inertia matrix, $\boldsymbol{C}_p = 2m_p\boldsymbol{S}(\boldsymbol{\omega}_t)$ is a Coriolis-like matrix, $\boldsymbol{D}_p\boldsymbol{\rho} = m_p[\boldsymbol{S}(\dot{\boldsymbol{\omega}}_t) + (\boldsymbol{S}(\boldsymbol{\omega}_t))^2 + (\mu/\rho_p^3)\boldsymbol{I}_3]\boldsymbol{\rho}$ can be viewed as a time-varying potential force, $\boldsymbol{n}_p = m_p\mu(\boldsymbol{\rho}_t/\rho_p^3 - \boldsymbol{\rho}_t/\rho_t^3)$ is a nonlinear term, and $\boldsymbol{f}^{\mathcal{T}}$ is the control force expressed in $\mathcal{T}$. Note that the position vectors $\boldsymbol{\rho}$, $\boldsymbol{\rho}_p$, and $\boldsymbol{\rho}_t$ are the projections of $\boldsymbol{r}^I$, $\boldsymbol{r}_p^I$, and $\boldsymbol{r}_t^I$ in the frame $\mathcal{T}$; moreover, we have $\rho_s = r_s$ and $\rho_t = r_t$. Furthermore, $\boldsymbol{\rho}_s = \boldsymbol{\rho}_t + \boldsymbol{\rho}$ and $\boldsymbol{\rho}_t = \boldsymbol{R}_{\mathcal{LT}}^\top[\rho_t, 0, 0]^\top$, where the scalar $\rho_t = a_o(1 - e_o^2)/(1 + e_o \cos v_o)$ with a_o, e_o and v_o the semimajor axis, orbital eccentricity and true anomaly of the target orbit, and $\boldsymbol{R}_{\mathcal{LT}} = \boldsymbol{R}_{\mathcal{LI}}\boldsymbol{R}_{\mathcal{TI}}^\top$ with $\boldsymbol{R}_{\mathcal{LI}}$ given by

$$\boldsymbol{R}_{\mathcal{LI}} = \begin{bmatrix} c_{\Omega_o}c_{\theta_o} - s_{\Omega_o}s_{\theta_o}c_{i_o} & s_{\Omega_o}c_{\theta_o} + c_{\Omega_o}s_{\theta_o}c_{i_o} & s_{\theta_o}s_{i_o} \\ -c_{\Omega_o}s_{\theta_o} - s_{\Omega_o}c_{\theta_o}c_{i_o} & -s_{\Omega_o}c_{\theta_o}c_{i_o} & c_{\theta_o}s_{i_o} \\ s_{\Omega_o}s_{i_o} & -c_{\Omega_o}s_{i_o} & c_{i_o} \end{bmatrix}$$

where $s_* \triangleq \sin(*)$ and $c_* \triangleq \cos(*)$ are the "sine" and "cosine" functions; $\theta_o = \omega_o + v_o$ is the argument of latitude; and Ω_o, ω_o, and i_o are the right ascension of the ascending node, argument of perigee, and orbit inclination, respectively.

Remark 2.1 The relative translational dynamics in (2.30) are, in essence, established by transforming the fully nonlinear CW equations in (2.26) from the frame $\mathcal{L}$ to the frame $\mathcal{T}$. Note, however, that an implicit assumption is that although the tumbling target is uncontrolled, its attitude information (i.e., attitude and angular velocity) is always available to the pursuer. For specific missions, this assumption is practically reasonable since the target's attitude information could be acquired by the onboard sensors of the pursuer or provided by the target itself.

2.4.3.2 Modified Relative Attitude Dynamics

We denote $\boldsymbol{q}_p = [\boldsymbol{q}_{pv}^\top, q_{p4}]^\top \in \mathbb{Q}$ and $\boldsymbol{\omega}_p \in \mathbb{R}^3$ as the attitude and angular velocity of the pursuer in the frame $\mathcal{P}$ with respect to (w.r.t.) the ECI frame $\mathcal{I}$, respectively. Let $\boldsymbol{q}_t = [q_{tv}^\top, q_{e4}]^\top \in \mathbb{Q}$ and $\boldsymbol{\omega}_t \in \mathbb{R}^3$ denote the inertial attitude and angular velocity of the target, expressed in the frame $\mathcal{T}$. Then, the attitude motion of the tumbling target is governed by

$$\dot{\boldsymbol{q}}_t = \frac{1}{2}\boldsymbol{Q}(\boldsymbol{q}_t)\boldsymbol{\omega}_t, \quad \boldsymbol{Q}(\boldsymbol{q}_t) = \begin{bmatrix} q_{t4}\boldsymbol{I}_3 + \boldsymbol{S}(\boldsymbol{q}_{tv}) \\ -\boldsymbol{q}_{tv}^\top \end{bmatrix}, \tag{2.31}$$

$$\boldsymbol{J}_t\dot{\boldsymbol{\omega}}_t = -\boldsymbol{S}(\boldsymbol{\omega}_t)\boldsymbol{J}_t\boldsymbol{\omega}_t, \tag{2.32}$$

where $\boldsymbol{J}_t \in \mathbb{R}^{3\times3}$ is the inertia matrix of the target. Similarly, the attitude motion of the pursuer can be described as

$$\dot{\boldsymbol{q}}_p = \frac{1}{2}\boldsymbol{Q}(\boldsymbol{q}_p)\boldsymbol{\omega}_p, \tag{2.33}$$

$$\boldsymbol{J}_p\dot{\boldsymbol{\omega}}_p = -\boldsymbol{S}(\boldsymbol{\omega}_p)\boldsymbol{J}_p\boldsymbol{\omega}_p + \boldsymbol{\tau}_c, \tag{2.34}$$

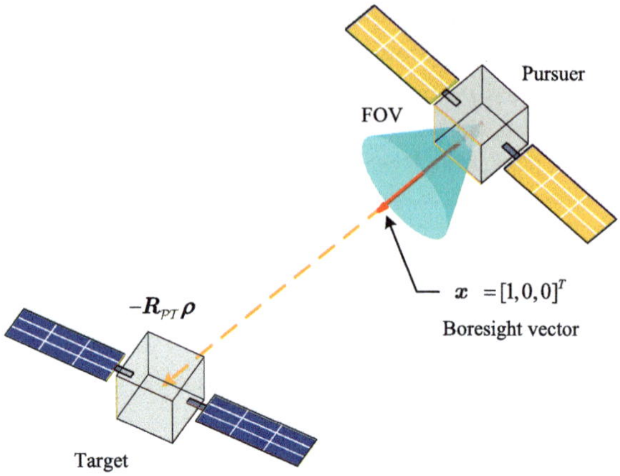

Fig. 2.3 Illustration of boresight pointing of visual sensor

where $\boldsymbol{J}_p \in \mathbb{R}^{3\times 3}$ is the inertia matrix of the pursuer; and $\boldsymbol{\tau}_c \in \mathbb{R}^3$ denotes the control torques exerting on the pursuer in $\mathcal{P}$.

Without loss of generality, we assume that the boresight axis of the visual sensor onboard the pursuer perfectly coincides with the $+\boldsymbol{X}_{\mathcal{P}}$ axis of the frame $\mathcal{P}$, that is, the boresight vector is $\boldsymbol{x}_{\mathcal{P}} = [1, 0, 0]^\top$, as shown in Fig. 2.3, where the cone with half-cone angle β represents the FOV of the visual sensor. Intuitively, the ultimate control goal for the relative attitude motion is to keep the boresight vector $\boldsymbol{x}_{\mathcal{P}}$ of the sensor oriented toward the target. This is actually equivalent to achieving $\boldsymbol{x}_{\mathcal{P}} = -\boldsymbol{R}_{\mathcal{PT}}\boldsymbol{\rho}/\|\boldsymbol{\rho}\|$. To achieve this, an LOS frame $\mathcal{D}$ is introduced such that the following holds (the attitude of the frame $\mathcal{D}$ relative to the ECI frame $\mathcal{I}$ is expressed as $\boldsymbol{q}_d = [\boldsymbol{q}_{dv}^\top, q_{d4}]^\top \in \mathbb{Q}$, and the corresponding rotation matrix $\boldsymbol{R}_{\mathcal{DT}}$):

$$\boldsymbol{R}_{\mathcal{DT}}\underbrace{(-\boldsymbol{\rho}/\|\boldsymbol{\rho}\|)}_{\boldsymbol{x}_\rho} = \boldsymbol{x}_{\mathcal{D}} \Longrightarrow \boldsymbol{R}_{\mathcal{DT}}\boldsymbol{x}_\rho = \boldsymbol{x}_{\mathcal{D}}, \tag{2.35}$$

where $\boldsymbol{x}_{\mathcal{D}}$ is the unit vector along the X-axis of $\mathcal{D}$, and $\|\boldsymbol{x}_\rho\| = \|\boldsymbol{x}_{\mathcal{D}}\| = 1$. To obtain $\boldsymbol{q}_d$, we need to extract the relative attitude $\bar{\boldsymbol{q}} = [\bar{\boldsymbol{q}}_v^\top, \bar{q}_4]^\top \in \mathbb{Q}$ of the LOS frame $\mathcal{D}$ with respect to the target's body frame $\mathcal{T}$. It should be noted that (2.35) can be rewritten as follows [8]:

$$\bar{\boldsymbol{q}}^{-1} \otimes \begin{bmatrix} \boldsymbol{x}_\rho \\ 0 \end{bmatrix} \otimes \bar{\boldsymbol{q}} = \begin{bmatrix} \boldsymbol{x}_{\mathcal{D}} \\ 0 \end{bmatrix}. \tag{2.36}$$

Let us rewrite $\bar{\boldsymbol{q}}$ as $\bar{\boldsymbol{q}} = (\hat{\boldsymbol{k}} \sin(\delta/2), \cos(\delta/2))$, where $\hat{\boldsymbol{k}}$ and $\delta \in [0, 2\pi)$ are Euler axis and Euler angle, respectively. A rotation scheme that satisfies (2.36) is $\hat{\boldsymbol{k}} = \boldsymbol{S}(\boldsymbol{x}_{\mathcal{D}})\boldsymbol{x}_\rho$. Then, $\bar{\boldsymbol{q}}$ can be determined as per the following lemma.

Lemma 2.1 *Under the condition that* $\|\boldsymbol{\rho}\| \neq 0$ *and* $\boldsymbol{x}_\rho \neq -\boldsymbol{x}_{\mathcal{D}}$*, the unit quaternion* $\bar{\boldsymbol{q}}$ *that satisfies* (2.35) *certainly exists, and a feasible solution that can minimize the rotation angle for its extraction is given by:*

$$\bar{\boldsymbol{q}}_v = \frac{\boldsymbol{S}(\boldsymbol{x}_{\mathcal{D}})\boldsymbol{x}_\rho}{\sqrt{2(1+\boldsymbol{x}_\rho^\top \boldsymbol{x}_{\mathcal{D}})}}, \quad \bar{q}_4 = \sqrt{\frac{1+\boldsymbol{x}_\rho^\top \boldsymbol{x}_{\mathcal{D}}}{2}}. \tag{2.37}$$

Proof Since the rotation axis is chosen as $\hat{\boldsymbol{k}} = \boldsymbol{S}(\boldsymbol{x}_{\mathcal{D}})\boldsymbol{x}_\rho$, one can get

$$\boldsymbol{x}_\rho^\top \boldsymbol{x}_{\mathcal{D}} = \|\boldsymbol{x}_\rho\| \|\boldsymbol{x}_{\mathcal{D}}\| \cos(\delta), \tag{2.38}$$

$$\boldsymbol{S}(\boldsymbol{x}_{\mathcal{D}})\boldsymbol{x}_\rho = \|\boldsymbol{x}_\rho\| \|\boldsymbol{x}_{\mathcal{D}}\| \sin(\delta)\hat{\boldsymbol{k}}. \tag{2.39}$$

As $\|\boldsymbol{x}_\rho\| = \|\boldsymbol{x}_{\mathcal{D}}\| = 1$, it follows from (2.38) that $\cos(\delta) = \boldsymbol{x}_\rho^\top \boldsymbol{x}_{\mathcal{D}}$. Further, using the fact that $\sin^2(\delta) + \cos^2(\delta) = 1$, one get $\sin(\delta) = (1 - (\boldsymbol{x}_\rho^\top \boldsymbol{x}_{\mathcal{D}})^2)^{1/2}$. Substituting $\sin(\delta)$ into (2.39) yields $\hat{\boldsymbol{k}} = \boldsymbol{S}(\boldsymbol{x}_{\mathcal{D}})\boldsymbol{x}_\rho/(1 - (\boldsymbol{x}_\rho^\top \boldsymbol{x}_{\mathcal{D}})^2)^{1/2}$. Then, according to the half angle formula $\cos(\delta) = 1 - 2\sin^2(\delta/2)$, one can obtain $\sin(\delta/2) = ((1 - \boldsymbol{x}_\rho^\top \boldsymbol{x}_{\mathcal{D}})/2)^{1/2}$, which together with the fact that $\sin^2(\delta/2) + \cos^2(\delta/2) = 1$ gives:

$$\bar{q}_4 = \cos(\delta/2) = \sqrt{\frac{1+\boldsymbol{x}_\rho^\top \boldsymbol{x}_{\mathcal{D}}}{2}}. \tag{2.40}$$

In addition, combining $\hat{\boldsymbol{k}}$ and $\sin(\delta/2)$ leads to

$$\bar{\boldsymbol{q}}_v = \sin(\delta/2)\hat{\boldsymbol{k}} = \frac{\boldsymbol{S}(\boldsymbol{x}_{\mathcal{D}})\boldsymbol{x}_\rho}{\sqrt{2(1+\boldsymbol{x}_\rho^\top \boldsymbol{x}_{\mathcal{D}})}}. \tag{2.41}$$

This completes the proof. □

Remark 2.2 The extraction algorithm sketched by Lemma 2.1 choses $\hat{\boldsymbol{k}} = \boldsymbol{S}(\boldsymbol{x}_{\mathcal{D}})\boldsymbol{x}_\rho$ as the rotation axis. Intuitively, such a rotation corresponds to a minimum angle rotation, which is instrumental for reducing fuel consumption and maneuvering time. Note, however, that it will suffer from a singularity problem when $\|\boldsymbol{\rho}\| = 0$ and/or $\boldsymbol{x}_\rho = -\boldsymbol{x}_{\mathcal{D}}$. Thus, to ensure that the extraction scheme is feasible, the controller design should guarantee that these two conditions do not hold.

In fact, $\bar{\boldsymbol{q}} \in \mathbb{Q}$ describes the discrepancy between $\boldsymbol{q}_d$ and $\boldsymbol{q}_t$ and, thus, can be calculated by $\bar{\boldsymbol{q}} = \boldsymbol{q}_t^{-1} \otimes \boldsymbol{q}_d$. According to the quaternion multiplication rule, one can get

$$\boldsymbol{q}_d = \boldsymbol{q}_t \otimes \bar{\boldsymbol{q}} = \begin{bmatrix} q_{t4}\bar{\boldsymbol{q}}_v + \bar{q}_4\boldsymbol{q}_{tv} + \boldsymbol{S}(\boldsymbol{q}_{tv})\bar{\boldsymbol{q}}_v \\ q_{t4}\bar{q}_4 - \boldsymbol{q}_{tv}^\top \bar{\boldsymbol{q}}_v \end{bmatrix}. \tag{2.42}$$

For $\bar{\boldsymbol{q}}$, its corresponding angular velocity (denoted by $\bar{\omega}$) can be obtained based on the well-known kinematic equation $\dot{\bar{\boldsymbol{q}}} = \frac{1}{2}\boldsymbol{Q}(\bar{\boldsymbol{q}})\bar{\omega}$. As $\boldsymbol{Q}(\bar{\boldsymbol{q}})^\top \boldsymbol{Q}(\bar{\boldsymbol{q}}) = \boldsymbol{I}_3$, the angular velocity $\bar{\omega}$ can then be deduced from

$$\bar{\omega} = \omega_d - \boldsymbol{R}_{\mathcal{DT}}\omega_t = 2\boldsymbol{Q}^\top(\bar{\boldsymbol{q}})\dot{\bar{\boldsymbol{q}}}, \tag{2.43}$$

where ω_d denotes the desired angular velocity, and $\dot{\bar{\boldsymbol{q}}}$ can be easily calculated from (2.37). It thus follows from (2.43) that

$$\omega_d = 2\boldsymbol{Q}^\top(\bar{\boldsymbol{q}})\dot{\bar{\boldsymbol{q}}} + \boldsymbol{R}_{\mathcal{DT}}\omega_t, \tag{2.44}$$

with $\boldsymbol{R}_{\mathcal{DT}}$ being given by (2.12) in terms of $\bar{\boldsymbol{q}}$.

To meet the boresight pointing requirement, we here define the attitude tracking error $\boldsymbol{q}_e = [\boldsymbol{q}_{ev}^\top, q_{e4}]^\top \in \mathbb{Q}$ as the relative orientation between the frames $\mathcal{D}$ and $\mathcal{P}$, which is computed as

$$\boldsymbol{q}_e = \boldsymbol{q}_d^{-1} \otimes \boldsymbol{q}_p = \begin{bmatrix} q_{d4}\boldsymbol{q}_{pv} - q_{p4}\boldsymbol{q}_{dv} + \boldsymbol{S}(\boldsymbol{q}_{pv})\boldsymbol{q}_{dv} \\ q_{d4}q_{p4} + \boldsymbol{q}_{dv}^\top \boldsymbol{q}_{pv} \end{bmatrix}. \tag{2.45}$$

The rotation matrix from $\mathcal{D}$ to $\mathcal{P}$ is given by $\boldsymbol{R}_{\mathcal{PD}} = (q_{e4}^2 - \boldsymbol{q}_{ev}^\top \boldsymbol{q}_{ev})\boldsymbol{I}_3 + 2\boldsymbol{q}_{ev}\boldsymbol{q}_{ev}^\top - 2q_{e4}\boldsymbol{S}(\boldsymbol{q}_{ev})$ according to (2.12). Then, the relative angular velocity of $\mathcal{P}$ with respect to $\mathcal{D}$ can be defined as $\omega_e = \omega_p - \boldsymbol{R}_{\mathcal{PD}}\omega_d$.

The rotational tracking error dynamics can be derived that

$$\dot{\boldsymbol{q}}_e = \frac{1}{2}\boldsymbol{Q}(\boldsymbol{q}_e)\omega_e, \tag{2.46}$$

$$\boldsymbol{J}_p\dot{\omega}_e = -\boldsymbol{S}(\omega_p)\boldsymbol{J}_p\omega_p + \boldsymbol{J}_p(\boldsymbol{S}(\omega_e)\boldsymbol{\Omega} - \bar{\boldsymbol{\Omega}}) + \boldsymbol{\tau}_c, \tag{2.47}$$

where $\boldsymbol{\Omega} = \boldsymbol{R}_{\mathcal{PD}}\omega_d$ and $\bar{\boldsymbol{\Omega}} = \boldsymbol{R}_{\mathcal{PD}}\dot{\omega}_d$. Let us define $\boldsymbol{P} = (0.5(q_{e4}\boldsymbol{I}_3 + \boldsymbol{S}(\boldsymbol{q}_{ev})))^{-1}$, then the dynamics described by (2.46) and (2.47) can be transformed into the Euler-Lagrange equation by following the exposition in [9]

$$\boldsymbol{M}_r\ddot{\boldsymbol{q}}_{ev} + \boldsymbol{C}_r\dot{\boldsymbol{q}}_{ev} + \boldsymbol{G}_r = \boldsymbol{P}^\top\boldsymbol{\tau}_c, \tag{2.48}$$

where $\boldsymbol{M}_r = \boldsymbol{P}^\top\boldsymbol{J}_p\boldsymbol{P}, \boldsymbol{C}_r = \boldsymbol{P}^\top\boldsymbol{J}_p\dot{\boldsymbol{P}} - \boldsymbol{P}^\top\boldsymbol{S}(\boldsymbol{J}_p\boldsymbol{P}\dot{\boldsymbol{q}}_{ev})\boldsymbol{P}$, and $\boldsymbol{G}_r = \boldsymbol{P}^\top[\boldsymbol{S}(\omega_e)\boldsymbol{J}_p \boldsymbol{\Omega} + \boldsymbol{S}(\boldsymbol{\Omega})\boldsymbol{J}_p(\omega_e + \boldsymbol{\Omega}) - \boldsymbol{J}_p(\boldsymbol{S}(\omega_e)\boldsymbol{\Omega} - \bar{\boldsymbol{\Omega}})]$.

Remark 2.3 To ensure the transformation from (2.46) and (2.47)–(2.48) is valid, the following condition must hold:

$$\det(q_{e4}\boldsymbol{I}_3 + \boldsymbol{S}(\boldsymbol{q}_{ev})) = q_{e4}(t) \neq 0, \quad \forall t \geq 0. \tag{2.49}$$

As such, it is required that the initial condition be restricted such that $q_{e4}(0) \neq 0$, and the controller be designed to guarantee $q_{e4}(t) \neq 0$ holds for all time.

Lemma 2.2 *The 2-norm of* $\boldsymbol{P}$ *satisfies* $\|\boldsymbol{P}\| = 2/|q_{e4}|$.

Proof According to the definition of 2-norm, $\|\boldsymbol{P}_2\|$ equals to the maximum singular value of $\boldsymbol{P}_2$. Considering

$$\boldsymbol{P}^\top \boldsymbol{P} = (\boldsymbol{Q}(\boldsymbol{q}_e)\boldsymbol{Q}^\top(\boldsymbol{q}_e))^{-1}. \tag{2.50}$$

To obtain $\|\boldsymbol{P}_2\|$, we calculate the eigenvalues of the matrix $\boldsymbol{Q}(\boldsymbol{q}_e)\boldsymbol{Q}^\top(\boldsymbol{q}_e)$ by using MATLAB symbolic computations. With a slight abuse of notations, we let x_1, x_2, x_3, and x_4 denote, respectively, q_{e1}, q_{e2}, q_{e3}, and q_{e4}. The code is given below.

```
syms x_1 x_2 x_3 x_4 real;
Q = 0.5*[x_4 -x_3 x_2; x_3 x_4 -x_1; -x_2 x_1 x_4];
assumeAlso(x_1^2+x_2^2+x_3^2+x_4^2==1);
simplify(eig(Q*Q'));
```

Executing the above codes yields the eigenvalues of $\boldsymbol{Q}(\boldsymbol{q}_e)\boldsymbol{Q}^\top(\boldsymbol{q}_e)$, i.e., $q_{e4}^2/4$, $1/4$, and $1/4$, from whose inverses we immediately get the maximum singular value of $\boldsymbol{P}$ as $2/|q_{e4}|$, due to $|q_{e4}| \leq 1$. It thus follows that $\|\boldsymbol{P}\| = 2/|q_{e4}|$. □

Remark 2.4 It is a time-consuming and sometimes troublesome task to obtain the analytical expression for $\dot{\boldsymbol{\omega}}_d$. To bypass this barrier, we can let $\boldsymbol{\omega}_d$ pass through a low-pass filter of the form $c\dot{\boldsymbol{z}} = -\boldsymbol{z} + \boldsymbol{\omega}_d$, where $c > 0$ is the filter time constant denoting the bandwidth of the filter. By choosing c sufficient small, $\boldsymbol{z}$ can be viewed as equivalent to $\boldsymbol{\omega}_d$, and at the same time, $\dot{\boldsymbol{z}} \approx \dot{\boldsymbol{\omega}}_d$. Therefore, $\dot{\boldsymbol{z}}$ can be used in lieu of $\dot{\boldsymbol{\omega}}_d$ in the subsequent control design.

2.4.3.3 Modified Relative Position-Attitude Coupled Dynamics

During the close-range proximity operations, the pursuer needs to achieve precise pose maneuvers so that the pursuer finally reaches a safe position above the target's receiving port with its docking port points to the target's receiving port. To this end, we introduce a desired relative position vector $\boldsymbol{\rho}_d = [\rho_d, 0, 0]^\top$ ($\rho_d < 0$), and define the relative position tracking error $\boldsymbol{\rho}_e = \boldsymbol{\rho} - \boldsymbol{\rho}_d$. Considering that the orbit control actuators of the pursuer is installed in the frame $\mathcal{T}$, the control force vector $\boldsymbol{d}^\mathcal{T}$ in (2.30) is equivalent to

$$\boldsymbol{f}^\mathcal{T} = \boldsymbol{R}_{\mathcal{PT}}^\top \boldsymbol{f}_c, \tag{2.51}$$

where $\boldsymbol{R}_{\mathcal{FT}}$ can be obtained by substituting $\boldsymbol{q}_e$ into (2.12), $\boldsymbol{f}_c$ is the control force expressed in the pursuer's body frame $\mathcal{P}$. In view of (2.30) and (2.51), the relative position tracking error dynamics can be expressed as

$$\boldsymbol{M}_p\ddot{\boldsymbol{\rho}}_e + \boldsymbol{C}_p\dot{\boldsymbol{\rho}}_e + \boldsymbol{G}_p = \boldsymbol{R}_{\mathcal{PT}}^\top \boldsymbol{f}_c, \tag{2.52}$$

where $\boldsymbol{G}_p = \boldsymbol{D}_p\boldsymbol{\rho} + \boldsymbol{n}_p$.

Define $\boldsymbol{e} = [\boldsymbol{\rho}_e^\top, \boldsymbol{q}_{ev}^\top]^\top \in \mathbb{R}^6$ as the pose tracking error. By integrating (2.48) and (2.52), the pose tracking error dynamics can be written as

$$\boldsymbol{M}\ddot{\boldsymbol{e}} + \boldsymbol{C}\dot{\boldsymbol{e}} + \boldsymbol{G} = \boldsymbol{A}\boldsymbol{u}, \tag{2.53}$$

where $\boldsymbol{M} = \text{blkdiag}\{\boldsymbol{M}_p, \boldsymbol{M}_r\}$, $\boldsymbol{C} = \text{blkdiag}\{\boldsymbol{C}_p, \boldsymbol{C}_r\}$, $\boldsymbol{G} = [\boldsymbol{G}_p^\top, \boldsymbol{G}_r^\top]^\top$, and $\boldsymbol{A} = \text{blkdiag}\{\boldsymbol{R}_{\mathcal{PT}}^\top, \boldsymbol{P}^\top\}$.

The dynamics (2.53) possesses the following properties:

Property 2.1 The matrix $\boldsymbol{M}$ is symmetric positive definite, and satisfies the following

$$\lambda_{\min}(\boldsymbol{M})\|\boldsymbol{x}\|^2 \le \boldsymbol{x}^\top \boldsymbol{M}\boldsymbol{x} \le \lambda_{\max}(\boldsymbol{M})\|\boldsymbol{x}\|^2, \quad \forall \boldsymbol{x} \in \mathbb{R}^3.$$

Property 2.2 $\dot{\boldsymbol{M}} - 2\boldsymbol{C}$ is skew-symmetric, that is, for any $\boldsymbol{x} \in \mathbb{R}^3$, one has

$$\boldsymbol{x}^\top(\dot{\boldsymbol{M}} - 2\boldsymbol{C})\boldsymbol{x} = 0.$$

Remark 2.5 The integrated relative position-attitude dynamics model (2.53) is a 6-DOF coupled model, and its coupling characteristics mainly lie in the following two aspects: (1) from (2.51), it can be seen that the control force vector for the relative position motion depends on the pursuer attitude, so the relative position motion will be affected by the relative attitude motion; (2) from formula (2.37), it is not difficult to find that the extraction of the desired attitude has a direct relationship with the relative position vector, thus the relative attitude motion is also affected by the relative position motion. In addition, the coupling properties also include the natural system coupling caused by various external disturbances. However, they are not explicitly considered in the dynamics modeling.

2.4.3.4 Thruster Configuration

It is common practice to endow the pursuer with a thrusters-only actuation system for the 6-DOF pose control, owing to its advantages of light and simple system design and of providing great agility in proximity operations. However, this inevitably induces dynamic coupling between translational and rotational motions. Since all the thrusters are fixed in $\mathcal{P}$, the control force in $\mathcal{T}$ can be expressed as $\boldsymbol{f} = \boldsymbol{R}_{\mathcal{PT}}^\top \boldsymbol{f}_c$, where $\boldsymbol{f}_c$ is the control force expressed in $\mathcal{P}$, and $\boldsymbol{R}_{\mathcal{PT}} = \boldsymbol{R}_{\mathcal{PI}}\boldsymbol{R}_{\mathcal{TI}}^\top$ with $\boldsymbol{R}_{\mathcal{PI}}$ and $\boldsymbol{R}_{\mathcal{TI}}$ given according to (2.12) in terms of $\boldsymbol{q}_p$ and $\boldsymbol{q}_t$, respectively. The thrusters used here work in an on-off mode with constant magnitudes. Thus, the control design for the pursuer with only thrusters is typically carried out in the following steps:

- Solve the control problem for the 6-DOF rototranslational dynamics, yielding continuous control laws $\boldsymbol{f}_c$ and $\boldsymbol{\tau}_c$.
- Map $\boldsymbol{f}_c$ and $\boldsymbol{\tau}_c$ into a continuous input command $\boldsymbol{u} \in \mathbb{R}^N$ (N is the number of thrusters), which provides the desired force outputs for the thrusters.

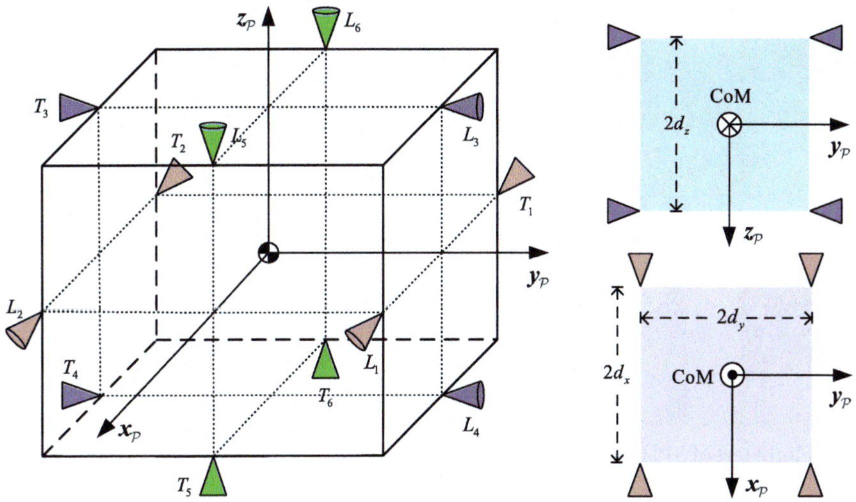

Fig. 2.4 Thruster configuration

- Convert u_i ($i = 1, 2, ..., N$) into on-off commands u_i^* by using the pulse modulation technique.

Following this line, we can write the control as

$$[\boldsymbol{f}_c^\top, \boldsymbol{\tau}_c^\top]^\top = \boldsymbol{D}\boldsymbol{u}, \tag{2.54}$$

where $\boldsymbol{D} \in \mathbb{R}^{6\times N}$ denotes the thrust distribution matrix related directly to the geometrical structure of N thrusters' placement on the pursuer. A generic thruster configuration is selected as the case study of this chapter, as shown in Fig. 2.4. The actuation system consists of 12 thrusters, each of which can only provide unidirectional thrust. The constants d_x, d_y, and d_z serve as the moment arms of the thrusters w.r.t. the CoM of the pursuer. Let us further arrange the thrusters in 6 thruster pairs, i.e., $\{T_i, L_i\}$, $i = 1, 2, ..., 6$, and each thruster pair can provide bidirectional thrust. Within this setting, $\boldsymbol{D}$ is as follows:

$$\boldsymbol{D} = \begin{bmatrix} 1 & 1 & 0 & 0 & 0 & 0 \\ 0 & 0 & 1 & 1 & 0 & 0 \\ 0 & 0 & 0 & 0 & 1 & 1 \\ 0 & 0 & -d_z & d_z & 0 & 0 \\ 0 & 0 & 0 & 0 & -d_x & d_x \\ -d_y & d_y & 0 & 0 & 0 & 0 \end{bmatrix}. \tag{2.55}$$

Notably, the structure of Fig. 2.4 is not based upon the particular geometry of the pursuer, but with a slight modification, it can be applied to many space vehicles, for example, the SPHERES testbed [10], as shown in Fig. 2.5.

(a) Flight test of SPHERES satellites in ISS (b) Structure of the SPHERES satellite

Fig. 2.5 SPHERES test satellite platform

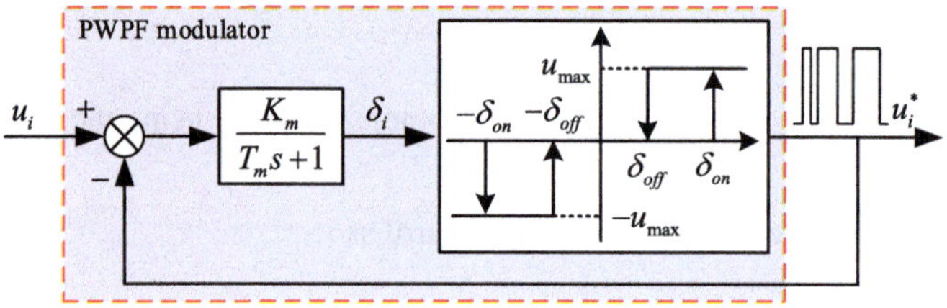

Fig. 2.6 PWPF modulator

Due to thruster's on-off nature, a modulation mechanism is therefore required to modulate the continuous input commands u_i $(i = 1, 2, ..., 6)$ in pulsed signals u_i^* suitable for controlling thruster firing. In practical engineering, some commonly used pulse modulators have Schmitt trigger, pulse-width modulator, pulse-width pulse-frequency (PWPF) modulator [11]. In this book, the default modulator is chosen as the PWPF modulator, whose structure is shown in Fig. 2.6. The parameters of interest are the prefilter coefficients K_m and T_m, the Schmidt Trigger parameters δ_{on} and δ_{off}, and the thrust magnitude $u_{\max}$. The interested readers may refer to [12] for more details about the PWPF modulator and its parameter setting.

2.4.3.5 Model Validation

In this subsection, the correctness of the modified relative position dynamics (2.30) and the effectiveness of the attitude extraction algorithm summarized in Lemma 2.1 will be demonstrated via simulations. Assume that the tumbling target orbits the Earth in a Molniya orbit with initial orbital elements listed in Table 2.1. The

Table 2.1 Initial orbital elements

Orbital elements	Vaules	Units
Semimajor axis	26628	km
Eccentricity	0.7417	–
Inclination	63.4	deg
RAAN	0	deg
Argument of perigee	–90	deg
True anomaly	0	deg

attitude evolution of the target is described by (2.31) and (2.32) with the inertia matrix $\boldsymbol{J}_t = \text{diag}[22, 20, 23]\ \text{kg} \cdot \text{m}^2$ and the initial conditions: $\boldsymbol{q}_t(0) = [0, 0, 0, 1]^\top$ and $\boldsymbol{\omega}_t(0) = [0.01, -0.01, 0.01]^\top$ rad/s. Moreover, the gravity-gradient torque is considered as the external disturbance, given by $\boldsymbol{\tau}_{td} = 3\mu(\boldsymbol{S}(\boldsymbol{\rho}_t)\boldsymbol{J}_t\boldsymbol{\rho}_t)/\|\boldsymbol{\rho}_t\|^5$.

The nominal mass and inertia of the pursuer are given as follows:

$$m_p = 200\ \text{kg}, \ \text{and}\ \boldsymbol{J}_p = \begin{bmatrix} 55 & 0.3 & 0.5 \\ 0.3 & 65 & 0.2 \\ 0.5 & 0.2 & 58 \end{bmatrix} \text{kg} \cdot \text{m}^2.$$

Initially, the relative position and velocity of the pursuer with respect to the target are $\boldsymbol{\rho}(0) = [150, -100, 80]^\top$ m and $\dot{\boldsymbol{\rho}}(0) = [-0.1, 0.5, -0.3]^\top$ m/s. First, we testify the correctness of the modified relative position dynamics (2.30). Rewrite (2.30) as

$$\boldsymbol{M}_p\ddot{\boldsymbol{\rho}} + \boldsymbol{C}_p\dot{\boldsymbol{\rho}} + \boldsymbol{D}_p\boldsymbol{\rho} + \boldsymbol{n}_p = \boldsymbol{0}. \tag{2.56}$$

Consider also the fully nonlinear CW equations (2.26)

$$\boldsymbol{M}_p^*\ddot{\boldsymbol{r}} + \boldsymbol{C}_p^*\dot{\boldsymbol{r}} + \boldsymbol{D}_p^*\boldsymbol{r} + \boldsymbol{n}_p^* = \boldsymbol{0}, \tag{2.57}$$

The relative position vector $\boldsymbol{\rho}$ is, in essence, the expression of $\boldsymbol{r}$ in $\mathcal{T}$. For fair comparison, the initial values of (2.54) is taken as $\boldsymbol{r}(0) = \boldsymbol{R}_{\mathcal{LT}}(0)\boldsymbol{\rho}(0)$ and $\dot{\boldsymbol{r}}(0) = \boldsymbol{R}_{\mathcal{LT}}(0)\boldsymbol{S}(\boldsymbol{\omega}_{tl}(0))\boldsymbol{\rho}(0) + \boldsymbol{R}_{\mathcal{LT}}(0)\dot{\boldsymbol{\rho}}(0)$, where $\boldsymbol{\omega}_{tl}(0) = \boldsymbol{\omega}_t(0) - \boldsymbol{R}_{\mathcal{LT}}^\top(0)\boldsymbol{\omega}_l(0)$, and $\boldsymbol{\omega}_l(0) = [0, 0, \dot{\upsilon}_o(0)]^\top$. Fig. 2.7 shows the comparison responses of equations (2.56) and (2.57) under the same initial conditions. To intuitively show the comparison results, we transform the output $\boldsymbol{\rho}$ to the LVLH frame $\boldsymbol{L}$ by using the relationship $\boldsymbol{\rho}^{\mathcal{L}} = \boldsymbol{R}_{\mathcal{LT}}\boldsymbol{\rho}$. From Fig. 2.7, it is clear that the two dynamics have the same responses of the relative position vector, indicating that the correctness of the dynamics model (2.30) is, in essence, derived by transforming the fully nonlinear CW equations (2.26) from $\mathcal{L}$ to $\mathcal{T}$.

Next, we will illustrate the validity of the attitude extraction algorithm outlined in Lemma 2.1. As a special case, the pursuer is assumed to perform a circular circumnavigation around the target with a radius of 100 m in the $\boldsymbol{Y}_{\mathcal{T}} - \boldsymbol{Z}_{\mathcal{T}}$ plane.

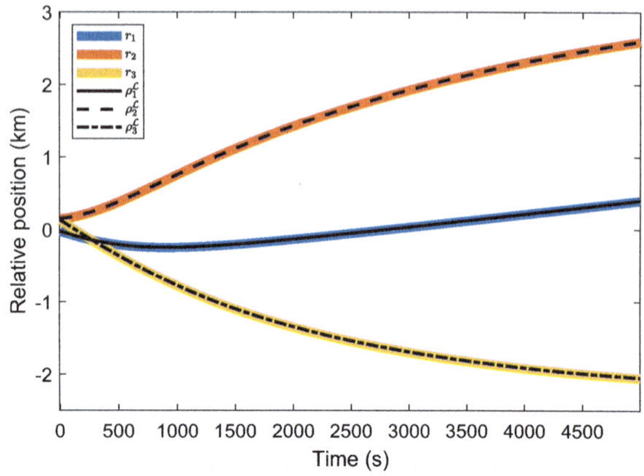

Fig. 2.7 Validation result for the derived relative translational dynamics

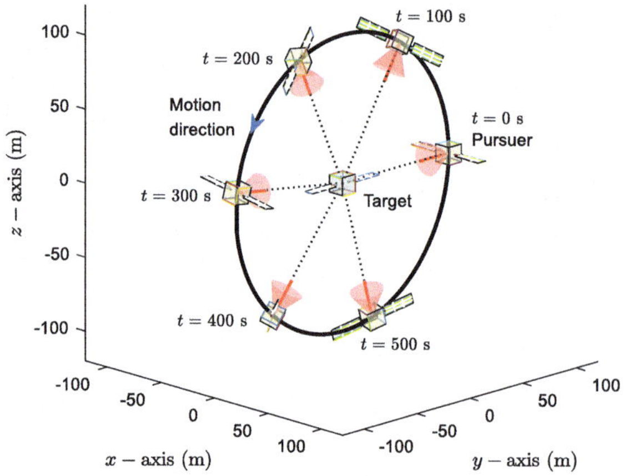

Fig. 2.8 Validation result for the derived attitude extraction algorithm

Specifically, the relative position vector is chosen as

$$\rho(t) = [0, 100\cos(0.01t), 100\sin(0.01t)]^\top \text{ m}.$$

The desired attitude is extracted using (2.35)–(2.42). The three-dimensional (3-D) motion trajectory of the pursuer and the snapshots of its desired attitude observed in $\mathcal{T}$ is depicted in Fig. 2.8. In the figure, the target and the pursuer are, respectively, portrayed as the yellow and grey cubes with solar panels, while the red arrow and cone binding on the pursuer are, respectively, the boresight axis and FOV of the onboard vision sensor. As illustrated in Fig. 2.8, the extracted desired attitude can indeed render the boresight axis of the onboard vision sensor oriented towards the target. Thus, the boresight pointing adjustment will be achieved when the desired attitude is tracked.

2.4.4 *Dual-Quaternion-Based Spacecraft Relative Motion Dynamics*

2.4.4.1 Dual Number and Dual Quaternion

The concept of dual numbers was originally introduced by Clifford [13] and further improved by Study [14]. Let $\hat{\mathbb{R}}$ denote the field of dual numbers. For a scalar dual number $\hat{a} \in \hat{\mathbb{R}}$, its definition is

$$\hat{a} = a_r + \varepsilon a_d, \tag{2.58}$$

where $a_r,\ a_d \in \mathbb{R}$ denote, respectively, the "real part" and "dual part" of $\hat{a}$, and ε is the dual unit satisfying

$$\varepsilon^2 = 0,\ \text{and}\ \varepsilon \neq 0, \tag{2.59}$$

A dual vector $\hat{\boldsymbol{a}} \in \hat{\mathbb{R}}^m$ is a dual number whose real and dual parts are both vectors, and can be defined as follows:

$$\hat{\boldsymbol{a}} = \boldsymbol{a}_r + \varepsilon \boldsymbol{a}_d, \tag{2.60}$$

where $\boldsymbol{a}_r,\ \boldsymbol{a}_d \in \mathbb{R}^m$. In fact, the dual number can be regarded as a special case of dual vector. Some commonly used algebraic operations and related properties of dual vectors are given as follows:

$$\lambda\hat{\boldsymbol{a}} = \lambda\boldsymbol{a}_r + \varepsilon\lambda\boldsymbol{a}_d,\ \lambda \in \mathbb{R}$$

$$\hat{\boldsymbol{a}}^\top = \boldsymbol{a}_r^\top + \varepsilon\boldsymbol{a}_d^\top,$$

$$\hat{\boldsymbol{a}}^s = \boldsymbol{a}_d + \varepsilon\boldsymbol{a}_r,\ (\hat{\boldsymbol{a}}^s)^s = \hat{\boldsymbol{a}},$$

$$\|\hat{\boldsymbol{a}}\| = \|\boldsymbol{a}_r\| + \varepsilon\|\boldsymbol{a}_d\|,$$

$$\mathrm{sgn}(\hat{\boldsymbol{a}}) = \mathrm{sgn}(\boldsymbol{a}_r) + \varepsilon\,\mathrm{sgn}(\boldsymbol{a}_d),$$

$$\hat{a}\hat{\boldsymbol{a}} = a_r\boldsymbol{a}_r + \varepsilon(a_r\boldsymbol{a}_d + a_d\boldsymbol{a}_r),\ \hat{a} \in \hat{\mathbb{R}}$$

$$\hat{a} \odot \hat{\boldsymbol{a}} = a_r\boldsymbol{a}_r + \varepsilon a_d\boldsymbol{a}_d,\ \hat{a} \in \hat{\mathbb{R}}$$

$$\hat{\boldsymbol{a}}_1 \pm \hat{\boldsymbol{a}}_2 = \boldsymbol{a}_{1r} \pm \boldsymbol{a}_{2r} + \varepsilon(\boldsymbol{a}_{1d} \pm \boldsymbol{a}_{2d}),$$

$$\hat{\boldsymbol{a}}_1 \cdot \hat{\boldsymbol{a}}_2 = \hat{\boldsymbol{a}}_1^\top\hat{\boldsymbol{a}}_2 = \boldsymbol{a}_{1r} \cdot \boldsymbol{a}_{2r} + \varepsilon(\boldsymbol{a}_{1r} \cdot \boldsymbol{a}_{2d} + \boldsymbol{a}_{1d} \cdot \boldsymbol{a}_{2r}),$$

$$\hat{\boldsymbol{a}}_1^\top \odot \hat{\boldsymbol{a}}_2 = \boldsymbol{a}_{1r}^\top\boldsymbol{a}_{2r} + \varepsilon\boldsymbol{a}_{1d}^\top\boldsymbol{a}_{2d},$$

$$\hat{\boldsymbol{a}}_1 \circ \hat{\boldsymbol{a}}_2 = \boldsymbol{a}_{1r}^\top\boldsymbol{a}_{2r} + \boldsymbol{a}_{1d}^\top\boldsymbol{a}_{2d},\ \hat{\boldsymbol{a}}_1^s \circ \hat{\boldsymbol{a}}_2^s = \hat{\boldsymbol{a}}_1 \circ \hat{\boldsymbol{a}}_2,$$

where $\lambda \in \mathbb{R}$, $\hat{\boldsymbol{a}}_1,\ \hat{\boldsymbol{a}}_2 \in \hat{\mathbb{R}}^m$, $\mathrm{sgn}(\cdot)$ denotes the standard sign function, and the notation $(\cdot)^s$ denotes the swap operation. Note that the dual vector dot multiplication (denoted by "$\cdot$") is not commutative, while the dual vector circle product (denoted by "$\circ$") is commutative. For $\hat{\boldsymbol{a}}_1,\ \hat{\boldsymbol{a}}_2 \in \hat{\mathbb{R}}^3$, the cross product operation can be defined by

$$\begin{aligned}\hat{\boldsymbol{a}}_1 \times \hat{\boldsymbol{a}}_2 &= -\,\hat{\boldsymbol{a}}_2 \times \hat{\boldsymbol{a}}_1 \\ &= \boldsymbol{a}_{1r} \times \boldsymbol{a}_{2r} + \varepsilon(\boldsymbol{a}_{1r} \times \boldsymbol{a}_{2d} + \boldsymbol{a}_{1d} \times \boldsymbol{a}_{2r}).\end{aligned} \tag{2.61}$$

Similar to real vectors, the cross product of dual vectors can be expressed as

$$\hat{\boldsymbol{a}}_1 \times \hat{\boldsymbol{a}}_2 = \boldsymbol{S}(\hat{\boldsymbol{a}}_1)\hat{\boldsymbol{a}}_2, \tag{2.62}$$

where $\boldsymbol{S}(\hat{\boldsymbol{a}}) = -\boldsymbol{S}^\top(\hat{\boldsymbol{a}})$ is a skew-symmetric matrix of the following form

$$\boldsymbol{S}(\hat{\boldsymbol{a}}) = \begin{bmatrix} 0 & -\hat{a}_3 & \hat{a}_2 \\ \hat{a}_3 & 0 & -\hat{a}_1 \\ -\hat{a}_2 & \hat{a}_1 & 0 \end{bmatrix} = \boldsymbol{S}(\boldsymbol{a}_r) + \varepsilon\boldsymbol{S}(\boldsymbol{a}_d), \tag{2.63}$$

where $\hat{\boldsymbol{a}} = [\hat{a}_1, \hat{a}_2, \hat{a}_3]^\top \in \hat{\mathbb{R}}^3$ with $\hat{a}_i = a_{ir} + \varepsilon a_{id}$ $(i = 1, 2, 3)$ being the i-th element of $\hat{\boldsymbol{a}}$. Some key properties of dual vectors are given in Lemma 2.3.

Lemma 2.3 *For any dual vectors $\hat{\boldsymbol{a}},\ \hat{\boldsymbol{b}},\ \hat{\boldsymbol{c}} \in \hat{\mathbb{R}}^3$, the following hold*

$$\hat{\boldsymbol{a}} \times (\hat{\boldsymbol{b}} \times \hat{\boldsymbol{c}}) + \hat{\boldsymbol{b}} \times (\hat{\boldsymbol{c}} \times \hat{\boldsymbol{a}}) + \hat{\boldsymbol{c}} \times (\hat{\boldsymbol{a}} \times \hat{\boldsymbol{b}}) = \hat{\boldsymbol{0}}_3, \tag{2.64}$$

$$\hat{\boldsymbol{a}} \times (\hat{\boldsymbol{b}} \times \hat{\boldsymbol{c}}) = \hat{\boldsymbol{b}}(\hat{\boldsymbol{a}} \cdot \hat{\boldsymbol{c}}) - \hat{\boldsymbol{c}}(\hat{\boldsymbol{a}} \cdot \hat{\boldsymbol{b}}), \tag{2.65}$$

$$(\hat{\boldsymbol{a}} \times \hat{\boldsymbol{b}}) \times \hat{\boldsymbol{a}} = \hat{\boldsymbol{0}}_3,\ \hat{\boldsymbol{a}} \cdot (\hat{\boldsymbol{a}} \times \hat{\boldsymbol{b}}) = \hat{\boldsymbol{0}}_3, \tag{2.66}$$

$$\hat{\boldsymbol{a}}^s \circ (\hat{\boldsymbol{a}} \times \hat{\boldsymbol{b}}) = \hat{\boldsymbol{0}}_3, \tag{2.67}$$

where $\hat{\boldsymbol{0}}_3 = \boldsymbol{0}_3 + \varepsilon\boldsymbol{0}_3$ is a three-dimensional dual zero vector.

Proof Please refer to [15] for the detailed proof. □

A dual quaternion $\hat{\boldsymbol{q}} \in \mathbb{DQ}$ ($\mathbb{DQ}$ is the set of dual quaternions) can be described as a quaternion with dual numbers as coefficients, and can be defined as

$$\hat{\boldsymbol{q}} = \boldsymbol{q}_r + \varepsilon\boldsymbol{q}_d, \tag{2.68}$$

where $\boldsymbol{q}_r, \boldsymbol{q}_r \in \mathbb{R}^4$ are quaternions. Similar to the vector notation for quaternions, a dual quaternion can also be expressed as a column vector by stacking its real and dual parts together as follows:

$$\hat{\boldsymbol{q}} = [\hat{\boldsymbol{q}}_v^\top, \hat{q}_4]^\top, \tag{2.69}$$

where $\hat{\boldsymbol{q}}_v = \boldsymbol{q}_{rv} + \varepsilon \boldsymbol{q}_{dv} \in \hat{\mathbb{R}}^3$ and $\hat{q}_4 = q_{r4} + \varepsilon q_{d4} \in \hat{\mathbb{R}}$ denote the vector and scalar parts of $\hat{\boldsymbol{q}}$. As a special case of dual vectors, the dual quaternions not only obey the aforementioned algebraic rules of dual vectors, but also inherit many definitions and algebraic rules of quaternions. The basic operations on dual quaternions are defined as follows [16]:

Addition and subtraction:

$$\hat{\boldsymbol{q}}_x \pm \hat{\boldsymbol{q}}_y = [\hat{\boldsymbol{q}}_{xv}^\top \pm \hat{\boldsymbol{q}}_{yv}^\top, \hat{q}_{x4} \pm \hat{q}_{y4}]^\top, \ \hat{\boldsymbol{q}}_x, \hat{\boldsymbol{q}}_y \in \mathbb{DQ}. \tag{2.70}$$

Multiplication by a scalar λ:

$$\lambda \hat{\boldsymbol{q}} = [\lambda \hat{\boldsymbol{q}}_v^\top, \lambda \hat{q}_4]^\top. \tag{2.71}$$

Multiplication:

$$\begin{aligned} \hat{\boldsymbol{q}}_x \otimes \hat{\boldsymbol{q}}_y &= \hat{\boldsymbol{q}}_{xr} \otimes \boldsymbol{q}_{yr} + \varepsilon(\boldsymbol{q}_{xr} \otimes \boldsymbol{q}_{yd} + \boldsymbol{q}_{xd} \otimes \boldsymbol{q}_{yr}) \\ &= [(\hat{q}_{x4}\hat{\boldsymbol{q}}_{yv} + \hat{q}_{y4}\hat{\boldsymbol{q}}_{xv} + \hat{\boldsymbol{q}}_{xv} \times \hat{\boldsymbol{q}}_{yv})^\top, \hat{q}_{x4}\hat{q}_{y4} - \hat{\boldsymbol{q}}_{xv}^\top \hat{\boldsymbol{q}}_{yv}]^\top \\ &= [\hat{\boldsymbol{q}}_x]_\otimes \hat{\boldsymbol{q}}_y, \end{aligned} \tag{2.72}$$

with

$$[\hat{\boldsymbol{q}}_x]_\otimes = \begin{bmatrix} \boldsymbol{S}(\hat{\boldsymbol{q}}_v) + \hat{q}_{x4}\boldsymbol{I}_3 & \hat{\boldsymbol{q}}_{xv} \\ -\hat{\boldsymbol{q}}_{xv}^\top & \hat{q}_{x4} \end{bmatrix}.$$

Conjugation:

$$\hat{\boldsymbol{q}}^* = [-\hat{\boldsymbol{q}}_v^\top, \hat{q}_4]^\top. \tag{2.73}$$

Vector part:

$$\mathrm{vec}(\hat{\boldsymbol{q}}) = \hat{\boldsymbol{q}}_v. \tag{2.74}$$

Scalar part:

$$\mathrm{sc}(\hat{\boldsymbol{q}}) = \hat{q}_4. \tag{2.75}$$

Dual norm:

$$\|\hat{\boldsymbol{q}}\|^2 = \hat{\boldsymbol{q}}\hat{\boldsymbol{q}}^* = \hat{\boldsymbol{q}}^*\hat{\boldsymbol{q}} = \hat{\boldsymbol{q}} \cdot \hat{\boldsymbol{q}} = (\boldsymbol{q}_r \cdot \boldsymbol{q}_r) + \varepsilon(2\boldsymbol{q}_r \cdot \boldsymbol{q}_d). \tag{2.76}$$

Gradient:

$$\hat{\nabla}_{\hat{q}} V(\hat{\boldsymbol{q}}) = \frac{\partial}{\partial \boldsymbol{q}_r} V(\hat{\boldsymbol{q}}) + \varepsilon \frac{\partial}{\partial \boldsymbol{q}_d} V(\hat{\boldsymbol{q}}), \ V(\hat{\boldsymbol{q}}) \in \mathbb{R}. \tag{2.77}$$

The set of unit dual quaternions can be defined as $\hat{\mathbb{Q}} = \{\hat{\boldsymbol{q}} \in \mathbb{DQ} \mid \hat{\boldsymbol{q}} \otimes \hat{\boldsymbol{q}}^* = \hat{\boldsymbol{q}}^* \otimes \hat{\boldsymbol{q}} = \hat{\boldsymbol{1}}\}$, where $\hat{\boldsymbol{1}} = \boldsymbol{q}_I + \varepsilon \boldsymbol{0}_4$. Thus, for unit dual quaternions $\hat{\boldsymbol{q}}, \hat{\boldsymbol{p}} \in \hat{\mathbb{Q}}$, we have $\hat{\boldsymbol{q}}^* \otimes \hat{\boldsymbol{q}} \otimes \hat{\boldsymbol{p}} = \hat{\boldsymbol{p}}$.

The position and orientation (i.e., pose) of a frame $\mathcal{A}$ w.r.t. a frame $\mathcal{B}$ can be described by a unit quaternion $\boldsymbol{q}_{ab} \in \mathbb{Q}$ and by a translation vector $\bar{\boldsymbol{r}}_{ab} \in \mathbb{R}^3$, respectively. Then, the pose of $\mathcal{A}$ w.r.t. $\mathcal{B}$ can be represented by the unit dual quaternion $\hat{\boldsymbol{q}}_{ab} \in \hat{\mathbb{Q}}$ in a compact form:

$$\hat{\boldsymbol{q}}_{ab} = \boldsymbol{q}_{ab} + \varepsilon \frac{1}{2} \boldsymbol{q}_{ab} \otimes \boldsymbol{r}^a_{ab} = \boldsymbol{q}_{ab} + \varepsilon \frac{1}{2} \boldsymbol{r}^b_{ab} \otimes \boldsymbol{q}_{ab}, \tag{2.78}$$

where $\boldsymbol{r}^c_{ab} = [(\bar{\boldsymbol{r}}^c_{ab})^\top, 0]^\top$ and $\bar{\boldsymbol{r}}^c_{ab}$ is the translation vector from the origin of the frame $\mathcal{B}$ to the origin of the frame $\mathcal{A}$ expressed in the frame C. Since $\boldsymbol{q}_{ab}$ is a unit quaternion, it is not difficult to check that $\hat{\boldsymbol{q}}_{ab} \in \hat{\mathbb{Q}}$.

For the sake of brevity, we will call unit dual quaternions as dual quaternions. From (2.78), it is clear that the real part of the dual quaternion $\hat{\boldsymbol{q}}_{ab}$ is an unit quaternion, which is used to describe the relative attitude between the two frames; while the dual part contains both the relative attitude and position information, so that $\hat{\boldsymbol{q}}_{ab}$ can also describe the relative position between the two frames at the same time. As such, dual quaternions can be used to describe six degrees-of-freedom (6-DOF) relative position and attitude motions of two rigid spacecraft.

2.4.4.2 Dual-Quaternion-Based Spacecraft Relative Dynamics

Consider the spacecraft PROs with a stationary target. Recalling Sect. 2.3 for the involved frames in RPOs: the pursuer body frame $\mathcal{P}$, the target body frame $\mathcal{T}$, and the inertial frame $\mathcal{I}$. The relation between frames are shown in Fig. 2.9. Denote by $\boldsymbol{q}_{pt} \in \mathbb{Q}$ and $\bar{\boldsymbol{r}}_{pt} \in \mathbb{R}^3$ the position and orientation of the frame $\mathcal{P}$ w.r.t. the frame $\mathcal{T}$, respectively. According to (2.78), the pose of $\mathcal{P}$ w.r.t. $\mathcal{T}$ can be more compactly represented by

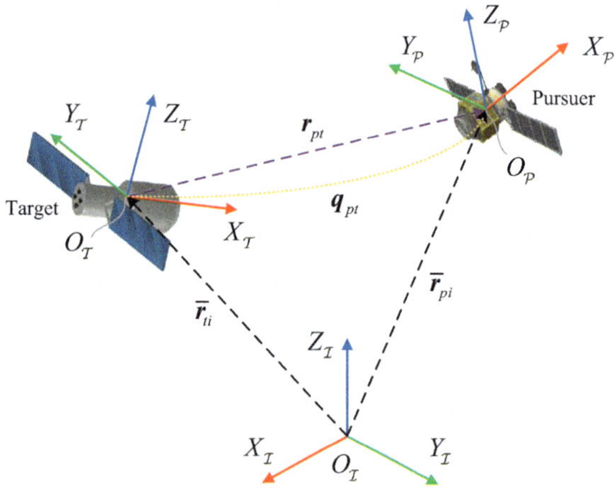

Fig. 2.9 Illustration of frames

$$\hat{\boldsymbol{q}}_{pt} = \boldsymbol{q}_{pt} + \varepsilon\frac{1}{2}\boldsymbol{q}_{pt} \otimes \boldsymbol{r}^p_{pt} = \boldsymbol{q}_{pt} + \varepsilon\frac{1}{2}\boldsymbol{r}^t_{pt} \otimes \boldsymbol{q}_{pt}, \tag{2.79}$$

where $\boldsymbol{r}^p_{pt} = [(\bar{\boldsymbol{r}}^p_{pt})^\top, 0]^\top$ and $\boldsymbol{r}^t_{pt} = [(\bar{\boldsymbol{r}}^t_{pt})^\top, 0]^\top$ with $\bar{\boldsymbol{r}}^p_{pt}$ (resp. $\bar{\boldsymbol{r}}^t_{pt}$) being the translation vector from the origin of the frame $\mathcal{P}$ to the frame $\mathcal{T}$ expressed in $\mathcal{P}$ (resp. $\mathcal{T}$). Then, based on the dual-quaternion formulation, the 6-DOF relative motion dynamics for the spacecraft RPOs with a stationary target are described by

$$\dot{\hat{\boldsymbol{q}}}_{pt} = \frac{1}{2}\hat{\boldsymbol{q}}_{pt} \otimes \hat{\boldsymbol{\omega}}^p_{pt}, \tag{2.80}$$

$$\hat{\boldsymbol{J}}_p\dot{\hat{\boldsymbol{\omega}}}^p_{pt} = -\hat{\boldsymbol{\omega}}^p_{pt} \times (\hat{\boldsymbol{J}}_p\hat{\boldsymbol{\omega}}^p_{pt}) + \hat{\boldsymbol{u}}, \tag{2.81}$$

where $\hat{\boldsymbol{\omega}}^p_{pt} = \boldsymbol{\omega}^p_{pt} + \varepsilon\boldsymbol{v}^p_{pt}$ is the dual relative angular velocity of the frame $\mathcal{P}$ w.r.t. the frame $\mathcal{T}$ expressed in the frame $\mathcal{P}$, whereas $\boldsymbol{\omega}^p_{pt} = [(\bar{\boldsymbol{\omega}}^p_{pt})^\top, 0]^\top$ and $\boldsymbol{v}^p_{pt} = [(\bar{\boldsymbol{v}}^p_{pt})^\top, 0]^\top$ with $\bar{\boldsymbol{\omega}}^p_{pt} \in \mathbb{R}^3$ and $\bar{\boldsymbol{v}}^p_{pt} \in \mathbb{R}^3$ being the angular and linear velocities of the frame $\mathcal{P}$ w.r.t. the frame $\mathcal{T}$ expressed in the frame $\mathcal{P}$, respectively; $\hat{\boldsymbol{u}} = \boldsymbol{f}^p + \varepsilon\boldsymbol{\tau}^p$ is called the dual control input expressed in the frame $\mathcal{P}$, and $\boldsymbol{f}^p, \boldsymbol{\tau}^p \in \mathbb{R}^3$ are the force and torque applied to the pursuer, respectively; and $\hat{\boldsymbol{J}}_p$ is the dual inertia of the pursuer, with the definition:

$$\hat{\boldsymbol{J}}_p = m_p\boldsymbol{I}_3\frac{d}{d\varepsilon} + \varepsilon\boldsymbol{J}_p, \tag{2.82}$$

where $m_p \in \mathbb{R}$ and $\boldsymbol{J}_p \in \mathbb{R}^{3\times3}$ represent the mass and inertia parameters of the pursuer. The inverse of $\hat{\boldsymbol{J}}_p$ is defined as follows:

$$\hat{\boldsymbol{J}}_p^{-1} = \boldsymbol{J}_p^{-1}\frac{d}{d\varepsilon} + \varepsilon\frac{1}{m_p}\boldsymbol{I}_3. \tag{2.83}$$

2.5 Lyapunov Stability Theory

Stability is a basic requirement for the AOCS design. At present, the stability analysis of linear systems has formed a complete theoretical framework, but there is no unified standard method for the stability analysis of nonlinear systems. For nonlinear systems (e.g., the spacecraft AOCS), the commonly used stability analysis methods mainly include phase portrait method, describing function method, and Lyapunov method, among which the Lyapunov method is more general and has been widely applied. This book mainly uses the Lyapunov method to analyze the closed-loop stability of the AOCS. In this section, we will give several important definitions and theorems of the Lyapunov stability theory.

Consider a nonlinear autonomous system of the following form:

$$\dot{\boldsymbol{x}} = \boldsymbol{f}(\boldsymbol{x}) \tag{2.84}$$

where $\boldsymbol{x} \in \mathbb{R}^n$ is the system state, whereas $\boldsymbol{f} : \mathbb{U} \to \mathbb{R}^n$ is a locally Lipschitz map from a domain $\mathbb{U}$ to $\mathbb{R}^n$. The system initial state is $\boldsymbol{x}(t_0) = \boldsymbol{x}_0$, and the initial time is $t_0 \geq 0$.

Definition 2.1 (*Equilibrium Point* [17]) For a point $\boldsymbol{x}_* \in \mathbb{U}$, if it satisfies $\boldsymbol{f}(\boldsymbol{x}_*) = \boldsymbol{0}$, then it is called an equilibrium point of (2.84). It is not difficult to check that one the system state reaches the equilibrium point ($\boldsymbol{x}(t) = \boldsymbol{x}_*$), it will remain on it thereafter. In general, the system's equilibrium point is the origin, i.e., $\boldsymbol{x}_* = \boldsymbol{0}$.

Definition 2.2 (*Stability* [17]) The equilibrium point $\boldsymbol{x}_* = \boldsymbol{0}$ of (2.84) is

- *stable* if, for any $\epsilon > 0$, there always exists $\delta = \delta(\epsilon) > 0$ such that

$$\|\boldsymbol{x}_0 - \boldsymbol{x}_*\| < \delta \Longrightarrow \|\boldsymbol{x}(t) - \boldsymbol{x}_*\| < \epsilon, \ \forall t \geq t_0;$$

- *asymptotically stable* if it is stable and there is $\delta > 0$ such that

$$\|\boldsymbol{x}_0 - \boldsymbol{x}_*\| < \delta \Longrightarrow \lim_{t\to\infty} \|\boldsymbol{x}(t) - \boldsymbol{x}_*\| = 0;$$

- *exponentially stable* if it is asymptotically stable and $\exists \alpha, \beta, \delta > 0$ such that

$$\|\boldsymbol{x}_0 - \boldsymbol{x}_*\| < \delta \Longrightarrow \|\boldsymbol{x}(t)\| \leq \alpha \|\boldsymbol{x}_0 - \boldsymbol{x}_*\| e^{-\beta t}, \ \forall t \geq t_0.$$

Theorem 2.1 (Lyapunov Stability Theorem *[17]) Let $\boldsymbol{x}_*$ be an equilibrium point of* (2.84) *and* $\mathbb{U} \in \mathbb{R}^n$ *be a domain containing* $\boldsymbol{x}_*$. *Consider a continuously differentiable function* $V : \mathbb{U} \to \mathbb{R}$ *satisfying* $V(\boldsymbol{x}_*) = 0$ *and* $V(\boldsymbol{x}) > 0$ *in* $\mathbb{U} - \{\boldsymbol{x}_*\}$. *Then, the equilibrium is*

- *stable if* $\dot{V}(\boldsymbol{x}) \leq 0, \ \forall \boldsymbol{x} \neq \boldsymbol{x}_*$;
- *asymptotically stable if* $\dot{V}(\boldsymbol{x}) < 0, \ \forall \boldsymbol{x} \neq \boldsymbol{x}_*$;
- *exponentially stable if* $\forall t \geq 0$ *and* $\forall \boldsymbol{x} \in \mathbb{U}$

$$k_1\|\boldsymbol{x}\|^\alpha \leq V(\boldsymbol{x}) \leq k_2\|\boldsymbol{x}\|^\alpha,$$

$$\frac{\partial V}{\partial t} + \frac{\partial V}{\partial \boldsymbol{x}} \boldsymbol{f}(\boldsymbol{x}) \leq -k_3\|\boldsymbol{x}\|^\alpha,$$

where k_1, k_2, k_3, *and* α *are positive constants.*

Lemma 2.4 (Barbalat Lemma [18]) *Suppose* $\boldsymbol{f} : [0, \infty) \to \mathbb{R}^n$ *is a continuously differentiable function, and the limit* $\lim_{t\to\infty} \boldsymbol{f}(t)$ *exists and is finite. If* $\dot{\boldsymbol{f}}(t)$ *is uniformly continuous (that is* $\ddot{\boldsymbol{f}}(t)$ *is bounded) on* $t \in [0, \infty)$, *then* $\lim_{t\to\infty} \dot{\boldsymbol{f}}(t) = 0$.

Lemma 2.5 (Corollary of Barbalat Lemma [18]) *If the function* $\boldsymbol{f} : [0, \infty) \to \mathbb{R}^n$ *is a uniformly continuous on* $t \in [0, \infty)$, *and the limit* $\lim_{t\to\infty} \int_0^t \boldsymbol{f}(s)ds$ *exists and is finite, then* $\lim_{t\to\infty} \boldsymbol{f}(t) = 0$.

References

1. Junkins JL, Schaub H (2009) Analytical mechanics of space systems. American Institute of Aeronautics and Astronautics, Reston, Virginia
2. Shuster MD (1993) A survey of attitude representations. The Journal of the Astronautical Sciences 41(4): 439–517
3. Lee U, Mesbahi M (2014) Feedback control for spacecraft reorientation under attitude constraints via convex potentials. IEEE Transactions on Aerospace and Electronic Systems 50(4): 2578–2592
4. Murray RM, Li Z, Sastry SS, Sastry SS (1994) A mathematical introduction to robotic manipulation. CRC press, Boca Raton, NY, USA
5. Tayebi A (2008) Unit quaternion-based output feedback for the attitude tracking problem. IEEE Transactions on Automatic Control 53(6): 1516–1520
6. Yamanaka K, Ankersen F (2002) New state transition matrix for relative motion on an arbitrary elliptical orbit. Journal of Guidance, Control, and Dynamics 25(1): 60–66
7. Kristiansen R, Nicklasson PJ (2009) Spacecraft formation flying: a review and new results on state feedback control. Acta Astronautica 65(11-12): 1537–1552
8. Roberts A, Tayebi A (2010) Adaptive position tracking of VTOL UAVs. IEEE Transactions on Robotics 27(1): 129–142
9. Shao X, Hu Q, Shi Y, Jiang B (2018) Fault-tolerant prescribed performance attitude tracking control for spacecraft under input saturation. IEEE Transactions on Control Systems Technology 28(2): 574–582
10. Saenz-Otero A, Miller D (2003) The spheres iss laboratory for rendezvous and formation flight. European Space Agency-Publications-ESA SP 516: 217–224
11. Anthony TC, Wie B, Carroll S (1990) Pulse-modulated control synthesis for a flexible spacecraft. Journal of Guidance, Control, and Dynamics 13(6): 1014–1022
12. Song G, Buck NV, Agrawal BN (1999) Spacecraft vibration reduction using pulse-width pulse-frequency modulated input shaper. Journal of Guidance, Control, and Dynamics 22(3): 433–440
13. Clifford WK (1873) Preliminary sketch of biquaternions. Proceedings of the London Mathematical Society s1-4(1): 381–395
14. Study E (1891) Von den bewegungen und umlegungen. Mathematische Annalen 39(4): 441–565
15. Salgueiro Filipe NR (2014) Nonlinear pose control and estimation for space proximity operations: An approach based on dual quaternions. PhD thesis, Georgia Institute of Technology
16. Filipe N, Tsiotras P (2015) Adaptive position and attitude-tracking controller for satellite proximity operations using dual quaternions. Journal of Guidance, Control, and Dynamics 38(4): 566–577
17. Khalil HK (2015) Nonlinear Control, vol 406. Pearson New York
18. Slotine JJE, Li W (1991) Applied nonlinear control, vol 199. Prentice hall Englewood Cliffs, NJ

Chapter 3
Data-Driven Adaptive Control for Spacecraft Constrained Reorientation

3.1 Introduction

Rigid-body attitude control dates back to early aeronautics and space applications that involve attitude maneuvers of aerial vehicles or spacecraft [1]. It is also motivated by applications of ground and underwater vehicles, and robotic systems. In recent years, with the ever-increasing demands on such engineering applications, the attitude control problem of rigid bodies has been attracting more and more attention from both academia and industrial sectors. Various attitude control methods have been presented in the literature, such as inverse optimal control [2], proportional-derivative plus feed-forward control [3], geometric control [4], disturbance observer-based control [5], output feedback control [6], etc. Although these methods can help to achieve high-performance attitude control in many cases, some underlying state constraints (as discussed later) are ignored, which may cause some safety issues or, even worse, lead to mission failure and severe economic losses.

Spacecraft can generally be viewed as a rigid body. For rigid spacecraft, a typical mission scenario is to reorient the spacecraft attitude to a desired orientation. In such a mission, executing attitude reorientation is usually subjected to state constraints that arise from two practical concerns. On the one hand, some on-board sensitive payloads (e.g., infrared telescopes or interferometers) should always be kept away from direct exposure to the Sun vector or other bright objects [7], in order to avoid functional damage. This is regarded as attitude constraints. On the other hand, the spacecraft angular velocities are restricted, due to limited measurement ranges of rate gyros (e.g., Rossi X-ray Timing Explorer (XTE) [8]). This is called here angular velocity constraints. Studying the spacecraft attitude control problem under multiple state constraints is of practical significance and has received widespread interest. In this respect, different methods have been proposed. In general, the methods for constrained attitude control problems can be categorized into two types: path planning based methods [7, 9–11] and artificial potential function (APF) based methods [12–17]. The former is somewhat computationally expensive and time-consuming

Q. Hu et al., *Intelligent Autonomous Control of Spacecraft with Multiple Constraints*,
https://doi.org/10.1007/978-981-99-0681-9_3

for implementation [11]. In contrast, the PF-based method utilizes an APF to form an admissible path, and provides an analytic control law easy for on-board implementation in the sense of Lyapunov. Typically, based on a logarithmic APF, Lee and Mesbahi [15] designed orientation control laws for spacecraft with multiple types of attitude constraints. Building upon the results in [15], Shen et al. [16] and Hu et al. [17] further explored the constrained attitude reorientation problems, while additionally taking into account angular rate limits and unwinding problem. In addition to the foregoing two types of methods, the model predictive control has also been employed in [18] to cope with attitude constraints. The PF-based attitude controllers reported in [16, 19] can account for attitude and angular rate constraints simultaneously. Recently, Dong et al. [20] proposed an approximate optimal control approach using reinforcement learning, which solves the attitude reorientation problem under both attitude and angular rate constraints.

Note that nearly every one of the afore-mentioned methods applicable for tackling attitude and/or angular rate constraints requires exact knowledge of the spacecraft inertia parameters. But, in practice, the inertia parameters may be uncertain due to for example, fuel consumption, payload motion, and appendage deployment, which makes most of those foregoing works inapplicable in theory. Adaptive control has emerged as an effective tool to deal with parameter uncertainties, and has been widely applied to the spacecraft attitude control problems subject to inertia uncertainties (see, indicatively, [21, 22] and references therein). One important caveat, however, is that most of the existing adaptive control methods obey the so-called certainty equivalence (CE) principle and require a certain "realizability" condition akin to that in [23]. Unfortunately, such a condition usually does not hold in the Lyapunov sense when angular rate constraints are taken into account. In recent years, several non-CE adaptive attitude controllers have been presented in [24–29] based upon the promising immersion and invariance (I&I) adaptive control methodology in [23]. These non-CE adaptive controllers can overcome some limitations inherent in the CE design structure, and possess the potential to circumvent the stringent realizability condition. But currently, the study of the non-CE adaptive attitude control problem for a spacecraft in the presence of both attitude and angular rate constraints has received less attention in the literature.

In addition, it is sometimes necessary to exactly identify the uncertain inertia parameters of the spacecraft. Note, however, that neither the CE nor the non-CE adaptive attitude controllers mentioned above can guarantee parameter convergence, unless the reference trajectories satisfy the persistent excitation (PE) condition [30]. But, the PE condition usually does not hold for attitude reorientation maneuvers of spacecraft. Promisingly, Chowdhary and Johnson [31] creatively proposed a data-driven adaptive method, known as concurrent learning (CL), with the aim of relaxing the PE condition for parameter convergence. Using CL in the adaptive process, parameter convergence can be guaranteed, only if the reference signals are exciting over a finite time interval (denoted as interval excitation (IE) condition). Later, Cho et al. [32] and Pan and Yu [33] remove the need for estimating the state derivatives in [31] using the regressor filtering technique. The CL technique has also been integrated into the adaptive dynamic programming algorithms to exactly learn the network weights in the absence of PE condition [34, 35]. It is noted that in these CL-based adaptive methods, the parameter convergence rates across all components are hard to be tuned in a well-balanced manner. Moreover, since they still follow the CE principle, the resulting closed-loop performance may be arbitrarily poor relative to the deterministic case, when the excitation is weak. Even though the method presented in [36] can partially overcome the above drawbacks, it cannot be readily extended to deal with multiple state constraints.

To the authors' best knowledge, there is no previous study on adaptive attitude control handling both attitude and angular rate constraints, whilst achieving on-line parameter identification. Toward this end, this chapter seeks to propose a data-driven I&I adaptive attitude control scheme to address all these considerations simultaneously. As a stepping stone, a dynamic scaling based I&I adaptive control framework is proposed to accommodate nonsatisfaction of the realizability condition, in which two PFs are artfully constructed to encode the information about attitude and angular rate constraints. It is proved that the derived I&I adaptive controller can enable both the attitude errors and angular rates to asymptotically converge to zero for most initial conditions in the accessible space, while obeying underlying state constraints. To further relax the dependence of parameter convergence on PE condition, the earlier designed I&I adaptive law is then extended to a data-driven counterpart through adding a learning term (driven by historical data) that is acquired by employing the regressor filtering technique in conjunction with the dynamic regressor extension and mixing (DREM) procedure recently presented in [37]. Such a term can ensure that the parameter estimates asymptotically converge to their true values, under the assumption of IE, not PE, thus significantly relaxing the stringent PE condition. In addition, benefiting from the DREM method and some special designs, the parameter convergence rates are not only independent of the excitation level, but they can also be flexibly tuned via an explicit and simple weights selection. Furthermore, the data-driven I&I adaptive controller preserves all the key features of the I&I adaptive control methodology, thus exhibiting better transient performance than the CE adaptive controllers.

This chapter is organized as follows. Section 3.2 contains some preliminaries and the problem formulation. The dynamically scaled I&I adaptive controller is derived in Sect. 3.3, whereas the data-driven adaptive extension is presented in Sect. 3.4. Further, the effectiveness of the proposed method is validated by numerical simulations in Sect. 3.5, and by HIL experiments in Sect. 3.6. Finally, some concluding remarks are given in Sect. 3.7.

3.2 Problem Statement

The attitude dynamics of a fully-actuated spacecraft are given in (2.13) and (2.14). Here we assume that $\boldsymbol{\tau}_d = \mathbf{0}$ in the theoretical analysis, and that the spacecraft inertia matrix $\boldsymbol{J}$ is a diagonal matrix. The primary emphasis of this chapter is focused on the rest-to-rest attitude reorientation issue. Let $\boldsymbol{q}_d = [\boldsymbol{q}_{dv}^\top, q_{d4}]^\top \in \mathbb{Q}_u$ (setpoint, that is, $\omega_d = \mathbf{0}$) be the desired attitude. Then, the unit-quaternion error is defined as $\boldsymbol{q}_e = [\boldsymbol{q}_{ev}^\top, q_{e4}]^\top = \boldsymbol{q}_d^{-1} \otimes \boldsymbol{q}$, which describes the discrepancy between $\boldsymbol{q}$ and $\boldsymbol{q}_d$. The governing differential equations for attitude error is given in (2.19) and (2.21). Note that in this chapter the control input $\boldsymbol{\tau}_c$ is simply written as $\boldsymbol{u}$.

3.2.1 Attitude Constraints

To protect the sensitive payloads (e.g., optical instruments) on-board the spacecraft from possible damage that arises from direct exposure to some certain unwanted objects, the forbidden zones (cone-shaped) are introduced, whose geometries are illustrated in Fig. 3.1. Suppose that the spacecraft is equipped with n sensitive payloads, and each has m forbidden zones. Let $\boldsymbol{n}_{bi}^{\mathcal{P}} \in \mathbb{R}^3$ be the unit boresight vector of the i-th sensitive payload in the spacecraft body-fixed frame $\mathcal{P}$, and $\boldsymbol{n}_{oj}^{\mathcal{I}} \in \mathbb{R}^3$ be the unit vector pointing toward the j-th undesired object in the inertial frame $\mathcal{I}$. Intuitively, the angle between $\boldsymbol{n}_{bi}^{\mathcal{P}}$ and $\boldsymbol{n}_{oj}^{\mathcal{I}}$ should remain larger than α_j^i, that is,

$$\bar{\alpha}_j^i = \boldsymbol{n}_{oj}^{\mathcal{I}} \cdot \boldsymbol{R}^\top \boldsymbol{n}_{bi}^{\mathcal{P}} < \cos(\alpha_j^i), \tag{3.1}$$

where $\boldsymbol{R} = (q_4^2 - \boldsymbol{q}_v^\top \boldsymbol{q}_v)\boldsymbol{I}_3 + 2\boldsymbol{q}_v\boldsymbol{q}_v^\top - 2q_4\boldsymbol{S}(\boldsymbol{q}_v)$ is the rotation matrix from $\mathcal{I}$ to $\mathcal{P}$. After some algebra, (3.1) can be compactly expressed as the following quadratic inequality:

$$\boldsymbol{q}^\top \boldsymbol{M}_j^i \boldsymbol{q} < 0, \tag{3.2}$$

with

$$\boldsymbol{M}_j^i = \begin{bmatrix} \boldsymbol{n}_{oj}^{\mathcal{I}}(\boldsymbol{n}_{bi}^{\mathcal{P}})^\top + \boldsymbol{n}_{bi}^{\mathcal{P}}(\boldsymbol{n}_{oj}^{\mathcal{I}})^\top - ((\boldsymbol{n}_{oj}^{\mathcal{I}})^\top \boldsymbol{n}_{bi}^{\mathcal{P}})\boldsymbol{I}_3 & \boldsymbol{n}_{bi}^{\mathcal{P}} \times \boldsymbol{n}_{oj}^{\mathcal{I}} \\ (\boldsymbol{n}_{bi}^{\mathcal{P}} \times \boldsymbol{n}_{oj}^{\mathcal{I}})^\top & (\boldsymbol{n}_{oj}^{\mathcal{I}})^\top \boldsymbol{n}_{bi}^{\mathcal{P}} \end{bmatrix} - \cos(\alpha_j^i)\boldsymbol{I}_4. \tag{3.3}$$

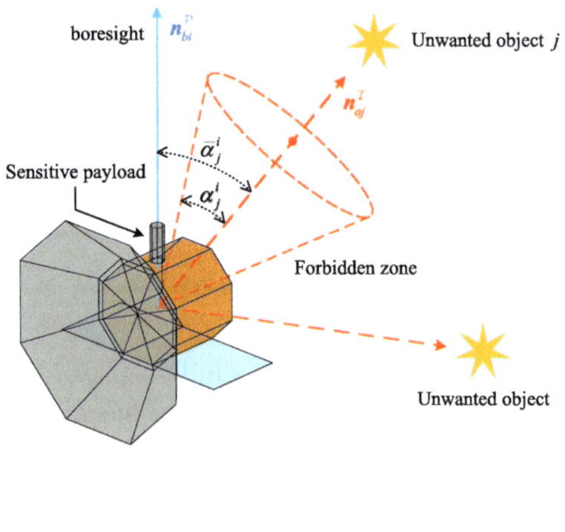

Fig. 3.1 Illustration of attitude constraints

Overall, the set of attitude orientations that lead to all boresight vectors lying outside the specified exclusion zones can be defined as $\mathbb{Q}_s = \{\boldsymbol{q} \in \mathbb{S} \mid \boldsymbol{q}^\top \boldsymbol{M}^i_j \boldsymbol{q} < 0,\ i = 1, 2, .., n,\ j = 1, 2, ..., m\}$.

Note additionally that the unwinding phenomenon may also occur due to the redundancy of the unit quaternion, causing an unnecessary rotation. A key idea of overcoming the unwinding phenomenon is to provide a discontinuous controller such that the attitude error is always regulated to the closest equilibrium point without passing through $q_{e4} = 0$. With this in mind, the permissible set of attitudes is further restricted as the subset $\mathbb{Q}_a = \{\boldsymbol{q} \in \mathbb{Q}_s \mid q_{e4} \neq 0\}$.

As per [15], $\boldsymbol{q}^\top \boldsymbol{M}^i_j \boldsymbol{q}$ usually satisfies $\boldsymbol{q}^\top \boldsymbol{M}^i_j \boldsymbol{q} > -2$. Thus, to achieve rest-to-rest reorientation maneuvers, while avoiding simultaneously forbidden pointing constraints and unwinding phenomenon, we consider the APF $V_a : \mathbb{Q}_a \to \mathbb{R}$ as:

$$V_a = \zeta \log(q_{e4}^2) \sum_{i=1}^{n} \sum_{j=1}^{m} \log\left(-\frac{\boldsymbol{q}^\top \boldsymbol{M}^i_j \boldsymbol{q}}{2}\right), \tag{3.4}$$

where $\zeta > 0$ is a weighting parameter, and the term $\log(q_{e4}^2)$ is introduced to avoid the unwinding phenomenon (see [17] for the details). Inspecting (3.4) reveals that $V_a = 0$ only when $\boldsymbol{q} = \pm \boldsymbol{q}_d$, and that $V_a \to \infty$ when $\boldsymbol{q}^\top \boldsymbol{M}^i_j \boldsymbol{q} \to 0$ and/or $q_{e4} \to 0$. The latter means that the spacecraft attitude would evolve strictly in the permitted set $\mathbb{Q}_a$, if we can guarantee that $V_a \in \mathcal{L}_\infty$ by designing properly an attitude controller under the condition of $\boldsymbol{q}(0) \in \mathbb{Q}_a$.

Evaluating the time derivative of V_a yields $\dot{V}_a = (\nabla V_a)^\top \dot{\boldsymbol{q}}$, where ∇V_a is the gradient of V_a with respect to (w.r.t.) $\boldsymbol{q}$, given by

$$\nabla V_a = 2\zeta \frac{\boldsymbol{q}_d}{q_{e4}} \sum_{i=1}^{n} \sum_{j=1}^{m} \log\left(-\frac{\boldsymbol{q}^\top \boldsymbol{M}^i_j \boldsymbol{q}}{2}\right) + \zeta \log(q_{e4}^2) \sum_{i=1}^{n} \sum_{j=1}^{m} \frac{2\boldsymbol{M}^i_j \boldsymbol{q}}{\boldsymbol{q}^\top \boldsymbol{M}^i_j \boldsymbol{q}}, \tag{3.5}$$

where $q_{e4} = \boldsymbol{q}_d^\top \boldsymbol{q}$ is used. Further, using (2.19) in $\dot{V}_a$ and noting the algebraic properties of quaternions in [15], we have

$$\dot{V}_a = -\frac{1}{2} \boldsymbol{\omega}^\top \mathrm{Vec}[\nabla V_a^* \otimes \boldsymbol{q}] = -\boldsymbol{\omega}^\top \boldsymbol{v}, \tag{3.6}$$

where $\mathrm{Vec}[\cdot]$ is the 3×1 vector part of the argument, whereas $\boldsymbol{v} = 0.5\mathrm{Vec}[\nabla V_a^* \otimes \boldsymbol{q}]$ is defined for notational simplicity.

3.2.2 Angular Velocity Constraints

The primary motivation for restricting the maximum angular velocity of the spacecraft lies in the limited measurement ranges of the equipped rate gyros. The permissible set of the angular velocities can be usually expressed as

$$\mathbb{W} = \{\boldsymbol{\omega} \in \mathbb{R}^3 \mid |\omega_i| < \omega_{\mathrm{m}},\ i = 1, 2, 3\},$$

where $\omega_{\mathrm{m}} > 0$ is the common magnitude limit on $\boldsymbol{\omega}$.

To deal with angular rate constraints, a commonly used APF $V_\omega : \mathbb{W} \to \mathbb{R}$ of the following form is introduced:

$$V_\omega = \frac{1}{2} \log \left(\prod_{i=1}^{3} \frac{\omega_{\mathrm{m}}^2}{\omega_{\mathrm{m}}^2 - \omega_i^2} \right). \tag{3.7}$$

Evidently, V_ω is positive definite and C^1 continuous in the set $\mathbb{W}$, and has a unique global minimum at $\boldsymbol{\omega} = \mathbf{0}$; moreover, if $|\omega_i| \to \omega_{\mathrm{m}}$, $V_\omega \to \infty$. Hence, if the controller to be developed can guarantee $V_\omega \in \mathcal{L}_\infty$ for all $t \geq 0$, then the hard constraints on the angular rates can be satisfied.

Taking the time derivative of V_ω leads to

$$\dot{V}_\omega = \boldsymbol{\omega}^\top \boldsymbol{N}_\omega \dot{\boldsymbol{\omega}} = \boldsymbol{\omega}^\top \boldsymbol{N}_\omega \boldsymbol{J}^{-1}(-\boldsymbol{S}(\boldsymbol{\omega})\boldsymbol{J}\boldsymbol{\omega} + \boldsymbol{u}), \tag{3.8}$$

where $\boldsymbol{N}_\omega = \mathrm{diag}_{i\in\{1,2,3\}}[N_{\omega i}]$ with $N_{\omega i} = 1/(\omega_{\mathrm{m}}^2 - \omega_i^2)$.

3.2.3 *Problem Statement and Challenges*

Formally, the problem statement reads as follows:

Problem 3.1 For the spacecraft attitude dynamics given by (2.19) and (2.21) with an uncertain inertia matrix $\boldsymbol{J}$, design an adaptive attitude control law $\boldsymbol{u}$ in the context of the APFs defined in (3.4) and (3.7) to render $\lim_{t\to\infty} \boldsymbol{q}(t) = \pm \boldsymbol{q}_d$ and $\lim_{t\to\infty} \boldsymbol{\omega}(t) = \mathbf{0}$, whilst ensuring that $\boldsymbol{q}(t) \in \mathbb{Q}_a$ and $\boldsymbol{\omega}(t) \in \mathbb{W}$ for all $t \geq 0$, and the parameter estimates converge to their true values.

Key challenges: Theoretically speaking, it is a non-trivial task to design an adaptive attitude controller using the existing adaptive control approaches to solve Problem 3.1. The technical barriers mainly lie in the following facts:

- As discussed in [29], the classical MRAC method [32] and its modified versions including the $\mathcal{L}_1$ adaptive control [38] and the simple adaptive control (SAC) [39] have a relatively fixed architecture with a lower flexibility, and can hardly be extended to handle state constraints due to involvement of a reference model to be tracked; the direct adaptive control [21, 22] and composite adaptive control [40] obey the CE principle, and require a "realizability" condition that may not hold in the Lyapunov sense when considering angular rate constraints; even though the robust adaptive control [41] has the potential to circumvent the realizability condition and ensure satisfaction of full-state constraints through integrating the element-wise and norm-wise adaptive estimations, it has a certain level of conservativeness and cannot recover the ideal closed-loop performance obtained in the deterministic case (that is, no uncertainties in inertia parameters).

- Most of the foregoing adaptive approaches can guarantee parameter convergence only if the regressor matrix satisfies the restrictive PE condition, which may not hold for rest-to-rest attitude reorientation maneuvers. Although the CL-based methods in [31–33] are recently developed to relax the stringent PE condition, they do not deviate from the CE framework and, therefore, cannot be used to solve Problem 1 either, as discussed above.

Therefore, it is imperative to tailor new adaptive solutions to Problem 3.1.

3.3 I&I Adaptive Attitude Control

As a stepping stone, a dynamic scaling based I&I adaptive control framework is proposed in this section to accommodate the nonsatisfaction of the realizability condition as discussed in Sect. 3.2, and providing an effective solution to Problem 3.1 (except for parameter convergence). Later, the I&I adaptive control scheme will be extended to a data-driven counterpart in Sect. 3.4 to further achieve parameter convergence.

3.3.1 Regressor Reconfiguration

Before proceeding, a linear operator $\boldsymbol{L}(\cdot): \mathbb{R}^3 \rightarrow \mathbb{R}^{3\times 3}$ is introduced such that $\boldsymbol{J}\boldsymbol{x} = \boldsymbol{L}(\boldsymbol{x})\boldsymbol{\theta}$ for any $\boldsymbol{x} \in \mathbb{R}^3$, where $\boldsymbol{\theta} = [J_{11}, J_{22}, J_{33}]^\top$, and $J_{ii}, i = 1, 2, 3$ (unknown but otherwise constant) are the principal moments of inertia. Based on the operator $\boldsymbol{L}(\cdot)$, the attitude dynamics (2.21) can be rewritten as

$$\dot{\boldsymbol{\omega}} = \boldsymbol{N}_{\omega}^{-1}\boldsymbol{v} - k(t)\boldsymbol{N}_{\omega}\boldsymbol{\omega} + \boldsymbol{J}^{-1}(\boldsymbol{\Phi}(k(t), \boldsymbol{v}, \boldsymbol{\omega})\theta + \boldsymbol{u}), \tag{3.9}$$

where the term $\boldsymbol{N}_{\omega}^{-1}\boldsymbol{v} - k(t)\boldsymbol{N}_{\omega}\boldsymbol{\omega}$ is judiciously added here to achieve the control objectives stated in Sect. 3.2.3, $k(t)$ (positive and independent of $\boldsymbol{\omega}$) is a time-varying gain to be determined, and $\boldsymbol{\Phi}(\cdot) \in \mathbb{R}^{3\times 3}$ is the regressor matrix satisfying $\boldsymbol{\Phi}(\cdot)\theta = -\boldsymbol{J}\boldsymbol{N}_{\omega}^{-1}\boldsymbol{v} + k(t)\boldsymbol{J}\boldsymbol{N}_{\omega}\boldsymbol{\omega} - \boldsymbol{S}(\boldsymbol{\omega})\boldsymbol{J}\boldsymbol{\omega}$. For the purpose of analysis, $\boldsymbol{\Phi}(\cdot)$ is further decomposed into two parts:

$$\boldsymbol{\Phi}(\cdot) = \boldsymbol{\Phi}_1(k(t), \boldsymbol{v}, \boldsymbol{\omega}) + \boldsymbol{\Phi}_2(\boldsymbol{\omega}), \tag{3.10}$$

with $\boldsymbol{\Phi}_1(\cdot)$ and $\boldsymbol{\Phi}_2(\cdot)$ determined by $\boldsymbol{\Phi}_1(\cdot) = -\boldsymbol{L}(\boldsymbol{N}_{\omega}^{-1}\boldsymbol{v}) + k(t)\boldsymbol{L}(\boldsymbol{N}_{\omega}\boldsymbol{\omega})$ and $\boldsymbol{\Phi}_2(\cdot) = -\boldsymbol{S}(\boldsymbol{\omega})\boldsymbol{L}(\boldsymbol{\omega})$, respectively.

Since, for all $i, j \in \{1, 2, 3\}$, $\partial \phi_{1i}/\partial \omega_j = \partial \phi_{1j}/\partial \omega_i$, where ϕ_{1i} and ϕ_{1j} denote the i-th and j-th columns of $\mathbf{\Phi}_1^\top(\cdot)$, respectively, there exists $\boldsymbol{\mu}_1$ (not unique) satisfying the following partial differential equation (PDE):

$$\frac{\partial \boldsymbol{\mu}_1}{\partial \boldsymbol{\omega}} = \mathbf{\Phi}_1^\top(\cdot). \tag{3.11}$$

By simple deductions, a feasible solution is given below

$$\boldsymbol{\mu}_1 = \begin{bmatrix} -v_1\omega_{\mathrm{m}}^2\omega_1 + \frac{1}{3}v_1\omega_1^3 - \frac{k(t)}{2}\log(N_{\omega 1}^{-1}) \\ -v_2\omega_{\mathrm{m}}^2\omega_2 + \frac{1}{3}v_2\omega_2^3 - \frac{k(t)}{2}\log(N_{\omega 2}^{-1}) \\ -v_3\omega_{\mathrm{m}}^2\omega_3 + \frac{1}{3}v_3\omega_3^3 - \frac{k(t)}{2}\log(N_{\omega 3}^{-1}) \end{bmatrix}. \tag{3.12}$$

However, unlike $\mathbf{\Phi}_1(\cdot)$, there exists no $\boldsymbol{\mu}_2$ such that the PDE $\partial \boldsymbol{\mu}_2/\partial \boldsymbol{\omega} = \mathbf{\Phi}_2^\top(\cdot)$ holds, owing to $\partial \phi_{2i}/\partial \omega_j \neq \partial \phi_{2j}/\partial \omega_i$, for all $i, j \in \{1, 2, 3\}$ other than $i = j$. This is the well-known "integrability obstacle" in the traditional I&I adaptive control. To overcome it, inspired by [25], we here construct a solvable PDE by introducing a 3×3 matrix $\mathbf{\Psi}(\boldsymbol{\omega})$ to render

$$\frac{\partial \phi_{2i}}{\partial \omega_j} + \frac{\partial \psi_i}{\partial \omega_j} = \frac{\partial \phi_{2j}}{\partial \omega_i} + \frac{\partial \psi_j}{\partial \omega_i}, \tag{3.13}$$

for all $i, j \in \{1, 2, 3\}$, where the subscripts i and j on ψ stand for the i-th and j-th columns of $\mathbf{\Psi}^\top(\cdot)$. A special choice of $\mathbf{\Psi}(\cdot)$ satisfying (3.13) is

$$\mathbf{\Psi}(\cdot) = -\mathbf{\Phi}_2(\cdot), \tag{3.14}$$

which allows us to solve the PDE:

$$\frac{\partial \boldsymbol{\mu}_2}{\partial \boldsymbol{\omega}} = \mathbf{\Phi}_2^\top(\cdot) + \mathbf{\Psi}^\top(\cdot), \tag{3.15}$$

and a direct solution is $\boldsymbol{\mu}_2 = \mathbf{0}_{3\times 1}$.

For notational compactness, we further define

$$\boldsymbol{\mu}(k(t), \boldsymbol{v}, \boldsymbol{\omega}) = \boldsymbol{\mu}_1 + \boldsymbol{\mu}_2. \tag{3.16}$$

3.3.2 I&I Adaptive Controller Design

At this point, we develop an I&I adaptive controller based on the solvable PDEs in (3.11) and (3.15) as well as the dynamic scaling method. For notational brevity, function arguments are dropped hereafter whenever no confusion can occur.

Design a smooth control law as

$$\boldsymbol{u} = -\boldsymbol{\Phi}(\hat{\boldsymbol{\theta}} + \boldsymbol{\beta}), \tag{3.17}$$

with $\hat{\boldsymbol{\theta}}$ and $\boldsymbol{\beta}$ determined by

$$\dot{\hat{\boldsymbol{\theta}}} = -\gamma[\dot{\bar{\boldsymbol{\mu}}} + (\boldsymbol{\Phi} + \boldsymbol{\Psi})^\top (\boldsymbol{N}_\omega^{-1} \boldsymbol{v} - k(t)\boldsymbol{N}_\omega \boldsymbol{\omega})], \tag{3.18}$$

$$\boldsymbol{\beta} = \gamma \boldsymbol{\mu}, \tag{3.19}$$

where $\gamma > 0$ is a design constant, and $\dot{\bar{\boldsymbol{\mu}}} = \dot{\boldsymbol{\mu}} - (\partial \boldsymbol{\mu}/\partial \boldsymbol{\omega})\dot{\boldsymbol{\omega}}$. It is noted that $\dot{\bar{\boldsymbol{\mu}}}$ is actually obtained by weeding out $(\partial \boldsymbol{\mu}/\partial \boldsymbol{\omega})\dot{\boldsymbol{\omega}}$ from $\dot{\boldsymbol{\mu}}$, and therefore the unmeasured $\dot{\boldsymbol{\omega}}$ will not be utilized in practical implementation. In this manner, the composite term $(\hat{\boldsymbol{\theta}} + \boldsymbol{\beta}) \in \mathbb{R}^3$ acts as the estimate of the unknown vector $\boldsymbol{\theta}$. Within this setting, the estimation error vector is defined as $\tilde{\boldsymbol{\theta}} = \hat{\boldsymbol{\theta}} + \boldsymbol{\beta} - \boldsymbol{\theta}$. Once the control law (3.17) is plugged in, (3.9) reduces to

$$\dot{\boldsymbol{\omega}} = \boldsymbol{N}_\omega^{-1} \boldsymbol{v} - k(t)\boldsymbol{N}_\omega \boldsymbol{\omega} - \boldsymbol{J}^{-1}\boldsymbol{\Phi}\tilde{\boldsymbol{\theta}}. \tag{3.20}$$

In addition, with (3.11), (3.15), and (3.18)–(3.20) in mind, the time derivative of $\tilde{\boldsymbol{\theta}}$ can be derived as

$$\dot{\tilde{\boldsymbol{\theta}}} = -\gamma \boldsymbol{\Phi}^\top \boldsymbol{J}^{-1} \boldsymbol{\Phi}\tilde{\boldsymbol{\theta}} - \gamma \boldsymbol{\Psi}^\top \boldsymbol{J}^{-1} \boldsymbol{\Phi}\tilde{\boldsymbol{\theta}}. \tag{3.21}$$

Actually, the second term of the right-hand side of (3.21) plays the role of "perturbation" to the adaptive parameter estimation, due to the involvement of $\boldsymbol{\Psi}$. The dynamic scaling technique, originally developed by Karagiannis et al. [42], will be used to deal with this term. To be specific, a dynamic scaling factor $r(t) \in \mathbb{R}$ satisfying $r(t) \geq 1$ $\forall t \geq 0$ is introduced to form the scaled estimation error [26]

$$z = \frac{e^{\frac{1}{2}(\frac{1}{J_\mathrm{m}^2}+1)}}{\sqrt{J_\mathrm{m}}} \cdot \frac{\tilde{\boldsymbol{\theta}}}{e^{\frac{\sqrt{\log r + 1}}{J_\mathrm{m}}}}, \tag{3.22}$$

where J_m stands for the minimum eigenvalue of $\boldsymbol{J}$ (note that J_m is introduced just for the subsequent stability analysis and will not be used in the control implementation), and r evolves over time along the differential equation below

$$\dot{r} = \gamma r \sqrt{\log r + 1} \|\boldsymbol{\Psi}\|_2^2, \; r(0) = 1. \tag{3.23}$$

It is noted that z is defined in (3.22), instead of the traditional form $z = \tilde{\theta}/r$ as in [25, 42], to eliminate the need of J_{m} in I&I adaptive design. Differentiating z w.r.t. time and using (3.21) and (3.23) give rise to

$$\dot{z} = -\gamma(\boldsymbol{\Phi} + \boldsymbol{\Psi})^{\top} \boldsymbol{J}^{-1} \boldsymbol{\Phi} z - \frac{\gamma}{2J_m} \|\boldsymbol{\Psi}\|_2^2 z. \tag{3.24}$$

Consider a positive-definite function $V_z = \frac{1}{2} z^{\top} z$. Taking the time derivative of V_z along (3.24) leads to

$$\begin{aligned} \dot{V}_z = & -\gamma z^{\top} (\boldsymbol{\Phi} + \boldsymbol{\Psi})^{\top} \boldsymbol{J}^{-1} \boldsymbol{\Phi} z - \frac{\gamma}{2J_m} \|\boldsymbol{\Psi}\|_2^2 \|z\|_2^2 \\ \leq & -\frac{\gamma J_m}{2} \|\boldsymbol{J}^{-1} \boldsymbol{\Phi} z\|_2^2 \leq 0, \end{aligned} \tag{3.25}$$

showing that the scaled parameter estimation error dynamics (3.24) has a globally stable equilibrium at $z = \mathbf{0}$.

Theorem 3.1 *Consider the spacecraft attitude dynamics given by (2.19) and (2.21) with unknown inertia matrix. Given the initial conditions satisfying* $\boldsymbol{q}(0) \in \mathbb{Q}_a$, *if the dynamic gain is chosen as* $k(t) = \kappa r(t)$, *where* $\kappa > 0$ *is a constant parameter, then the implementation of the control law* (3.17) *in conjunction with the adaptive law* (3.18) *and the nonlinear function* (3.19) *can guarantee that:*

1. V_a *and* V_ω *remain bounded for all* $t \geq 0$, *indicating the satisfaction of double-level state constraints;*
2. *The gradient-related vector* $\boldsymbol{v}$ *and angular rate* $\boldsymbol{\omega}$ *converge asymptotically to zero, i.e.,* $\lim_{t\to\infty}[\boldsymbol{v}(t), \boldsymbol{\omega}(t)] = \mathbf{0}$.

Proof Choose the overall Lyapunov-like function as

$$V = V_a + V_\omega + \frac{\eta}{2\gamma} z^{\top} z, \tag{3.26}$$

where $\eta > \frac{1}{\kappa} + \varrho$ with ϱ a positive constant is chosen just for stability analysis. Now taking the time derivative of V along (3.6), (3.20) and (3.24) and noting (3.21) yield

$$\begin{aligned} \dot{V} = & -\boldsymbol{\omega}^{\top} \boldsymbol{v} + \boldsymbol{\omega}^{\top} \boldsymbol{N}_\omega \dot{\boldsymbol{\omega}} + \frac{\eta}{\gamma} z^{\top} \dot{z} \\ = & -k(t) \|\boldsymbol{N}_\omega \boldsymbol{\omega}\|_2^2 - \frac{e^{\frac{\sqrt{\log r + 1}}{J_{\mathrm{m}}}} \sqrt{J_{\mathrm{m}}}}{e^{\frac{1}{2}(\frac{1}{J_{\mathrm{m}}^2} + 1)}} \boldsymbol{\omega}^{\top} \boldsymbol{N}_\omega \boldsymbol{J}^{-1} \boldsymbol{\Phi} z \\ & -\eta z^{\top} (\boldsymbol{\Phi} + \boldsymbol{\Psi})^{\top} \boldsymbol{J}^{-1} \boldsymbol{\Phi} z - \frac{\eta}{2J_m} \|\boldsymbol{\Psi}\|_2^2 \|z\|_2^2. \end{aligned} \tag{3.27}$$

By Young's inequality, we have the following inequalities:

$$e^{\frac{\sqrt{\log r+1}}{J_{\mathrm{m}}}} \leq e^{(\frac{\log r+1}{2}+\frac{1}{2J_{\mathrm{m}}^2})} = e^{\frac{1}{2}(\frac{1}{J_{\mathrm{m}}^2}+1)}\sqrt{r},$$

$$\sqrt{J_{\mathrm{m}}r}\left|\boldsymbol{\omega}^\top \boldsymbol{N}_\omega \boldsymbol{J}^{-1}\boldsymbol{\Phi} z\right| \leq \frac{\kappa r}{2}\|\boldsymbol{N}_\omega \boldsymbol{\omega}\|_2^2 + \frac{J_{\mathrm{m}}}{2\kappa}\|\boldsymbol{J}^{-1}\boldsymbol{\Phi} z\|_2^2,$$

$$-\eta z^\top \boldsymbol{\Psi}^\top \boldsymbol{J}^{-1}\boldsymbol{\Phi} z \leq \frac{\eta}{2J_{\mathrm{m}}}\|\boldsymbol{\Psi}\|_2^2\|z\|_2^2 + \frac{J_{\mathrm{m}}\eta}{2}\|\boldsymbol{J}^{-1}\boldsymbol{\Phi} z\|_2^2.$$

Then, by the definitions of $k(t)$ and η, we further have

$$\begin{aligned}\dot{V} &\leq -\frac{\kappa r}{2}\|\boldsymbol{N}_\omega \boldsymbol{\omega}\|_2^2 - \frac{J_{\mathrm{m}}}{2}\left(\eta - \frac{1}{\kappa}\right)\|\boldsymbol{J}^{-1}\boldsymbol{\Phi} z\|_2^2 \\ &\leq -\frac{\kappa}{2}\|\boldsymbol{N}_\omega \boldsymbol{\omega}\|_2^2 - \frac{J_{\mathrm{m}}\varrho}{2}\|\boldsymbol{J}^{-1}\boldsymbol{\Phi} z\|_2^2.\end{aligned} \tag{3.28}$$

The following analyses are two-fold:

(1) Inspecting (3.28) reveals that $\dot{V} \leq 0$ and accordingly $V(t)$ and hence $V_a(t)$ and $V_\omega(t)$ are uniformly bounded for all $t \geq 0$. Since $\boldsymbol{q}(0) \in \mathbb{Q}_a$ and $\boldsymbol{\omega}(0) \in \mathbb{W}$, the boundedness of V_a and V_ω indicates that $\boldsymbol{q}(t) \in \mathbb{Q}_a$ and $\boldsymbol{\omega}(t) \in \mathbb{W}$ for all time. As a result, both attitude and angular velocity constraints are satisfied during the entire maneuver process.

(2) As $\dot{V} \leq 0$, $V(t)$ is upper bounded by $V(0)$, from which we conclude that $\int_0^\infty \dot{V}(t)\,dt$ exists and is finite. As a result, $\boldsymbol{N}_\omega \boldsymbol{\omega}$ and $\boldsymbol{J}^{-1}\boldsymbol{\Phi} z \in \mathcal{L}_2 \cap \mathcal{L}_\infty$. However, at this point, there is still no way to show the convergence of $\boldsymbol{v}$ and $\boldsymbol{\omega}$ since the boundedness of r is not yet established. In view of this, let us first show $r \in \mathcal{L}_\infty$ before proceeding. Solving (3.23) yields

$$\sqrt{\log r(t) + 1} = 1 + \frac{\gamma}{2}\int_0^t \|\boldsymbol{\Psi}(\sigma)\|_2^2 d\sigma. \tag{3.29}$$

For further analysis, one could rewrite $\boldsymbol{\Psi}$ as

$$\boldsymbol{\Psi} = [\boldsymbol{H}(\mathbf{I}_3 \otimes (\boldsymbol{N}_\omega \boldsymbol{\omega}))]^\top, \tag{3.30}$$

where "⊗" represents the Kronecker product and $\boldsymbol{H} \in \mathbb{R}^{3\times 9}$ is defined as $\boldsymbol{H} = [\,\boldsymbol{H}_1\ \boldsymbol{H}_2\ \boldsymbol{H}_3\,]$ with the block matrices $\boldsymbol{H}_i \in \mathbb{R}^{3\times 3}$, $i = 1, 2, 3$ detailed below

$$\boldsymbol{H}_1 = \begin{bmatrix} 0 & 0 & 0 \\ 0 & -\omega_3 N_{\omega 2}^{-1} & 0 \\ 0 & 0 & \omega_2 N_{\omega 3}^{-1} \end{bmatrix},$$

$$\boldsymbol{H}_2 = \begin{bmatrix} \omega_3 N_{\omega 1}^{-1} & 0 & 0 \\ 0 & 0 & 0 \\ 0 & 0 & -\omega_1 N_{\omega 3}^{-1} \end{bmatrix},$$

$$\boldsymbol{H}_3 = \begin{bmatrix} -\omega_2 N_{\omega 1}^{-1} & 0 & 0 \\ 0 & \omega_1 N_{\omega 2}^{-1} & 0 \\ 0 & 0 & 0 \end{bmatrix}.$$

As it has been shown previously that $\boldsymbol{\omega}(t) \in \mathbb{W}$ for all $t \geq 0$, it follows that $\boldsymbol{H}$ is bounded and there exists a positive constant L_h such that $\|\boldsymbol{H}\|_2 \leq L_h$ for all $t \geq 0$. On the other hand, it is not difficult to check from (3.28) that

$$\int_0^t \|\boldsymbol{N}_\omega(\sigma)\boldsymbol{\omega}(\sigma)\|_2^2 d\sigma \leq \frac{2}{\kappa} V(0) < \infty. \tag{3.31}$$

Thus, from (3.30) and (3.31), we can further deduce

$$\begin{aligned} \int_0^t \|\boldsymbol{\Psi}(\sigma)\|_2^2 d\sigma &\leq \int_0^t \|\boldsymbol{H}(\sigma)\|_2^2 \|\boldsymbol{N}_\omega(\sigma)\boldsymbol{\omega}(\sigma)\|_2^2 d\sigma \\ &\leq \frac{2L_h^2}{\kappa} V(0) < \infty, \end{aligned} \tag{3.32}$$

for all $t \geq 0$, where the facts that $\|\boldsymbol{\Psi}\|_2 = \|\boldsymbol{\Psi}^\top\|_2$ and $\|\mathbf{I}_3 \otimes (\boldsymbol{N}_\omega \boldsymbol{\omega})\|_2 = \|\boldsymbol{N}_\omega \boldsymbol{\omega}\|_2$ have been used. With (3.29) and (3.32) in mind, it can be concluded that $r(t) \in \mathcal{L}_\infty$. From the boundedness of $\boldsymbol{J}^{-1}\boldsymbol{\Phi} z$ and r aided by (3.22), it is evident that $\boldsymbol{J}^{-1}\boldsymbol{\Phi}\tilde{\boldsymbol{\theta}}$ and $k(t) \in \mathcal{L}_\infty$. Additionally, invoking 1) shows $\boldsymbol{v}$, $\boldsymbol{\omega}$, $\boldsymbol{N}_\omega$, and $\boldsymbol{N}_\omega^{-1} \in \mathcal{L}_\infty$, whereby it can further be inferred from (3.10), (3.14), (3.20) and (3.21) that $\boldsymbol{\Phi}$, $\boldsymbol{\Psi}$, $\dot{\boldsymbol{\omega}}$, and hence $\dot{\tilde{\boldsymbol{\theta}}} \in \mathcal{L}_\infty$. With $\boldsymbol{\omega}$ and $\dot{\boldsymbol{\omega}}$ bounded, it follows that $\dot{\boldsymbol{v}}$, $\dot{\boldsymbol{N}}_\omega$, and $\dot{\boldsymbol{N}}_\omega^{-1} \in \mathcal{L}_\infty$. Further, by (3.23), we can infer that $\dot{k}(t) \in \mathcal{L}_\infty$, which together with the above suggest that $\dot{\boldsymbol{\Phi}} \in \mathcal{L}_\infty$.

Based on the above discussion, we conclude that $\boldsymbol{N}_\omega \boldsymbol{\omega}$ and $\boldsymbol{J}^{-1}\boldsymbol{\Phi} z$ are square integrable; moreover, from the following

$$\begin{cases} \dfrac{d}{dt}(\boldsymbol{N}_\omega \boldsymbol{\omega}) = \dot{\boldsymbol{N}}_\omega \boldsymbol{\omega} + \boldsymbol{N}_\omega \dot{\boldsymbol{\omega}} \\ \dfrac{d}{dt}(\boldsymbol{J}^{-1}\boldsymbol{\Phi}\tilde{\boldsymbol{\theta}}) = \boldsymbol{J}^{-1}(\dot{\boldsymbol{\Phi}}\tilde{\boldsymbol{\theta}} + \boldsymbol{\Phi}\dot{\tilde{\boldsymbol{\theta}}}) \end{cases}, \tag{3.33}$$

we can further conclude that $\boldsymbol{N}_\omega \boldsymbol{\omega}$ and $\boldsymbol{J}^{-1}\boldsymbol{\Phi} z$ are uniformly continuous. Then, applying Barbalat's lemma establishes

$$\lim_{t \to \infty} [\boldsymbol{N}_\omega(t)\boldsymbol{\omega}(t), \boldsymbol{J}^{-1}\boldsymbol{\Phi}(t)\tilde{\boldsymbol{\theta}}(t)] = \mathbf{0}. \tag{3.34}$$

Note that $\lim_{t\to\infty} \boldsymbol{N}_\omega(t)\boldsymbol{\omega}(t) = \mathbf{0}$ is, in essence, equivalent to $\lim_{t\to\infty} \boldsymbol{\omega}(t) = \mathbf{0}$.

Next, let us show the asymptotic convergence of $\boldsymbol{v}$. Actually, according to the above conclusions, it is very easy to establish the boundedness of $\ddot{\boldsymbol{\omega}}$, which implies the uniformly continuous of $\dot{\boldsymbol{\omega}}$. Then, together with the convergence of $\boldsymbol{\omega}$ to the origin, it can be claimed from Barbalat's Lemma that $\lim_{t\to\infty}\dot{\boldsymbol{\omega}}(t) = \mathbf{0}$. Thus, from (3.20), it follows that $\lim_{t\to\infty}\boldsymbol{v}(t) = \mathbf{0}$, which completes the proof.

Remark 3.1 An important caveat is that, from $\boldsymbol{v} = \mathbf{0}$, we may not necessarily draw the conclusion that $\boldsymbol{q} = \pm\boldsymbol{q}_d$, since there may exist critical points, like saddle points and local minima. This is an inherent drawback of the PF-based methods. But, such an undesirable behavior occurs rarely in practice, and the spacecraft attitude could converge to the desired setpoint from most initial free configurations. Even if the spacecraft attitude gets indeed trapped into a critical point, as suggested in [17], a small additive torque with a direction orthogonal to $\boldsymbol{q}_{ev}$ can be evoked to help it escape from the critical point.

Remark 3.2 In the proof of Theorem 1, it has been shown that $\lim_{t\to\infty}\boldsymbol{J}^{-1}\boldsymbol{\Phi}(t)\tilde{\boldsymbol{\theta}}(t) = \mathbf{0}$, which indicates the establishment of an attracting manifold $\mathcal{M}$ defined by

$$\mathcal{M} = \{\tilde{\boldsymbol{\theta}} \in \mathbb{R}^3 \mid \boldsymbol{\Phi}\tilde{\boldsymbol{\theta}} = \mathbf{0}\}. \tag{3.35}$$

In theory, all the closed-loop trajectories will end up inside $\mathcal{M}$. It can, therefore, be concluded from (3.20) that the closed-loop dynamics ultimately recovers to the ideal case (no parameter uncertainties), that is,

$$\dot{\boldsymbol{\omega}} = \boldsymbol{N}_\omega^{-1}\boldsymbol{v} - k(t)\boldsymbol{N}_\omega\boldsymbol{\omega}, \tag{3.36}$$

without resorting to any fragile cancellation operation and PE condition. In addition, as clearly seen from (3.21), the estimation error dynamics is linear w.r.t. the estimation error $\tilde{\boldsymbol{\theta}}$. Hence, once $\tilde{\boldsymbol{\theta}}(t) = \mathbf{0}$ occurs at any instance of time t^*, the adaptation will then stop thereafter and, as a consequence, the parameter estimates $\hat{\theta}_i + \beta_i$, $i = 1, 2, 3$ will stay locked at the true values θ_i, $i = 1, 2, 3$. The foregoing two features (i.e., performance recovery and parameter locking) can hardly be obtained in the traditional adaptive control methods.

Remark 3.3 As can be clearly seen from (3.23), the dynamic scaling factor $r(t)$ will grow monotonically with time in the presence of disturbances and/or measurement noises, due to lack of damping. This may give rise to a high-gain parameter $k(t)$, which, in turn, may lead to degradation of the control system robustness. To eliminate such a detrimental effect, we can revise $k(t)$ in Theorem 1 as $k(t) = \kappa\varphi(t)$, where $\varphi(t) = \rho(t)r(t)$ and $\rho(t)$ is a non-increasing function defined by

$$\dot{\rho} = -\gamma[\kappa_1 + (\kappa_2 + 1)\sqrt{\log r + 1}\|\boldsymbol{\Psi}\|_2^2]\left(\rho - \frac{1}{r} - \rho^\star\right), \tag{3.37}$$

where $\rho(0) = \frac{1}{r(0)} + \rho^\star$ with $\rho^\star > 0$ a sufficiently small scalar, and $\kappa_1, \kappa_2 > 0$. Then, from (3.23) and (3.37), it follows that

$$\dot{\varphi} = -\gamma \Big\{ (\kappa_1 + \kappa_2 \sqrt{\log r + 1} \|\boldsymbol{\Psi}\|_2^2)\varphi \\ -[\kappa_1 + (\kappa_2 + 1)\sqrt{\log r + 1}\|\boldsymbol{\Psi}\|_2^2](1 + \rho^\star r) \Big\}. \tag{3.38}$$

A close observation of (3.38) shows that $\dot{\varphi} > 0$ when $\varphi = 1$ and $\dot{\varphi} < 0$ when $\varphi = (1 + \frac{1}{\kappa_2})(1 + \rho^\star r)$, indicating that

$$1 \le \varphi(t) \le \frac{\kappa_2 + 1}{\kappa_2}(1 + \rho^\star r), \tag{3.39}$$

for all $t \ge 0$. Thus, $\varphi(t)$ can be rendered not too large or even close to 1, through properly choosing $\rho^\star$ and κ_2. In this way, the potential issue of robustness degradation can be countered to some extent, when practically implementing the proposed I&I adaptive controller. It should also be emphasized that the above revision will not break the original stability analysis.

Remark 3.4 Although the filter-based methods presented in [25, 26, 29] provide alternative ways to derive the dynamically scaled I&I adaptive controller, it requires a velocity filter to design the parameter updating law and the nonlinear term $\boldsymbol{\beta}$, which inevitably leads to increased complexity of the adaptive design. In contrast, the filter-free method employed in this chapter has a simple structure and only requires lower-dimensional dynamic extensions.

Remark 3.5 It should be noted that, for the case in which $\boldsymbol{J}$ is a diagonal matrix, V_ω in (3.7) can be revised as

$$V_\omega = \frac{1}{2}\sum_{i=1}^{3} J_{ii} \log\left(\frac{\omega_{\mathrm{m}}^2}{\omega_{\mathrm{m}}^2 - \omega_i^2}\right),$$

its time derivative satisfies

$$\dot{V}_\omega = \boldsymbol{\omega}^\top \boldsymbol{N}_\omega \boldsymbol{J}\dot{\boldsymbol{\omega}} = \boldsymbol{\omega}^\top \boldsymbol{N}_\omega(-\boldsymbol{S}(\boldsymbol{\omega})\boldsymbol{J}\boldsymbol{\omega} + \boldsymbol{u}).$$

It is clear from the above equality that the CE-based adaptive control algorithms can be readily adopted to solve Problem 1 (except for parameter convergence), without encountering the unrealizability problem. However, when $\boldsymbol{J}$ is in a general from that has non-zero inertia products, they would be inapplicable for Problem 1, theoretically speaking, due to the fact that the realizability condition (i.e., $\partial V_\omega/\partial\boldsymbol{\omega}\boldsymbol{J}^{-1}$ is known) no longer holds, as discussed in Sect. 3.2.3. Although the proposed I&I adaptive control strategy in its current from also requires that $\boldsymbol{J}$ be a diagonal matrix (see Remark 3.6 for more details), it provides a crucial first step toward overcoming the unrealizability obstacle, and is shown to outperform the CE-based adaptive solutions in transient performance and robustness.

Remark 3.6 If $\boldsymbol{J}$ has nonzero inertia products, the proposed I&I adaptive control strategy will be inapplicable for Problem 3.1, due to the following technical barriers. Specifically, the sub-regressor matrix $\boldsymbol{\Phi}_1^\star = k(t)\boldsymbol{L}(\boldsymbol{N}_\omega\boldsymbol{\omega})$ will turn to be a 3×6 matrix and does no longer satisfy $\partial\phi_{1i}^\star/\partial\omega_j = \partial\phi_{1j}^\star/\partial\omega_i$ for all $i, j \in \{1, 2, 3\}$, where $\phi_{1i}^\star$ and $\phi_{1j}^\star$ denote the ith and jth columns of $(\boldsymbol{\Phi}_1^\star)^\top$. Thus, it is necessary to introduce a matrix $\boldsymbol{\Psi}_1^\star$ to construct a solvable PDE $\partial\boldsymbol{\mu}^\star/\partial\boldsymbol{\omega} = (\boldsymbol{\Phi}_1^\star)^\top + (\boldsymbol{\Psi}_1^\star)^\top$. However, it is stressed that $\boldsymbol{\Psi}_1^\star$ inevitably involves $k(t)$ and hence $r(t)$, so does $\boldsymbol{\Psi}$. Under such a condition, we cannot claim directly that $\boldsymbol{\Psi} \in \mathcal{L}_\infty$ as shown in (3.32), and moreover, the local Lipschitz condition for $\boldsymbol{\Psi}$ will also be invalid. As a consequence, one cannot conclude the boundedness of $r(t)$ and hence the asymptotic convergence of $\boldsymbol{v}$ and $\boldsymbol{\omega}$. Although, at first glance, the filter-based methods in [25, 26, 29] may be effective to show the boundedness of $r(t)$, through choosing properly the dynamic filter gain with an extra term, the filtered states $\hat{\omega}_i$, $i = 1, 2, 3$ might be beyond the specified limit ω_m on $\boldsymbol{\omega}$, leading to the singularity of $\boldsymbol{\Psi}_1^\star(\hat{\boldsymbol{\omega}})$ due to the existence of $1/(\omega_\mathrm{m}^2 - \hat{\omega}_i^2)$, $i = 1, 2, 3$ in $\boldsymbol{\Psi}_1^\star(\hat{\boldsymbol{\omega}})$. Up to now, how to extend the current result to the general case still remains open and calls for further investigation.

3.4 Data-Driven I&I Adaptive Control

To achieve high-precision attitude control or meet specific mission demands, it is sometimes necessary to identify on-line the uncertain inertia parameters of the spacecraft. However, the I&I adaptive control method proposed in Sect. 3.3 cannot ensure that the parameter estimates converge to the true values, unless the regressor matrix $\boldsymbol{\Phi}$ satisfies the PE condition. However, the PE condition is restrictive and usually does not hold for the rest-to-rest attitude maneuvers. To relax the PE condition for parameter convergence, in this section, we will propose a data-driven I&I adaptive control scheme, which is complementary to the theoretical findings of Sect. 3.3. The block diagram of the resulted closed-loop system is shown in Fig. 3.2.

During the past two decades, relevant researches [43] have studied the importance of PE condition in adaptive control. However, the work in this chapter aims to relax the requirement of PE condition for parameter convergence. To proceed, two necessary definitions associated with signal excitation are given [31, 44], which is the basis for design of adaptive laws and stability analysis.

Definition 3.1 (*IE of a Signal*) A bounded signal $\mathbf{v}(t) \in \mathbb{R}^{m \times n}$, where $t \in [0, \infty)$, is of (t_s, T, α)-IE (also denoted as $\mathbf{v}(t) \in \mathrm{IE}$) over a finite time interval $[t_s, t_s + T]$ if, there exist $t_s \geq 0$, $T, \alpha > 0$ such that the following equation holds:

$$\int_{t_s}^{t_s+T} \mathbf{v}(\tau)\mathbf{v}^\top(\tau)d\tau \geq \alpha \boldsymbol{I}_m > \mathbf{0}.$$

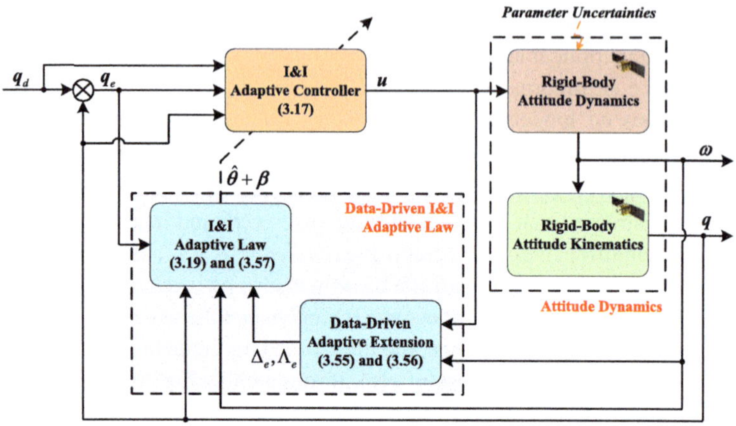

Fig. 3.2 Block diagram of the closed-loop system

Definition 3.2 (*PE of a Signal*) A bounded signal $\mathbf{v}(t) \in \mathbb{R}^{m\times n}$, where $t \in [0, \infty)$, is of (T, α)-PE (also denoted as $\mathbf{v}(t) \in \text{PE}$) if, there exist $T, \alpha > 0$ such that the following holds:

$$\int_t^{t+T} \mathbf{v}(\tau)\mathbf{v}^\top(\tau)d\tau \geq \alpha \boldsymbol{I}_m > \mathbf{0},\ \forall t \geq 0.$$

Note that if $t_s = 0$, the (t_s, T, α)-IE is also called the initial excitation. It is not difficult to check that the IE condition is strictly weaker than PE, since PE implies the satisfaction of IE for all $t_s \geq 0$.

3.4.1 Filtered System Dynamics

In fact, the data-driven I&I adaptive law to be designed later is partially inspired by the composite adaptive control methods [33, 40], the key idea of which is to extract information about the uncertain parameters from the prediction errors. However, in general, the involved information extraction requires using unmeasured state derivatives, like $\dot{\omega}$ in our scenario. To eliminate the need of unavailable signals in parameter adaptation, a regressor filtering scheme is adopted, forming a filtered system dynamics, as similarly done in [32].

Let us first rewrite (2.21) as follows:

$$\boldsymbol{J}\dot{\boldsymbol{\omega}} = \boldsymbol{W}\boldsymbol{\theta} + \boldsymbol{u}, \tag{3.40}$$

where $\boldsymbol{W} = -\boldsymbol{S}(\boldsymbol{\omega})\boldsymbol{L}(\boldsymbol{\omega})$ is a new regressor matrix. Then, as the central part of the regressor filtering scheme, the following stable low-pass filters are introduced:

$$\dot{\omega}_f = -c\omega_f + \omega,\ \omega_f(0) = \omega(0)/c, \tag{3.41}$$
$$\dot{\boldsymbol{W}}_f = -c\boldsymbol{W}_f + \boldsymbol{W},\ \boldsymbol{W}_f(0) = \boldsymbol{0}, \tag{3.42}$$
$$\dot{\boldsymbol{u}}_f = -c\boldsymbol{u}_f + \boldsymbol{u},\ \boldsymbol{u}_f(0) = \boldsymbol{0}, \tag{3.43}$$

where $c > 0$ is the time constant of the filters. Taking the time derivative of (3.41) and noting (3.40), (3.42), and (3.43) lead to

$$\dot{\boldsymbol{\delta}} = -c\boldsymbol{\delta},\ \boldsymbol{\delta} = \dot{\omega}_f - \boldsymbol{J}^{-1}(\boldsymbol{W}_f\boldsymbol{\theta} + \boldsymbol{u}_f), \tag{3.44}$$

which immediately renders $\dot{\omega}_f = \boldsymbol{J}^{-1}(\boldsymbol{W}_f\boldsymbol{\theta} + \boldsymbol{u}_f) + \boldsymbol{\delta}(0)e^{-ct}$. According to the initial conditions of the filters defined in (3.41)-(3.43), it can be derived that $\boldsymbol{\delta}(0) = \boldsymbol{0}$. Thus, we have

$$\dot{\omega}_f = \boldsymbol{J}^{-1}(\boldsymbol{W}_f\boldsymbol{\theta} + \boldsymbol{u}_f). \tag{3.45}$$

Notably, $\dot{\omega}_f$ in the filtered system dynamics (3.45) is obtainable from (3.41). This fact contributes to eliminating the need of the state derivation $\dot{\omega}$ in estimator design. Reorganizing (3.45) results in a linear regressor equation (LRE) of the form:

$$\boldsymbol{u}_f = \boldsymbol{W}_a\boldsymbol{\theta}, \tag{3.46}$$

where $\boldsymbol{W}_a = \boldsymbol{L}(\dot{\omega}_f) - \boldsymbol{W}_f$ is a known regressor. It is obvious that the filtered input $\boldsymbol{u}_f$ contains information about the ideal parameter $\boldsymbol{\theta}$, and hence will be used as the measurement for parameter estimation.

Assumption 3.1. *There exist positive constants t_s and T such that $\boldsymbol{W}_a^\top$ is of IE over the time interval $[t_s, t_s + T]$.*

Remark 3.7 If the directions of the three row vectors of the regressor $\boldsymbol{W}_a$ vary sufficiently in a certain finite time interval, then Assumption 1 would be satisfied. In fact, the IE of $\boldsymbol{W}_a^\top$ is a rather mild condition, which is almost always satisfied in practice, due to initial transient, system noise, etc.

3.4.2 Data-Driven Adaptive Extension

At this point, the Dynamic Regressor Extension and Mixing (DREM) procedure, recently reported in [37], is tactfully used to extend the previously proposed parameter update law (3.18) to a data-driven counterpart, which can achieve parameter convergence under an IE condition. It gets started by carrying out the *dynamic regressor extension* step. For that, we premultiply both sides of (3.46) by $\boldsymbol{W}_a$, yielding

$$\boldsymbol{W}_a^\top\boldsymbol{u}_f = \boldsymbol{W}_a^\top\boldsymbol{W}_a\boldsymbol{\theta}, \tag{3.47}$$

to which the Kreisselmeier's regressor extension method introduced by Kreisselmeier [45] and recently employed in [46, 47] is used, which proceeds as follows:

$$\dot{\boldsymbol{N}} = -a\boldsymbol{N} + \boldsymbol{W}_a^\top \boldsymbol{u}_f,\ \ \boldsymbol{N}(0) = \mathbf{0}, \tag{3.48}$$

$$\dot{\boldsymbol{\Omega}} = -a\boldsymbol{\Omega} + \boldsymbol{W}_a^\top \boldsymbol{W}_a,\ \ \boldsymbol{\Omega}(0) = \mathbf{0}, \tag{3.49}$$

where $a > 0$ is the time constant of the filters, $\boldsymbol{N} \in \mathbb{R}^3$ and $\boldsymbol{\Omega} \in \mathbb{R}^{3\times 3}$ are an auxiliary matrix and an information matrix, respectively. Solving (3.48) and (3.49) directly obtains

$$\boldsymbol{N}(t) = \int_0^t e^{-a(t-\tau)} \boldsymbol{W}_a^\top(\tau)\boldsymbol{u}_f(\tau)d\tau, \tag{3.50}$$

$$\boldsymbol{\Omega}(t) = \int_0^t e^{-a(t-\tau)} \boldsymbol{W}_a^\top(\tau)\boldsymbol{W}_a(\tau)d\tau. \tag{3.51}$$

Using (3.47) in (3.50) and (3.51), one gets

$$\boldsymbol{N}(t) = \boldsymbol{\Omega}(t)\boldsymbol{\theta}. \tag{3.52}$$

Subsequently, the *mixing* step is performed to generate a set of scalar LREs. All we can recall that, for any square and possibly singular matrix $\boldsymbol{A} \in \mathbb{R}^{n\times n}$, the formula $\text{adj}(\boldsymbol{A})\boldsymbol{A} = \det(\boldsymbol{A})\mathbf{I}_n$ always holds, where $\text{adj}(\boldsymbol{A})$ is the adjoint matrix of $\boldsymbol{A}$. Premultiplying both sides of (3.52) by $\text{adj}(\boldsymbol{\Omega})$, we have

$$\boldsymbol{\Lambda}(t) = \Delta(t)\boldsymbol{\theta}, \tag{3.53}$$

where $\boldsymbol{\Lambda}(t) = \text{adj}(\boldsymbol{\Omega}(t))\boldsymbol{N}(t)$[1] and $\Delta(t) = \det(\boldsymbol{\Omega}(t))$. With the help of the DREM procedure, a set of 3 scalar LREs that share the same regressor Δ are obtained, as seen in (3.53).

Inspecting (3.51), we find that the information matrix $\boldsymbol{\Omega}$ at a time instance t is calculated through a weighted accumulation (via forward integration) of all incoming data from 0 up to t. As such, the rank of $\boldsymbol{\Omega}$ can be gradually populated over time such that $\boldsymbol{\Omega}$ has full rank (equivalent to $\Delta > 0$) after a certain moment, if the regressor matrix $\boldsymbol{W}_a^\top$ satisfies the IE condition (see Assumption 1). This can be interpreted below. Assumption 3.1 implies that $\int_{t_s}^{t_s+T} \boldsymbol{W}_a^\top(\tau)\boldsymbol{W}_a(\tau)d\tau \geq \alpha\mathbf{I}_3$, where $\alpha > 0$ denotes the excitation level. Given this fact, one can deduce that

[1] As highlighted in [46], numerical computation of $\text{adj}(\boldsymbol{\Omega})$ is not necessary for obtaining $\boldsymbol{\Lambda}$. Actually, the elements Λ_i $(i = 1, 2, 3)$ of $\boldsymbol{\Lambda}$ can be computed applying the Cramer's rule as $\Lambda_i = \det(\boldsymbol{\Omega}_{N,i})$, where $\boldsymbol{\Omega}_{N,i}$ is the matrix $\boldsymbol{\Omega}$ with its i-th column replaced with the vector $\boldsymbol{N}$.

$$\begin{aligned}\boldsymbol{\Omega}(t_s+T) &= \int_0^{t_s+T} e^{-a(t_s+T-\tau)}\boldsymbol{W}_a^\top(\tau)\boldsymbol{W}_a(\tau)d\tau \\ &\geq e^{-a(t_s+T)}\int_{t_s}^{t_s+T} e^{a\tau}\boldsymbol{W}_a^\top(\tau)\boldsymbol{W}_a(\tau)d\tau \\ &\geq e^{-aT}\int_{t_s}^{t_s+T} \boldsymbol{W}_a^\top(\tau)\boldsymbol{W}_a(\tau)d\tau \\ &\geq \alpha e^{-aT}\boldsymbol{I}_3 > 0,\end{aligned} \tag{3.54}$$

from which it can be observed that $\boldsymbol{\Omega}$ will become a full rank matrix after $t = t_s + T$. However, it should be pointed out that, due to the exponential forgetting design in (3.49), if $\boldsymbol{W}_a^\top$ is only of IE, Δ will decay exponentially after the end of IE. Consequently, in the case of IE, direct use of Δ to design the adaptive laws as in [37] will lead to a conspicuous decrease in parameter convergence rate over time. To avoid this problem, an interception mechanism is introduced as follows:

$$t_e \triangleq \min\left\{\arg_{\tau\in(0,t]} \Delta(\tau) \geq \Delta_{thr}\right\}, \tag{3.55}$$

$$\Delta_e \triangleq \Delta(t_e),\ \boldsymbol{\Lambda}_e \triangleq \boldsymbol{\Lambda}(t_e) = \Delta_e\boldsymbol{\theta}, \tag{3.56}$$

where $\Delta_{thr} > 0$ is a user-defined threshold. In general, Δ_{thr} should be chosen sufficiently small to accommodate a possible low level of excitation.

Then, under the control law $\boldsymbol{u} = -\boldsymbol{\Phi}(\hat{\boldsymbol{\theta}} + \boldsymbol{\beta})$ of same form as (3.17), $\boldsymbol{\beta}$ is still given by (3.19), whereas $\hat{\boldsymbol{\theta}}$ is updated by the following learning law

$$\dot{\hat{\boldsymbol{\theta}}} = \underbrace{-\gamma[\dot{\boldsymbol{\mu}} + (\boldsymbol{\Phi}+\boldsymbol{\Psi})^\top(\boldsymbol{N}_\omega^{-1}\boldsymbol{v} - k(t)\boldsymbol{N}_\omega\boldsymbol{\omega})]}_{\text{current data}}\underbrace{-\gamma\Delta_e^{-1}\boldsymbol{\Gamma}\boldsymbol{\Theta}}_{\text{historical data}}, \tag{3.57}$$

where $\boldsymbol{\Gamma} \in \mathbb{R}^{3\times 3}$ is a positive-definite diagonal weighting matrix, and $\boldsymbol{\Theta}$ is a prediction error vector given by

$$\boldsymbol{\Theta} = \begin{cases} \boldsymbol{0}, & \text{for } t < t_e \\ \Delta_e(\hat{\boldsymbol{\theta}} + \boldsymbol{\beta}) - \boldsymbol{\Lambda}_e, & \text{for } t \geq t_e \end{cases}. \tag{3.58}$$

In (3.57), Δ_e^{-1} is used instead of Δ_e to make the convergence rates of parameter adaptation independent of excitation levels, which will become clear later. By "data-driven", it is here meant that information-rich historical data is used concurrently with the standard I&I adaptive law (3.18) (driven by the current data) to update the parameter estimates.

Likewise, the estimation error vector is defined as $\tilde{\boldsymbol{\theta}} = \hat{\boldsymbol{\theta}} + \boldsymbol{\beta} - \boldsymbol{\theta}$. Then, according to (3.19) and (3.57), we derive that

$$\dot{\tilde{\boldsymbol{\theta}}} = -\gamma(\boldsymbol{\Phi} + \boldsymbol{\Psi})^\top \boldsymbol{J}^{-1}\boldsymbol{\Phi}\tilde{\boldsymbol{\theta}} - \gamma C_\Theta(t)\boldsymbol{\Gamma}\tilde{\boldsymbol{\theta}}, \tag{3.59}$$

where $C_\Theta(t) \in \mathbb{R}$ is defined as

$$C_\Theta(t) = \begin{cases} 0, & \text{for } t < t_e \\ 1, & \text{for } t \geq t_e \end{cases}. \tag{3.60}$$

A scaled estimation error z that takes the same form as (3.22) is also defined to deal with the "perturbation" term $\boldsymbol{\Psi}^\top \boldsymbol{J}^{-1}\boldsymbol{\Phi}\tilde{\boldsymbol{\theta}}$ in (3.59). Substituting (3.23) and (3.59) into $\dot{z}$ gives

$$\dot{z} = -\gamma(\boldsymbol{\Phi} + \boldsymbol{\Psi})^\top \boldsymbol{J}^{-1}\boldsymbol{\Phi}z - \frac{\gamma}{2J_m}\|\boldsymbol{\Psi}\|_2^2 z - \gamma C_\Theta(t)\boldsymbol{\Gamma}z. \tag{3.61}$$

Theorem 3.2 *Consider the spacecraft attitude dynamics in (2.19) and (2.21). Given the initial conditions satisfying $\boldsymbol{q}(0) \in \mathbb{Q}_a$, if the dynamic gain is chosen as $k(t) = \kappa r(t)$, and the regressor matrix $\mathbf{W}_a^\top$ satisfies the IE condition as stated in Assumption 3.1, then the control law $\boldsymbol{u} = -\boldsymbol{\Phi}(\hat{\boldsymbol{\theta}} + \boldsymbol{\beta})$ with $\hat{\boldsymbol{\theta}}$ and $\boldsymbol{\beta}$ defined respectively by* (3.57) *and* (3.19) *can guarantee that:*

1. *All the results of Theorem 3.1 hold on $t \in [0, \infty)$;*
2. *The parameter estimation error $\tilde{\boldsymbol{\theta}}(t)$ converges asymptotically to zero on $t \in [t_e, \infty)$.*

Proof Consider the same Lyapunov-like function V as (3.26). Under the data-driven I&I adaptive controller proposed in this section, the inequality (3.62) can be obtained following a similar procedure for deriving (3.28).

$$\dot{V} \leq -\frac{\kappa}{2}\|\boldsymbol{N}_\omega\boldsymbol{\omega}\|_2^2 - \frac{J_{\text{m}}\varrho}{2}\|\boldsymbol{J}^{-1}\boldsymbol{\Phi}z\|_2^2 - \eta C_\Theta(t)z^\top\boldsymbol{\Gamma}z. \tag{3.62}$$

Since $\eta C_\Theta(t)z^\top\boldsymbol{\Gamma}z \geq 0, \forall t \geq 0$, it therefore follows from (3.62) that $\dot{V} \leq -\frac{\kappa}{2}\|\boldsymbol{N}_\omega \boldsymbol{\omega}\|_2^2 - \frac{J_{\text{m}}\varrho}{2}\|\boldsymbol{J}^{-1}\boldsymbol{\Phi}z\|_2^2$ holds for all $t \geq 0$, which has the same form as (3.28). This implies that all theoretical results of Theorem 1 can be preserved for all $t \geq 0$. Thus, all the key features (see Remark 3.2) of the I&I adaptive control framework presented in Sect. 3.3 can be preserved.

We now assume that the IE condition as stated in Assumption 3.1 is met such that there exists a moment t_e satisfying (3.55), and emphatically consider the stability analysis on $t \in [t_e, \infty)$. From (3.60), it is known that $C_\Theta(t) = 1$ for all $t \geq t_e$. Thus, (3.62) reduces to

$$\dot{V} \leq -\frac{\kappa}{2}\|\boldsymbol{N}_\omega\boldsymbol{\omega}\|_2^2 - \frac{J_{\text{m}}\varrho}{2}\|\boldsymbol{J}^{-1}\boldsymbol{\Phi}z\|_2^2 - \eta\Gamma_m\|z\|^2, \tag{3.63}$$

where $\Gamma_m > 0$ denotes the minimum eigenvalue of $\boldsymbol{\Gamma}$. From (3.63) and the definition of V, it is evident that $z \in \mathcal{L}_2 \cap \mathcal{L}_\infty$. In addition, Theorem 1 has shown the boundedness of $\boldsymbol{J}^{-1}\boldsymbol{\Phi}z$, $\boldsymbol{\Phi}$, and $\boldsymbol{\Psi}$. Based on these two facts, it can be concluded from (3.61) that z is uniformly continuous. Then, by using Barbalat's lemma, it can be claimed that z converges asymptotically to zero on $t \in [t_e, \infty)$. Recalling the definition of z given in (3.22), we further conclude that $\tilde{\boldsymbol{\theta}}(t)$ also converges asymptotically to zero on $t \in [t_e, \infty)$. This completes the proof.

A close inspection of the proposed data-driven I&I adaptive control scheme illuminates the following discussions:

(1) The data-driven learning law (3.57) is an immediate extension of the previously proposed parameter update law (3.18) by adding a prediction-error-driven term $-\gamma\Delta_e^{-1}\boldsymbol{\Gamma}\boldsymbol{\Theta}$, which can take full advantage of stored historical data to inject damping into the parameter estimation error dynamics (3.59). Benefiting from the inclusion of such a term in the parameter update law (3.57), the parameter estimate $\hat{\boldsymbol{\theta}} + \boldsymbol{\beta}$ is able to accurately learn $\boldsymbol{\theta}$ under a strictly weak IE condition, thus relaxing the stringent PE condition for parameter convergence. In fact, the construction of the prediction error $\boldsymbol{\Theta}$ is partially inspired by the composite learning in [33] and its complement in [32], but still remains a substantial difference. To be specific, the DREM procedure is introduced to generate a set of scalar LREs sharing the same regressor Δ, as compactly expressed in (3.53). This contributes directly to a new prediction error vector $\boldsymbol{\Theta}(t) = \Delta_e\tilde{\boldsymbol{\theta}}\ \ \forall t \geq t_e$ that has a linear relationship with $\tilde{\boldsymbol{\theta}}$. This property can notably improve the transient performance of the estimator. To provide intuitive insights into this point, the estimation error dynamics (3.59) on $t \in [t_e, \infty)$ is rewritten in an element-wise manner

$$\dot{\tilde{\theta}}_i = -\gamma[(\boldsymbol{\Phi} + \boldsymbol{\Psi})^\top \boldsymbol{J}^{-1}\boldsymbol{\Phi}\tilde{\boldsymbol{\theta}}]_i - \gamma\Gamma_i\tilde{\theta}_i,\ i \in 1, 2, 3, \tag{3.64}$$

where $[\cdot]_i$ represents the i-th element of "$\cdot$". The first term on the right-hand side of (3.64) plays the role of establishing the attracting manifold $\mathcal{M}$ defined by (3.35), while the second term serves as a damping injecting term for parameter convergence. Evidently, given a value for γ in advance, adjusting the weight Γ_i only affects the convergence rate of $\tilde{\theta}_i$. Such an element-wise tuning can help improve the transient responses of $\tilde{\boldsymbol{\theta}}$, and make the weights selection transparent and flexible.

Another prominent advantage of the prediction-error-driven term $-\gamma\Delta_e^{-1}\boldsymbol{\Gamma}\boldsymbol{\Theta}$ is that Δ_e^{-1} is added instead of Δ_e to offset Δ_e in $\boldsymbol{\Theta}$ for $t \geq t_e$. This renders the convergence rates of $\tilde{\theta}_i$, $i = 1, 2, 3$ independent of the specific value of Δ_e and hence the excitation level, as seen in (3.64).

(2) *Comparison with the Concurrent Learning-Based Adaptive Control Methods*: The proposed data-driven I&I adaptive control algorithm shares some similarities with the CL-based adaptive methods (see, indicatively, [31–33]), but nonetheless exhibits outperformance over them in transient performance of parameter estimation, as dictated by the first bullet point. Apart from this, we note that these CL-based methods are developed based upon the CE principle, and the closed-loop performance obtained from them will be arbitrarily poor relative to the ideal deterministic control

case, if the IE conditions are not satisfied and/or the parameter convergence rates are slow caused by low levels of excitation. In contrast, the proposed method deviates significantly from the CE design framework, and yields a non-CE adaptive controller which owns all the key features of the I&I adaptive control methodology, as summarized in Remark 3.2. Such a controller can overcome the detrimental performance degradation of the CE controllers through introducing a stable attracting manifold $\mathcal{M}$ as defined by (3.35), thus improving the transient performance of the closed-loop system.

(3) *Comparison with the DREM-Based Adaptive Control Methods*: It is noteworthy that to achieve asymptotic parameter convergence, the traditional DREM-based adaptive estimators (see, e.g., [37, 46, 47]) require a non-square integrability condition on the scalar regressor Δ, which is often infeasible to monitor online similar to the PE condition, since it depends on how Δ behaves in the future. However, in this section, the Kreisselmeier's regressor extension is utilized to generate the information matrix $\boldsymbol{\Omega}$ through a weighted accumulation of all incoming data. So, the rank of $\boldsymbol{\Omega}$ can be populated to the full rank over time under the IE condition. On the other hand, an interception mechanism is proposed to avoid the information degradation caused by the exponential forgetting design (see (3.49) and (3.51)). The above two properties enable the novel data-driven estimator to achieve accurate parameter learning under a detectable IE condition, which is strictly weaker than the condition of $\Delta \notin \mathcal{L}_2$.

3.5 Numerical Simulations

In this section, numerical simulations are presented to show the effectiveness and superiority of the proposed method. For a spacecraft carrying an optical instrument with a fixed boresight aligned with the z-axis of the body-fixed frame $\mathcal{P}$, we consider a mission scenario where the spacecraft is required to perform a rest-to-rest attitude reorientation maneuver to reorient its on-board optical instrument to the desired direction, while evading certain forbidden zones.

Simulation setup: The spacecraft inertia matrix is $\boldsymbol{J} = \text{diag}[10, 7, 5]\ \text{kg} \cdot \text{m}^2$, and its initial attitude is given as $\boldsymbol{q}(0) = [0.33, 0.66, -0.62, -0.26665]^\top$ with $\boldsymbol{\omega}(0) = \mathbf{0}\,\text{rad/s}$. The desired attitude is $\boldsymbol{q}_d = [0.2, -0.5, -0.5, -0.67823]^\top$, which is evidently a non-PE reference. During attitude reorientation maneuvers, there are four forbidden zones that are required to be avoided. Their geometrical details are listed in Table 3.1. The maximum allowed angular rate is set to $\omega_\text{m} = 0.05\,\text{rad/s}$. Besides, for practical implementation, the control input torques are saturated by $u_\text{m} = 0.5\,\text{Nm}$. Please note that all subsequent simulations are executed using the fixed-step ODE 4 (Runge-Kutta) solver with a sample step of 0.01 s.

Table 3.1 Geometrical details of the forbidden zones

Attitude constraints	Axis vector	Angle
FZ 1	$[0.437, -0.783, 0.442]^\top$	35°
FZ 2	$[0, 0.707, 0.707]^\top$	25°
FZ 3	$[-0.853, 0.436, -0.286]^\top$	25°
FZ 4	$[0.413, -0.474, -0.783]^\top$	20°

3.5.1 Performance Validation

To verify the effectiveness and performance of the proposed data-driven I&I adaptive control algorithm, simulation results obtained by implementing the designed controller are provided in this subsection. The control parameters are selected as: $\zeta = 15$, $\kappa = 0.004$, $c = 3$, $a = 0.1$, $\gamma = 0.1$, $\Gamma = \text{diag}[1, 1, 1]$ and $\hat{\boldsymbol{\theta}}(0) = [12, 13, 25]^\top$.

The resulting closed-loop responses are documented in Figs. 3.3, 3.4 and 3.5, from which it can be seen that both the attitude errors and angular velocities converge asymptotically to zero, and the control torques remain bounded by u_m for all times. Note that, in Fig. 3.3, an unconventional attitude maneuver is observed at $40 - 100$ s (the time interval marked by the shadow). This is caused by the fact that the proposed controller allows the spacecraft to perform a circuitous slew maneuver to avoid the forbidden zones, which will soon be witnessed in Fig. 3.6. Also, as shown in Fig. 3.4, the angular velocity constraints are strictly satisfied over the entire maneuver period. The 3-D reorientation trajectory is further provided in Fig. 3.6 for a better illustration. Intuitively, the designed controller successfully retargets the boresight of the optical instrument to the desired orientation, while evading all forbidden zones.

The time histories of $\|\boldsymbol{\Phi}\tilde{\boldsymbol{\theta}}\|$ and $\tilde{\boldsymbol{\theta}}$ are depicted in Fig. 3.7. It is evident that $\boldsymbol{\Phi}\tilde{\boldsymbol{\theta}}$ converges asymptotically to zero, indicating the establishment of the attracting manifold $\mathcal{M}$. This result is consistent with the theoretical analysis (by recalling Theorem 3.2 and Remark 3.2). Besides, from the right subfigures of Fig. 3.7, one can observe that the parameter estimation errors $\tilde{\theta}_i$, $i = 1, 2, 3$ converge to zero, despite the nonsatisfaction of the PE condition. The above two observations imply that the data-driven I&I adaptive control method proposed in this work not only inherits all the key features of the I&I adaptive control methodology, but also relaxes the PE condition (for parameter convergence) to IE. These two features can hardly be obtained simultaneously by the existing CE- and non-CE-based adaptive controllers. To illustrate the element-wise tuning property of the parameter estimator (3.57), we simulate three cases in each of which we deliberately choose two additional values for one weight (Γ_i), while fixing the other two. The simulation results are plotted in Fig. 3.8. As can be seen, adjusting Γ_i $(i = 1, 2, 3)$ can only affect the convergence rate of $\tilde{\theta}_i$, and moreover, the convergence rate is independent of the excitation level. This fact renders the selection of the weighting values Γ_i, $i = 1, 2, 3$ a transparent and flexible procedure. As a result, the parameter convergence rates across all parameter vector components can be flexibly adjusted in an explicit way.

3.5.2 Comparison Results

In order to show the superiority of the proposed data-driven I&I adaptive controller (termed here as D-I&IAC), two other controllers are also simulated for comparison.

(1) *APF-based controller in* [17] *(termed as APFC)*: It is a model-free controller, and can ensure that the spacecraft achieves the reorientation maneuvers, while avoiding the attitude constriants and the unwinding phenomenon. The design parameters of this controller are chosen as: $\alpha = 3$, $\beta = 0.1$, and $l_1 = 50$.

(2) *CL-based adaptive controller in* [32] *(termed as CLAC)*: The CLAC can not only achieve the reorientation objectives (i.e., $\lim_{t\to\infty} \boldsymbol{q}(t) = \boldsymbol{q}_d$ and $\lim_{t\to\infty} \boldsymbol{\omega}(t) = \mathbf{0}$), but is capable of identifying on-line the unknown inertia parameters, even if the PE condition is not satisfied. We should emphasize that, although the CLAC is designed following the approach in [32], a slight modification is made to simplify the control design. The control law is detailed below:

$$\boldsymbol{u} = -\bar{\boldsymbol{\Phi}}\hat{\boldsymbol{\theta}} - k_1 \boldsymbol{s},$$

where $k_1 > 0$ is the control gain, $\boldsymbol{s} = \boldsymbol{\omega} + k\boldsymbol{q}_{ev}$ with $k > 0$ a constant, $\bar{\boldsymbol{\Phi}}$ is the regressor matrix satisfying $\bar{\boldsymbol{\Phi}}\boldsymbol{\theta} = -\boldsymbol{S}(\boldsymbol{\omega})\boldsymbol{J}\boldsymbol{\omega} + (k/2)\boldsymbol{J}(\boldsymbol{S}(\boldsymbol{q}_{ev}) + q_{e4}\mathbf{I}_3)\boldsymbol{\omega}$, and $\hat{\boldsymbol{\theta}}$ is updated by the following learning law:

$$\dot{\hat{\boldsymbol{\theta}}} = \begin{cases} \eta\bar{\boldsymbol{\Phi}}^\top \boldsymbol{s}, & \text{for } t < t_e \\ \eta[\bar{\boldsymbol{\Phi}}^\top \boldsymbol{s} - \iota(\boldsymbol{\Omega}_e\hat{\boldsymbol{\theta}} - \boldsymbol{N}_e)], & \text{for } t \geq t_e \end{cases},$$

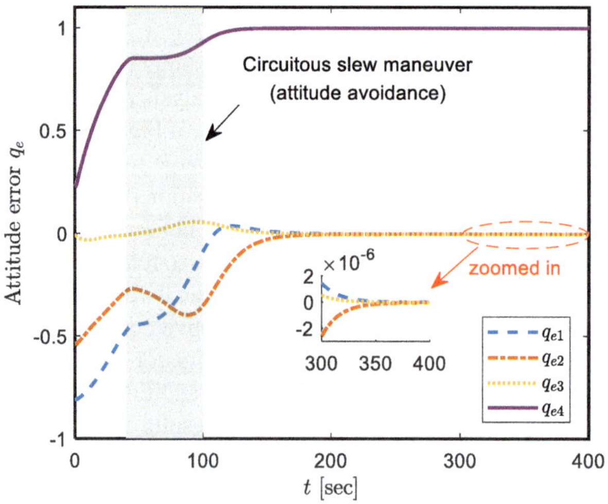

Fig. 3.3 Time histories of the attitude errors

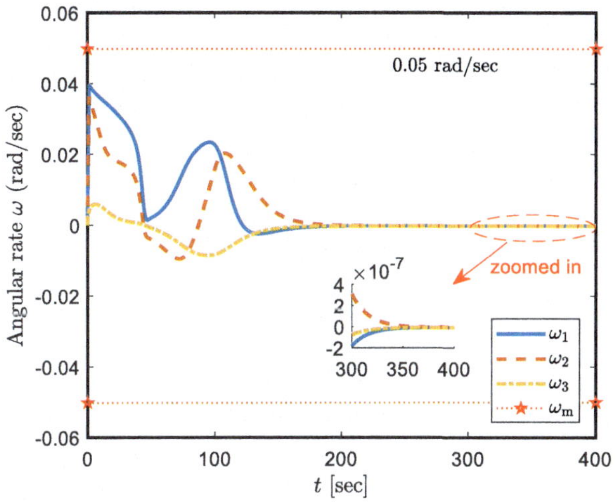

Fig. 3.4 Time histories of the angular velocities

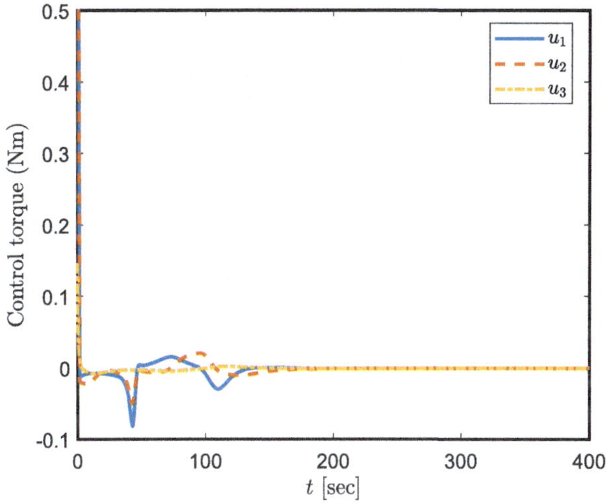

Fig. 3.5 Time histories of the control torques

where $\eta,\ \iota > 0$ are constants, while $\boldsymbol{\Omega}_e$ and $\boldsymbol{N}_e$ are $\boldsymbol{\Omega}(t)$ and $\boldsymbol{N}(t)$ (see (3.50) and (3.51)) taking values at t_e defined in (3.55), respectively. The control parameters are chosen to be $k = 0.07$, $k_1 = 10$, $\eta = 5$, and $\iota = 5000$.

It is noteworthy that, for fair comparison, the control parameters of the above two controllers have been judiciously chosen by trial and error to render almost the same settling times of attitude errors as the D-I&IAC proposed in this work. The comparison results are shown in Fig. 3.9. Inspecting Fig. 3.9a and b reveals that all the three controllers achieve the asymptotic convergence of the attitude errors and

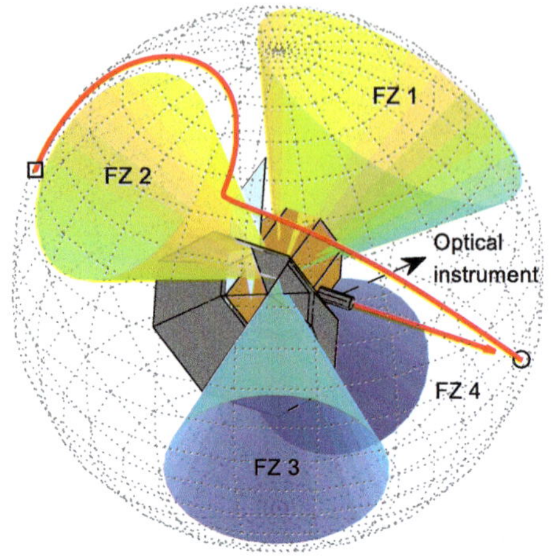

Fig. 3.6 3-D reorientation trajectory of optical instrument boresight pointing on unit sphere. The initial and desired orientations are marked by "circle" and "square", respectively

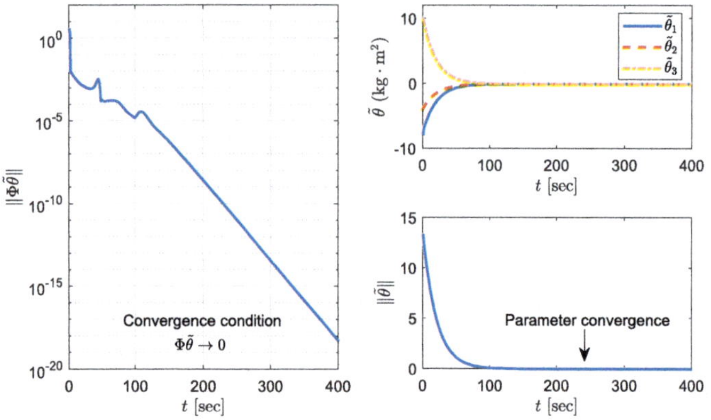

Fig. 3.7 Time histories of $\|\boldsymbol{\Phi}\tilde{\boldsymbol{\theta}}\|$ and $\tilde{\boldsymbol{\theta}}$

angular rates, but, quantitatively speaking, the derived D-I&IAC delivers the best transient performance. In addition, from Fig. 3.9b (infinity norm of $\boldsymbol{\omega}$), we found that the D-I&IAC complies strictly with the angular rate constraints, whereas the other two controllers fail. The comparison of control efforts is depicted in Fig. 3.9c, where the energy index is defined as $Energy = \int_0^{top} \|\boldsymbol{u}(\tau)\|^2 d\tau$. From Fig. 3.9c, it is evident that the D-I&IAC and the APFC require more control efforts than the CLAC especially during the initial phase (0-20 s). This is because they require extra control efforts to steer the spacecraft to perform a circuitous slew maneuver to evade the attitude forbidden zones (as will be witnessed in Fig. 3.9d). Figures 3.9d and e plot the trajectories of the optical instrument boresight pointing on 3-D unit sphere and in 2-D cylindrical projection, respectively. As can be seen, the D-I&IAC and

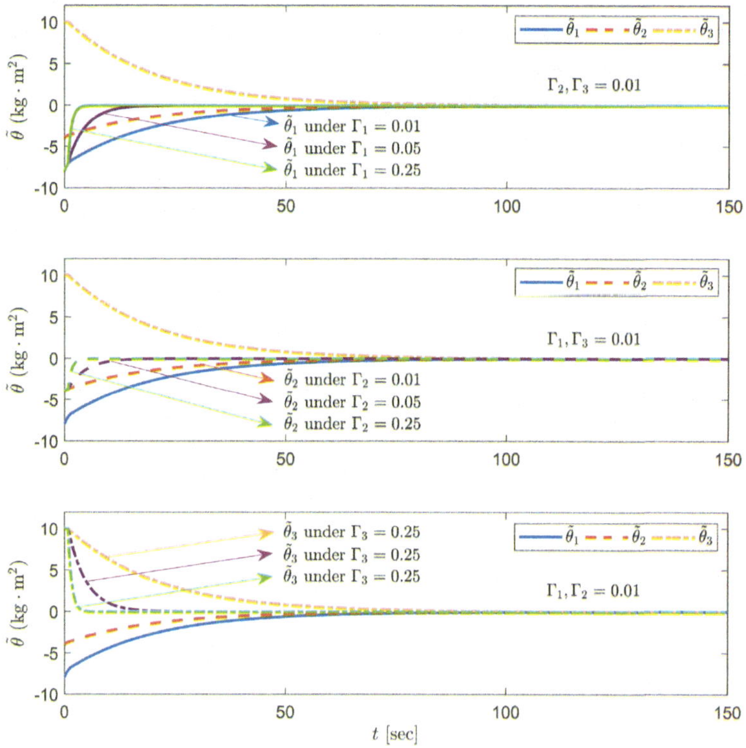

Fig. 3.8 Time histories of $\tilde{\boldsymbol{\theta}}$ under different weights

the APFC evade all the forbidden zones, but the CLAC fails (legend: CLAC for Case 1). Specially, in this case, the trajectory generated by the CLAC traverses FZ 2. To further show the anti-unwinding ability of the three controllers, we deliberately reset the initial attitude vector to $-\boldsymbol{q}(0)$ such that $q_{e4}(0) < 0$, and keep all the other conditions unchanged (This case is called Case 2). As no changes occur in the motion trajectories of the D-I&IAC and the APFC in Case 2, we here directly plot the attitude trajectory generated by the CLAC in Figs. 3.9d, e (legend: CLAC for Case 2). Notably, when $q_{e4}(0) < 0$, the CLAC experiences the unwinding issue (intuitively, the rotation angle is larger than 180°, as shown in Fig. 3.9e), and moreover, the resulted path traverses FZ 4. In contrast, the D-I&IAC and the APFC can regulate the attitude error to the closest equilibrium $\boldsymbol{q}_e = [0, 0, 0, -1]^\top$, rather than $\boldsymbol{q}_e = [0, 0, 0, 1]^\top$, thus remaining shorter paths.

To show outperformance of the proposed data-driven parameter estimator in (3.57) w.r.t. the traditional CL-based estimators, we further examine and compare the dynamic responses of the parameter estimation errors between the D-I&IAC and the CLAC. The comparison results are detailed in Fig. 3.10. The baseline I&I adaptive controller (without data-driven extension term) presented in Sect. 3.3 is also simu-

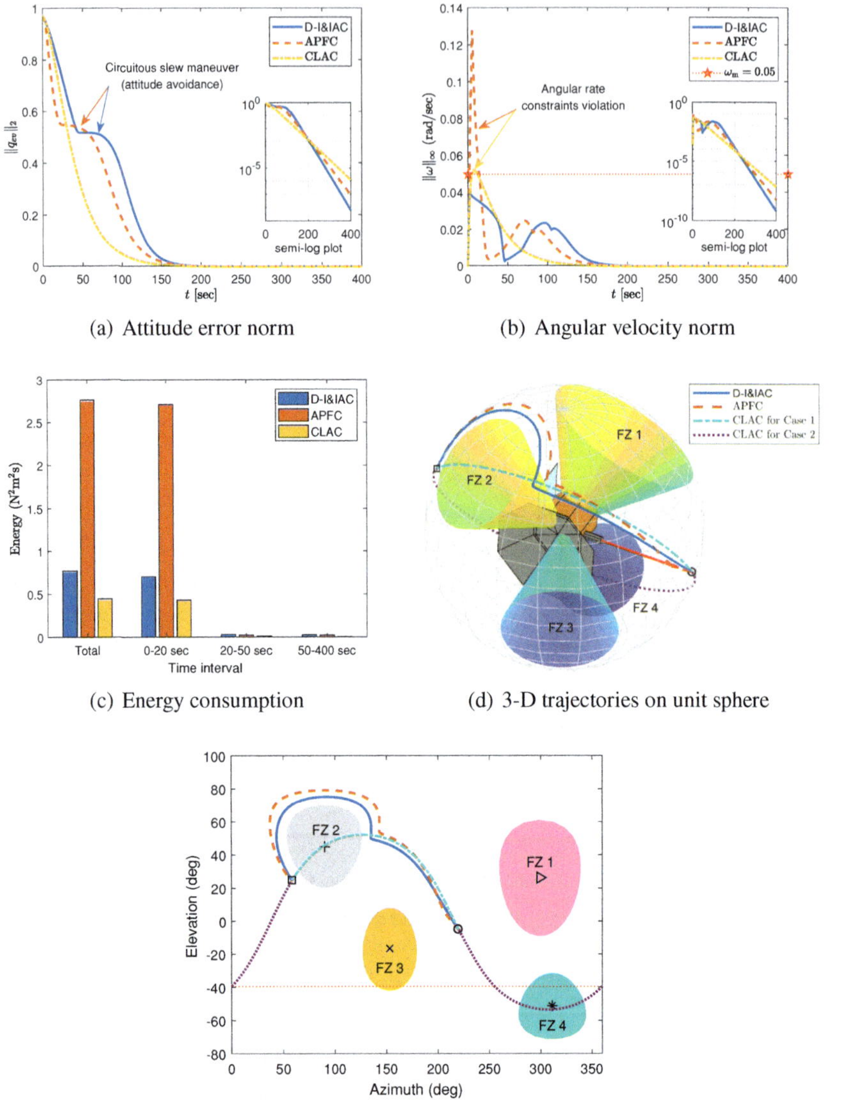

(a) Attitude error norm

(b) Angular velocity norm

(c) Energy consumption

(d) 3-D trajectories on unit sphere

(e) 2-D trajectories in cylindrical projection

Fig. 3.9 Comparison results. In (**d**) and (**e**), the initial and desired orientations are marked by "circle" and "square", respectively

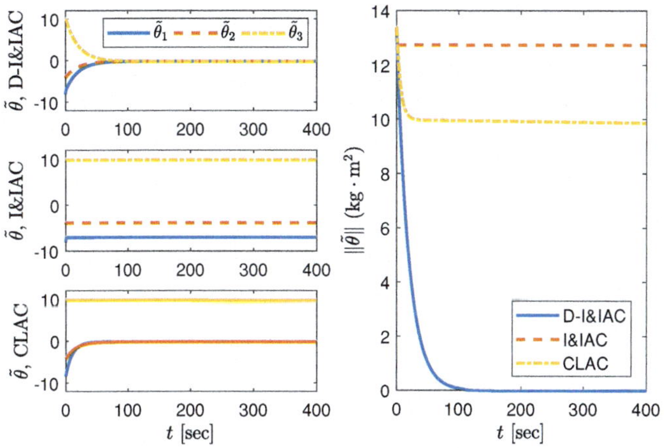

Fig. 3.10 Simulation results of $\tilde{\boldsymbol{\theta}}$ under different controllers

lated to highlight the crucial role of the data-driven term $-\gamma\Delta_e^{-1}\boldsymbol{\Gamma}\boldsymbol{\Theta}$ in relaxing the dependence of parameter convergence on PE condition. This controller is termed as I&IAC hereafter. From Fig. 3.10, one can check that the D-I&IAC and the CLAC can achieve parameter convergence in the absence of PE condition, while the baseline I&IAC which is driven only by the current data fails to obtain such a result. This indicates that the introduction of historical data is instrumental for relaxing the PE condition. However, it should be stressed that although the CLAC exhibits asymptotic parameter convergence, the convergence rates across all three parameter vector components are ill-balanced (indeed $\tilde{\theta}_3$ converges rather slowly). The main reason for this problem is certainly that the parameter estimates are interactively driven by the instantaneous information matrix $\boldsymbol{\Omega}_e$, which renders the parameter convergence rates among all the components of $\tilde{\boldsymbol{\theta}}$ not only highly dependent on the excitation levels in different regressor channels, but also strongly coupled to each other. In contrast, the proposed data-driven method can fully overcome such a problem through introducing the DREM procedure and the inverse of Δ_e in the estimator design, as discussed below the proof of Theorem 3.2. In reality, for the proposed D-I&IAC, the parameter convergence rates of $\tilde{\theta}_i$, $i = 1, 2, 3$ can be easily balanced by adjusting the components of the diagonal $\boldsymbol{\Gamma}$.

In summary, the above comparison simulation results show outperformance of the proposed data-driven I&I adaptive control algorithm in transient performance, constraints guarantee, anti-unwinding ability, and parameter identification.

3.5.3 Robustness Tests

We next examine the robustness of the proposed data-driven I&I adaptive control scheme against external disturbances and measurement noise. The disturbances of the following form ($w = 0.02$) are considered

$$\boldsymbol{\tau}_d = 10^{-4} \times \begin{bmatrix} 3\cos(10wt) + 4\sin(3wt) - 10 \\ -1.5\sin(2wt) + 3\cos(5wt) + 15 \\ 3\sin(10wt) - 8\sin(4wt) + 5 \end{bmatrix} \text{Nm},$$

The measurement noises of attitude is generated following the line of [48]. Specifically, rewrite the unit quaternion $\boldsymbol{q}$ as

$$\boldsymbol{q}_v = \hat{\boldsymbol{n}} \sin(\frac{\varepsilon}{2}), \; q_4 = \cos(\frac{\varepsilon}{2}),$$

where $\hat{\boldsymbol{n}}$ and ε denote the eigenaxis and eigenangle associated with $\boldsymbol{q}$, respectively. Then, the noisy measurements are given by randomly perturbing the true $\hat{\boldsymbol{n}}$ within a specified spherical cone (with uniform distribution) centered around it. Here, the cone half-angle is taken as 0.5 deg. The measurement noises with mean zero and variance 1×10^{-6} rad/s are also added to the angular rates. With explicit consideration of both external disturbances and measurement noises, the simulation scenario in Sect. 3.5.2 is repeated. In order to clearly illustrate the steady-state behaviors of different controllers, the simulation duration is prolonged to 1000 s.

The performance comparison of different controllers under perturbed and noisy conditions are plotted on semilogarithmic scales in Fig. 3.11. By comparing Fig. 3.11 with Figs. 3.9 and 3.10, it can be seen that all the three controllers suffer from obvious performance degradation when both external disturbances and measurement noises are present. The asymptotic convergence of attitude errors and angular rates has been broken. Instead, they can only converge to small residual sets around the origin. Moreover, the proposed D-I&IAC and the APFC yield almost the same steady-state behaviors, which are superior than that obtained from the CE-based CLAC. Note that the steady-state performance of the D-I&IAC and the APFC is still acceptable for general attitude reorientation missions. From Fig. 3.11c, we find that the D-I&IAC can still ensure that the parameter estimation error $\tilde{\boldsymbol{\theta}}$ rapidly converges to a small residual set, in the presence of external disturbances and measurement noises. In summary, the proposed control scheme has a certain level of robustness against external disturbances and measurement noises. However, we emphasize that, in the present framework, it is hard to quantitatively evaluate the robustness.

Fig. 3.11 Performance comparison of different controllers under disturbed and noisy circumstance

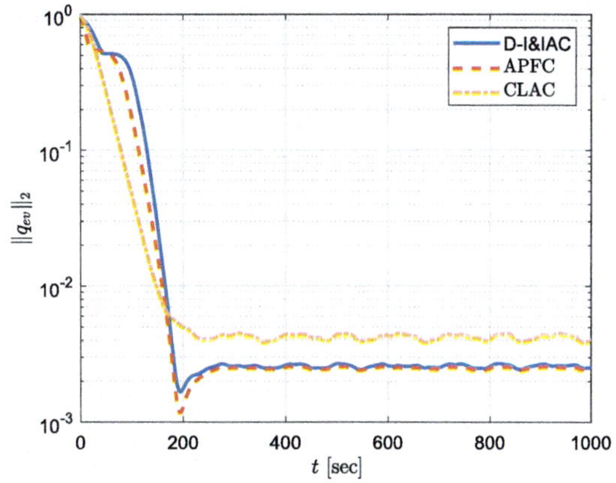

(a) Attitude error norm

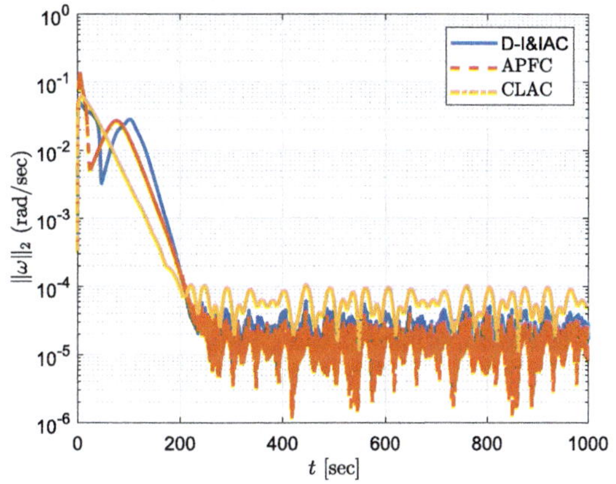

(b) Angular velocity norm

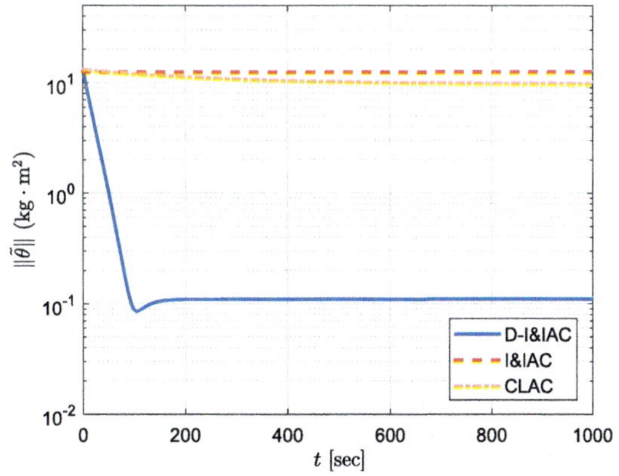

(c) Parameter estimation error norm

3.6 Hardware-in-Loop Experiments

In the section, we further carry out a hardware-in-the-loop (HIL) experiment to validate the practical applicability of the proposed adaptive control method as in (3.17)–(3.19) without integrating the data-driven part. The experimental setup is shown in Fig. 3.12, and consists of the following parts:

- A three-axis turntable used to simulate the attitude motion of the spacecraft. Three rasters and four fiber-optic gyroscopes are mounted on the turntable to measure its Euler angles and angular velocities, respectively. The measured values $\hat{\boldsymbol{q}}$ and $\hat{\boldsymbol{\omega}}$ (where Euler angles are transformed to unit quaternions) are delivered to the simulation computer for control calculation.
- A high-performance real-time VxWorks simulator computing the command control signals $\boldsymbol{\tau}$ as per (3.17), and then performing actuator redundancy management based on control allocation and transferring the allocated signals $\boldsymbol{u}_{cmd}$ to the underlying control module.
- An Arduino Mega2560 based underlying control module, which is connected with reaction wheels (RWs) via RS-422 and provides actuating commands $\hat{\boldsymbol{u}}_{cmd}$ for the RWs.
- Four RWs (each has a maximum torque output 0.1 Nm and a maximum rotation speed of 5000 rpm) act as an actuator simulator. They issue the control torques $\boldsymbol{\tau}_{act}$ through the configuration matrix $\boldsymbol{D}$ to the spacecraft dynamics running in the VxWorks simulator.

In the experiments, the configuration matrix of the four RWs is chosen as $\boldsymbol{D} = [\boldsymbol{I}_3, \text{col}(1, 1, 1)/\sqrt{3}]$, and the pseudo-inverse method is used for control allocation.

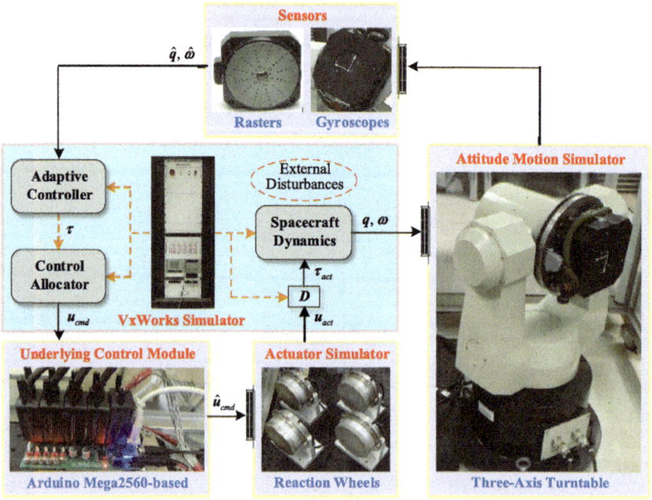

Fig. 3.12 HIL experimental setup

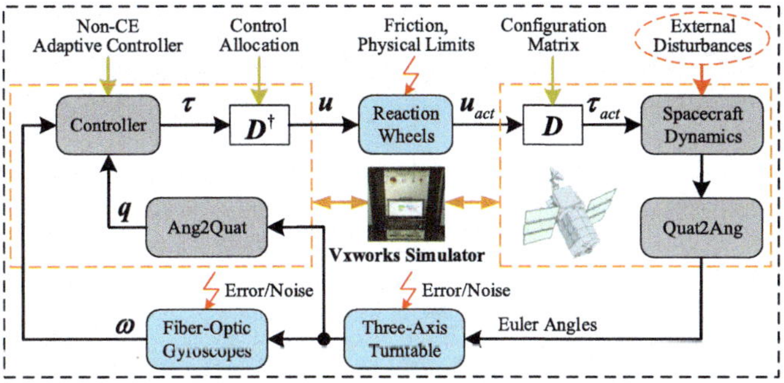

Fig. 3.13 Block diagram of the experimental setup

The disturbance torque τ_d is also considered, and its form is the same as that in Sect. 3.5.3. The block diagram of the HIL experimental system is shown in Fig. 3.13. All the parameters and conditions are the same as Sect. 3.5. The sample frequency is 20 Hz. The video showing the experimental results is available at https://youtu.be/pFiPr_wwvJs.

As can be seen in Fig. 3.14a, b, the rest-to-rest attitude reorientation is achieved with steady-state accuracy of the attitude error and angular rate higher than 5×10^{-3} (relative Euler angles less than 0.5 deg) and 6×10^{-4} rad/s; moreover, the angular rate limit $\omega_{\mathrm{m}} = 0.05$ rad/s is not violated. Note that the primary reason why asymptotic results as observed in the numerical simulations are no longer obtained in the experiment lies in the adverse affects from external disturbances, control delay, bearing friction in the RWs, measurement error/noise, communication breakpoint, etc. In addition, in Fig. 3.14a, and b, an unconventional attitude maneuver is notably observed at 40-120 s (the interval marked by the shadow). Actually, this is caused by the proposed controller rendering a circuitous slew maneuver of spacecraft to avoid FZ 2, as will be seen in Figs. 3.15 and 3.16. The torque outputs and speeds of the RWs are plotted in Fig. 3.14c and d, from which we can observe that all the four RWs function normally within rated torques and speeds. Light oscillations in RWs' torque outputs are observed in Fig. 3.14c at 0-10 s. This phenomenon is caused by the fact that larger control torques are required at the initial phase, but the RWs cannot timely respond such demands due to the physical limitation of the motors (in reality, the initial torque output of each RW can only increase from zero). The 3-D reorientation trajectory and 2-D cylindrical projection are provided, respectively, in Figs. 3.15 and 3.16 for illustration. Intuitively, the derived controller retargets the instrument's boresight to the desired orientation, while evading all forbidden zones. In summary, the proposed control scheme exhibits an acceptable performance in the HIL setting, and thus is practically applicable.

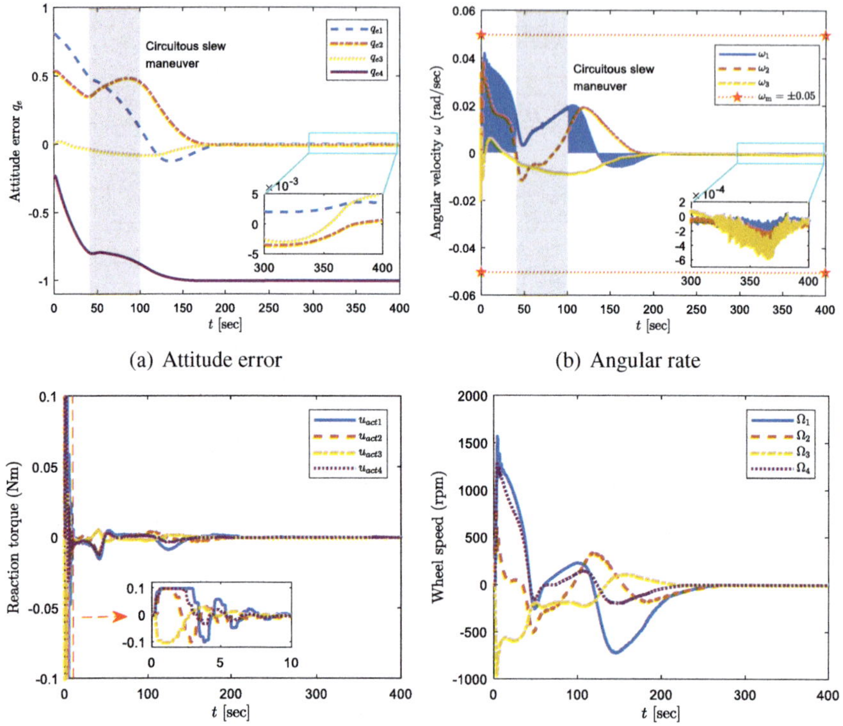

(a) Attitude error (b) Angular rate

(c) Reaction torque (d) Wheel speed

Fig. 3.14 Control performance illustration

Fig. 3.15 HIL experimental setup

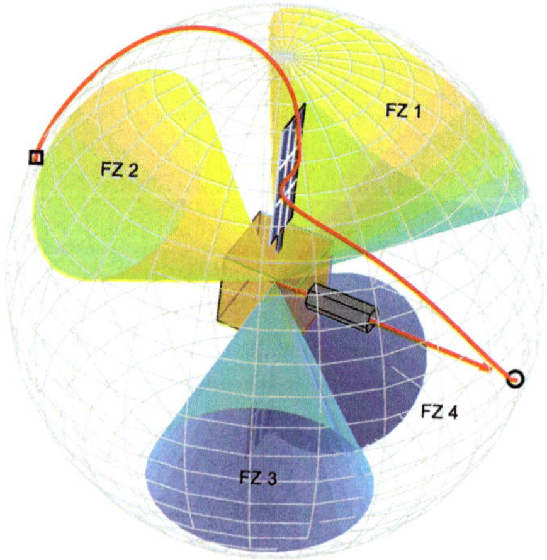

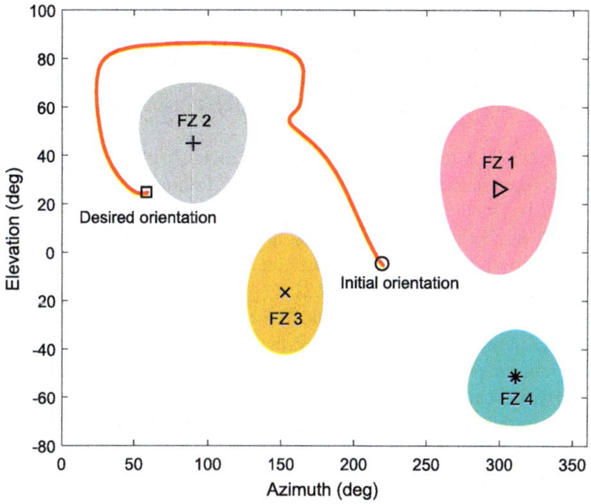

Fig. 3.16 HIL experimental setup

3.7 Summary

A novel data-driven I&I adaptive control scheme for spacecraft attitude reorientation under attitude and angular velocity constraints, as well as inertia uncertainties, has been proposed in this chapter. The basic framework of the developed algorithm is built upon the I&I adaptive control methodology, which helps remove the restrictive realizability condition that does not hold in the Lyapunov sense when angular velocity constraints are taken into account. The excellent features of the designed controller lie in the following several aspects: It can (i) enable the spacecraft attitude to arrive at the desired setpoint from most initial free configurations, whilst satisfying both attitude and angular velocity constraints; (ii) preserve all the key properties of the I&I adaptive control methodology, thus exhibiting better transient behaviors than the traditional CE-based adaptive controllers; (iii) guarantee asymptotic parameter convergence under a strictly weak IE condition, inducing a significant relaxation of PE. In addition, the parameter convergence rates across all entries can be tuned in an easily-balanced way, and moreover, they are independent of the excitation level. Finally, simulation results show the efficiency and superiority the proposed method.

References

1. Chaturvedi NA, Sanyal AK, McClamroch NH (2011) Rigid-body attitude control. IEEE Control Systems Magazine 31(3): 30–51
2. Krstic M, Tsiotras P (1999) Inverse optimal stabilization of a rigid spacecraft. IEEE Transactions on Automatic Control 44(5): 1042–1049

3. Arjun Ram S, Akella MR (2020) Uniform exponential stability result for the rigid-body attitude tracking control problem. Journal of Guidance, Control, and Dynamics 43(1): 39–45
4. Lee T (2011) Geometric tracking control of the attitude dynamics of a rigid body on SO(3). In: Proceedings of the 2011 American Control Conference, San Francisco, CA, USA, pp 1200–1205
5. Sun L, Zheng Z (2017) Disturbance-observer-based robust backstepping attitude stabilization of spacecraft under input saturation and measurement uncertainty. IEEE Transactions on Industrial Electronics 64(10): 7994–8002
6. Peng X, Geng Z, Sun J (2020) The specified finite-time distributed observers-based velocity-free attitude synchronization for rigid bodies on SO(3). IEEE Transactions on Systems, Man, and Cybernetics: Systems 50(4): 1610–1621
7. Frazzoli E, Dahleh M, Feron E, Kornfeld R (2001) A randomized attitude slew planning algorithm for autonomous spacecraft. In: Proceedings of AIAA Guidance, Navigation, and Control Conference and Exhibit, Montreal, Quebec, Canada, pp AIAA 2001–4155
8. Wie B, Lu J (1995) Feedback control logic for spacecraft eigenaxis rotations under slew rate and control constraints. Journal of Guidance, Control, and Dynamics 18(6): 1372–1379
9. Biggs JD, Colley L (2016) Geometric attitude motion planning for spacecraft with pointing and actuator constraints. Journal of Guidance, Control, and Dynamics 39(7): 1672–1677
10. Kjellberg HC, Lightsey EG (2016) Discretized quaternion constrained attitude pathfinding. Journal of Guidance, Control, and Dynamics 39(3): 710–715
11. Tan X, Berkane S, Dimarogonas DV (2020) Constrained attitude maneuvers on SO(3): Rotation space sampling, planning and low-level control. Automatica 112: 108659
12. McInnes CR (1994) Large angle slew maneuvers with autonomous sun vector avoidance. Journal of Guidance, Control, and Dynamics 17(4): 875–877
13. Ramos MD, Schaub H (2018) Kinematic steering law for conically constrained torque-limited spacecraft attitude control. Journal of Guidance, Control, and Dynamics 41(9): 1990–2001
14. Kulumani S, Lee T (2017) Constrained geometric attitude control on SO(3). International Journal of Control, Automation and Systems 15(6): 2796–2809
15. Lee U, Mesbahi M (2014) Feedback control for spacecraft reorientation under attitude constraints via convex potentials. IEEE Transactions on Aerospace and Electronic Systems 50(4): 2578–2592
16. Shen Q, Yue C, Goh CH, Wu B, Wang D (2018) Rigid-body attitude stabilization with attitude and angular rate constraints. Automatica 90: 157–163
17. Hu Q, Chi B, Akella MR (2019) Anti-unwinding attitude control of spacecraft with forbidden pointing constraints. Journal of Guidance, Control, and Dynamics 42(4): 822–835
18. Lee DY, Gupta R, Kalabić UV, Di Cairano S, Bloch AM, Cutler JW, Kolmanovsky IV (2017) Geometric mechanics based nonlinear model predictive spacecraft attitude control with reaction wheels. Journal of Guidance, Control, and Dynamics 40(2): 309–319
19. Hu Q, Chi B, Akella MR (2019) Reduced attitude control for boresight alignment with dynamic pointing constraints. IEEE/ASME Transactions on Mechatronics 24(6): 2942–2952
20. Dong H, Zhao X, Yang H (2020) Reinforcement learning-based approximate optimal control for attitude reorientation under state constraints. IEEE Transactions on Control Systems Technology 29(4): 1664–1673
21. Thakur D, Srikant S, Akella MR (2015) Adaptive attitude-tracking control of spacecraft with uncertain time-varying inertia parameters. Journal of Guidance, Control, and Dynamics 38(1): 41–52
22. Shao X, Hu Q, Shi Y, Jiang B (2018) Fault-tolerant prescribed performance attitude tracking control for spacecraft under input saturation. IEEE Transactions on Control Systems Technology 28(2): 574–582
23. Astolfi A, Ortega R (2003) Immersion and invariance: A new tool for stabilization and adaptive control of nonlinear systems. IEEE Transactions on Automatic control 48(4): 590–606
24. Seo D, Akella MR (2008) High-performance spacecraft adaptive attitude-tracking control through attracting-manifold design. Journal of Guidance, Control, and Dynamics 31(4): 884–891

25. Yang S, Akella MR, Mazenc F (2017) Dynamically scaled immersion and invariance adaptive control for euler–lagrange mechanical systems. Journal of Guidance, Control, and Dynamics 40(11): 2844–2856
26. Wen H, Yue X, Yuan J (2018) Dynamic scaling–based noncertainty-equivalent adaptive spacecraft attitude tracking control. Journal of Aerospace Engineering 31(2): 04017098
27. Zou Y, Meng Z (2019) Immersion and invariance-based adaptive controller for quadrotor systems. IEEE Transactions on Systems, Man, and Cybernetics: Systems 49(11): 2288–2297
28. Shao X, Hu Q, Shi Y (2021) Adaptive pose control for spacecraft proximity operations with prescribed performance under spatial motion constraints. IEEE Transactions on Control Systems Technology 29(4): 1405–1419
29. Shao X, Hu Q (2021) Immersion and invariance adaptive pose control for spacecraft proximity operations under kinematic and dynamic constraints. IEEE Transactions on Aerospace and Electronic Systems 57(4): 2183–2200
30. Jenkins BM, Annaswamy AM, Lavretsky E, Gibson TE (2018) Convergence properties of adaptive systems and the definition of exponential stability. SIAM Journal on Control and Optimization 56(4): 2463–2484
31. Chowdhary G, Johnson E (2010) Concurrent learning for convergence in adaptive control without persistency of excitation. In: Proceedings of 49th IEEE Conference on Decision and Control (CDC), Atlanta, GA, USA, pp 3674–3679
32. Cho N, Shin HS, Kim Y, Tsourdos A (2017) Composite model reference adaptive control with parameter convergence under finite excitation. IEEE Transactions on Automatic Control 63(3): 811–818
33. Pan Y, Yu H (2018) Composite learning robot control with guaranteed parameter convergence. Automatica 89: 398–406
34. Zhang Q, Zhao D, Zhu Y (2016) Event-triggered H_∞ control for continuous-time nonlinear system via concurrent learning. IEEE Transactions on Systems, Man, and Cybernetics: Systems 47(7): 1071–1081
35. Xue S, Luo B, Liu D, Yang Y (early access, 2020, Constrained event-triggered H_∞ control based on adaptive dynamic programming with concurrent learning. IEEE Transactions on Systems, Man, and Cybernetics: Systems, https://doi.org/10.1109/TSMC.2020.2997559
36. Dong H, Hu Q, Akella MR, Yang H (2019) Composite adaptive attitude-tracking control with parameter convergence under finite excitation. IEEE Transactions on Control Systems Technology 28(6): 2657–2664
37. Aranovskiy S, Bobtsov A, Ortega R, Pyrkin A (2016) Performance enhancement of parameter estimators via dynamic regressor extension and mixing. IEEE Transactions on Automatic Control 62(7): 3546–3550
38. Zuo Z, Ru P (2014) Augmented $\mathcal{L}_1$ adaptive tracking control of quad-rotor unmanned aircrafts. IEEE Transactions on Aerospace and Electronic Systems 50(4): 3090–3101
39. Ulrich S, Saenz-Otero A, Barkana I (2016) Passivity-based adaptive control of robotic spacecraft for proximity operations under uncertainties. Journal of Guidance, Control, and Dynamics 39(6): 1444–1453
40. Slotine JJE, Li W (1989) Composite adaptive control of robot manipulators. Automatica 25(4): 509–519
41. Sun L, Huo W, Jiao Z (2016) Adaptive backstepping control of spacecraft rendezvous and proximity operations with input saturation and full-state constraint. IEEE Transactions on Industrial Electronics 64(1): 480–492
42. Karagiannis D, Sassano M, Astolfi A (2009) Dynamic scaling and observer design with application to adaptive control. Automatica 45(12): 2883–2889
43. Boyd S, Sastry SS (1986) Necessary and sufficient conditions for parameter convergence in adaptive control. Automatica 22(6): 629–639
44. Tao G (2003) Adaptive control design and analysis. John Wiley & Sons, Hoboken, NJ, USA
45. Kreisselmeier G (1977) Adaptive observers with exponential rate of convergence. IEEE Transactions on Automatic Control 22(1): 2–8

46. Korotina M, Aranovskiy S, Ushirobira R, Vedyakov A (2020) On parameter tuning and convergence properties of the DREM procedure. In: Proceedings of European Control Conference, Saint Petersburg, Russia, pp 53–58
47. Yi B, Ortega R (2022, Conditions for convergence of dynamic regressor extension and mixing parameter estimators using lti filters. IEEE Transactions on Automatic Control, https://doi.org/10.1109/TAC.2022.3149964
48. Akella MR, Thakur D, Mazenc F (2015) Partial Lyapunov strictification: Smooth angular velocity observers for attitude tracking control. Journal of Guidance, Control, and Dynamics 38(3): 442–451

Chapter 4
Learning-Based Fault-Tolerant Control for Spacecraft Constrained Reorientation Maneuvers

4.1 Introduction

Over the past decades, the research on spacecraft attitude control has attracted extensive attention, owing to its significance in observation, communication, power supply, and many other space tasks [1]. Various advanced control algorithms have been developed and successfully applied to the strongly coupled and nonlinear attitude control system of spacecraft, such as sliding mode control [2, 3], inverse optimal control [4], adaptive control [5, 6], etc. However, most of the traditional control methods tend to ignore the underlying attitude constraints, which often arise in practical missions. For example, the onboard infrared telescope should avoid direct exposure to bright objects, while the antenna should remain within a specific zone towards transmission stations [7]. These kinds of requirements towards onboard instruments can be regarded as attitude constraints, which reduce the feasible zones of spacecraft attitude motion and pose a great challenge for the control design.

State constraint is an important issue for safe and autonomous spacecraft attitude control. For example, some sensitive payloads should avoid direct exposure to bright objects, and maneuver velocity should be kept within the limited measurement range of gyros. These requirements lead to attitude and angular velocity constraints, respectively. Lee and Mesbahi [8] proposed a novel reorientation control algorithm in the presence of multiple types of attitude-constrained zones using a logarithmic barrier potential. Shen et al. [9] proposed a velocity-free attitude reorientation control law for flexible spacecraft, wherein attitude constraints are addressed using artificial potential function (APF) method. Cui et al. [10] employed barrier Lyapunov function (BLF) approach to cope with the constraints of tracking errors, and then derived a finite-time tracking control scheme for a class of uncertain nonlinear systems. Shao and Hu [11] employed the BLF method to cope with angular velocity constraints of spacecraft attitude maneuver, and designed an immersion and invariance adaptive controller to overcome nonsatisfaction of the realizability condition. The APF and BLF can also be incorporated into the ADP framework to achieve performance

Q. Hu et al., *Intelligent Autonomous Control of Spacecraft with Multiple Constraints*,
https://doi.org/10.1007/978-981-99-0681-9_4

optimality and guarantee constraint satisfaction. For example, by combining barrier functions and ADP technique, Dong et al. [12] proposed an optimal control law with constraint-handling abilities on attitude forbidden zones and angular velocity limits. Yang et al. [13] designed an ADP-based spacecraft attitude controller under actuator misalignment and attitude constraints, wherein the undesired states are encoded into the reward function using APF method.

Note that the unwinding problem caused by the redundancy of the quaternion-based representation has not been emphasized enough in the literature on the constrained attitude problem (see, indicatively, [8, 9, 14]). Since the state space of unit-quaternion representation is a double cover of SO(3), each attitude in the physical three-dimensional space corresponds to two opposite quaternions (i.e., $\boldsymbol{q}$ and $-\boldsymbol{q}$). It indicates that there are two equilibria for the attitude control system described by unit-quaternion. However, in many related works, only one of the two equilibria is designed to be stable. This may lead to an unnecessary large angle rotation even if the initial orientation is close to desired one, which in turn results in longer path and higher fuel consumption. To deal with the unwinding problem, Kristiansen and Nicklasson [15] regulated the attitude of satellite to the closest equilibrium point by a tactful choice of integrator backstepping variables, which avoids unwanted rotation to reach the desired attitude. Hu et al. [16, 17] proposed a class of attitude error vectors to ensure that both two equilibria are locally asymptotically stable and accordingly solved the unwinding problem. It is important to underscore that most of the aforementioned works that concentrate on achieving constrained attitude control or unwinding avoidance require the exact knowledge of inertia parameters (e.g., see [9, 14, 16]), which however are often uncertain in practice due to fuel consumption and payload variation of the rigid body. Adaptive control methods have been widely developed to address parameter uncertainties and successfully applied to attitude control problem subject to unknown inertia parameters. Thakur et al. [18] designed an adaptive attitude tracking control algorithm that compensates for inertia variations for a spacecraft with both rigid and nonrigid (time-varying) inertia components. Benallegue et al. [19] presented a new adaptive controller for rigid body with unknown inertia and unknown gyro-bias by using inertial vector measurements, which guarantees the asymptotic convergence of the attitude and angular velocity to their desired values.

A safe and reliable controller should also be able to tolerate potential actuator faults, which may cause performance degradation or even system instability [20, 21]. In this respect, fault-tolerant control (FTC) serves as an efficient technology. Active FTC [22–25] uses fault detection and diagnosis strategy to compensate the adverse influence caused by actuator faults, while the passive FTC [26–29] accounts for all faulty conditions in advance and makes the controller robust against the preconsidered faults. In addition, some researches further consider the optimality problem of FTC design. Maziar et al. [30] proposed an optimal FTC approach for multiagent systems, and offline RL is used to learn the optimal control law for each agent. Zhao et al. [31] designed an RL algorithm to iteratively learn the optimal control policy for nonlinear quadrotors. Then, an adaptive fault-tolerant controller involving the optimal control policy is proposed to restrain the effects of actuator faults.

Meng et al. [32] employed the incremental nonlinear control technology to simplify spacecraft attitude system into an incremental nominal model with a synthetic uncertainty/fault term. Based on a sliding mode disturbance observer, the original optimal FTC problem is then transformed into a guaranteed performance optimal problem of the incremental model. However, how to deal with velocity constraints in the framework of passive FTC still remains an open problem, because most passive FTC design requires a strong controllability assumption which may not hold with unknown multiplicative faults and inertia matrix. Shao et al. [33] presented a feasible region for the uniform strong controllability assumption, based on which a passive FTC law is developed with constraint-handling capability.

Motivated by the above concerns, this chapter is devoted to design fault-tolerant control laws for spacecraft attitude reorientation under complex state constraints. Firstly, an adaptive FTC scheme is proposed for constrained spacecraft reorientation, wherein APF and integral BLF (iBLF) are employed to handle attitude and angular velocity constraints. Adaptive approach is further used to deal with actuator fault, uncertain inertia, and external disturbance. Secondly, an approximate optimal FTC scheme is derived for spacecraft attitude reorientation under both attitude and angular velocity constraints. A specially designed cost function is developed, which is online approximated by a single-critic neural network (NN) in the framework of adaptive dynamic programming (ADP) method. Simulations are presented for both of the two control laws.

The remainder of the chapter is structured as follows. Section 4.2[1] proposes two adaptive FTC schemes for spacecraft reorientation under attitude and/or angular velocity constraints. In Sect. 4.3, an RL-based optimal fault-tolerant control scheme is further developed to achieve optimal attitude reorientation under both attitude and angular velocity constraints, despite the presence of actuator faults. Finally, this chapter is wrapped up with some concluding remarks in Sect. 4.4.

4.2 Adaptive FTC for Spacecraft Constrained Reorientation

In this section, we develop an adaptive robust FTC scheme for spacecraft reorientation under attitude constraints, in order to show the basic design idea for addressing actuator faults and state constraints simultaneously. Later, this controller is extended to handle both attitude and angular velocity constraints, while sufficient conditions and feasibility analysis for controllability are given at the same time.

[1] Reproduced from Yuan Tian, Qinglei Hu, and Xiaodong Shao. Adaptive fault-tolerant control for attitude reorientation under complex attitude constraints. Aerospace Science and Technology 2022; 121: 107332.

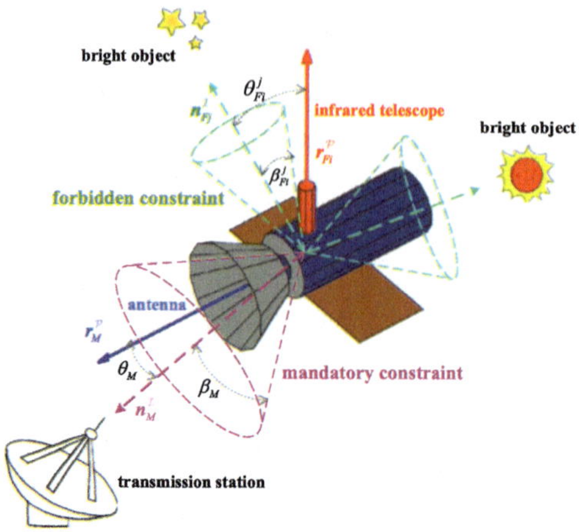

Fig. 4.1 Illustration of attitude constraints

4.2.1 Problem Formulation

4.2.1.1 Attitude and Angular Velocity Constraints

Various mission-oriented attitude constraints imposed on onboard payload should be taken into consideration for spacecraft attitude control. Suppose that the spacecraft is equipped with n optical instruments, each of which has m forbidden zones. Let $\boldsymbol{r}^{\mathcal{P}}_{Fi}$ denote the unit boresight vector of the i-th instrument expressed in $\mathcal{P}$, and $\boldsymbol{n}^{\mathcal{I}}_{Fj}$ denote the unit vector pointing towards the j-th unwanted object expressed in $\mathcal{I}$, as shown in (Fig. 4.1). The angle between $\boldsymbol{r}^{\mathcal{P}}_{Fi}$ and $\boldsymbol{n}^{\mathcal{I}}_{Fj}$ should be greater than β^{j}_{Fi}, that is

$$\cos(\theta^{j}_{Fi}) = \boldsymbol{n}^{\mathcal{I}}_{Fj} \cdot \boldsymbol{R} \cdot \boldsymbol{r}^{\mathcal{P}}_{Fi} < \cos(\beta^{j}_{Fi}), \tag{4.1}$$

where $\boldsymbol{R} = (q_4^2 - \boldsymbol{q}_v^\top \boldsymbol{q}_v)\boldsymbol{I}_3 + 2\boldsymbol{q}_v\boldsymbol{q}_v^\top - 2q_4\boldsymbol{S}(\boldsymbol{q}_v)$ is the rotation matrix from $\mathcal{I}$ to $\mathcal{P}$. By some formula calculation, (4.1) can be rewritten concisely as

$$\boldsymbol{q}^\top \boldsymbol{M}^{j}_{Fi}\boldsymbol{q} < 0, (i = 1, 2, \ldots, n, j = 1, 2, \ldots, m), \tag{4.2}$$

where $\boldsymbol{M}^{j}_{Fi}$ is a symmetric matrix, given by

$$\boldsymbol{M}^{j}_{Fi} = \begin{bmatrix} \boldsymbol{n}^{\mathcal{I}}_{Fj}(\boldsymbol{r}^{\mathcal{P}}_{Fi})^\top + \boldsymbol{r}^{\mathcal{P}}_{Fi}(\boldsymbol{n}^{\mathcal{I}}_{Fj})^\top - ((\boldsymbol{n}^{\mathcal{I}}_{Fj})^\top \boldsymbol{r}^{\mathcal{P}}_{Fi})\boldsymbol{I}_3 & \boldsymbol{r}^{\mathcal{P}}_{Fi} \times \boldsymbol{n}^{\mathcal{I}}_{Fj} \\ (\boldsymbol{r}^{\mathcal{P}}_{Fi} \times \boldsymbol{n}^{\mathcal{I}}_{Fj})^\top & (\boldsymbol{n}^{\mathcal{I}}_{Fj})^\top \boldsymbol{r}^{\mathcal{P}}_{Fi} \end{bmatrix} - \cos(\beta^{j}_{Fi})\boldsymbol{I}_4. \tag{4.3}$$

As for an attitude-mandatory constraint, the angle between $\boldsymbol{r}^{\mathcal{P}}_{M}$ and $\boldsymbol{n}^{\mathcal{I}}_{M}$ should be less than β_M, that is

$$\cos(\theta_M) = \boldsymbol{n}^{\mathcal{I}}_{M} \cdot \boldsymbol{R} \cdot \boldsymbol{r}^{\mathcal{P}}_{M} > \cos(\beta_M), \tag{4.4}$$

which can also be rewritten concisely as follows:

$$\boldsymbol{q}^\top \boldsymbol{M}_M \boldsymbol{q} > 0, \tag{4.5}$$

where $\boldsymbol{M}_M$ has the same form as $\boldsymbol{M}^j_{Fi}$ in (4.3), and substitutes $\boldsymbol{n}^{\mathcal{I}}_M, \boldsymbol{r}^{\mathcal{P}}_M, \beta_M$ for $\boldsymbol{n}^{\mathcal{I}}_{Fj}$, $\boldsymbol{r}^{\mathcal{P}}_{Fi}$, and β^j_{Fi}. Based on (4.2) and (4.5), we define the constrained set of attitude feasible zone as $\mathbb{Q}_f = \{\boldsymbol{q} \in \mathbb{Q} \mid \boldsymbol{q}^\top \boldsymbol{M}^j_{Fi}\boldsymbol{q} < 0 \text{ and } \boldsymbol{q}^\top \boldsymbol{M}_M \boldsymbol{q} > 0,\ i = 1, 2, \ldots, n,\ j = 1, 2, \ldots, m\}$.

Due to the limited measurement range of rate gyros or certain performance requirements, the spacecraft angular velocity is restricted in a constrained set, which can be described as

$$\mathbb{W} = \{\boldsymbol{\omega} \in \mathbb{R}^3 \mid |\omega_i| < \omega_m,\ i = 1, 2, 3\}, \tag{4.6}$$

where ω_m is the maximum allowable angular velocity.

4.2.1.2 Actuator Fault

The actuator faults are modeled as

$$\boldsymbol{\tau}_c = \boldsymbol{D}(\boldsymbol{E}\boldsymbol{u} + \bar{\boldsymbol{u}}), \tag{4.7}$$

where $\boldsymbol{D} \in \mathbb{R}^{3\times n}$ (n is the number of actuators) is the reaction wheel distribution matrix, $\boldsymbol{E} = \text{diag}[e_1, e_2, \ldots, e_n]$ is the effectiveness matrix with $0 \le e_i \le 1$, $i = 1, 2, \ldots n$, $\boldsymbol{u} \in \mathbb{R}^n$ denotes the command torques, and $\bar{\boldsymbol{u}} \in \mathbb{R}^3$ is the additive fault. In this way, four faulty modes of the reaction wheels can be considered, and their corresponding mathematical description can be summarized as follows [34]:

(1) Decreased reaction torque: $0 < e_i < 1$, $\bar{u}_i = 0$;
(2) Increased bias torque: $e_i = 1$, $\bar{u}_i \neq 0$;
(3) Failure to response to control signal: $e_i = 0$, $\bar{u}_i = 0$;
(4) Continuous generation of torque: $e_i = 0$, $\bar{u}_i \neq 0$.

To proceed, several reasonable assumptions are made:

Assumption 4.1 The inertia matrix $\boldsymbol{J}$ is uncertain but otherwise remains symmetric and positive definite. It is practical to assume that both $\boldsymbol{J}$ is bounded, so that there exists a positive constant c_1, such that $\|\boldsymbol{J}\| \le c_1$ holds.

Assumption 4.2 Both the external disturbances $\boldsymbol{\tau}_d$ and torque deviation $\bar{\boldsymbol{u}}$ are unknown but bounded, that is, $\|\boldsymbol{D}\bar{\boldsymbol{u}} + \boldsymbol{\tau}_d\| \le c_2$ for some positive constant c_2.

Assumption 4.3 For all faulty scenarios under consideration, the spacecraft attitude control system always remains either over-actuated or fully-actuated. Thus, $\text{rank}(\boldsymbol{D}\boldsymbol{E}) = 3$ holds, which according to Lemma 1 of [35] indicates that $\boldsymbol{D}\boldsymbol{E}\boldsymbol{D}^\top$ is positive definite.

4.2.1.3 Control Objective

The control objective of this section is to design an adaptive robust FTC law for spacecraft reorientation subject to attitude and angular velocity constraints, in the presence of inertia uncertainties, external disturbances, and actuator faults. The controller should ensure the following:

(1) Both the gradient-related vector and angular velocity converge to zero, i.e., $\lim_{t\to\infty}[\boldsymbol{v}(t), \boldsymbol{\omega}(t)] = \mathbf{0}$.

(2) Both attitude and angular velocity constraints are satisfied during the reorientation maneuver, i.e., $\boldsymbol{q}(t) \in \mathbb{Q}_f$ and $\boldsymbol{\omega}(t) \in \mathbb{W}$ for all $t \geq 0$.

4.2.2 *Adaptive FTC Under Attitude Constraints*

The following barrier function is constructed as an attitude error function to guide the spacecraft to the desired orientation, while avoiding the unwinding problem

$$\phi = -(q_{e4}^2 - 1)/q_{e4}^2. \tag{4.8}$$

Since $\phi \to \infty$ as $q_{e4} \to 0$, we can ensure that q_{e4} would not approach zero if $\phi \in L_\infty$. This implies that the rotation angle can be limited within π by ensuring $\phi \in L_\infty$, so that the unwinding phenomenon can be avoided.

Recalling the parameterized attitude constraints in (4.2) and (4.5), we consider the following APF $V_p : \mathbb{Q}_f \to \mathbb{R}$

$$V_p = \underbrace{\left(\frac{q_{e4}^2 - 1}{q_{e4}^2}\right)}_{\text{anti-unwinding}} \left[\underbrace{\alpha_1 \sum_{j=1}^{m}\sum_{i=1}^{n} \ln\left(-\frac{\boldsymbol{q}^\top \boldsymbol{M}_{Fi}^j \boldsymbol{q}}{2}\right)}_{\text{forbidden constraints}} + \underbrace{\alpha_2 \ln\left(\frac{\boldsymbol{q}^\top \boldsymbol{M}_M \boldsymbol{q}}{2}\right)}_{\text{mandatory constraint}}\right], \tag{4.9}$$

where $\alpha_1, \alpha_2 > 0$ are design constants.

Lemma 4.1 *The potential function V_p in* (4.9) *has the following properties:*

(P1) $V_p(\pm\boldsymbol{q}_d) = 0$;

(P2) $V_p(\boldsymbol{q}) > 0$, $\forall \boldsymbol{q} \in \mathbb{Q}_f \backslash \{\pm\boldsymbol{q}_d\}$;

(P3) $V_p(\boldsymbol{q}) \to \infty$, *if* $q_{e4} \to 0$;

(P4) $V_p(\boldsymbol{q}) \to \infty$, *if* $\boldsymbol{q}^\top \boldsymbol{M}_{Fi}^j \boldsymbol{q} \to 0$ *or* $\boldsymbol{q}^\top \boldsymbol{M}_M \boldsymbol{q} \to 0$.

Proof Since $\boldsymbol{q} = \pm\boldsymbol{q}_d$ is equivalent to $q_{e4} = \pm 1$, it is obvious to get P1. As in [8], $-2 < \boldsymbol{q}^\top \boldsymbol{M}_{Fi}^j \boldsymbol{q} < 0$ and $0 < \boldsymbol{q}^\top \boldsymbol{M}_M \boldsymbol{q} < 2$ holds, thus making it clear of P2 along with the property that $q_{e4}^2 \leq 1$. We can arrive at P3 and P4 using the properties of the APF in (4.8) and the last term in (4.9), respectively. □

Taking the time derivative of V_p leads to $\dot{V}_p = (\nabla V_p)^\top \dot{\boldsymbol{q}}$, where ∇V_p is given by

$$\begin{aligned}\nabla V_p = &\left[\frac{2\alpha_1}{q_{e4}^3}\sum_{j=1}^{m}\sum_{i=1}^{n}\ln\left(-\frac{\boldsymbol{q}^\top \boldsymbol{M}_{Fi}^j \boldsymbol{q}}{2}\right) + \frac{2\alpha_2}{q_{e4}^3}\ln\left(\frac{\boldsymbol{q}^\top \boldsymbol{M}_M \boldsymbol{q}}{2}\right)\right]\boldsymbol{q}_d \\ &+ \left(\frac{q_{e4}^2 - 1}{q_{e4}^2}\right)\cdot\left(\alpha_1 \sum_{j=1}^{m}\sum_{i=1}^{n}\frac{2\boldsymbol{M}_{Fi}^j \boldsymbol{q}}{\boldsymbol{q}^\top \boldsymbol{M}_{Fi}^j \boldsymbol{q}} + \alpha_2 \frac{2\boldsymbol{M}_M \boldsymbol{q}}{\boldsymbol{q}^\top \boldsymbol{M}_M \boldsymbol{q}}\right).\end{aligned} \tag{4.10}$$

Then recalling the algebraic property of quaternion, it yields

$$\dot{V}_p = \frac{1}{2} V_p^\top (\boldsymbol{q} \odot \tilde{\boldsymbol{\omega}}) = -\frac{1}{2}\boldsymbol{\omega}^\top \operatorname{vec}(\nabla V_p^* \odot \boldsymbol{q}) = -\boldsymbol{v}^\top \boldsymbol{\omega}, \tag{4.11}$$

where vec(·) denotes the vector part of a four dimensional vector, and the gradient-related term $\boldsymbol{v} = 0.5\operatorname{vec}(\nabla V_p^* \odot \boldsymbol{q})$ is defined for the sake of brevity.

A virtual input $\boldsymbol{\omega}_c$ that can stabilize (4.11) is designed as

$$\boldsymbol{\omega}_c = k_1 \operatorname{Tanh}(\boldsymbol{v}), \tag{4.12}$$

where k_1 is a positive parameter, and $\operatorname{Tanh}(\boldsymbol{v}) = [\tanh(v_1), \tanh(v_2), \tanh(v_3)]^\top$. By adding and subtracting $\boldsymbol{\omega}_c$ on the right-hand side of (4.11), we have

$$\begin{aligned}\dot{V}_p &= -\boldsymbol{v}^\top \boldsymbol{\omega} \\ &= -\boldsymbol{v}^\top \boldsymbol{\omega}_c - \boldsymbol{v}^\top (\boldsymbol{\omega} - \boldsymbol{\omega}_c) \\ &= -k_1 \boldsymbol{v}^\top \operatorname{Tanh}(\boldsymbol{v}) - \boldsymbol{v}^\top \boldsymbol{z}.\end{aligned} \tag{4.13}$$

In (4.13), $\boldsymbol{z} = \boldsymbol{\omega} - \boldsymbol{\omega}_c$. We further consider the following Lyapunov function candidate

$$V_1 = \frac{1}{2}\boldsymbol{z}^\top \boldsymbol{J}\boldsymbol{z}, \tag{4.14}$$

whose time derivative is given by

$$\begin{aligned}\dot{V}_1 &= \boldsymbol{z}^\top \boldsymbol{J}\dot{\boldsymbol{z}} \\ &= \boldsymbol{z}^\top[-\boldsymbol{S}(\boldsymbol{\omega})\boldsymbol{J}\boldsymbol{\omega} + \boldsymbol{\tau} + \boldsymbol{\tau}_d - \boldsymbol{J}\dot{\boldsymbol{\omega}}_c] \\ &= \boldsymbol{z}^\top[-\boldsymbol{S}(\boldsymbol{\omega})\boldsymbol{J}\boldsymbol{\omega} + \boldsymbol{A}(\boldsymbol{E}\boldsymbol{u} + \bar{\boldsymbol{u}}) + \boldsymbol{\tau}_d - \boldsymbol{J}\dot{\boldsymbol{\omega}}_c] \\ &= \boldsymbol{z}^\top(\boldsymbol{D}\boldsymbol{E}\boldsymbol{u} + \boldsymbol{T}_d),\end{aligned} \tag{4.15}$$

where $\boldsymbol{T}_d = -\boldsymbol{S}(\boldsymbol{\omega})\boldsymbol{J}\boldsymbol{\omega} + \boldsymbol{A}\bar{\boldsymbol{u}} + \boldsymbol{\tau}_d - \boldsymbol{J}\dot{\boldsymbol{\omega}}_c$ is the lumped disturbances containing the unknown time-varying inertia matrix $\boldsymbol{J}$. By Assumptions 4.1 and 4.2, we further conclude that $\exists d_1, d_2, d_3, d_4 > 0$ such that

$$\|\boldsymbol{T}_d\| \le d_1 + d_2\|\boldsymbol{\omega}\| + d_3\|\boldsymbol{\omega}\|^2 + d_4\|\dot{\boldsymbol{\omega}}_c\| = \boldsymbol{d}^\top \boldsymbol{\Phi}, \tag{4.16}$$

where $\boldsymbol{d} = [d_1, d_2, d_3, d_4]^\top$, and $\boldsymbol{\Phi} = \left[1, \|\boldsymbol{\omega}\|, \|\boldsymbol{\omega}\|^2, \|\dot{\boldsymbol{\omega}}_c\|\right]^\top$.

Denote $\delta = \lambda_{\min}(\boldsymbol{DED}^\top)$, then there exists some positive constant δ_m such that $\delta > \delta_m$ holds since $\boldsymbol{DED}^\top$ is positive definite. We further define $\lambda = 1/\delta$ and $\boldsymbol{d}_m = \lambda \boldsymbol{d}$, then it yields that $\lambda < \lambda_m$, with $\lambda_m = 1/\delta_m$. And $\hat{\lambda}$, $\hat{\boldsymbol{d}}$, $\hat{\lambda}_m$, $\hat{\boldsymbol{d}}_m$ are denoted as the estimation of λ, $\boldsymbol{d}$, λ_m, $\boldsymbol{d}_m$ respectively.

The adaptive fault-tolerant controller is designed as

$$\boldsymbol{u} = -k_2 \boldsymbol{D}^\top \boldsymbol{z} - k(t)\boldsymbol{D}^\top \frac{\boldsymbol{z}}{\|\boldsymbol{z}\| + \rho^2 \varepsilon}, \tag{4.17}$$

with

$$k(t) = \hat{\lambda}\|\boldsymbol{v}\| + \hat{\boldsymbol{d}}^\top \boldsymbol{\Phi}. \tag{4.18}$$

The accompanying adaptive laws are designed as

$$\dot{\hat{\lambda}} = \zeta_1 \left[\|\boldsymbol{v}\|\|\boldsymbol{z}\| - \mu_1(\hat{\lambda} - \hat{\lambda}_m)\right], \quad \dot{\hat{\lambda}}_m = \beta_1(\hat{\lambda} - \hat{\lambda}_m), \tag{4.19}$$

$$\dot{\hat{\boldsymbol{d}}} = \zeta_2 \left[\boldsymbol{\Phi}\|\boldsymbol{z}\| - \mu_2(\hat{\boldsymbol{d}} - \hat{\boldsymbol{d}}_m)\right], \quad \dot{\hat{\boldsymbol{d}}}_m = \beta_2(\hat{\boldsymbol{d}} - \hat{\boldsymbol{d}}_m), \tag{4.20}$$

where k_2, ζ_i, μ_i, and β_i $(i = 1, 2)$ are positive constants, and the time-varying parameter ρ is updated by

$$\dot{\rho} = -\gamma \varepsilon \rho k(t) \frac{\|\boldsymbol{z}\|}{\|\boldsymbol{z}\| + \rho^2 \varepsilon}, \tag{4.21}$$

where γ, ε are positive constants.

Theorem 4.1 *Consider the spacecraft attitude error system given by (2.19) and (2.21) with actuator faults modeled in* (4.7) *under Assumptions 4.1–4.3. Suppose that the initial state satisfies* $\boldsymbol{q}_0 \in \mathbb{Q}_f$, *then the proposed controller* (4.17)-(4.18) *with adaptive laws* (4.19)-(4.21) *can ensure that:*

- V_p *is bounded, so that attitude constraints are always satisfied;*
- *The gradient-related vector asymptotically converges to zero, i.e.,* $\lim_{t\to\infty} \boldsymbol{v}(t) = \boldsymbol{0}$;
- *The angular rate asymptotically converges to zero, i.e.,* $\lim_{t\to\infty} \boldsymbol{\omega}(t) = \boldsymbol{0}$.

Proof Consider the following Lyapunov function candidate

$$\begin{aligned} V = {} & V_p + V_1 + \frac{\delta}{2\gamma}\rho^2 + \frac{\delta}{2\zeta_1}(\hat{\lambda} - \lambda_m)^2 + \frac{\delta\mu_1}{2\beta_1}(\hat{\lambda}_m - \lambda_m)^2 \\ & + \frac{\delta}{2\zeta_2}\|\hat{\boldsymbol{d}} - \boldsymbol{d}_m\|^2 + \frac{\delta\mu_2}{2\beta_2}\|\hat{\boldsymbol{d}}_m - \boldsymbol{d}_m\|^2. \end{aligned} \tag{4.22}$$

□

Taking the time derivative of V yields

$$\begin{aligned}\dot{V} = & -k_1 \boldsymbol{v}^\top \text{Tanh}(\boldsymbol{v}) - \boldsymbol{v}^\top \boldsymbol{z} + \boldsymbol{z}^\top (\boldsymbol{DEu} + \boldsymbol{T}_d) \\ & + \frac{\delta}{\gamma}\rho\dot{\rho} + \frac{\delta}{\zeta_1}(\hat{\lambda} - \lambda_m)\dot{\hat{\lambda}} + \frac{\delta\mu_1}{\beta_1}(\hat{\lambda}_m - \lambda_m)\dot{\hat{\lambda}}_m \\ & + \frac{\delta}{\zeta_2}(\hat{\boldsymbol{d}} - \boldsymbol{d}_m)^\top \dot{\hat{\boldsymbol{d}}} + \frac{\delta\mu_2}{\beta_2}(\hat{\boldsymbol{d}}_m - \boldsymbol{d}_m)^\top \dot{\hat{\boldsymbol{d}}}_m.\end{aligned} \tag{4.23}$$

Substituting (4.17)-(4.18) and (4.21) into (4.23), one gets

$$\begin{aligned}\dot{V} \leq & -k_1 \boldsymbol{v}^\top \text{Tanh}(\boldsymbol{v}) - k_2\delta\|\boldsymbol{z}\|^2 - k(t)\delta\frac{\|\boldsymbol{z}\|^2}{\|\boldsymbol{z}\| + \rho^2\varepsilon} - \delta\rho k(t)\frac{\rho\varepsilon\|\boldsymbol{z}\|}{\|\boldsymbol{z}\| + \rho^2\varepsilon} \\ & - \boldsymbol{z}^\top \boldsymbol{v} + \boldsymbol{z}^\top \boldsymbol{T}_d + \frac{\delta}{\zeta_1}(\hat{\lambda} - \lambda_m)\dot{\hat{\lambda}} + \frac{\delta\mu_1}{\beta_1}(\hat{\lambda}_m - \lambda_m)\dot{\hat{\lambda}}_m \\ & + \frac{\delta}{\zeta_2}(\hat{\boldsymbol{d}} - \boldsymbol{d}_m)^\top \dot{\hat{\boldsymbol{d}}} + \frac{\delta\mu_2}{\beta_2}(\hat{\boldsymbol{d}}_m - \boldsymbol{d}_m)^\top \dot{\hat{\boldsymbol{d}}}_m \\ = & -k_1 \boldsymbol{v}^\top \text{Tanh}(\boldsymbol{v}) - k_2\delta\|\boldsymbol{z}\|^2 - (\hat{\lambda}\|\boldsymbol{v}\| + \hat{\boldsymbol{d}}^\top \boldsymbol{\Phi})\delta\|\boldsymbol{z}\| - \boldsymbol{z}^\top \boldsymbol{v} \\ & + \boldsymbol{z}^\top \boldsymbol{T}_d + \frac{\delta}{\zeta_1}(\hat{\lambda} - \lambda_m)\dot{\hat{\lambda}} + \frac{\delta\mu_1}{\beta_1}(\hat{\lambda}_m - \lambda_m)\dot{\hat{\lambda}}_m \\ & + \frac{\delta}{\zeta_2}(\hat{\boldsymbol{d}} - \boldsymbol{d}_m)^\top \dot{\hat{\boldsymbol{d}}} + \frac{\delta\mu_2}{\beta_2}(\hat{\boldsymbol{d}}_m - \boldsymbol{d}_m)^\top \dot{\hat{\boldsymbol{d}}}_m.\end{aligned} \tag{4.24}$$

Then, substituting the adaptive laws (4.19)-(4.20) into (4.24) leads to

$$\begin{aligned}\dot{V} \leq & -k_1 \boldsymbol{v}^\top \text{Tanh}(\boldsymbol{v}) - k_2\delta\|\boldsymbol{z}\|^2 - (\hat{\lambda}\|\boldsymbol{v}\| + \hat{\boldsymbol{d}}^\top \boldsymbol{\Phi})\delta\|\boldsymbol{z}\| - \boldsymbol{z}^\top \boldsymbol{v} + \boldsymbol{z}^\top \boldsymbol{T}_d \\ & + \delta(\hat{\lambda} - \lambda_m)\left[\|\boldsymbol{v}\|\|\boldsymbol{z}\| - \mu_1(\hat{\lambda} - \hat{\lambda}_m)\right] + \delta\mu_1(\hat{\lambda}_m - \lambda_m)(\hat{\lambda} - \hat{\lambda}_m) \\ & + \delta(\hat{\boldsymbol{d}} - \boldsymbol{d}_m)^\top \left[\boldsymbol{\Phi}\|\boldsymbol{z}\| - \mu_2(\hat{\boldsymbol{d}} - \hat{\boldsymbol{d}}_m)\right] + \delta\mu_2(\hat{\boldsymbol{d}}_m - \boldsymbol{d}_m)^\top (\hat{\boldsymbol{d}} - \hat{\boldsymbol{d}}_m) \\ = & -k_1 \boldsymbol{v}^\top \text{Tanh}(\boldsymbol{v}) - k_2\delta\|\boldsymbol{z}\|^2 - \boldsymbol{z}^\top \boldsymbol{v} + \boldsymbol{z}^\top \boldsymbol{T}_d - \delta\hat{\lambda}_m\|\boldsymbol{v}\|\|\boldsymbol{z}\|| \\ & - \delta\hat{\boldsymbol{d}}_m^\top \boldsymbol{\Phi}\|\boldsymbol{z}\| - \delta\mu_1(\hat{\lambda} - \hat{\lambda}_m)^2 - \delta\mu_2\|\hat{\boldsymbol{d}} - \hat{\boldsymbol{d}}_m\|^2 \\ \leq & -k_1 \boldsymbol{v}^\top \text{Tanh}(\boldsymbol{v}) - k_2\delta\|\boldsymbol{z}\|^2 - \delta\mu_1(\hat{\lambda} - \hat{\lambda}_m)^2 - \delta\mu_2\|\hat{\boldsymbol{d}} - \hat{\boldsymbol{d}}_m\|^2.\end{aligned} \tag{4.25}$$

As an immediate consequence of (4.25), V is negative semidefinite, from which it follows that $V_p, \boldsymbol{z}, \hat{\lambda}, \hat{\boldsymbol{d}} \in \mathcal{L}_\infty$. As V_p is bounded, we can easily verify from (4.10) that $\nabla V_p \in \mathcal{L}_\infty$, which implies that $\boldsymbol{v} \in \mathcal{L}_\infty$. Since $\boldsymbol{z}$ is bounded, it is clear that $\boldsymbol{\omega} \in \mathcal{L}_\infty$ and hence $\dot{\boldsymbol{z}} \in \mathcal{L}_\infty$. Consequently, $\boldsymbol{z}$ is uniformly continuous. Since $\boldsymbol{v}, \nabla V_p, \boldsymbol{\omega}$ are bounded, it is clear that $\dot{\boldsymbol{v}} \in \mathcal{L}_\infty$, showing that $\boldsymbol{v}$ is uniformly continuous. On the other hand, integrating $\dot{V}$ in (4.25) yields

$$V(0) - V(\infty) \geq k_1 \int_0^\infty \boldsymbol{v}^\top(t)\mathrm{Tanh}(\boldsymbol{v}(t))dt + k_2\delta \int_0^\infty \|\boldsymbol{z}(t)\|^2 dt. \quad (4.26)$$

Hence, $\int_0^\infty \boldsymbol{v}^\top(t)\mathrm{Tanh}(\boldsymbol{v}(t))dt$ and $\int_0^\infty \|\boldsymbol{z}(t)\|^2 dt$ are bounded. According to Barbalat's Lemma, one can easily infer that $\lim_{t\to\infty}[\boldsymbol{v}(t), \boldsymbol{z}(t)] = \mathbf{0}$. Further, it can be concluded that $\lim_{t\to\infty} \boldsymbol{\omega}(t) = \mathbf{0}$. This completes the proof.

Remark 4.1 The virtual input $\boldsymbol{\omega}_c$ in (4.12) is well designed using the hyperbolic tangent function, so that the boundness of $\boldsymbol{z}$ can ensure the boundness of $\boldsymbol{\omega}$. In addition, a large k_2 guarantees that $\boldsymbol{z}$ has a high convergence rate and small steady-state error, so that $\boldsymbol{\omega}$ can quickly and closely follow its saturated virtual trajectory $\boldsymbol{\omega}_c$. This implies that k_1 can be set as a loose upper bound to limit the magnitude of $\boldsymbol{\omega}$ to some extent since $\|\omega_{ci}\| \leq k_1$ always holds.

Remark 4.2 The time varying parameter ρ is introduced into controller (4.17) because mere use of $\frac{\boldsymbol{z}}{\|\boldsymbol{z}\|}$ will lead to system chattering. Most research would otherwise use $\frac{\boldsymbol{z}}{\|\boldsymbol{z}\|+\varepsilon}$ (ε is a small positive constant) or boundary layer method to avoid this problem, which can only guarantee $\boldsymbol{z}$ converges to a small neighborhood around the origin. Since $\dot{\rho} \leq -\gamma\varepsilon k(t)\rho$ according to (4.21), the proposed controller can ensure a low convergence rate of ρ by choosing the product of $\gamma\varepsilon$ as a small value. Hence, ρ will not approach zero in a limited mission period with a low convergence rate and a proper initial value, thus not only avoiding chattering but also making $\boldsymbol{z}$ asymptotically converge to origin.

Remark 4.3 It should be noted that $\lim_{t\to\infty} \boldsymbol{q}(t) = \pm\boldsymbol{q}_d$ cannot be strictly guaranteed from $\lim_{t\to\infty} \boldsymbol{v}(t) = \mathbf{0}$, since V_p may converge to a local minimum or saddle point. However, this problem rarely happens in practice and can be avoided by adjusting the values of the parameters α_1, α_2 in (4.9).

4.2.3 *Adaptive FTC Under Attitude and Angular Velocity Constraints*

To further handle angular velocity constraint, the adaptive FTC scheme proposed in Sect. 4.2.2 is extended. Consider the following integral BLF (iBLF):

$$V_\omega(\boldsymbol{z}, \boldsymbol{\omega}_c) = \sum_{i=1}^{3} V_{\omega i}(z_i, \omega_{ci}), \quad (4.27)$$

with

$$V_{\omega i} = \int_0^{z_i} \frac{\sigma\omega_m^2}{\omega_m^2 - (\sigma + \omega_{ci})^2} d\sigma. \quad (4.28)$$

Lemma 4.2 *The iBLF in* (4.27) *satisfies:*

- $0.5z_i^2 \leq V_{\omega i} \leq [\omega_m^2/(\omega_m^2 - \omega_i^2)]z_i^2$, *for* $|\omega_i| < \omega_m$;
- *With initial condition* $\boldsymbol{\omega}(0) \in \mathbb{W}$ *and* $V_{\omega i}$ *bounded, the state* $\boldsymbol{\omega}(t)$ *remains in the constrained set* $\mathbb{W}$ *for all* $t > 0$.

Proof See the proof of Theorem 1 in [36]. □

Remark 4.4 Many existing BLF-based methods enforce the original state constraint indirectly by imposing constraints on z. In contrast, the proposed iBLF in (4.27) incorporates the original state $\boldsymbol{\omega}$ with the error vector z so that the angular velocity constraint is enforced directly, thus reducing conservatism.

Taking the time derivative of V_ω along with (4.28) yields

$$\dot{V}_\omega = \sum_{i=1}^{3} \left(\frac{\partial V_\omega}{\partial z_i} \dot{z}_i + \frac{\partial V_\omega}{\partial \omega_{ci}} \dot{\omega}_{ci} \right). \tag{4.29}$$

The two partial derivatives in (4.29) are given by

$$\frac{\partial V_\omega}{\partial z_i} = \frac{z_i \omega_m^2}{\omega_m^2 - \omega_i^2}, \quad \frac{\partial V_\omega}{\partial \omega_{ci}} = z_i \left(\frac{\omega_m^2}{\omega_m^2 - \omega_i^2} - \eta_i \right), \tag{4.30}$$

with

$$\eta_i = \frac{\omega_m}{2z_i} \ln \frac{(\omega_m + \omega_i)(\omega_m - \omega_{ci})}{(\omega_m - \omega_i)(\omega_m + \omega_{ci})}. \tag{4.31}$$

By L'Hopital's rule, it can be obtained that

$$\lim_{z_i \to 0} \eta_i = \frac{\omega_m^2}{\omega_m^2 - \omega_{ci}^2}, \tag{4.32}$$

which implies that η_i is well designed in the neighborhood of $z_i = 0$, thus avoiding the singularity problem of (4.31). For the sake of brevity, we define diagonal matrices as $\boldsymbol{N}_z = \text{diag}\{N_{zi}\}$, $(i = 1, 2, 3)$ with $N_{zi} = \omega_m^2/(\omega_m^2 - \omega_i^2)$, and $\boldsymbol{N}_\omega = \text{diag}\{N_{\omega i}\}$, $(i = 1, 2, 3)$ with $N_{\omega i} = \omega_m^2/(\omega_m^2 - \omega_i^2) - \eta_i$. Then, (4.29) can be rewritten as

$$\dot{V}_\omega = \boldsymbol{z}^\top (\boldsymbol{N}_z \dot{\boldsymbol{z}} + \boldsymbol{N}_\omega \dot{\boldsymbol{\omega}}_c) = \boldsymbol{z}^\top \boldsymbol{N}_z (\dot{\boldsymbol{z}} + \boldsymbol{N}_z^{-1} \boldsymbol{N}_\omega \dot{\boldsymbol{\omega}}_c), \tag{4.33}$$

where the time derivative of z is given by

$$\dot{\boldsymbol{z}} = \boldsymbol{J}^{-1} (-\boldsymbol{S}(\boldsymbol{\omega}) \boldsymbol{J} \boldsymbol{\omega} + \boldsymbol{D} \boldsymbol{E} \boldsymbol{u} + \boldsymbol{D} \bar{\boldsymbol{u}} + \boldsymbol{\tau}_d) - \dot{\boldsymbol{\omega}}_c. \tag{4.34}$$

Combining two Lyapunov candidates V_p and V_ω, we have

$$\dot{V}_p + \dot{V}_\omega = -k_1 \boldsymbol{v}^\top \text{Tanh}(\boldsymbol{v}) + \boldsymbol{z}^\top \boldsymbol{N}_z (\boldsymbol{J}^{-1} \boldsymbol{D} \boldsymbol{E} \boldsymbol{u} + \boldsymbol{u}_d + \boldsymbol{T}_d), \tag{4.35}$$

where $\boldsymbol{u}_d = -\dot{\boldsymbol{\omega}}_c + \boldsymbol{N}_z^{-1}\boldsymbol{N}_\omega\dot{\boldsymbol{\omega}}_c - \boldsymbol{N}_z^{-1}\boldsymbol{v}$ is a combination of known terms, and $\boldsymbol{T}_d = -\boldsymbol{J}^{-1}\boldsymbol{S}(\boldsymbol{\omega})\boldsymbol{J}\boldsymbol{\omega} + \boldsymbol{J}^{-1}(\boldsymbol{D}\bar{\boldsymbol{u}} + \boldsymbol{\tau}_d)$. It follows that,

$$\|\boldsymbol{T}_d\| \le d_1 + d_2\|\boldsymbol{\omega}\|^2 = \boldsymbol{d}^\top\boldsymbol{\Phi}, \tag{4.36}$$

where $\boldsymbol{d} = [d_1, d_2]^\top$ with d_1 and d_2 being some position constants, and $\boldsymbol{\Phi} = [1, \|\boldsymbol{\omega}\|^2]^\top$.

Challenges arise when dealing with (4.35) using the traditional FTC methods, since the uniform strong controllability assumption may not hold in the presence of multiplicative actuator faults. By "uniform strong controllability", it is meant that $\boldsymbol{P}^* = \frac{1}{2}[\boldsymbol{D}\boldsymbol{E}\boldsymbol{D}^\top + (\boldsymbol{D}\boldsymbol{E}\boldsymbol{D}^\top)^\top]$ remains positive definite for all $t \ge 0$. However, given the positive-definite matrix $\boldsymbol{D}\boldsymbol{E}\boldsymbol{D}^\top$ (a fundamental requirement of the existing FTC methods), its product with another positive-definite matrix $\boldsymbol{J}^{-1}$ may not remain positive-definite. Let us define $\boldsymbol{P} = \frac{1}{2}[\boldsymbol{J}^{-1}\boldsymbol{D}\boldsymbol{E}\boldsymbol{D}^\top + (\boldsymbol{J}^{-1}\boldsymbol{D}\boldsymbol{E}\boldsymbol{D}^\top)^\top]$ and give the following assumption.

Assumption 4.4 The matrix $\boldsymbol{P}$ remains positive definite for all faulty scenarios under consideration.

The above assumption does not always hold in practical engineering, thus we provide in this chapter sufficient conditions and feasibility analysis for two typical control distribution matrix $\boldsymbol{D}_1$ and $\boldsymbol{D}_2$.

$$\boldsymbol{D}_1 = \begin{bmatrix} 1 & 0 & 0 & 1/\sqrt{3} \\ 0 & 1 & 0 & 1/\sqrt{3} \\ 0 & 0 & 1 & 1/\sqrt{3} \end{bmatrix}, \quad \boldsymbol{D}_2 = \frac{1}{\sqrt{2}}\begin{bmatrix} 1 & -1 & 1 & -1 \\ 1 & 0 & -1 & 0 \\ 0 & -1 & 0 & 1 \end{bmatrix} \tag{4.37}$$

To proceed, define $r = \max(\boldsymbol{J}_{ii})/\min(\boldsymbol{J}_{ii})$ $(i = 1, 2, 3)$ to describe degree of similarity between the inertia of three principal axes. In addition, actuator faults are divided into three types according to the effective factor e_i:

(F1) total failure: $0 \le e_i < e_t$ (e_t is a tiny value);

(F2) severe failure: $e_t \le e_i < e_s$;

(F3) partial failure: $e_s \le e_i < 1$.

It is noteworthy that the spacecraft is assumed to equip with four actuators. Hence, at most one actuator can suffer from F1 fault, otherwise the attitude control system will become an under-actuated one. Concerning the control distribution matrix $\boldsymbol{D}_1$, sufficient conditions for Assumption 4.4 are given under four typical faulty scenarios in Table 4.1. It can be seen that the positive definiteness of $\boldsymbol{P}$ can be guaranteed by limiting the value of r. That is, Assumption 4.4 always holds for spacecraft with similar inertia on three principal axes.

In order to give a concise expression, the sufficient conditions given in Table 4.1 are relatively conservative. Therefore, we further generate a large number of actuator fault data by Monte Carlo method to verify the feasibility of Assumption 4.4. The envelop for non positive-definite $\boldsymbol{P}$ are presented in Fig. 4.2, with Fig. 4.2a for the first two faulty scenario in Table 4.1, and Fig. 4.2b for the last scenario.

Table 4.1 Sufficient conditions for Assumption 4.4 ($\boldsymbol{D}_1$)

Faulty scenario	Sufficient conditions	Parameters set as $e_t = 0.01, e_s = 0.1$
One actuator with F1 and others with F2	$r + \frac{1}{r} < 2 + 6e_t$	$r < 1.22$
One actuator with F1 and others with F3	$r + \frac{1}{r} < 2 + 6e_s$	$r < 1.44$
All actuators with F2	$r + \frac{1}{r} < 2 + 12e_t$	$r < 2.08$
All actuators with F3	$r + \frac{1}{r} < 2 + 12e_s$	$r < 4.32$

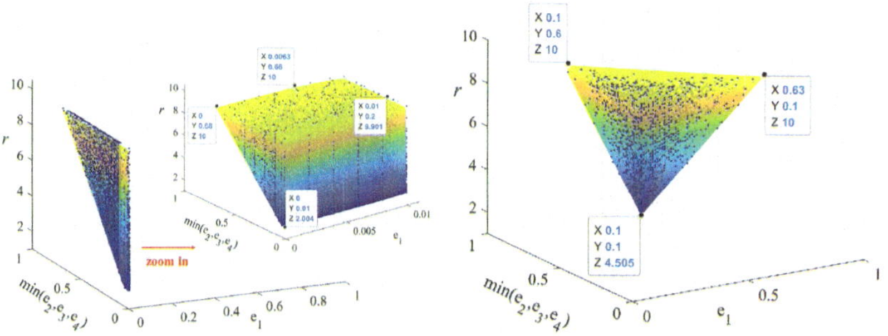

(a) Faulty scenarios 1 and 2 in Tab. 4.1

(b) Faulty scenario 4 in Tab. 4.1

Fig. 4.2 Envelop for non positive-definite $\boldsymbol{P}$ of $\boldsymbol{D}_1$

In Fig. 4.3a, e_1 is the total failure factor valuing from 0 to e_t (set as 0.01), and e_2, e_3, e_4 value from 0.01 to 1. It is shown that the minimum value of r in the envelope is 2.004, which suggests that Assumption 4.4 holds if $r < 2.004$. In addition, r decreases as $\min(e_2, e_3, e_4)$ goes down. In other words, as the fault goes slighter, the restrictions on r can be relaxed accordingly. To sum up, we can ensure $\boldsymbol{P}$ is positive-definite by setting a limitation on r. Follow a similar line, sufficient conditions and Monte Carlo analysis for $\boldsymbol{D}_2$ are presented in Table 4.2 and Fig. 4.3.

Remark 4.5 To simplify the tedious calculation of deriving sufficient conditions, $\boldsymbol{J}$ is assumed to be a diagonal matrix in Assumption 4.3. However, for a general $\boldsymbol{J}$ with non-zero inertia products, the above conclusion and analysis still almost hold, since the inertia products are usually much smaller than the principal tensors.

Hereafter, an adaptive controller is derived to further cope with (4.35), so as to achieve system stability. Since $\boldsymbol{P}$ is positive-definite according to Assumption 4.4, there exist two positive constants δ and δ_m such that $\delta = \lambda_{\min}(\boldsymbol{P})$ and $\delta_m < \delta$ always hold. We further define $\lambda = 1/\delta$ and $\boldsymbol{d}_m = \lambda \boldsymbol{d}$, which follows that $\lambda < \lambda_m$ with $\lambda_m = 1/\delta_m$. Adaptive method is employed to estimate λ, $\boldsymbol{d}$, λ_m, $\boldsymbol{d}_m$, with their estimation denoted as $\hat{\lambda}$, $\hat{\boldsymbol{d}}$, $\hat{\lambda}_m$, $\hat{\boldsymbol{d}}_m$, respectively.

Table 4.2 Sufficient conditions for Assumption 4.4 ($\boldsymbol{D}_2$)

Faulty scenario	Sufficient conditions	Parameters set as $e_t = 0.01, e_s = 0.1$
One actuator with F1 and others with F2	$r + \frac{1}{r} < 2 + 4/\left(e_t + \frac{1}{e_t} - 2\right)$	$r < 1.22$
One actuator with F1 and others with F3	$r + \frac{1}{r} < 2 + 4/\left(e_s + \frac{1}{e_s} - 2\right)$	$r < 1.99$
All actuators with F2	$r + \frac{1}{r} < 2 + 16/\left(e_t + \frac{1}{e_t} - 2\right)$	$r < 1.49$
All actuators with F3	$r + \frac{1}{r} < 2 + 16/\left(e_s + \frac{1}{e_s} - 2\right)$	$r < 3.71$

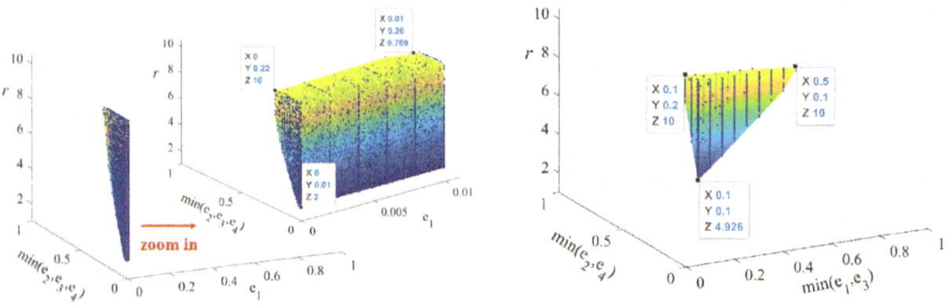

(a) Faulty scenarios 1 and 2 in Table 4.2 (b) Faulty scenario 4 in Table 4.2

Fig. 4.3 Envelop for non positive-definite $\boldsymbol{P}$ of $\boldsymbol{D}_2$

The adaptive control law is designed as

$$\boldsymbol{u} = -k_2 \boldsymbol{D}^\top \boldsymbol{N}_z \boldsymbol{z} - k(t) \boldsymbol{D}^\top \frac{\boldsymbol{N}_z \boldsymbol{z}}{\|\boldsymbol{N}_z \boldsymbol{z}\| + \rho^2 \varepsilon}, \tag{4.38}$$

with

$$k(t) = \hat{\lambda} \|\boldsymbol{u}_d\| + \hat{\boldsymbol{d}}^\top \boldsymbol{\Phi}. \tag{4.39}$$

The adaptive laws are designed as

$$\dot{\hat{\lambda}} = \zeta_1 \left[\|\boldsymbol{u}_d\| \|\boldsymbol{N}_z \boldsymbol{z}\| - \mu_1 (\hat{\lambda} - \hat{\lambda}_m) \right], \tag{4.40}$$

$$\dot{\hat{\lambda}}_m = \beta_1 (\hat{\lambda} - \hat{\lambda}_m), \tag{4.41}$$

$$\dot{\hat{\boldsymbol{d}}} = \zeta_2 \left[\boldsymbol{\Phi} \|\boldsymbol{N}_z \boldsymbol{z}\| - \mu_2 (\hat{\boldsymbol{d}} - \hat{\boldsymbol{d}}_m) \right], \tag{4.42}$$

$$\dot{\hat{\boldsymbol{d}}}_m = \beta_2 (\hat{\boldsymbol{d}} - \hat{\boldsymbol{d}}_m), \tag{4.43}$$

where k_2, ζ_i, μ_i, and β_i $(i = 1, 2)$ are positive constants, whereas the time-varying parameter ρ is updated by

$$\dot{\rho} = -\gamma\varepsilon\rho k(t)\frac{\|\boldsymbol{N}_z\boldsymbol{z}\|}{\|\boldsymbol{N}_z\boldsymbol{z}\| + \rho^2\varepsilon}, \tag{4.44}$$

where γ, ε are positive constants.

Theorem 4.2 *For the system described by (2.19) and (2.21) with the initial states satisfying* $\boldsymbol{q}(0) \in \mathbb{Q}_f$ *and* $\boldsymbol{\omega}(0) \in \mathbb{W}$, *the proposed fault-tolerant controller* (4.38)-(4.39) *with adaptive laws* (4.40)–(4.44) *can guarantee:*

- V_p *and* V_ω *are bounded, implying that both the attitude and angular velocity constraints are never violated;*
- *Gradient-related vector and angular velocity asymptotically converge to zero, i.e.,* $\lim_{t\to\infty}[\boldsymbol{v}(t), \boldsymbol{\omega}(t)] = \boldsymbol{0}$.

Proof Choose the following Lyapunov function candidate

$$\begin{aligned} V =& V_p + V_\omega + \frac{\delta}{2\gamma}\rho^2 + \frac{\delta}{2\zeta_1}(\hat{\lambda} - \lambda_m)^2 + \frac{\delta\mu_1}{2\beta_1}(\hat{\lambda}_m - \lambda_m)^2 \\ &+ \frac{\delta}{2\zeta_2}\|\hat{\boldsymbol{d}} - \boldsymbol{d}_m\|^2 + \frac{\delta\mu_2}{2\beta_2}\|\hat{\boldsymbol{d}}_m - \boldsymbol{d}_m\|^2. \end{aligned} \tag{4.45}$$

Evaluating its time derivative, we obtain that

$$\dot{V} \le -k_1\boldsymbol{v}^\top\mathrm{Tanh}(\boldsymbol{v}) - k_2\delta\|\boldsymbol{N}_z\boldsymbol{z}\|^2 - \delta\mu_1(\hat{\lambda} - \hat{\lambda}_m)^2 - \delta\mu_2\|\hat{\boldsymbol{d}} - \hat{\boldsymbol{d}}_m\|^2. \tag{4.46}$$

Similar to the proof of Theorem 4.1, we can further conclude that $\lim_{t\to\infty}[\boldsymbol{v}(t), \boldsymbol{\omega}(t)] = \boldsymbol{0}$ and V_p, V_ω are bounded. This completes the proof. □

4.2.4 Numerical Simulations

In this section, numerical simulations are carried out to show the effectiveness and superiority of the proposed controller. It is assumed that the spacecraft carries an infrared telescope and an antenna, whose boresight are fixed in body axes along the Z-axis and X-axis of the frame $\mathcal{P}$, respectively. The infrared telescope has to avoid four bright objects and the antenna has to stay in a transmission area, their attitude constraints detailed in Table 4.3 as four forbidden zones (termed as FZ) and a mandatory zone (termed as MZ). Consider a rest-to-rest reorientation mission, where the spacecraft is required to maneuver from initial attitude $\boldsymbol{q}_0 = [0.6,\ 0.3,\ -0.2,\ -0.7141]^\top$ to desired attitude $\boldsymbol{q}_d = [0.5,\ 0.3,\ 0.5,\ 0.6403]^\top$ while satisfying complex attitude constraints and tolerating potential actuator faults simultaneously. The inertia matrix of the spacecraft is given by

Table 4.3 Detailed attitude constraints

Attitude constraints	Axis vector	Angle
FZ1	$[-0.5,\ 0.7071,\ -0.5]^{\top}$	20°
FZ2	$[0,\ 0.7071,\ 0.7071]^{\top}$	40°
FZ3	$[0.4,\ -0.6,\ 0.6928]^{\top}$	30°
FZ4	$[0.8944,\ 0.2,\ 0.4]^{\top}$	20°
MZ	$[0.6,\ 0.8,\ 0]^{\top}$	50°

$$\boldsymbol{J} = \begin{bmatrix} 20 & 1.2 & 0.9 \\ 1.2 & 17 & 1.4 \\ 0.9 & 1.4 & 15 \end{bmatrix} \text{kg} \cdot \text{m}^2.$$

The reaction wheel distribution matrix is $\boldsymbol{D} = [1, 0, 0, 1/\sqrt{3}; 0, 1, 0, 1/\sqrt{3}; 0, 0, 1, 1/\sqrt{3}]$, and the external disturbances are as follows

$$\boldsymbol{\tau}_d = 2 \times 10^{-4} \times \begin{bmatrix} 3\cos(0.01t) + 1 \\ 1.5\sin(0.02t) + 3\cos(0.025t) \\ 3\sin(0.01t) - 1 \end{bmatrix} \text{Nm}.$$

The fault of four reactions wheels are set as

$$\begin{aligned} e_1(t) &= 0.3 + 0.02\sin(0.01t) + 0.02\,\text{rand}(0, 1), & \bar{u}_1 &= 0.01 \\ e_2(t) &= 0.5 + 0.05\cos(0.01t) + 0.03\,\text{rand}(0, 1), & \bar{u}_2 &= -0.02 + 0.01\sin(0.01t) \\ e_3(t) &= 0.6(t < 30),\ 0(t > 30), & \bar{u}_3 &= 0.01 \\ e_4(t) &= 0.4 + 0.05\sin(0.02t) + 0.02\,\text{rand}(0, 1), & \bar{u}_4 &= -0.015 + 0.01e^{-0.1t} \end{aligned}$$

The design parameters of proposed controller are: $\alpha_1 = 0.5$, $\alpha_2 = 0.1$, $k_1 = 0.03$, $k_2 = 100$, $\gamma = 0.1$, $\varepsilon = 1$, $\zeta_i = \beta_i = \mu_i = 0.1$ $(i = 1, 2)$. The initial values of λ, λ_m, d_i, d_{mi} are set to 0.1, and the initial value of ρ is taken as 10.

4.2.4.1 Simulation Scenario Under Attitude Constraints

Here we consider a simulation scenario with attitude constraints. As can be seen from Figs. 4.4 and 4.5, both the attitude error and angular velocity converge to a small neighborhood around the origin despite the presence of actuator faults. Figure 4.6 shows the time histories of control torque. It is noteworthy that u_3 and u_4 have a sudden change at $t = 30$ s. This is because the 3-th reaction wheel occurs F4 fault. Intuitively, Figs. 4.7 and 4.8 illustrate the 3D and 2D attitude maneuver trajectory of the spacecraft, respectively, which show the successful accomplishment of spacecraft

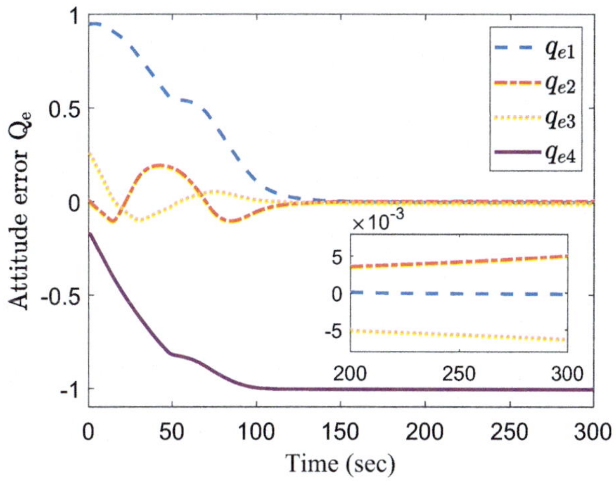

Fig. 4.4 Time responses of attitude error

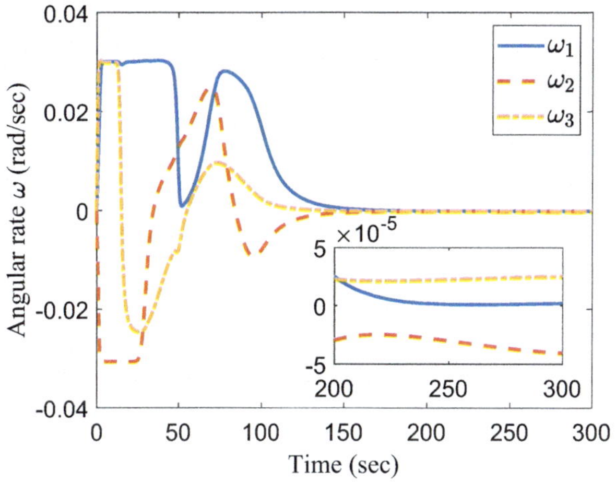

Fig. 4.5 Time responses of angular velocity

constrained reorientation, despite the presence of actuator faults. The initial and desired orientations are marked by "circle" and "star", respectively.

In addition, two groups of comparison simulations are performed to show the advantages of the proposed controller in dealing with both attitude constraints and actuator faults. Three other controllers are also simulated, with parameters given in Table 4.4. The controller in [8] only takes attitude constraints into consideration, while controller in [7] further addresses unwinding problem, but neither of them accounts for actuator faults, external disturbances or inertia uncertainties. For fair

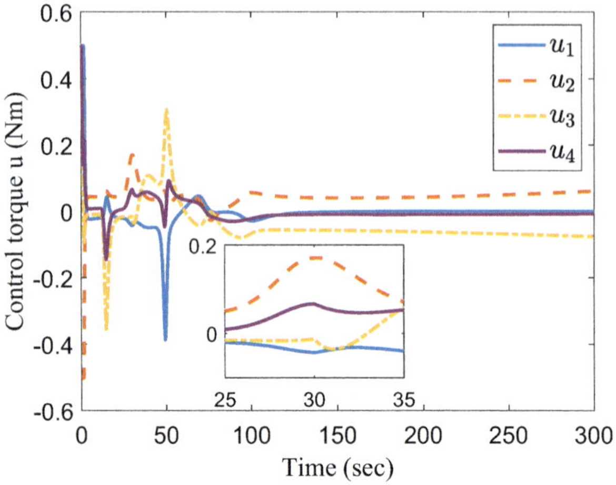

Fig. 4.6 Time responses of control torque

Fig. 4.7 3D trajectory

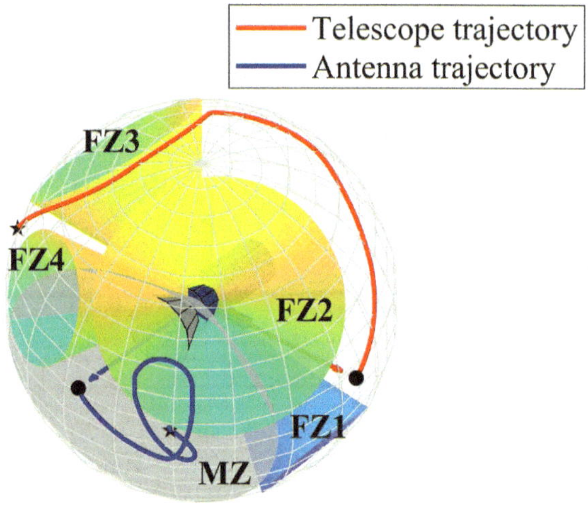

comparison, all parameters are specially chosen to ensure similar settling time of the attitude error and angular velocity as the proposed controller, as shown later in Figs. 4.9a, b and 4.10a.

Figure 4.9 shows outperformance of the proposed controller in terms of addressing attitude constraints and its anti-unwinding ability compared with PD controller and controller in [8]. It is shown in Fig. 4.9a and b that all the three controllers render convergence of attitude error and angular velocity with almost the same settling time, while the proposed controller achieves highest precision because of the consideration of external disturbances. In addition, we can see that PD controller enables the attitude error norm to decrease rapidly, because it ignores the attitude constrained zones.

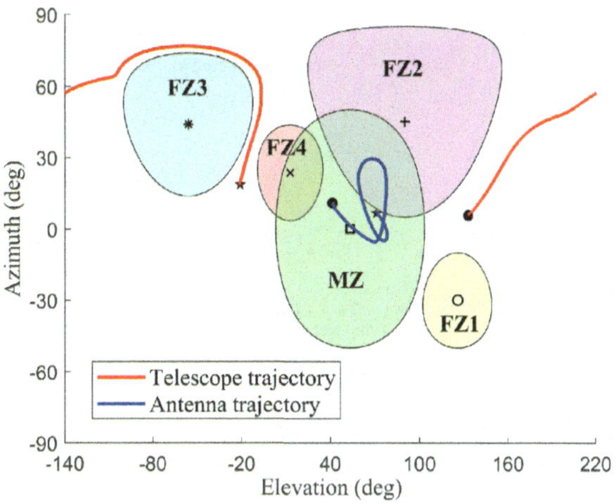

Fig. 4.8 2D trajectory

Table 4.4 Parameters of three other controllers

Controller	Parameters
PD controller $\boldsymbol{\tau} = -k_p \boldsymbol{q}_e - k_d \boldsymbol{\omega}$	$k_p = -1,\ k_d = 14$
Controller in [7]	$k_1 = 0.03,\ k_2 = 0.01,\ \alpha = 1.8$
Controller in [8]	$\alpha = 1,\ \beta = 0.01,\ l_1 = 40$

The controller in [8] cannot guarantee a monotone reduction of the attitude error norm, on the other hand, the attitude error norm goes up to 1 and then converges to 0, which implies the unwinding problem. Then from Fig. 4.9c, the PD controller consumes the least energy due to its shortest trajectory, but the shortcut fails to satisfy attitude constraints. Controller in [8] consumes the most energy because of the unwinding trajectory. The above analysis is more intuitive in Figs. 4.9d, e with the 3D and 2D reorientation trajectories. Compared with these two controllers, the proposed controller not only ensures high-precision maneuver, but also achieves anti-unwinding reorientation against attitude constraints.

Comparison results are shown in Fig. 4.10. The controller in [7] performs lower precision reorientation when suffering from the same fault as the proposed controller, as can be seen in Fig. 4.10a. By contrast, the proposed controller has a stronger fault-tolerant ability and can still perform high-precision attitude maneuver in the presence of actuator faults. As for energy consumption, it is shown in Fig. 4.10b that the proposed controller consumes less energy in both healthy and faulty cases. In addition, it also needs less additional energy from healthy to fault case, which is due to consideration of actuator faults in controller design. However, the controller in [7] requires more energy consumption to compensate for the declined performance of control torque. Figures 4.10c, d further depict the attitude maneuver trajectory under trhe controller in [7].

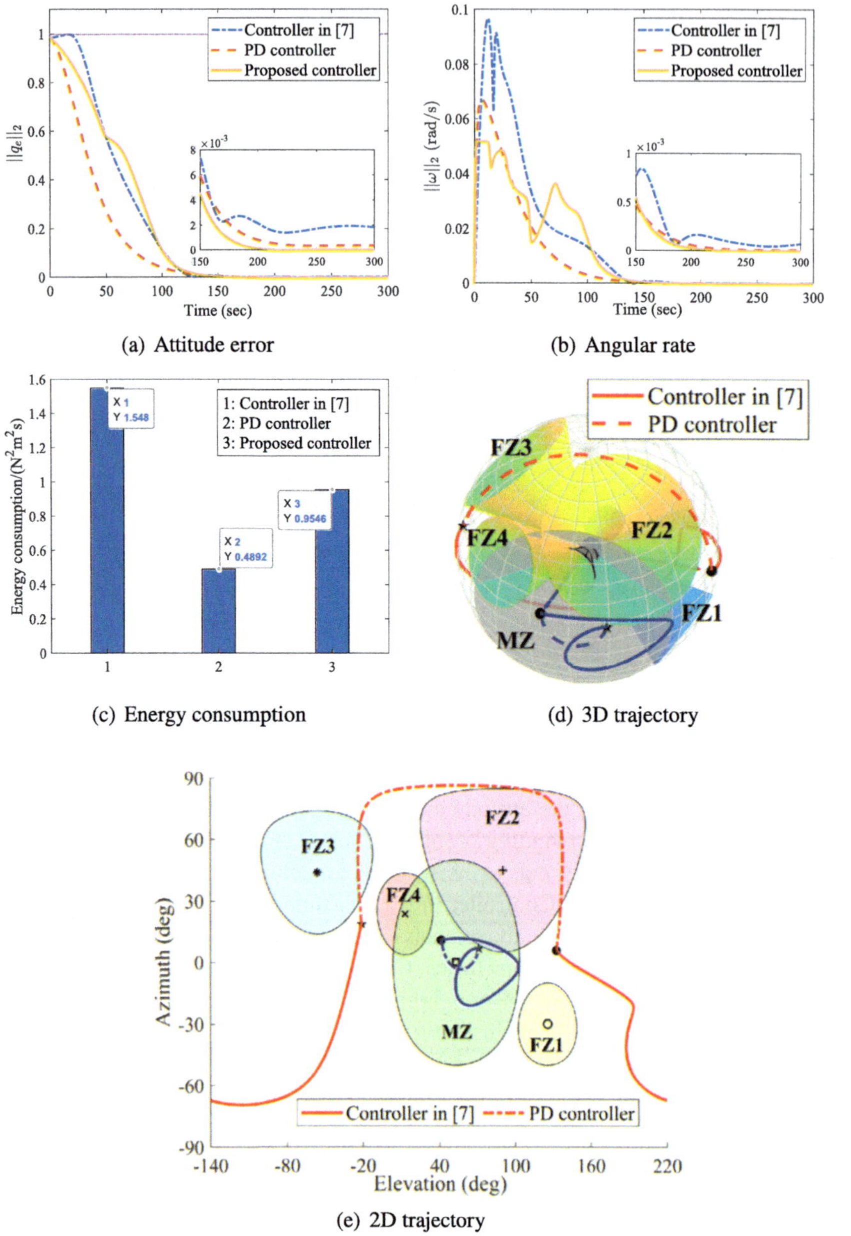

Fig. 4.9 Comparison results in terms of addressing attitude constraints. Initial and desired orientations are marked by "circle" and "star" respectively while telescope and antenna trajectory are plotted in red and blue curves. Solid and dotted line stand for trajectory under controller in [8] and PD controller

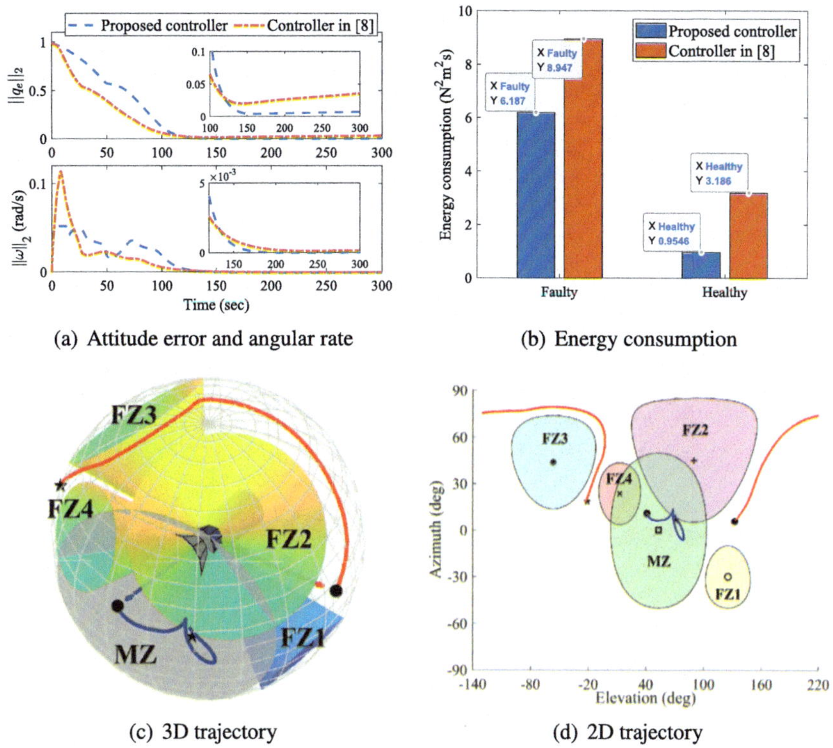

(a) Attitude error and angular rate

(b) Energy consumption

(c) 3D trajectory

(d) 2D trajectory

Fig. 4.10 Comparison results in terms of addressing actuator faults. Initial and desired orientations are marked by "circle" and "star" respectively while telescope and antenna trajectory are plotted in red and blue curves

4.2.4.2 Simulation Scenario Under Attitude and Angular Velocity Constraints

We further consider angular velocity constraint, with ω_m taken as 0.03 rad/s. It is shown in Figs. 4.11a, b that both the vector part of attitude error and the angular velocity ω converge asymptotically to zero, while angular velocity remains below 0.03 rad/s. In addition, q_{e4} converges to -1 monotonously, which suggests that the reorientation avoids the unwinding problem. Figures 4.11d, e illustrate the trajectory from 3D and 2D view, respectively, where the sensitive payload successfully evades four forbidden zones and reorients to the desired attitude.

4.2.4.3 Robustness Test

The robustness of proposed controller against measurement noise is examined in this subsection. Simulation cases of faulty actuators are repeated with measurement

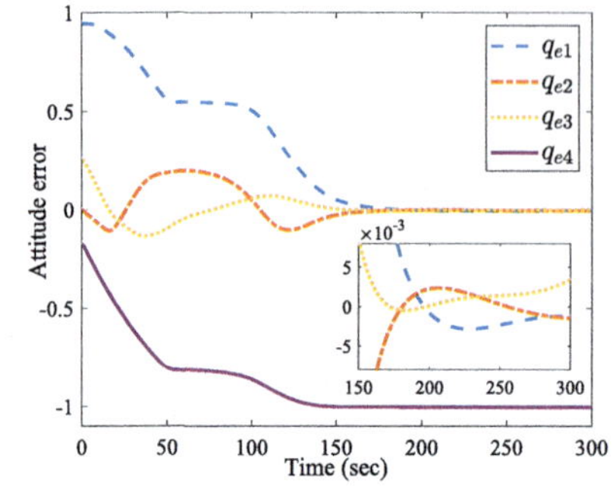

(a) Attitude error

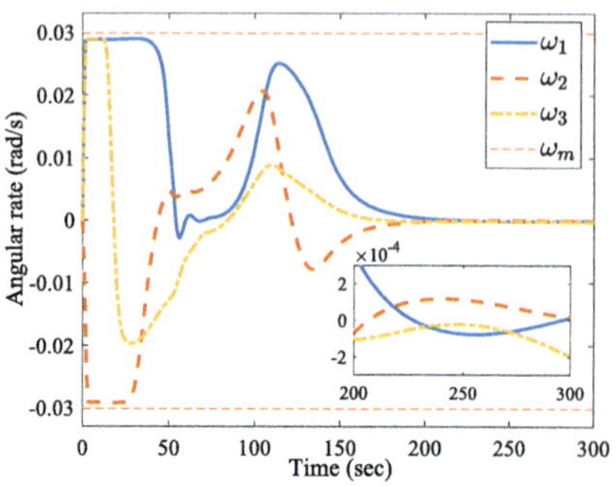

(b) Angular velocity

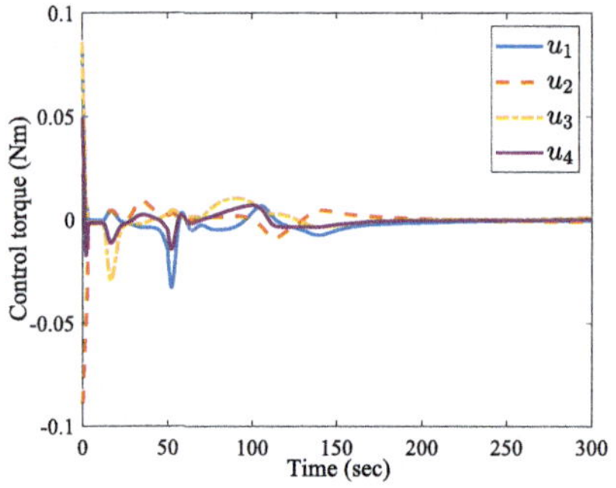

(c) Control torque

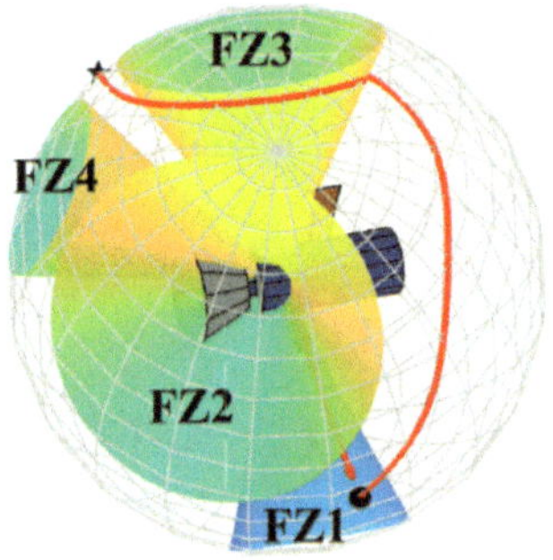

(d) 3D Trajectory

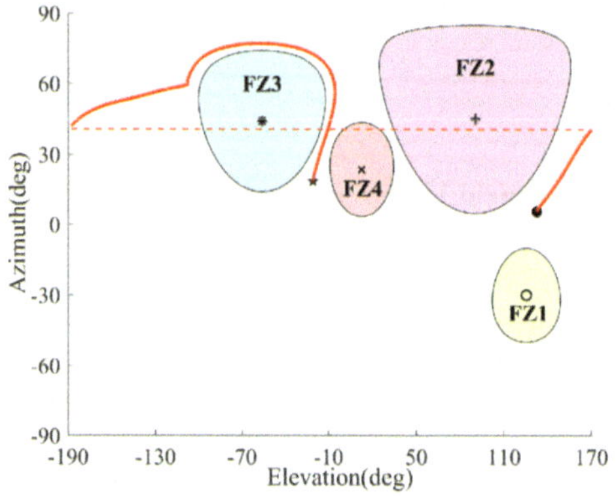

(e) 2D Trajectory

Fig. 4.11 Simulation results for attitude and angular constraints. The initial and desired orientations are marked by "circle" and "star" respectively while trajectory are plotted in a red curve

noises added to the attitude and angular velocity of spacecraft. Rewrite the unit quaternion $\boldsymbol{q} = [\boldsymbol{q}_v^\top, q_4]^T$ as

$$\boldsymbol{q}_v = \boldsymbol{e}\sin(\frac{\psi}{2}),\ q_4 = \cos(\frac{\psi}{2}),$$

where $\boldsymbol{e}$ and ψ represent the Euler eigenaxis and eigenangle with respect to (w.r.t.) $\boldsymbol{q}$, respectively. Then the measurement noise of $\boldsymbol{q}$ is set by randomly perturbing the true $\boldsymbol{e}$ with uniform distribution in a spherical cone centered around it. The cone half-angle is 1 deg. And the measurement noise with mean 0 and variance 10^{-6} rad/sec is also added to angular rate. Figure 4.12 shows the performance with measurement

Fig. 4.12 Simulation results with measurement noise

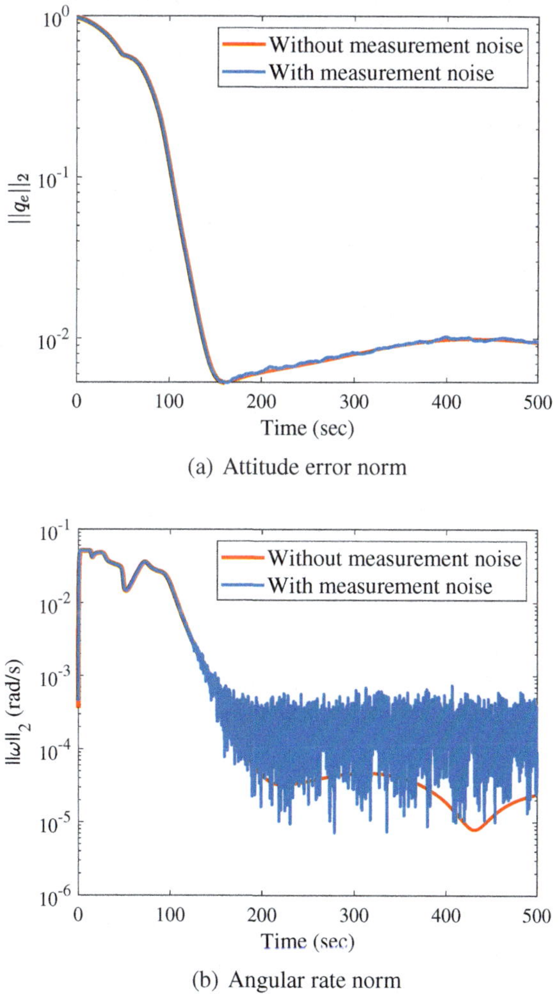

noises on a semilogarithmic scale, where the simulation time is extended to 500 s It is seen that the performance of the controller is degraded, especially the increasing steady-state value of the angular rate. However, the proposed controller shows a certain degree of robustness against measurement noises since both the attitude error and angular rate still converge to a small residual set around the origin.

4.3 Learning-Based Optimal FTC for Spacecraft Constrained Reorientation

In this section, we investigates the fault-tolerant optimal attitude control problem for a rigid spacecraft subject to both attitude and angular velocity constraints. A special cost function is proposed to balance control consumption and performance, wherein a judiciously designed term is introduced to accommodate actuator faults. In addition, the constraint information of attitude and angular velocity is encoded into the cost function using the concept of artificial potential field. Then, a single-critic NN is developed to online approximate the cost function, within the framework of reinforcement learning. Lyapunov stability theory proves that the derived approximate optimal control policy can guarantee the boundedness of states and NN estimation errors while satisfying attitude and angular velocity constraints, despite the presence of actuator faults. Finally, numerical simulations show the effectiveness of the proposed control scheme.

4.3.1 Problem Formulation

This subsection aims to design an RL-based approximate optimal control policy for spacecraft attitude reorientation by minimizing the specific cost function. In addition, the control scheme should also tolerate actuator faults and ensure attitude and angular velocity constraints are never violated.

4.3.2 Constrained Optimal FTC Design

Let us rewrite the spacecraft attitude dynamics described by (2.19) and (2.21) into the following form

$$\begin{aligned} \dot{\boldsymbol{x}} &= \boldsymbol{f}(\boldsymbol{x}) + \boldsymbol{g}\boldsymbol{u} && \text{(nominal system)}, \\ \dot{\boldsymbol{x}} &= \boldsymbol{f}(\boldsymbol{x}) + \boldsymbol{g}(\boldsymbol{I}_r - \boldsymbol{\Lambda})\boldsymbol{u} && \text{(faulty system)}, \end{aligned} \tag{4.47}$$

where $\boldsymbol{x} = [(\boldsymbol{q} - \boldsymbol{q}_d)^\top, \boldsymbol{\omega}^\top]^\top \in \mathbb{R}^7$, and

$$f(\boldsymbol{x}) = \begin{bmatrix} 0.5\boldsymbol{Q}(\boldsymbol{q})\omega \\ -\boldsymbol{J}^{-1}\boldsymbol{S}(\omega)\boldsymbol{J}\omega \end{bmatrix}, \ g = \begin{bmatrix} \boldsymbol{0}_{4\times 4} \\ \boldsymbol{J}^{-1}\boldsymbol{D}_{3\times 4} \end{bmatrix}. \tag{4.48}$$

Assumption 4.5 The fault information $\boldsymbol{\Lambda}$ is unavailable. However, the upper bound of fault matrix is supposed to be known, which is defined as $\bar{\boldsymbol{\Lambda}} = \text{diag}(\bar{\Lambda}_i)$, $i = 1, 2, \ldots r$.

To encode attitude constraints into the optimal control problem, the following cost function is specially chosen

$$V_p = -||\boldsymbol{q} - \boldsymbol{q}_d||^2 \sum_{i=1}^{n} \sum_{j=1}^{m} k_{ij} \log \left(-\frac{\boldsymbol{q}^\top \boldsymbol{M}_i^j \boldsymbol{q}}{2} \right), \tag{4.49}$$

where $\boldsymbol{q}_d$ is the desired attitude, and k_{ij} is a positive weighting parameter. Furthermore, a BLF is developed to encode angular velocity constraint

$$V_\omega = k_\omega \sum_{i=1}^{3} \log \left(\frac{\omega_m^2}{\omega_m^2 - \omega_i^2} \right), \ k_\omega > 0. \tag{4.50}$$

Remark 4.6 It is clear that $V_p(\boldsymbol{q}_d) = 0$ and $V_\omega(\boldsymbol{\omega}_d) = 0$ $(\boldsymbol{\omega}_d = 0)$. In addition, V_p and V_ω are positive-definite w.r.t. the relevant states and go to infinity when approaching the forbidden boundary. These properties facilitate the design of cost function and satisfaction of state constraints, as will be shown in the following.

Consider the following cost function for the nominal system in (4.47)

$$V(\boldsymbol{x}(t)) = \int_t^\infty R(\boldsymbol{x}(\sigma), \boldsymbol{u}(\sigma)) d\sigma, \tag{4.51}$$

where $R(\boldsymbol{x}, \boldsymbol{u}) = \boldsymbol{x}^\top \boldsymbol{Q} \boldsymbol{x} + \boldsymbol{u}^\top \boldsymbol{R} \boldsymbol{u} + V_p + V_\omega - V_x^\top \boldsymbol{g} \bar{\boldsymbol{\Lambda}} \boldsymbol{u}$, and $V_x = \partial V / \partial \boldsymbol{x}$. Compared with the generally used quadratic performance index, V_p and V_ω are incorporated into (4.51) to handle state constraints. In addition, the term $-V_x^\top \boldsymbol{g} \bar{\boldsymbol{\Lambda}} \boldsymbol{u}$ is designed to deal with actuator faults [30]. Then, by taking the time derivative of (4.51), we obtain the HJB equation

$$H(\boldsymbol{x}, \boldsymbol{u}^\star, V_x^\star) \triangleq R(\boldsymbol{x}, \boldsymbol{u}^\star) + (V_x^\star)^\top (\boldsymbol{f} + \boldsymbol{g}\boldsymbol{u}^\star) = 0. \tag{4.52}$$

The optimal control policy $\boldsymbol{u}^\star$ can be obtained by taking partial differential of both sides in (4.52) w.r.t. $\boldsymbol{u}^\star$

$$\boldsymbol{u}^\star = -0.5\boldsymbol{R}^{-1}(\boldsymbol{I}_r - \bar{\boldsymbol{\Lambda}}) g^\top V_x^\star. \tag{4.53}$$

Lemma 4.3 *Consider the spacecraft attitude error system given by (2.19) and (2.21) with actuator faults modeled in* (4.7). *The optimal control law* (4.53) *is fault-tolerant and can guarantee that attitude and angular velocity constraints are satisfied as long as initial states are in the admissible set* $\mathbb{Q}_f$ *and* $\mathbb{W}$.

Proof Taking $V^\star$ as the Lyapunov candidate and recalling (4.52), the time derivative along the faulty system trajectory is evaluated as

$$\begin{aligned}\dot{V}^\star &= (V_x^\star)^\top\left[\boldsymbol{f}(\boldsymbol{x}) + \boldsymbol{g}(\boldsymbol{I}_r - \boldsymbol{\Lambda})\boldsymbol{u}^\star\right] \\ &= -\boldsymbol{x}^\top\boldsymbol{Q}\boldsymbol{x} - (\boldsymbol{u}^\star)^\top\boldsymbol{R}\boldsymbol{u}^\star - V_p - V_\omega \\ &+ (V_x^\star)^\top(\bar{\boldsymbol{\Lambda}} - \boldsymbol{\Lambda})\boldsymbol{g}\boldsymbol{u}^\star \\ &\le -\boldsymbol{x}^\top\boldsymbol{Q}\boldsymbol{x} - V_p - V_\omega \le 0,\end{aligned} \tag{4.54}$$

which indicates that the optimal control $\boldsymbol{u}^\star$ can ensure $\lim_{t\to\infty}\boldsymbol{x}(t) = 0$ against actuator faults. In addition, since $\boldsymbol{q}(0), \boldsymbol{\omega}(0)$ belongs to $\mathbb{Q}_f$ and $\mathbb{W}$, V_p and V_ω remains bounded for all time. Thus, both attitude and angular velocity constraints are never violated. □

However, it is difficult or even impossible to derive the optimal control policy by analytical approach, because the HJB equation is a nonlinear partial differential equation which is hard to solve. Towards this end, ADP serves as a significant method which uses the online learning NNs to estimate the cost function and derive an approximate optimal solution.

4.3.3 Single-Critic NN Design and Stability Analysis

By employing the reinforcement learning idea, ADP approach generally utilizes actor-critic framework which builds two NNs to approximate the control policy and the optimal cost function respectively. Hereafter, we develop an online learning algorithm in the framework of ADP to obtain an approximate optimal control scheme. Specially, the proposed control policy employs a single critic NN and thus has a simple structure for implementation.

The optimal cost function $V^\star(x)$ can be approximated by an NN which includes plenty basis functions, that is

$$V^\star(x) = \boldsymbol{W}^\top\boldsymbol{\sigma}(\boldsymbol{x}) + \epsilon(\boldsymbol{x}), \tag{4.55}$$

where $\boldsymbol{W} \in \mathbb{R}^N$ and $\boldsymbol{\sigma} \in \mathbb{R}^N$ (N is the number of neurons) are the NN weight and basic function respectively, while $\epsilon \in \mathbb{R}$ is the approximation error. Then the residual error of (4.52) caused by approximation error is

$$\begin{aligned}\varepsilon_H = H(\boldsymbol{x}, \boldsymbol{u}, \boldsymbol{W}) &= \boldsymbol{x}^\top\boldsymbol{Q}\boldsymbol{x} + \boldsymbol{u}^\top\boldsymbol{R}\boldsymbol{u} + V_p + V_\omega \\ &- \boldsymbol{W}^\top\boldsymbol{\sigma}_x\boldsymbol{g}\bar{\boldsymbol{\Lambda}}\boldsymbol{u} + \boldsymbol{W}^\top\boldsymbol{\sigma}_x(\boldsymbol{f} + \boldsymbol{g}\boldsymbol{u}),\end{aligned} \tag{4.56}$$

where $\|\varepsilon_H\| = \varepsilon_x^\top \left(\boldsymbol{f} + \boldsymbol{g}(\boldsymbol{I}_r - \bar{\boldsymbol{\Lambda}})\boldsymbol{u}\right)$ is bounded, which is a basic assumption of ADP design. The Hamiltonian error is

$$\begin{aligned} e = H(\boldsymbol{x}, \boldsymbol{u}, \hat{\boldsymbol{W}}) &= \boldsymbol{x}^\top \boldsymbol{Q}\boldsymbol{x} + \boldsymbol{u}^\top \boldsymbol{R}\boldsymbol{u} + V_p + V_\omega \\ &\quad - \hat{\boldsymbol{W}}^\top \boldsymbol{\sigma}_x \boldsymbol{g}\bar{\boldsymbol{\Lambda}}\boldsymbol{u} + \hat{\boldsymbol{W}}^\top \boldsymbol{\sigma}_x(\boldsymbol{f} + \boldsymbol{g}\boldsymbol{u}) \\ &= -\tilde{\boldsymbol{W}}^\top \boldsymbol{\phi} + \varepsilon_H, \end{aligned} \tag{4.57}$$

where $\tilde{\boldsymbol{W}} = \boldsymbol{W} - \hat{\boldsymbol{W}}$, $\hat{\boldsymbol{W}}$, $\tilde{\boldsymbol{W}}$ are the estimate and estimation error of $\boldsymbol{W}$, respectively, and $\boldsymbol{\phi} = \boldsymbol{\sigma}_x \left(\boldsymbol{f} + \boldsymbol{g}(\boldsymbol{I}_r - \bar{\boldsymbol{\Lambda}})\boldsymbol{u}\right)$ is defined for notation brevity. Inspired by the idea of concurrent learning (CL) which uses both instantaneous and historical data for adaptation [37], we also record the Hamiltonian error regarding the past data as

$$e(t, t_j) = -\tilde{\boldsymbol{W}}(t)^\top \boldsymbol{\phi}(t_j) + \varepsilon_H(t_j), \tag{4.58}$$

where $\boldsymbol{\phi}(t_j)$ uses the past state $\boldsymbol{x}(t_j)$ and input $\boldsymbol{u}(t_j)$. The following error function is chosen to be minimized,

$$E(t) = \frac{1}{2}\frac{e(t)^2}{(1 + \boldsymbol{\phi}(t)^\top \boldsymbol{\phi}(t))^2} + \frac{1}{2}\sum_{j=1}^{p} \frac{e(t, t_j)^2}{(1 + \boldsymbol{\phi}(t_j)^\top \boldsymbol{\phi}(t_j))^2}, \tag{4.59}$$

where $(1 + \boldsymbol{\phi}^\top \boldsymbol{\phi})^2$ is used for normalization. Define $\boldsymbol{\phi}_1 = \boldsymbol{\phi}/(1 + \boldsymbol{\phi}^\top \boldsymbol{\phi})$ and $\boldsymbol{\phi}_2 = \boldsymbol{\phi}_1/(1 + \boldsymbol{\phi}^\top \boldsymbol{\phi})$ for ease of notation. According to the gradient descent rule, the NN update law is designed as

$$\begin{aligned} \dot{\hat{\boldsymbol{W}}} &= -\alpha \frac{\partial E}{\partial \hat{\boldsymbol{W}}} \\ &= -\alpha \frac{e(t)\boldsymbol{\phi}(t)}{(1 + \boldsymbol{\phi}(t)^\top \boldsymbol{\phi}(t))^2} - \alpha \sum_{i=1}^{p} \frac{e(t, t_j)\boldsymbol{\phi}(t_j)}{(1 + \boldsymbol{\phi}(t_j)^\top \boldsymbol{\phi}(t_j))^2} \\ &= -\alpha \left(\frac{\boldsymbol{\phi}(t)\boldsymbol{\phi}(t)^\top}{(1 + \boldsymbol{\phi}(t)^\top \boldsymbol{\phi}(t))^2} + \sum_{i=1}^{p} \frac{\boldsymbol{\phi}(t_j)\boldsymbol{\phi}(t_j)^\top}{(1 + \boldsymbol{\phi}(t_j)^\top \boldsymbol{\phi}(t_j))^2} \right) \tilde{\boldsymbol{W}}(t) \\ &\quad + \alpha \frac{\boldsymbol{\phi}(t)\varepsilon_H(t)}{(1 + \boldsymbol{\phi}(t)^\top \boldsymbol{\phi}(t))^2} + \alpha \sum_{i=1}^{p} \frac{\boldsymbol{\phi}(t_j)\varepsilon_H(t_j)}{(1 + \boldsymbol{\phi}(t_j)^\top \boldsymbol{\phi}(t_j))^2}, \end{aligned} \tag{4.60}$$

where α is a positive learning rate.

Now, the near-optimal control policy is derived as

$$\boldsymbol{u} = -0.5\boldsymbol{R}^{-1}(\boldsymbol{I}_r - \bar{\boldsymbol{\Lambda}})\boldsymbol{g}^\top \boldsymbol{\sigma}_x^\top \hat{\boldsymbol{W}}. \tag{4.61}$$

The excitation conditions are involved in the following design and analysis. The reader is referred to Sect. 3.4 for the definitions of IE and PE conditions. Note that

IE condition is much weaker than PE, and we will show later that by using the stored historical data, the signal excitation condition can be greatly relaxed. Assume that ϕ_1 satisfies the IE condition, which further indicates that there exist a moment t_p that $\sum_{i=1}^{p} \phi_1(t_j)\phi_1(t_j)^\top$ becomes positive definite. This means the recorded data has as many linearly independent elements as the dimension of $\phi_1(t)$, which can be easily checked by $\text{rank}(Z) = N$ with $Z = [\phi_1(t_1), \ldots, \phi_1(t_p)]$.

Lemma 4.4 *Given the update law* (4.60), *the critic NN approximation error* $\tilde{\boldsymbol{W}}$ *is uniformly ultimately bounded (UUB) under the assumption that* ϕ_1 *satisfies IE condition. In addition,* $\tilde{\boldsymbol{W}}$ *can exponentially converge to zero if* $\varepsilon_H = 0$.

Proof Let us denote $\boldsymbol{\Omega} = \phi_1(t)\phi_1(t)^\top + \sum_{i=1}^{p} \phi_1(t_j)\phi_1(t_j)^\top$, and $\boldsymbol{\Delta} = \phi_2(t)\varepsilon_H(t) + \sum_{i=1}^{p} \phi_2(t_j)\varepsilon_H(t_j)$, then the error dynamics is

$$\dot{\tilde{\boldsymbol{W}}} = -\alpha\boldsymbol{\Omega}\tilde{W} + \alpha\boldsymbol{\Delta}. \tag{4.62}$$

It should be noted that $\boldsymbol{\Omega}$ is positive definite since ϕ_1 satisfies the IE condition. Then, let us consider the Lyapunov function candidate $V_1 = (0.5\beta/\alpha)\tilde{\boldsymbol{W}}^\top\tilde{\boldsymbol{W}}$, whose time derivative is evaluated as

$$\dot{V}_1 = -\beta\tilde{\boldsymbol{W}}^\top\boldsymbol{\Omega}\tilde{W} + \beta\tilde{\boldsymbol{W}}\boldsymbol{\Delta}, \tag{4.63}$$

where $\|\boldsymbol{\Delta}\|$ is bounded, as the residual error ε_H is bounded. With this in mind, $\dot{V}_1$ is negative definite if the following condition holds

$$\|\tilde{\boldsymbol{W}}\| \geq \frac{\|\boldsymbol{\Delta}\|}{\lambda_{\min}(\boldsymbol{\Omega})}. \tag{4.64}$$

In addition, if there is no NN construction error, i.e. $\varepsilon_H = 0$, we have $\dot{\tilde{\boldsymbol{W}}} = -\alpha\boldsymbol{\Omega}\tilde{\boldsymbol{W}}$, indicating that $\tilde{\boldsymbol{W}}$ converges exponentially to zero. □

Remark 4.7 The traditional critic NN update law is designed to minimize the error function $0.5e^2(t)$, and thus $\tilde{\boldsymbol{W}}$ can only converge to zero only if ϕ_1 satisfies PE condition. However, PE condition is mostly unavailable in a rest-to-rest attitude maneuver, so the second term in (4.59) is introduced to relax the signal excitation condition to IE.

Theorem 4.3 *Consider the spacecraft model given by* (4.47) *and* (4.48) *with initial states* $\boldsymbol{q}(0), \boldsymbol{\omega}(0) \in \mathbb{Q}_f, \mathbb{W}$. *The proposed control policy* (4.53) *in conjunction with the NN update law* (4.60) *can guarantee that* $\boldsymbol{x}$, $\tilde{\boldsymbol{W}}$ *are UUB, and* V_p, V_ω *are bounded, despite the presence of actuator faults.*

Proof Choose the overall Lyapunov function candidate as $L = V^\star + V_1$. Taking its time derivative gets

$$
\begin{aligned}
\dot{L} &= (V_x^\star)^\top \left[f + g(I_r - \Lambda)u\right] + \beta\tilde{W}^\top(-\Omega\tilde{W} + \Delta) \\
&= -x^\top Qx - V_p - V_\omega - (u^\star)^\top Ru^\star + (V_x^\star)^\top g(\bar{\Lambda} - \Lambda)u^\star \\
&\quad + (V_x^\star)^\top g(I_r - \Lambda)(u - u^\star) + \beta\tilde{W}^\top(-\Omega\tilde{W} + \Delta) \\
&= -x^\top Qx - V_p - V_\omega + \beta\tilde{W}^\top(-\Omega\tilde{W} + \Delta) \\
&\quad - \frac{1}{4}W^\top\sigma_x g(I_r + \bar{\Lambda} - 2\Lambda)R^{-1}(I_r - \bar{\Lambda})g^\top\sigma_x^\top W \\
&\quad - \frac{1}{2}W^\top\sigma_x g(\bar{\Lambda} - \Lambda)R^{-1}(I_r - \bar{\Lambda})g^\top\epsilon_x \\
&\quad + \frac{1}{2}W^\top\sigma_x g(I - \Lambda)R^{-1}(I_r - \bar{\Lambda})g^\top\sigma_x^\top\tilde{W} \\
&\quad + \frac{1}{2}\epsilon_x^\top g(I_r - \Lambda)R^{-1}(I_r - \bar{\Lambda})g^\top\sigma_x^\top\tilde{W} \\
&\quad + \frac{1}{4}\epsilon_x^\top g(I_r - \bar{\Lambda})R^{-1}(I_r - \bar{\Lambda})g^\top\epsilon_x.
\end{aligned}
\tag{4.65}
$$

Then it yields that

$$
\begin{aligned}
\dot{L} \le & -x^\top Qx - V_p - V_\omega + \beta\tilde{W}^\top(-\Omega\tilde{W} + \Delta) \\
& + \frac{1}{2}\tilde{W}^\top\sigma_x g(I_r - \Lambda)R^{-1}(I_r - \bar{\Lambda})g^\top\sigma_x^\top\tilde{W} \\
& + \frac{1}{2}\epsilon_x^\top g(I_r - \Lambda)R^{-1}(I_r - \bar{\Lambda})g^\top\epsilon_x \\
= & -x^\top Qx - V_p - V_\omega - \tilde{W}^\top Y\tilde{W} + \tilde{W}\Delta + \epsilon_L,
\end{aligned}
\tag{4.66}
$$

where $Y = \beta\Omega - 0.5M$, $M = \sigma_x g(I_r - \Lambda)R^{-1}(I_r - \bar{\Lambda})g^\top\sigma_x^\top$, and $\epsilon_L = 0.5\epsilon_x^\top g(I_r - \Lambda)R^{-1}(I_r - \bar{\Lambda})g^\top\epsilon_x$. Therefore, by taking $\beta\lambda_{\min}(\Omega) > 0.5\lambda_{\max}(M)$, we have $Y > 0$. Then, it can be concluded that, x, $\tilde{W}$, V_p, $V_\omega \in \mathcal{L}_\infty$. This completes the proof. □

4.3.4 Numerical Simulations

In this section, numerical simulations are presented to demonstrate the effectiveness of the proposed control scheme and its advantages compared with other controllers. Consider a rest-to-rest attitude reorientation scenario, in which the spacecraft is required to maneuver from $q(0) = [0.4356, -0.6597, -0.5303, 0.3062]^\top$ to $q_d = [0, 0, 0, 1]^\top$. Actuator control allocation is set the same as that in Sect. 4.2.4. The attitude forbidden zones are detailed in Table 4.5, and the maximum allowable angular velocity is taken as $\omega_m = 0.03\,\mathrm{rad/s}$. Actuator faults are set as

Table 4.5 Attitude forbidden zones

Forbidden zones	Axis vector	Angle
FZ1	$[-0.9,\ 0,\ 0.4359]^{\mathrm{T}}$	20°
FZ2	$[-0.1,\ -0.5,\ 0.8602]^{\mathrm{T}}$	20°
FZ3	$[-0.4,\ -0.6,\ 0.6928]^{\mathrm{T}}$	20°

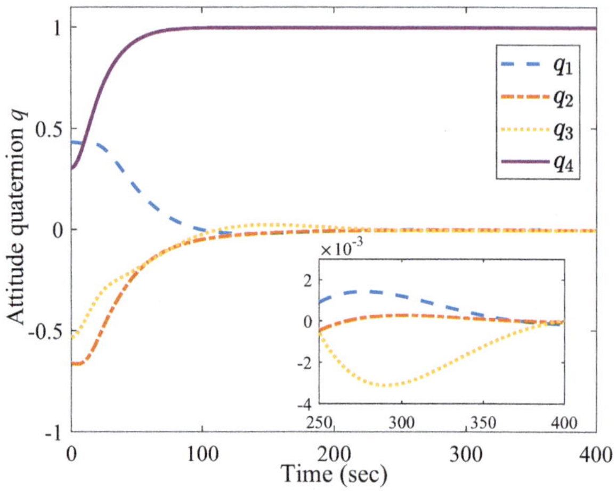

Fig. 4.13 Time responses of attitude quaternion

$$\Lambda_1 = 0.7 + 0.02\sin(0.01t);$$
$$\Lambda_2 = 0.5 + 0.05\cos(0.01t);$$
$$\Lambda_3 = 0.4\ (t < 30),\ 0.75\ (t \geq 30);$$
$$\Lambda_4 = 0.6 + 0.05\sin(0.02t).$$

Control parameters are taken as $k_{11} = 10$, $k_{12} = 30$, $k_{13} = 5$, $k_\omega = 1$, $\boldsymbol{Q} = 10\boldsymbol{I}_7$, $\boldsymbol{R} = 0.2\boldsymbol{I}_4$, $\alpha = 1$, and $\bar{\boldsymbol{\Lambda}} = 0.8\boldsymbol{I}_4$.

Simulation results of the proposed control scheme are presented in Figs. 4.13, ,4.14 and 4.15. As can be seen, the attitude quaternion converges to the desired state, and angular velocity converges to zero and remains below $\omega_m = 0.03$ rad/s.

The potential function controller (denoted as PFC) in Sect. 4.2.2 is used for comparison, and the settling time for PFC and the proposed optimal controller (denoted as OPC) are tuned similar for fairness. The PFC addresses actuator fault and attitude constraint problem in the presence of inertia uncertainties and external disturbances. However, it does not consider the angular velocity problem, and can not optimize system performance. Figure 4.16 shows time responses of PFC in the above attitude reorientation scenario, in which the angular velocity exceeds the maximum allowable value. Figure 4.17 illustrates the three-dimensional attitude maneuver trajectory of two controllers, with PFC and OPC marked by blue and red curves respectively. Both of the two control laws can regulate the spacecraft maneuver bypassing three attitude

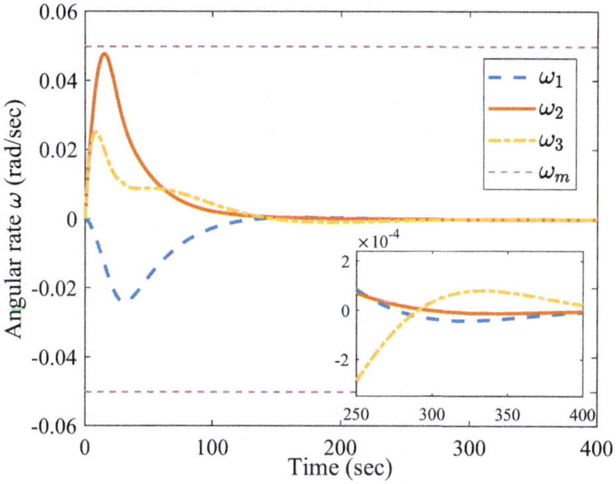

Fig. 4.14 Time responses of angular velocity

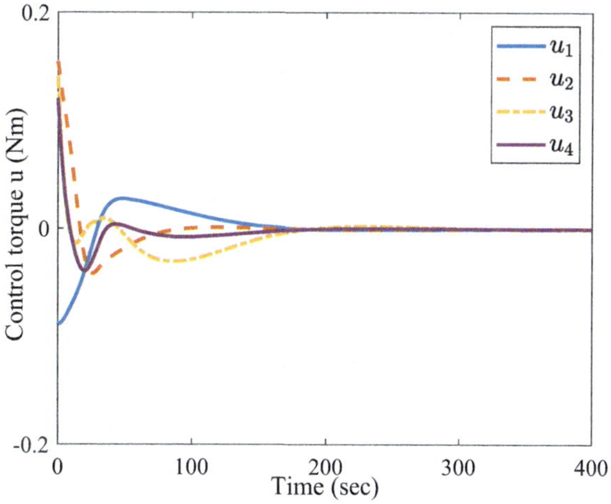

Fig. 4.15 Time responses of control torques

forbidden zones, and it is shown that the proposed control scheme derives a more smooth reorientation trajectory. In addition, energy consumption is also taken as a comparison index, which is defined as Energy$= \int_0^t \|\boldsymbol{u}\| dt$. As shown in Fig. 4.18, the proposed OPC scheme consumes less energy during the whole maneuver process, showing its optimization capability.

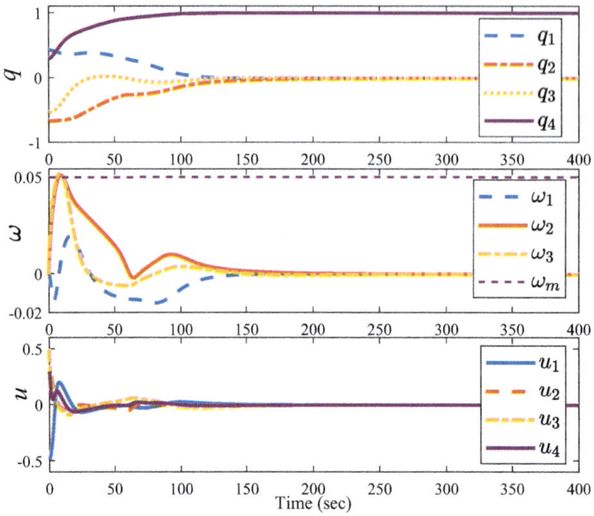

Fig. 4.16 Time responses of potential function based controller

Fig. 4.17 Maneuver trajectory comparison

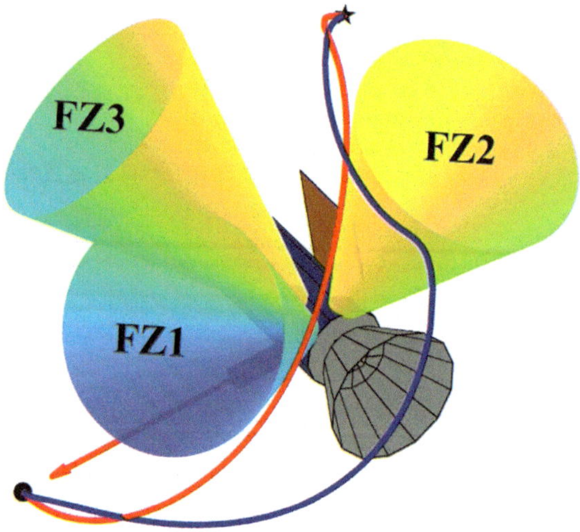

4.4 Summary

This chapter proposes two FTC laws for spacecraft constrained attitude reorientation. Firstly, an adaptive FTC scheme is proposed for constrained spacecraft reorientation, wherein attitude and angular velocity constraints are addressed using two barrier functions. Adaptive approach is further used to deal with actuator fault, uncertain inertia, and external disturbance. Secondly, an RL-based approximate optimal FTC

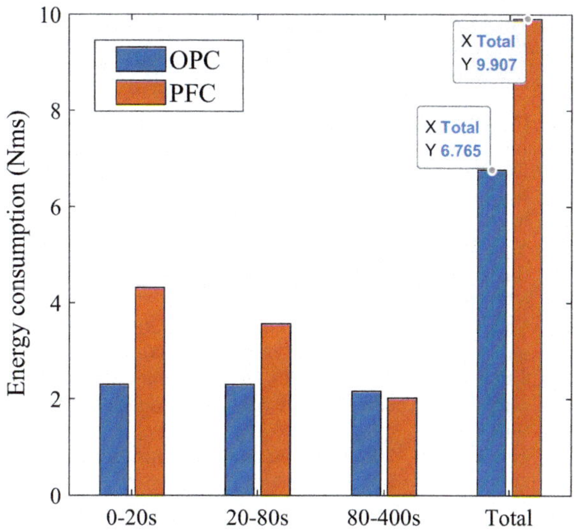

Fig. 4.18 Comparison of energy consumption

scheme is derived for spacecraft attitude reorientation under both attitude and angular velocity constraints. A specially designed cost function is developed, which is online approximated by a single-critic neural network by using reinforcement learning idea. Simulations are presented for both of the two control laws.

References

1. Chaturvedi NA, Sanyal AK, McClamroch NH (2011) Rigid-body attitude control. IEEE Control Systems Magazine 31(3): 30–51
2. Chen YP, Lo SC (1993) Sliding-mode controller design for spacecraft attitude tracking maneuvers. IEEE Transactions on Aerospace and Electronic Systems 29(4): 1328–1333
3. Tiwari PM, Janardhanan S, un Nabi M (2015) Rigid spacecraft attitude control using adaptive integral second order sliding mode. Aerospace Science and Technology 42: 50–57
4. Krstic M, Tsiotras P (1999) Inverse optimal stabilization of a rigid spacecraft. IEEE Transactions on Automatic Control 44(5): 1042–1049
5. Sun L, Huo W (2015) Robust adaptive relative position tracking and attitude synchronization for spacecraft rendezvous. Aerospace Science and Technology 41: 28–35
6. Junkins JL, Akella MR, Robinett RD (1997) Nonlinear adaptive control of spacecraft maneuvers. Journal of Guidance, Control, and Dynamics 20(6): 1104–1110
7. Hu Q, Chi B, Akella MR (2019) Anti-unwinding attitude control of spacecraft with forbidden pointing constraints. Journal of Guidance, Control, and Dynamics 42(4): 822–835
8. Lee U, Mesbahi M (2014) Feedback control for spacecraft reorientation under attitude constraints via convex potentials. IEEE Transactions on Aerospace and Electronic Systems 50(4): 2578–2592
9. Shen Q, Yue C, Goh CH (2017) Velocity-free attitude reorientation of a flexible spacecraft with attitude constraints. Journal of Guidance, Control, and Dynamics 40(5): 1293–1299

10. Cui B, Xia Y, Liu K, Shen G (2020) Finite-time tracking control for a class of uncertain strict-feedback nonlinear systems with state constraints: A smooth control approach. IEEE Transactions on Neural Networks and Learning Systems 31(11): 4920–4932
11. Shao X, Hu Q (2021) Immersion and invariance adaptive pose control for spacecraft proximity operations under kinematic and dynamic constraints. IEEE Transactions on Aerospace and Electronic Systems 57(4): 2183–2200
12. Dong H, Zhao X, Yang H (2020) Reinforcement learning-based approximate optimal control for attitude reorientation under state constraints. IEEE Transactions on Control Systems Technology 29(4): 1664–1673
13. Yang H, Hu Q, Dong H, Zhao X (2021) ADP-based spacecraft attitude control under actuator misalignment and pointing constraints. IEEE Transactions on Industrial Electronics 69(9): 9342–9352
14. Shen Q, Yue C, Goh CH, Wu B, Wang D (2018) Rigid-body attitude stabilization with attitude and angular rate constraints. Automatica 90: 157–163
15. Kristiansen R, Nicklasson PJ (2005) Satellite attitude control by quaternion-based backstepping. In: Proceedings of the American Control Conference, Portland, OR, United states, pp 907–912
16. Hu Q, Tan X (2017) Unified attitude control for spacecraft under velocity and control constraints. Aerospace Science and Technology 67: 257–264
17. Hu Q, Li L, Friswell MI (2015) Spacecraft anti-unwinding attitude control with actuator nonlinearities and velocity limit. Journal of Guidance, Control, and Dynamics 38(10): 2042–2050
18. Thakur D, Srikant S, Akella MR (2015) Adaptive attitude-tracking control of spacecraft with uncertain time-varying inertia parameters. Journal of Guidance, Control, and Dynamics 38(1): 41–52
19. Benallegue A, Chitour Y, Tayebi A (2018) Adaptive attitude tracking control of rigid body systems with unknown inertia and gyro-bias. IEEE Transactions on Automatic Control 63(11): 3986–3993
20. Liang X, Wang Q, Hu C, Dong C (2020) Fixed-time observer based fault tolerant attitude control for reusable launch vehicle with actuator faults. Aerospace Science and Technology 107: 106314
21. Tian Y, Hu Q, Shao X (2022) Adaptive fault-tolerant control for attitude reorientation under complex attitude constraints. Aerospace Science and Technology 121: 107332
22. Shen Q, Wang D, Zhu S, Poh EK (2015) Inertia-free fault-tolerant spacecraft attitude tracking using control allocation. Automatica 62: 114–121
23. Shen Q, Wang D, Zhu S, Poh EK (2017) Robust control allocation for spacecraft attitude tracking under actuator faults. IEEE Transactions on Control Systems Technology 25(3): 1068–1075
24. Hu H, Liu L, Wang Y, Cheng Z, Luo Q (2020) Active fault-tolerant attitude tracking control with adaptive gain for spacecrafts. Aerospace Science and Technology 98: 105706
25. Hu H, Wang B, Cheng Z, Liu L, Wang Y, Luo X (2021) A novel active fault-tolerant control for spacecrafts with full state constraints and input saturation. Aerospace Science and Technology 108: 106368
26. Zhang C, Wang J, Zhang D, Shao X (2018) Fault-tolerant adaptive finite-time attitude synchronization and tracking control for multi-spacecraft formation. Aerospace Science and Technology 73: 197–209
27. Hu Q, Shi Y, Shao X (2018) Adaptive fault-tolerant attitude control for satellite reorientation under input saturation. Aerospace Science and Technology 78: 171–182
28. Shao X, Hu Q, Shi Y, Jiang B (2020) Fault-tolerant prescribed performance attitude tracking control for spacecraft under input saturation. IEEE Transactions on Control Systems Technology 28(2): 574–582
29. Cai W, Liao X, Song Y (2008) Indirect robust adaptive fault-tolerant control for attitude tracking of spacecraft. Journal of Guidance, Control, and Dynamics 31(5): 1456–1463
30. Dehshalie ME, Menhaj MB, Karrari M (2019) Fault tolerant cooperative control for affine multi-agent systems: An optimal control approach. Journal of the Franklin Institute 356(3): 1360–1378

31. Zhao W, Liu H, Lewis FL (2021) Data-driven fault-tolerant control for attitude synchronization of nonlinear quadrotors. IEEE Transactions on Automatic Control 66(11): 5584–5591
32. Meng Q, Yang H, Jiang B (2022) Fault-tolerant optimal spacecraft attitude maneuver: An incremental model approach. Journal of Guidance, Control, and Dynamics 45(9): 1676–1691
33. Shao X, Hu Q, Zhu ZH, Zhang Y (2022) Fault-tolerant reduced-attitude control for spacecraft constrained boresight reorientation. Journal of Guidance, Control, and Dynamics 45(8): 1481–1495
34. Shen Q, Wang D, Zhu S, Poh K (2015) Finite-time fault-tolerant attitude stabilization for spacecraft with actuator saturation. IEEE Transactions on Aerospace and Electronic Systems 51(3): 2390–2405
35. Shao X, Hu Q, Shi Y, Jiang B (2018) Fault-tolerant prescribed performance attitude tracking control for spacecraft under input saturation. IEEE Transactions on Control Systems Technology 28(2): 574–582
36. Tee KP, Ge SS (2012) Control of state-constrained nonlinear systems using integral barrier lyapunov functionals. In: Proceedings of the IEEE Conference on Decision and Control, Maui, HI, United states, pp 3239–3244
37. Chowdhary G, Johnson E (2010) Concurrent learning for convergence in adaptive control without persistency of excitation. In: Proceedings of 49th IEEE Conference on Decision and Control (CDC), Atlanta, GA, USA, pp 3674–3679

Chapter 5
Intelligent Fault Diagnosis and Fault-Tolerant Control of Spacecraft

5.1 Introduction

Control Moment Gyroscopes (CMGs) are usually used as the actuator for large spacecraft due to their torque amplification characteristics. For example, the International Space Station (ISS) launched in 1998, and the Tianhe core module launched in 2021 are all equipped with CMGs. Compared with other momentum exchange devices like reaction wheels, the structure and working principle of CMGs are more complex. In practical applications, CMGs are prone to various types of faults and are difficult to implement fault diagnosis. Due to long-term high-speed rotating in harsh environments, the rotating mechanism in momentum exchange devices is vulnerable to damage. These faults may lead to control performance degradation or, even worse, render mission abortion and hence bring a series of severe problems. For example, it has been reported that two of the four CMGs used in the ISS failed and shut down, and the ISS then operated with only two of the four CMGs. From the lessons learned from these occurred mishaps, it is known that the spacecraft Attitude Control System (ACS) should possess strong fault diagnosis and fault-tolerance ability against actuator faults.

Fault-tolerant control (FTC) provides an effective tool to deal with actuator faults. In general, the existing FTC schemes can be classified into two types: passive FTC [1–4] and active FTC [5, 6]. Passive FTC needs neither fault diagnosis schemes nor controller reconfiguration, but it has limited fault-tolerant capabilities [7]. On the contrary, the active FTC is to reconfigure the control system according to the fault diagnosis results from fault diagnosis and detection mechanism and hence can achieve graceful performance degradation [8]. In this chapter, a neural network-based fault diagnosis scheme is proposed to perform active FTC. However, the majority of the observers design for fault diagnosis do not perform fault isolation [5, 6]. As a result, the torque deviation caused by the fault actuator could only be compensated by healthy actuators instead of rectified directly, which may lead to energy loss and overuse of actuators. When multiple time-varying faults occur simultaneously, it is generally impossible to achieve fault isolation by simply using the attitude informa-

Q. Hu et al., *Intelligent Autonomous Control of Spacecraft with Multiple Constraints*,
https://doi.org/10.1007/978-981-99-0681-9_5

tion. To solve this problem, Fonod et al. [9] proposed a layered fault isolation scheme for spacecraft equipped with redundant propulsion devices, where both attitude and position information are used. For spacecraft ACS with redundant actuators, the state information of each actuator must be used to perform fault isolation in the case of simultaneous occurrence of multiple time-varying faults. In [10, 11], the value of frame angles or fly-wheel spinning speeds were used to perform fault estimation for Single-Gimbal CMGs (SGCMGs) and flywheels, respectively, but there is still a room for improvement in estimation speed and accuracy.

In recent years, although considerable effort has been devoted to the fault diagnosis problem of reaction wheels, the study of fault diagnosis on SGCMGs has received less attention in the literature. Yue et al. [12] established the fault models of CMGs by analyzing their mechanical structure. Shen et al. [10] used adaptive observers to diagnose the faults of gimbals in SGCMGs, but this method suffers from low accuracy. Farahani and Rahimi [13] developed a data-driven fault diagnosis scheme for CMGs using Support Vector Machine (SVM), which cannot implement fault estimation. One caveat here is that all the above-mentioned fault diagnosis methods require the establishment of complex observers. As stated in [14], the analogue of Luenberger observers needs to be done in a case by case approach, the Kalman method needs to solve the HJB equation and nonlinear optimal observer often leads to dimensional disasters. Recently, neural networks-based intelligent algorithms have been widely used in various fields. Benefiting from its powerful fitting and memory capabilities, some researchers introduced the neural networks into the observer design and established neural network observers [15–20]. Wu and Saif replaced a part of the traditional observer with the output of neural network and developed neural network observers [15]. Talebi et al. [16, 17] designed a neural network fault diagnosis observer for sensor and actuator faults with detection, isolation, and diagnosis functions, and proved its stability in detail. An active FTC system integrating a neural network controller was developed by Shen et al. [19]. However, none of the above schemes considers the situation where multiple actuators fault occur at the same time, which limits their practical applications.

External disturbances are the key factors that affect the fault diagnosis results, due to the coupling of torques caused by external disturbances and actuator faults. The disturbance observer can be employed to efficiently estimate and compensate the unknown disturbances. Some disturbance observers are used in the control design to improve control performance or perform fault diagnosis [21–23]. Neural networks are also used to estimate and compensate disturbances actively. For example, Cheng et al. [23] proposed a cycle neural network (NN) to fit and compensate for the residual of the fault detection observer, which helps improve the detection ability for small faults in the presence of external disturbances, but they ignored the case in which faults take place in the disturbance observer training phase.

In this chapter, a Neural Network Disturbance Observer (NDO) is proposed to estimate and compensate the external disturbances actively. According to the estimation results, a fault diagnosis and FTC algorithm for SGCMGs based on neural network were developed. The contributions of this chapter are two-fold:

- An NDO is proposed to enhance the fault detection and fault diagnosis ability for on-orbit spacecraft. Compared with the existing disturbance observers, it has the memory capacity and therefore can decouple the deviation torques caused by the actuator faults and periodic disturbances. Moreover, unlike most neural network-based observers, the proposed scheme considers the case in which faults take place during neural network training.
- For spacecraft equipped with SGCMGs, fault isolation and high precision fault estimation of time-varying faults are realized by the proposed scheme. Several local observers were developed to achieve fault isolation and preliminary fault estimation. And estimation deviation is compensated by neural networks by combine spacecraft attitude data so that the accuracy of fault estimation is improved.

The remainder of this chapter is organized as follows. Section 5.2 introduces the spacecraft attitude dynamics and fault models of the SGCMGs. The NN-based disturbance observer and fault diagnosis schemes are designed in Sects. 5.3 and 5.4, respectively. Then, an adaptive FTC controller is proposed in Sect. 5.5. Subsequently, numerical simulations are carried out in Sect. 5.6 to verify the effectiveness of the proposed method. Finally, concluding remarks are given in Sect. 5.7.

5.2 Preliminaries

SGCMGs are composed of several Single Gimbal Control Moment Gyros, while an SGCMG consists of a gimbal that can perform one-dimensional movement and a constant speed rotor, as shown in Fig. 5.1. In the working process: the rotor rotates at a constant speed to generate a constant angular moment, and the gimbal changes the rotor's direction and hence the direction of the angular moment, whereby a control torque is generated. The control torque generated by a healthy SGCMG is

$$\boldsymbol{u}_i = -h_i\dot{\delta}_i\boldsymbol{t}_i,\ i = 1, 2, ..., N, \tag{5.1}$$

where N is the number of gyros, h_i represents the magnitude of constant angular moment generated by one rotor ($h_i = J_{\Omega i}\Omega_i$, $J_{\Omega i}$ is the inertia moment of the rotor, Ω_i is the rotor speed), $\dot{\delta}_i$ is the rotation rate of the gimbal, and $\boldsymbol{t}_i$ gives the unit vector of the output torque's direction.

Each SGCMG can only produce control torque in one direction. Considering the spacecraft requires three-dimensional control torque and issues of redundancy and singularity (SGCMGs cannot produce the desired control torque when their gimbals combined into a specific structure), a spacecraft is generally equipped with 4–6 SGCMGs. In practical engineering, to achieve better control performance, the pentagonal pyramid configuration of 6 gyros is generally adopted. However, in academic research, in order to better highlight the characteristics of the SGCMGs, a pyramid configuration of 4 gyros is generally adopted. Figures 5.2 and 5.3 display two classi-

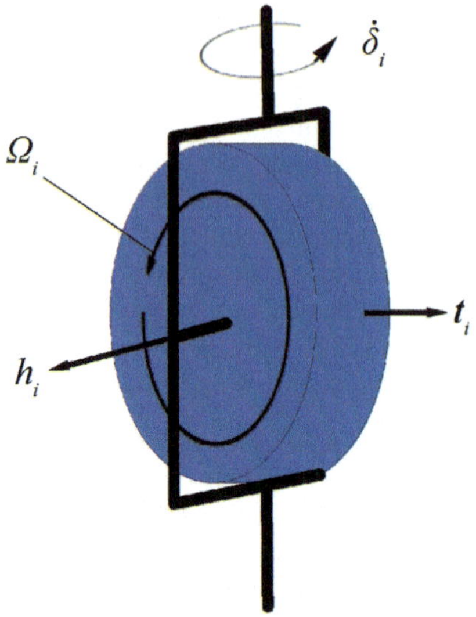

Fig. 5.1 Structure diagram of an SGCMG

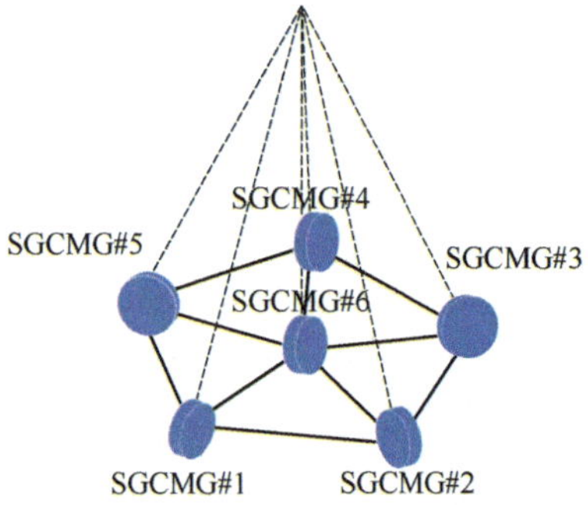

Fig. 5.2 SGCMG of pentagonal pyramid configuration

cal structure diagrams of SGCMGs. More specifically, Fig. 5.2 shows a pentagonal pyramid configuration, while Fig. 5.3 shows a pyramid configuration.

The control torque $\boldsymbol{\tau} \in \mathbb{R}^3$ generated by SGCMGs with N healthy gyros is:

$$\boldsymbol{\tau} = -h_0 \boldsymbol{A}_s \dot{\boldsymbol{\delta}} - \boldsymbol{S}(\boldsymbol{\omega})\boldsymbol{h}, \tag{5.2}$$

where h_0 denotes the magnitude of constant angular moment generated by each rotor, $\boldsymbol{A}_s \in \mathbb{R}^{3\times N}$ is the derivative Jacobian matrix of $\boldsymbol{h}$ ($\boldsymbol{A} \in \mathbb{R}^3$ is the angular moment generated by N gyros when their gimbals are static), $\dot{\boldsymbol{\delta}} \in \mathbb{R}^N$ is the rotational rate of gimbals, and $\boldsymbol{S}(\boldsymbol{\omega})\boldsymbol{h}$ is the gyroscopic moment produced by the rotation of spacecraft. The gimbal and rotor in an SGCMG can be regarded as an Electric Motor-Variable Speed Drive (EM-VSD), so the SGCMG can be regarded as a cascaded EM-VSD. In general, faults in the EM-VSD are caused by mechanical wear, harsh working

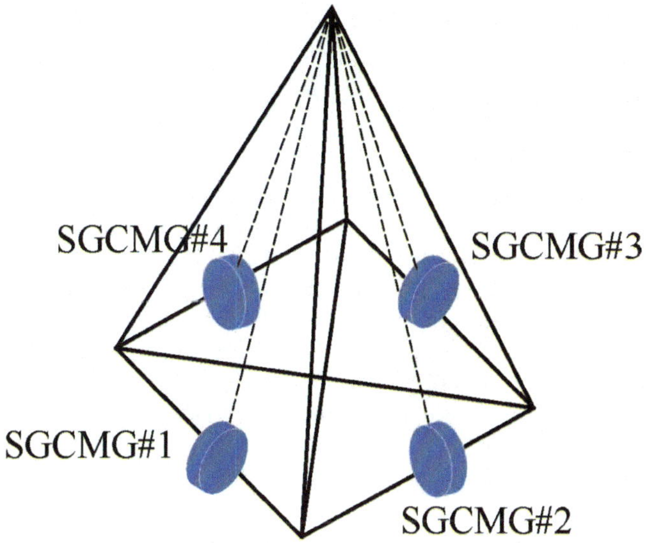

Fig. 5.3 SGCMG of pyramid configuration

environment, aging, voltage load, and some other facts. The fault models of this cascaded EM-VSD system are:

$$\begin{cases} \boldsymbol{\Omega} = \boldsymbol{\eta}_{\Omega}\boldsymbol{\Omega}_c + \boldsymbol{\Omega}_a, & \text{Rotor speed control loop} \\ \dot{\boldsymbol{\delta}} = \boldsymbol{\eta}_{\dot{\delta}}\dot{\boldsymbol{\delta}}_c + \dot{\boldsymbol{\delta}}_a, & \text{Gimbal rate control loop} \end{cases} \tag{5.3}$$

where $\boldsymbol{\Omega} \in \mathbb{R}^N$ and $\dot{\boldsymbol{\delta}} \in \mathbb{R}^N$ denote the rotation speed of rotors and gimbals, respectively, $\boldsymbol{\Omega}_c \in \mathbb{R}^N$, $\dot{\boldsymbol{\delta}}_c \in \mathbb{R}^N$ are the commend rotation speed of them, $\boldsymbol{0}_N \leq \boldsymbol{\eta}_{\Omega} \leq \boldsymbol{I}_N$, $\boldsymbol{0}_N \leq \boldsymbol{\eta}_{\delta} \leq \boldsymbol{I}_N$ ($\boldsymbol{0}_N \in \mathbb{R}^{N\times N}$ is a matrix with all entries being zero) give the efficiency matrix of them, while $\boldsymbol{\Omega}_a \in \mathbb{R}^N$, $\dot{\boldsymbol{\delta}}_a \in \mathbb{R}^N$ denote the deviation faults of them.

According to [10, 12], the above fault models can also be written as:

$$\begin{cases} \boldsymbol{\Omega} = \boldsymbol{\Omega}_c + \boldsymbol{f}_{\Omega}, & \text{Rotor speed control loop} \\ \dot{\boldsymbol{\delta}} = \dot{\boldsymbol{\delta}}_c + \boldsymbol{f}_{\dot{\delta}}, & \text{Gimbal rate control loop} \end{cases} \tag{5.4}$$

where $\boldsymbol{f}_{\Omega} = (\boldsymbol{\eta}_{\Omega} - \boldsymbol{I}_N)\boldsymbol{\Omega}_c + \boldsymbol{\Omega}_A$, and $\boldsymbol{f}_{\dot{\delta}} = (\boldsymbol{\eta}_{\dot{\delta}} - \boldsymbol{I}_N)\dot{\boldsymbol{\delta}}_c + \dot{\boldsymbol{\delta}}_a$.

As the rotor is supposed to rotate at a constant speed, it is easy to recognize its fault via rotational speed measurements. Consequently, only the gimbal fault was considered in this chapter.

5.3 Disturbance Observation Scheme

In the proposed scheme, as illustrated in Fig. 5.4, several adaptive estimators (their specific form would be discussed in Sect. 5.4.1) were set to judge whether a fault occurs in SGCMGs. The NDO would fit and memory the disturbance when no fault alarm produced by adaptive estimators. The training phase t_{train} would take the time of an orbital period T_{orb}, and after that, it would turn to the offline shape and output the estimated value of disturbances to the controller. If a fault alarm is produced by adaptive estimators in the training phase, the system would turn into fault diagnosis phase and ignore the observer results from NDO. This scheme could enhance fault diagnosis accuracy if no fault occurs in the first orbital period and retain low-level fault diagnosis ability otherwise.

Based on the spacecraft attitude dynamics, the NDO is established by

$$\dot{\hat{\omega}} = \boldsymbol{F}(\hat{\omega}) + \boldsymbol{B}\boldsymbol{\tau}_c + \boldsymbol{B}\hat{\boldsymbol{T}}_{dNN}, \tag{5.5}$$

where $\omega \in \mathbb{R}^3$ is the estimated angular velocity, $\boldsymbol{F}(\hat{\omega}) = -\boldsymbol{J}^{-1}\boldsymbol{S}(\hat{\omega})\boldsymbol{J}\hat{\omega}$, $\boldsymbol{B}$ represents the inverse of $\boldsymbol{J}$, $\boldsymbol{\tau}_c$ is the command control torque (included the gyroscopic moment produced by the rotation of the satellite), $\hat{\boldsymbol{T}}_{dNN} = \hat{\boldsymbol{W}}_d\sigma(\hat{\boldsymbol{V}}_d\boldsymbol{d})$ is the estimated value of disturbances from NNs, $\sigma(\cdot) = 2(1 - \exp^{-2(\cdot)})^{-1} - 1$ is the sigmoidal function, $\hat{\boldsymbol{W}}_d$ and $\hat{\boldsymbol{V}}_d$ are the estimated values of neural network weights, and $\boldsymbol{d} = [\omega^\top, \boldsymbol{\tau}_c^\top]^\top$.

The spacecraft attitude dynamics (2.14) can be rewritten as

$$\dot{\omega} = \boldsymbol{F}(\omega) + \boldsymbol{B}\boldsymbol{\tau}_c + \boldsymbol{B}\boldsymbol{\tau}_d. \tag{5.6}$$

To proceed, let us rewrite (5.6) as

$$\dot{\omega} = \boldsymbol{A}\omega + \boldsymbol{g}(\omega) + \boldsymbol{B}\boldsymbol{\tau}_c + \boldsymbol{B}\boldsymbol{\tau}_d, \tag{5.7}$$

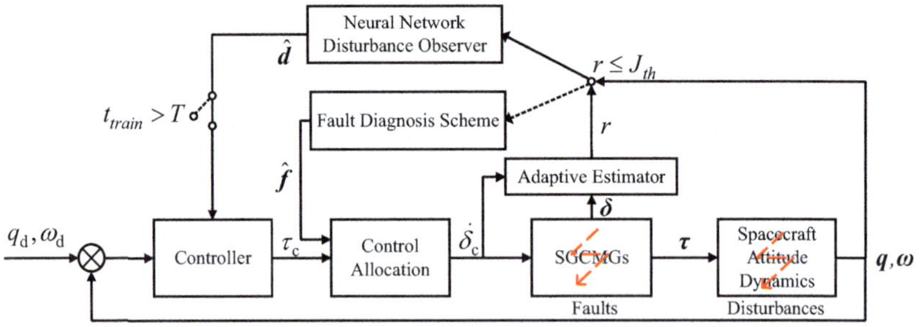

Fig. 5.4 Architecture of the disturbance observation and fault diagnosis scheme

where $\boldsymbol{g}(\boldsymbol{\omega}) = \boldsymbol{F}(\boldsymbol{\omega}) - \boldsymbol{A}\boldsymbol{\omega}$, and $\boldsymbol{A}$ is a Hurwitz matrix. According to the characteristic that neural network could represent any continuous function, $\boldsymbol{\tau}_d$ can be represented as

$$\boldsymbol{\tau}_d = \boldsymbol{W}_d\sigma(\boldsymbol{V}_d\boldsymbol{d}) + \varepsilon(\boldsymbol{d}), \tag{5.8}$$

where $\boldsymbol{W}_d$ and $\boldsymbol{V}_d$ are the ideal weight of the neural network and $\varepsilon(\boldsymbol{d})$ represents the approximation error.

Denoting the state estimation error as $\tilde{\boldsymbol{\omega}} = \boldsymbol{\omega} - \hat{\boldsymbol{\omega}}$, the weight error as $\tilde{\boldsymbol{W}}_d = \boldsymbol{W}_d - \hat{\boldsymbol{W}}_d$ and $\tilde{\boldsymbol{V}}_d = \boldsymbol{V}_d - \hat{\boldsymbol{V}}_d$. Using (5.6), (5.7), and (5.8), the error dynamics is stated as follows:

$$\dot{\tilde{\boldsymbol{\omega}}} = \boldsymbol{A}\tilde{\boldsymbol{\omega}} + \boldsymbol{g}(\boldsymbol{\omega}) - \boldsymbol{g}(\hat{\boldsymbol{\omega}}) + \boldsymbol{B}(\tilde{\boldsymbol{W}}_d\sigma(\hat{\boldsymbol{V}}_d\hat{\boldsymbol{d}}) + \boldsymbol{P}_d) + \boldsymbol{T}, \tag{5.9}$$

where $\boldsymbol{P}_d = \boldsymbol{W}_d[\sigma(\boldsymbol{V}_d\boldsymbol{d}) - \sigma(\hat{\boldsymbol{V}}_d\hat{\boldsymbol{d}})] + \varepsilon_d(\boldsymbol{d})$, whereas $\boldsymbol{T}$ is related to the faults and only used in Sect. 5.4.2. Let us define $\boldsymbol{S}_d = \boldsymbol{B}\sigma(\hat{\boldsymbol{V}}_d\hat{\boldsymbol{d}})$, $\boldsymbol{\theta}_d = \boldsymbol{B}\boldsymbol{P}_d$. According to the boundedness of sigmoidal function and approximation of neural network, both $\boldsymbol{S}_d$ and $\boldsymbol{\theta}_d$ are bounded, i.e., $\|\boldsymbol{S}_d\| \leq \bar{S}_d$ and $\|\boldsymbol{\theta}_d\| \leq \bar{\theta}_d$. With the above in mind, (5.9) can be expressed as

$$\dot{\tilde{\boldsymbol{\omega}}} = \boldsymbol{A}\tilde{\boldsymbol{\omega}} + \boldsymbol{g}(\boldsymbol{\omega}) - \boldsymbol{g}(\hat{\boldsymbol{\omega}}) + \tilde{\boldsymbol{W}}_d\boldsymbol{S}_d + \boldsymbol{\theta}_d + \boldsymbol{T}. \tag{5.10}$$

Assumption 5.1 The function $\boldsymbol{g}(\boldsymbol{\omega})$ satisfies the Lipschitz condition with the bound l_g in $\boldsymbol{\omega}$, that is

$$\|\boldsymbol{g}(\boldsymbol{\omega}) - \boldsymbol{g}(\hat{\boldsymbol{\omega}})\| \leq l_g\|\boldsymbol{\omega} - \hat{\boldsymbol{\omega}}\|. \tag{5.11}$$

Assumption 5.2 The nonlinear term $\boldsymbol{T}$ is bounded, and $\exists \bar{T} > 0$ such that

$$\|\boldsymbol{T}\| \leq \bar{T}. \tag{5.12}$$

Assumption 5.3 There exists an unknown constant ε such that $|\varepsilon_d(\boldsymbol{d})| \leq \varepsilon$.

In fact, for the spacecraft ACS, the above conditions are easy to meet.

Theorem 5.1 *Consider the spacecraft attitude dynamics in (5.7) and the observer model in (5.5). Given Assumptions 5.1–5.3, if the weights of the neural network are updated according to (for brevity, the subscript d is ignored in this part)*

$$\dot{\hat{\boldsymbol{W}}} = -\eta_1\frac{\partial V_{\tilde{\omega}}}{\partial\hat{\boldsymbol{W}}} - \rho_1\|\tilde{\boldsymbol{\omega}}\|\hat{\boldsymbol{W}}, \tag{5.13}$$

$$\dot{\hat{\boldsymbol{V}}} = -\eta_2\frac{\partial V_{\tilde{\omega}}}{\partial\hat{\boldsymbol{V}}} - \rho_2\|\tilde{\boldsymbol{\omega}}\|\hat{\boldsymbol{V}}, \tag{5.14}$$

where η_1, η_2 are learning rate, ρ_1, ρ_2 are small positive constant, and $V_{\tilde{\omega}} = 1/2(\tilde{\boldsymbol{\omega}}^\top\tilde{\boldsymbol{\omega}})$ is the cost function of the neural network, then $\tilde{\boldsymbol{\omega}}$, $\tilde{\boldsymbol{W}}$ and $\tilde{\boldsymbol{V}}$ are uni-

formly ultimately bounded, i.e., the disturbance observation error of the proposed scheme is uniformly ultimately bounded.

Proof By using the chain rule and static gradient approximation, as shown in Appendix, the error of neural network weight can be presented as

$$\dot{\tilde{W}} = l_1\tilde{\omega}S_d^{\top} + \rho_1\|\tilde{\omega}\|\hat{W}, \tag{5.15}$$

$$\dot{\tilde{V}} = S_2^{\top}\hat{W}^{\top}l_2\tilde{\omega}\hat{d}^{\top} + \rho_2\|\tilde{\omega}\|\hat{V}, \tag{5.16}$$

where $l_1 = \eta_1 J_0^{-\top}, l_2 = \eta_2 J_0^{-\top}, S_2 = \mathbf{I} - \mathrm{diag}(\sigma^2(\hat{V}\hat{d})), J_0 = \frac{\partial(F(\omega)+B\tau)}{\partial\omega}|_{\omega=0}$.

Consider the Lyapunov function candidate

$$L = \frac{1}{2}\tilde{\omega}^{\top}P\tilde{\omega} + \frac{1}{2}\mathrm{tr}(\tilde{W}^{\top}\rho_1^{-1}\tilde{W}), \tag{5.17}$$

where $P = P^{\top}$ is a positive-definite matrix satisfying the following condition

$$A^{\top}P + PA = -Q, \tag{5.18}$$

where Q is a positive-definitive matrix. The time derivative of (5.17) is

$$\dot{L} = \frac{1}{2}\dot{\tilde{\omega}}^{\top}P\tilde{\omega} + \frac{1}{2}\tilde{\omega}^{\top}P\dot{\tilde{\omega}} + \mathrm{tr}(\tilde{W}^{\top}\rho_1^{-1}\dot{\tilde{W}}). \tag{5.19}$$

By substituting (5.10), (5.15) and (5.18) into (5.19), it can be shown that

$$\begin{aligned}\dot{L} = &-\frac{1}{2}\tilde{\omega}^{\top}Q\tilde{\omega} + \tilde{\omega}^{\top}P(g(\omega) - g(\hat{\omega}) + BT + \tilde{W}S_d + \theta_d)\\ &+ \mathrm{tr}(\tilde{W}^{\top}(l_1\rho_1^{-1}\tilde{\omega}S_d^{\top}) + \tilde{W}^{\top}\|\tilde{\omega}\|(W - \tilde{W})).\end{aligned} \tag{5.20}$$

Note the following inequalities

$$\mathrm{tr}(\tilde{W}^{\top}(W - \tilde{W})) \le W_M\|\tilde{W}\| - \|\tilde{W}\|^2, \tag{5.21}$$

$$\mathrm{tr}(\tilde{W}^{\top}(l_1\rho_1^{-1}\tilde{\omega}S_d^{\top})) \le \sigma_m\rho_1^{-1}\|\tilde{W}\|\|l_1\|\|\tilde{\omega}\|\|B\|, \tag{5.22}$$

where σ_m and W_M denote the upper bounds of sigmoidal function $\sigma(\cdot)$ and ideal weight W, respectively.

Consequently, (5.11), (5.12), (5.20), (5.21) and (5.22) imply that

$$\dot{L} \le -\beta_1\|\tilde{\omega}\|^2 + \beta_2\|\tilde{\omega}\| + \beta_3\|\tilde{\omega}\|\|\tilde{W}\| - \|\tilde{W}\|^2\|\tilde{\omega}\|, \tag{5.23}$$

where $\beta_1 = 1/2\lambda_{\min}(\boldsymbol{Q}) - l_g\|\boldsymbol{P}\|$, $\beta_2 = (\bar{\theta} + \|\boldsymbol{B}\|\bar{\boldsymbol{T}})\|\boldsymbol{P}\|$ and $\beta_3 = \sigma_m(\|\boldsymbol{P}\|\|\boldsymbol{B}\| + \rho_1^{-1}\|\boldsymbol{l}_1\|\|\boldsymbol{B}\|) + W_M$, and $\lambda_{\min}(\boldsymbol{Q})$ denotes the minimum eigenvalue of $\boldsymbol{Q}$.

By completing the squares involving $\tilde{\boldsymbol{W}}$, we get

$$\begin{aligned}\dot{L} &\le -\beta_1\|\tilde{\omega}\|^2 + \beta_2\|\tilde{\omega}\| + (-(\|\tilde{\boldsymbol{W}}\| - \frac{\beta_3}{2})^2 + \frac{\beta_3{}^2}{4})\|\tilde{\omega}\| \\ &\le -\beta_1\|\tilde{\omega}\|^2 + (\beta_2 + \frac{\beta_3{}^2}{4})\|\tilde{\omega}\|.\end{aligned} \tag{5.24}$$

It can be readily obtained from (5.24) that $\dot{L} \le 0$ with $\lambda_{\min}(\boldsymbol{Q}) > 2l_g\|\boldsymbol{P}\|$ and $\|\tilde{\omega}\| \ge -(4\beta_2 + \beta_3^2)/8\beta_1$. This shows that $\|\tilde{\omega}\|$ is bounded.

To show the boundedness of $\tilde{\boldsymbol{W}}$, we rewrite (5.15) as

$$\dot{\tilde{\boldsymbol{W}}} = \boldsymbol{l}_1\tilde{\omega}\boldsymbol{S}_d^{\top} + \rho_1\|\tilde{\omega}\|\boldsymbol{W} - \rho_1\|\tilde{\omega}\|\tilde{\boldsymbol{W}}, \tag{5.25}$$

where $\boldsymbol{l}_1\tilde{\omega}\boldsymbol{S}_d^{\top}$ is bounded for the reason that $\tilde{\omega}$ and $\boldsymbol{S}_d$ are bounded and $\boldsymbol{J}_0$ is a Hurwitz matrix. Given the fact that the ideal weight $\boldsymbol{W}$ is fixed, (5.25) can be regarded as a linear system. The system is stable since $\rho_1\|\tilde{\omega}\|$ is positive and its input is bounded. This shows $\tilde{\boldsymbol{W}}$ is bounded.

Further consider (5.16) and rewrite it as follows:

$$\dot{\tilde{\boldsymbol{V}}} = \boldsymbol{S}_2^{\top}\hat{\boldsymbol{W}}^{\top}\boldsymbol{l}_2\tilde{\omega}\hat{\boldsymbol{d}}^{\top} + \rho_2\|\tilde{\omega}\|\boldsymbol{V} - \rho_2\|\tilde{\omega}\|\tilde{\boldsymbol{V}}. \tag{5.26}$$

Similar to the analysis of $\tilde{\boldsymbol{W}}$, it is easy to show that $\tilde{\boldsymbol{V}}$ is bounded. □

Remark 5.1 Generally speaking, when the disturbance models are unknown, it is difficult to decouple the disturbances and faults, which may decrease the accuracy of fault diagnosis. The proposed NDO uses the fitting and memory capabilities of neural networks to fit and memorize on-orbit periodic disturbances, which can decouple part of the disturbances. Here it is assumed that actuator faults do not occur in the NDO training time. A disadvantage of the proposed scheme is that it needs a certain storage space to store health period data.

5.4 Fault Diagnosis Scheme

When disturbances are observed and compensated, it can be considered that the deviation torque is caused by actuator faults. So, fault diagnosis can be performed. As illustrated in Fig. 5.5, several local observers were developed to achieve fault isolation and preliminary fault estimation, and the estimation deviation is compensated by neural networks, so that the accuracy of fault estimation is improved.

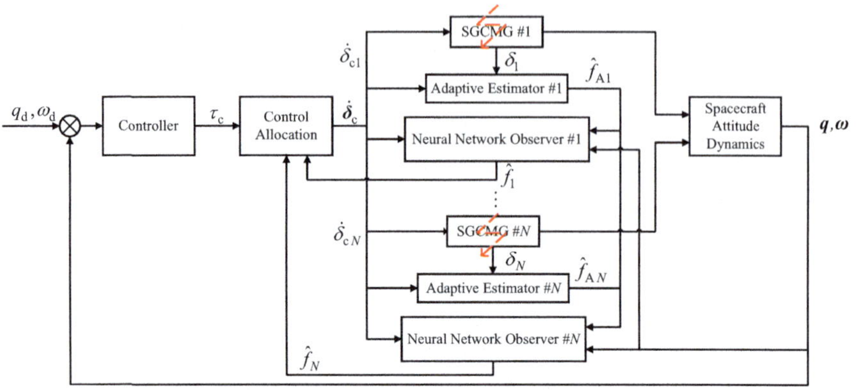

Fig. 5.5 Architecture of fault diagnosis scheme

5.4.1 Fault Diagnosis Using Addaptive Estimator

As shown in (5.4), the fault model of gimble in SGCMG is

$$\dot{\delta}_i = \dot{\delta}_{ci} + f_i, \tag{5.27}$$

where $\dot{\delta}_i$ and $\dot{\delta}_{ci}$ ($i = 1, 2, \ldots, N$, $N > 0$ is the number of gyros) denotes the real and except spinning speed of gimbal respectively, and f_i is the value of fault. Inspired by [10], N separate adaptive fault estimators (AE) are built for gimbals. First, we define an auxiliary variable as

$$\xi_i = f_i - k\delta_i, \tag{5.28}$$

where k is a positive constant. Then, the AE can be established as

$$\dot{\hat{\delta}}_i = \dot{\delta}_{ci} + \alpha\delta_i - \alpha\hat{\delta}_i + \hat{f}_{Ai}, \tag{5.29}$$

$$\dot{\hat{\xi}}_i = -k\dot{\delta}_{ci} - k\hat{\xi}_i - k^2\hat{\delta}_i, \tag{5.30}$$

where α is a positive constant. The estimated result can be written as

$$\hat{f}_{Ai} = \hat{\xi}_i + k\hat{\delta}_i, \tag{5.31}$$

and the observation residual is $r_i = \delta_i - \hat{\delta}_i$ ($i = 1, 2, \ldots, N$).

Assumption 5.4 The value of fault for each SGCMG is differentiable and the derivative is bounded, i.e., $|\dot{f}_i| \leq \bar{f}_i$ where $\bar{f}_i$ is a positive constant.

Lemma 5.1 *For the fault model in* (5.27) *satisfying Assumption 5.4, we use the AE given by* (5.28)–(5.31). *If* $k - \alpha < 0$ *and* $k^4 + 2k^2 - 2\alpha k + \alpha + 1 < 0$, *then the estimated error would be convergent.*

Proof The proof of Lemma 5.1 can be found in [10]. □

5.4.2 Fault Diagnosis Using Neural Network

It is very difficult to achieve rapid and accurate fault diagnosis by simply using the state information of actuators. The proposed scheme combines attitude information to improve the accuracy of fault estimation.

The fault estimation result from Sect. 5.1 can be given as $\hat{\boldsymbol{f}}_A = [\hat{f}_{A1}, \hat{f}_{A2}, \dots, \hat{f}_{AN}]^\top$. According to (5.4) the spacecraft attitude dynamics function under SGCMGs fault can be represented as

$$\dot{\omega} = \boldsymbol{F}(\omega) + \boldsymbol{B}\boldsymbol{D}\dot{\boldsymbol{\delta}}_c - \boldsymbol{B}\boldsymbol{S}(\omega)\boldsymbol{h} + \boldsymbol{B}\boldsymbol{\tau}_d + \boldsymbol{B}\boldsymbol{D}\boldsymbol{f}, \tag{5.32}$$

where $\boldsymbol{D} = -\mathrm{h}_0\boldsymbol{A}_s$, $\dot{\boldsymbol{\delta}}_c = -\boldsymbol{G}(\boldsymbol{\tau}_c - \boldsymbol{S}(\omega)\boldsymbol{h})/\mathrm{h}_0$ ($\boldsymbol{G}$ is the control allocation matrix). For each gimbal in SGCMGs, design a Neural network Observer (NO) as

$$\begin{aligned}\dot{\hat{\omega}} &= \boldsymbol{F}(\hat{\omega}) + \boldsymbol{B}\boldsymbol{D}\dot{\boldsymbol{\delta}}_c - \boldsymbol{B}\boldsymbol{S}(\omega)\boldsymbol{h} \\ &\quad + \boldsymbol{B}\boldsymbol{D}\hat{\boldsymbol{f}}_A + \boldsymbol{B}\boldsymbol{T}_{dNN} + \boldsymbol{B}\boldsymbol{D}[\underbrace{0, \cdots,}_{i-1} \hat{f}_{NNi}, \cdots 0]^\top,\end{aligned} \tag{5.33}$$

where $\boldsymbol{T}_{dNN}$ is the estimated value of disturbance from the NDO, $\hat{f}_{NNi} = \hat{\boldsymbol{W}}_i\sigma(\hat{\boldsymbol{V}}_i\boldsymbol{x})$ denotes the estimation result of the neural network on the diagnosis bias $(f_i - \hat{f}_{Ai})$ of the AE. $\hat{\boldsymbol{W}}_i$ and $\hat{\boldsymbol{V}}_i$ is the estimated weights of the neural network, $\boldsymbol{x} = [\omega^\top, (\dot{\boldsymbol{\delta}}_c + \hat{\boldsymbol{f}}_A)^\top, \boldsymbol{T}_{dNN}^\top]^\top$. The fault diagnosis result is $\hat{\boldsymbol{f}} = \hat{\boldsymbol{f}}_A + \hat{\boldsymbol{f}}_{NN}$.

Theorem 5.2 *Consider the spacecraft attitude dynamics described by (5.32) and the observer model in (5.33). Given Assumptions 5.1–5.4, if the weights of the neural network are updated according to (5.13) and (5.14), then* $\tilde{\omega}$, $\tilde{\boldsymbol{W}}_i$ *and* $\tilde{\boldsymbol{V}}_i$ *are uniformly ultimately bounded, that is, the fault diagnosis error is uniformly ultimately bounded.*

Proof Let $\tilde{\boldsymbol{f}}_A = \boldsymbol{f} - \hat{\boldsymbol{f}}_A$ represent the diagnosis bias of the AE, and then (5.32) can be rewritten as

$$\dot{\omega} = \boldsymbol{F}(\omega) + \boldsymbol{B}\boldsymbol{D}(\dot{\boldsymbol{\delta}}_c + \hat{\boldsymbol{f}}_A) + \boldsymbol{B}\boldsymbol{\tau}_d + \boldsymbol{B}\boldsymbol{D}\tilde{\boldsymbol{f}}_A. \tag{5.34}$$

Further arranging the above equation leads to

$$\dot{\omega} = \boldsymbol{A}\omega + \boldsymbol{g}_f(\omega) + \boldsymbol{B}\boldsymbol{\tau}_d + \boldsymbol{B}\boldsymbol{D}\tilde{\boldsymbol{f}}_A, \tag{5.35}$$

where $\boldsymbol{A}$ is a Hurwitz matrix, and $\boldsymbol{g}_f(\boldsymbol{\omega}) = \boldsymbol{F}(\boldsymbol{\omega}) + \boldsymbol{B}\boldsymbol{D}\dot{\boldsymbol{\delta}}_c - \boldsymbol{B}\boldsymbol{S}(\boldsymbol{\omega})\boldsymbol{h} + \boldsymbol{B}\boldsymbol{D}\hat{\boldsymbol{f}}_A - \boldsymbol{A}\boldsymbol{\omega}$. Let $\tilde{\boldsymbol{f}}_A = [\tilde{f}_{A1}, \tilde{f}_{A2}, \cdots, \tilde{f}_{AN}]^\top$. We decompose (5.35) as

$$\begin{aligned}\dot{\boldsymbol{\omega}} &= \boldsymbol{A}\boldsymbol{\omega} + \boldsymbol{g}_f(\boldsymbol{\omega}) + \boldsymbol{B}\boldsymbol{\tau}_d + \boldsymbol{B}\boldsymbol{D}[\underbrace{\tilde{f}_{A1}, \cdots, 0}_{i-1}, \cdots, \tilde{f}_{AN}]^\top \\ &\quad + \boldsymbol{B}\boldsymbol{D}[\underbrace{0, \cdots, 0}_{i-1}, \tilde{f}_{Ai}, 0, \cdots, 0]^\top.\end{aligned} \tag{5.36}$$

As the neural network can approximate any continuous function, $\tilde{f}_{Ai}$ can be represented by

$$\tilde{f}_{Ai} = \boldsymbol{W}_i\sigma(\boldsymbol{V}_i\boldsymbol{x}) + \varepsilon_i(\boldsymbol{x}), \tag{5.37}$$

where $\varepsilon_i(\boldsymbol{x})$ is approximation error. According to (5.34) and (5.35), (5.33) can be rewritten as

$$\dot{\hat{\boldsymbol{\omega}}} = \boldsymbol{A}\hat{\boldsymbol{\omega}} + \boldsymbol{g}_f(\hat{\boldsymbol{\omega}}) + \boldsymbol{B}\boldsymbol{T}_{dNN} + \boldsymbol{B}\boldsymbol{D}[0, \cdots, 0, \hat{\boldsymbol{W}}_i\sigma(\hat{\boldsymbol{V}}_i\hat{\boldsymbol{x}}), 0, \cdots, 0]^\top. \tag{5.38}$$

Let $\tilde{\boldsymbol{\omega}} = \boldsymbol{\omega} - \hat{\boldsymbol{\omega}}$ and $\tilde{\boldsymbol{W}}_i = \boldsymbol{W}_i - \hat{\boldsymbol{W}}_i$. The error dynamics can be presented by

$$\dot{\tilde{\boldsymbol{\omega}}} = \boldsymbol{A}\tilde{\boldsymbol{\omega}} + \boldsymbol{g}_f(\boldsymbol{\omega}) - \boldsymbol{g}_f(\hat{\boldsymbol{\omega}}) + \boldsymbol{B}\boldsymbol{T} + \boldsymbol{B}\boldsymbol{D}[0, \cdots, 0, Z_i, 0, \cdots, 0]^\top, \tag{5.39}$$

where $Z_i = \tilde{\boldsymbol{W}}_i\sigma(\hat{\boldsymbol{V}}_i\hat{\boldsymbol{x}}) + P_i$, $P_i = \boldsymbol{W}_i[\sigma(\boldsymbol{V}_i\boldsymbol{x}) - \sigma(\hat{\boldsymbol{V}}_i\hat{\boldsymbol{x}})] + \varepsilon_i(\boldsymbol{x})$, and $\boldsymbol{T} = \boldsymbol{\tau}_d - \boldsymbol{T}_{dNN} + \boldsymbol{D}[\tilde{f}_{A1}, \cdots, 0, \cdots, \tilde{f}_{AN}]^\top$. It is noted that $\boldsymbol{T}$ is bounded since both $\boldsymbol{\tau}_d - \boldsymbol{T}_{dNN}$ and $\tilde{f}_{Aj}$ $(j \neq i)$ are bounded. Let us define $\boldsymbol{S}_i = \boldsymbol{B}\boldsymbol{D}[0, \cdots, 0, \sigma(\hat{\boldsymbol{V}}_i\hat{\boldsymbol{x}}), 0, \cdots, 0]^\top$, and $\boldsymbol{\theta}_i = \boldsymbol{B}\boldsymbol{D}[0, \cdots, 0, P_i, 0, \cdots, 0]^\top$. According to the boundedness of sigmoidal function and approximation characteristics of neural network, both $\boldsymbol{S}_i$ and $\boldsymbol{\theta}_i$ are bounded, i.e., $\|\boldsymbol{S}_i\| \leq \bar{S}_i$, $\|\boldsymbol{\theta}_i\| \leq \bar{\theta}_i$. Then, (5.39) can be rewritten as

$$\dot{\tilde{\boldsymbol{\omega}}} = \boldsymbol{A}\tilde{\boldsymbol{\omega}} + \boldsymbol{g}_f(\boldsymbol{\omega}) - \boldsymbol{g}_f(\hat{\boldsymbol{\omega}}) + \boldsymbol{B}\boldsymbol{T} + \tilde{\boldsymbol{W}}_i\boldsymbol{S}_i + \boldsymbol{\theta}_i. \tag{5.40}$$

According to Theorem 5.1, it is easy to conclude that the observer is sable and, therefore, $\tilde{\boldsymbol{\omega}}$, $\hat{\boldsymbol{W}}_i$, and $\hat{\boldsymbol{V}}_i$ are uniformly ultimately bounded. □

Remark 5.2 Note that the number of actuators is greater than the spacecraft attitude dimension, it is thus unable to realize a complete mapping from spacecraft attitude data to control torques deviation. It can be seen from (5.36) that the estimation error of other actuators $\tilde{f}_{Aj}$, $j \neq i$ would reduce the estimation accuracy of the fault diagnosis observer. Therefore, when time-varying faults occur in multiple actuators at the same time (which happens very rarely in actual physical systems), the fault diagnosis accuracy of the NO would lower than AE possibly, in spite of the proposed scheme can ensure that the estimation error is bounded. To avoid this disadvantage, this article assumes that when multiple residuals $r_i = \delta_i - \hat{\delta}_i$ $(i = 1, 2, \ldots, N)$ of AE exceed the threshold, the results from NO observer would be abandoned and only the AE is used to perform fault diagnosis.

5.5 Fault-Tolerant Control

To verify the effectiveness of the fault diagnosis scheme proposed in this chapter, a closed-loop control scheme was formed by the NDO proposed in Sect. 5.3, the NO proposed in Sect. 5.4, and an adaptive sliding mode fault-tolerant controller.

Considering the disturbance estimated value $\boldsymbol{T}_{dNN}$ from NDO, we design the control law $\boldsymbol{\tau}_c$ as

$$\boldsymbol{\tau}_c = \boldsymbol{u}_c - \boldsymbol{T}_{dNN}, \tag{5.41}$$

where $\boldsymbol{u}_c\mathbb{R}^3$ denotes the normal control input.

According to the gimbal fault model in (5.4), the actual gimbal rate $\dot{\boldsymbol{\delta}}$, and the gimbal rate command $\dot{\boldsymbol{\delta}}_c$ have the following relationship:

$$\dot{\boldsymbol{\delta}} = \dot{\boldsymbol{\delta}}_c + \boldsymbol{f}. \tag{5.42}$$

Considering the estimated value of gimbals faults is $\hat{\boldsymbol{f}} = [\hat{f}_1, \hat{f}_2, \cdots, \hat{f}_N]^\top$, to compensate the effect of the faults, the commanded gimbal rate $\dot{\boldsymbol{\delta}}_c$ is replaced by $\dot{\boldsymbol{\delta}}_c^* = \dot{\boldsymbol{\delta}}_c - \hat{\boldsymbol{f}}$. Then, we have

$$\begin{aligned}\boldsymbol{\tau} =& - h_0\boldsymbol{A}_s(\boldsymbol{G}(\boldsymbol{\tau}_c + \boldsymbol{S}(\boldsymbol{\omega})\boldsymbol{h})/h_0 + \boldsymbol{f} - \hat{\boldsymbol{f}}) - \boldsymbol{S}(\boldsymbol{\omega})\boldsymbol{h} \\ =& - h_0\boldsymbol{A}_s(\boldsymbol{f} - \hat{\boldsymbol{f}}) + \boldsymbol{\tau}_c,\end{aligned} \tag{5.43}$$

where $\boldsymbol{G}$ is the control allocation matrix.

Consider the attitude tracking problem and recall the attitude tracking error dynamics described by (2.19) and (2.20). Define the sliding mode variable as $\boldsymbol{s} = \boldsymbol{\omega}_e + \beta_e\boldsymbol{q}_e$, where $\beta_e > 0$ is a design constant. Given this, the open-loop tracking error dynamics can be expressed as

$$\boldsymbol{J}\dot{\boldsymbol{s}} = -\boldsymbol{S}(\boldsymbol{\omega})\boldsymbol{J}\boldsymbol{\omega} + \boldsymbol{J}(\boldsymbol{S}(\boldsymbol{\omega}_e)\boldsymbol{C}\boldsymbol{\omega}_d - \boldsymbol{C}\dot{\boldsymbol{\omega}}_d) + \boldsymbol{\tau} + \boldsymbol{\tau}_d + \beta_e\boldsymbol{J}\dot{\boldsymbol{q}}_{ev}, \tag{5.44}$$

where $\boldsymbol{\tau}$ represents the moment generate by SGCMGs (included the gyroscopic moment produced by the rotation of spacecraft).

According to (5.41) and (5.43), (5.44) can be rewritten as:

$$\begin{aligned}\boldsymbol{J}\dot{\boldsymbol{s}} =& - \boldsymbol{S}(\boldsymbol{\omega})\boldsymbol{J}\boldsymbol{\omega} + \boldsymbol{J}(\boldsymbol{S}(\boldsymbol{\omega}_e)\boldsymbol{C}\boldsymbol{\omega}_d - \boldsymbol{C}\dot{\boldsymbol{\omega}}_d) + \boldsymbol{T}_d - \boldsymbol{T}_{dNN} \\ &+ \boldsymbol{u}_c - h_0\boldsymbol{A}(\boldsymbol{f} - \hat{\boldsymbol{f}}) + \beta_e\boldsymbol{J}\dot{\boldsymbol{q}}_{ev} \\ =& \ \boldsymbol{F} + \boldsymbol{\tau}_d,\end{aligned} \tag{5.45}$$

where $\boldsymbol{F} = -\boldsymbol{S}(\boldsymbol{\omega})\boldsymbol{J}\boldsymbol{\omega} + \boldsymbol{J}(\boldsymbol{S}(\boldsymbol{\omega}_e)\boldsymbol{C}\boldsymbol{\omega}_d - \boldsymbol{C}\dot{\boldsymbol{\omega}}_d) + \boldsymbol{T}_d - \boldsymbol{T}_{dNN} - h_0\boldsymbol{A}(\boldsymbol{f} - \hat{\boldsymbol{f}}) + \beta_e\boldsymbol{J}\dot{\boldsymbol{q}}_{ev}$. By simple algebraic operations [3], one can deduce that

$$\|\boldsymbol{F}\| \le b(1 + \|\boldsymbol{\omega}\| + \|\boldsymbol{\omega}\|^2) = b\Phi, \tag{5.46}$$

where $b > 0$ is an unknown constant, and Φ is a known regressor matrix.

Design the adaptive fault-tolerant control law as:

$$\boldsymbol{u}_c = -k_c \boldsymbol{s} - \frac{\hat{b}\Phi \boldsymbol{s}}{\|\boldsymbol{s}\| + \varepsilon}, \tag{5.47}$$

with

$$\dot{\hat{b}} = \eta \left(\frac{\|\boldsymbol{s}\|^2 \Phi}{\|\boldsymbol{s}\| + \varepsilon} - r\hat{b} \right), \ \varepsilon = \frac{\mu}{1 + \Phi}.$$

Consider the following Lyapunov function candidate:

$$V = \frac{1}{2}\boldsymbol{s}^\top \boldsymbol{J} \boldsymbol{s} + \frac{1}{2\eta}\tilde{b}^2, \tag{5.48}$$

where $\eta > 0$ is a design constant, and $\tilde{b} = b - \hat{b}$ ($\hat{b}$ is the estimate of b) denotes the parameter estimation error. Taking the time derivative of V leads to

$$\dot{V} \leq \|\boldsymbol{s}\| b\Phi + \boldsymbol{s}^\top \boldsymbol{\tau}_d - \frac{1}{\eta}\tilde{b}\dot{\hat{b}}. \tag{5.49}$$

Inserting the control law (5.47) into (5.49), we get

$$\dot{V} \leq -k_c \|\boldsymbol{s}\|^2 + b\mu + r\tilde{b}\hat{b}. \tag{5.50}$$

Using Young's inequality, it can be deduced that

$$\begin{aligned} r\tilde{b}\hat{b} &= -\frac{1}{2}r\tilde{b}^2 + \frac{1}{2}r(b^2 - \hat{b}^2) \\ &\leq -\frac{1}{2}r\tilde{b}^2 + \frac{1}{2}rb^2. \end{aligned} \tag{5.51}$$

It is readily obtained from (5.48) that

$$\begin{aligned} \dot{V} &\leq -k_c \|\boldsymbol{s}\|^2 + b\mu + (-\frac{1}{2}r\tilde{b}^2 + \frac{1}{2}rb^2) \\ &\leq -\lambda_0 V + \varepsilon_0, \end{aligned} \tag{5.52}$$

which implies that the ACS is stable, and V is uniformly ultimately bounded. Consequently, we have $\boldsymbol{s}$, $\boldsymbol{q}_e$, and $\hat{b}$ are uniformly ultimately bounded.

5.6 Numerical Simulation

5.6.1 Disturbances Model

Most of the existing works use the combination of sine and cosine functions to represent the external disturbances. But such a model may deviate from the actual environment [17, 18]. To provide a high-fidelity model, we establish an disturbance model for on-orbit spacecraft. In the following, the gravity gradient torque, aerodynamic damping torque, and sunlight pressure torque are discussed in order.

5.6.1.1 Gravity Gradient Torque

The gravity gradient torque $\boldsymbol{T}_g \in \mathbb{R}^3$ can be expressed as:

$$\boldsymbol{T}_g = \frac{3g_c}{R_c^5}\boldsymbol{R}_c \times \boldsymbol{J}\boldsymbol{R}_c, \tag{5.53}$$

where $\boldsymbol{R}_c$ denotes the vector from the spacecraft to the center of earth, and g_c is the gravitational constant.

5.6.1.2 Aerodynamic Damping Torque

To calculate the aerodynamic damping torque, several basic assumptions are made:

Assumption 5.5 Atmospheric molecules that reach the surface of the spacecraft give all of their momenta to the surface.

Assumption 5.6 The average velocity of atmospheric thermal motion is about 1 km/s when calculated using Maxwell's maximum probability, which is less than the velocity of the spacecraft.

Assumption 5.7 The momentum exchange generated by atmospheric molecules leaving the surface of the spacecraft is omitted.

Based on the above assumptions, for free molecular flow, the following formula can be used to calculate the aerodynamic damping torque $\boldsymbol{T}_a \in \mathbb{R}^3$

$$\boldsymbol{T}_a = \rho_a \boldsymbol{C}_D \boldsymbol{A}_\rho \boldsymbol{l} \times (v\boldsymbol{v}/2), \tag{5.54}$$

$$\boldsymbol{v} = \boldsymbol{v}_a - \boldsymbol{v}_s, \tag{5.55}$$

where ρ_a denotes the density of the atmosphere; $\boldsymbol{C}_D$ gives the drag coefficient; $\boldsymbol{A}_\rho$ represents the area of the oncoming surface and can be obtained by simple geometric operation; $\boldsymbol{l}$ is the vector from the center of mass to the center of air pressure; $\boldsymbol{v}$ denotes the velocity vector of spacecraft relative to incoming flow; while $\boldsymbol{v}_a = \boldsymbol{R}_c \times \omega_0$ and $\boldsymbol{v}_s$ represent the velocity of the air and spacecraft, respectively.

5.6.1.3 Sunlight Pressure Torque

The sunlight pressure torque is expressed as $\boldsymbol{T}_s = \sum_{i=1}^{m} \boldsymbol{l}_i \times \boldsymbol{F}_i$, where m is the number of spacecraft's surfaces, $\boldsymbol{l}_i$ denotes the force arm of each surface relative to the center of mass, and

$$\boldsymbol{F}_i = -\rho_{sun} S_{si} \cos\theta_{si}[(1-\eta)\boldsymbol{L} + 2\eta\cos\theta_{si}\boldsymbol{n}_i], \tag{5.56}$$

where ρ_{sun} is the pressure intensity of light, S_{si} denotes the area of surface, θ_{si} represents the angle between the normal direction $\boldsymbol{n}_i$ of the illuminated surface and the radiation source vector $\boldsymbol{L}$, η gives the reflection coefficient. It is worth noting that the occluded planes do not generate light pressure moment, which includes occluded by earth or other surfaces.

In this chapter, the disturbance model can be written as $\boldsymbol{\tau}_d = \boldsymbol{T}_g + \boldsymbol{T}_a + \boldsymbol{T}_s$.

5.6.2 *Simulation Conditions*

The inertia matrix of the spacecraft is as follows:

$$\boldsymbol{J} = \begin{bmatrix} 20 & 0 & 0.9 \\ 0 & 17 & 0 \\ 0.9 & 0 & 15 \end{bmatrix} \text{kg} \cdot \text{m}^2. \tag{5.57}$$

The apogee height of the spacecraft is 3000 km, the eccentricity $e = 0.1$, the gravity constant $g_c = 398600\,\text{km}^3/\text{s}^3$, the drag coefficient $\boldsymbol{C}_D$ is 2.6, the intensity of sunlight pressure $\rho_{sun} = 4.56 \times 10^{-6}$, the angular velocity of the earth's rotation $\omega_0 = 7.292 \times 10^{-5}$ rad/s, the vector from the center of gravity to the center of air pressure $\boldsymbol{l} = [0.001, 0.002, 0]^\top$ m, and the spacecraft is regarded as a cube with the area of each surface is $S = 5\,\text{m}^2$.

The SGCMGs adopt the pyramid scheme, and the rotation moment of each rotor is 1 Nm. The configuration matrix of the pyramid structure $\boldsymbol{A}_s$ is:

$$\boldsymbol{A}_s = \begin{bmatrix} -\cos\beta\cos\delta_1 & \sin\delta_2 & \cos\beta\cos\delta_3 & -\sin\delta_4 \\ -\sin\delta_1 & -\cos\beta\cos\delta_2 & \sin\delta_3 & \cos\beta\cos\delta_4 \\ \sin\beta\cos\delta_1 & \sin\beta\cos\delta_2 & \sin\beta\cos\delta_3 & \sin\beta\cos\delta_4 \end{bmatrix}, \tag{5.58}$$

where β is the installation angle of the gimbals, which is 53.13° in the pyramid configuration. Here the pseudo-inverse ($\boldsymbol{A}_s^{\dagger} = \boldsymbol{A}_s^{\top}(\boldsymbol{A}_s\boldsymbol{A}_s^{\top})^{-1}$) is adopted as the control allocation method.

The initial attitude and the target attitude of the spacecraft are:

$$\boldsymbol{q}(0) = [0.35, -0.525, -0.70, 0.334]^{\top}, \tag{5.59}$$

$$\boldsymbol{q}(d) = [0, 0, 0, 1]^{\top}. \tag{5.60}$$

The control parameters are taken as: $k_c = 20$, $\beta_e = 0.05$, $\mu = 0, 1$, $\eta = 100$, and $r = 0.0001$. The network parameters are taken as: $\eta_1 = \eta_2 = 20$ and $\rho_1 = \rho_2 = 10^{-6}$.

5.6.3 Simulation of Disturbance Observation Scheme

The training time of the proposed NDO is one orbital period (about 9028 s). Assuming that no fault occurs during this time, the actual and estimated disturbances are shown in Fig. 5.6. As shown in Fig. 5.6, the estimated disturbances are pretty close to the actual ones. After one orbit period, the neural network stops training and turned to offline mode, and the controller uses the estimated value to actively resist the disturbances. It can be seen from Figs. 5.7 and 5.8 that due to the influence of disturbances, the actual attitude will deviate from the expected value $[0, 0, 0, 1]^{\top}$. The use of NDO can suppress the effects of disturbance. Define the attitude error index $\sqrt{\boldsymbol{q}_e^{\top}(t)\boldsymbol{q}_e(t)}$, whose response is depicted in Fig. 5.9. As can be

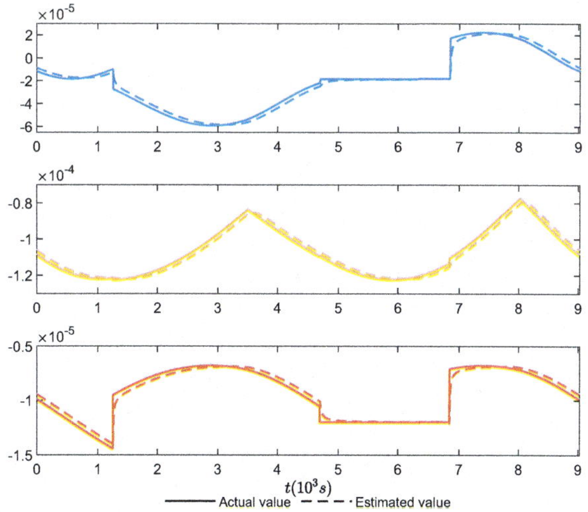

Fig. 5.6 Actual and estimated disturbances for spacecraft in an orbital period

Fig. 5.7 Comparison of attitude tracking error without and with NDO

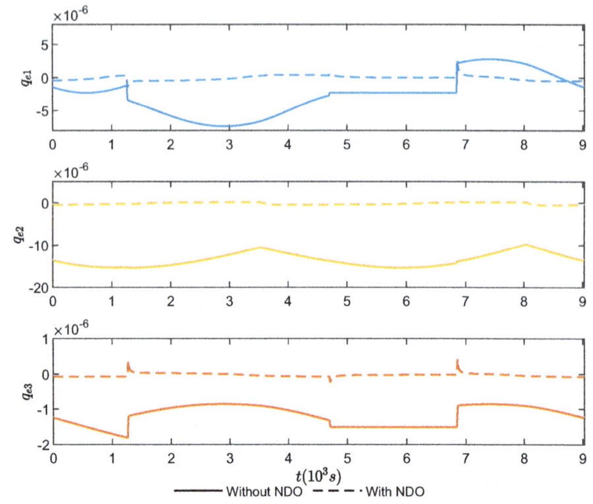

Fig. 5.8 Comparison of attitude stabilization results without and with NDO

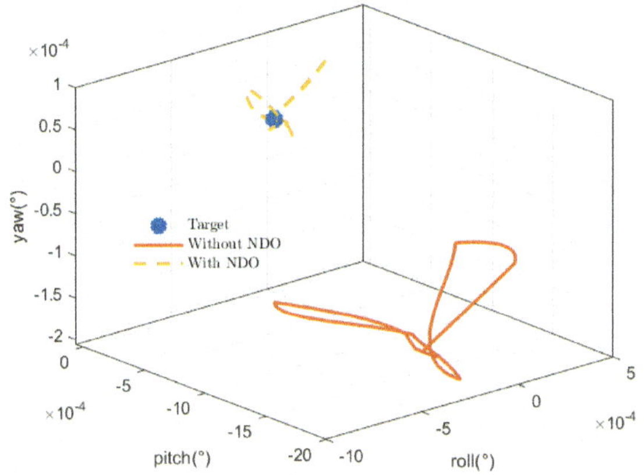

seen, $\int_0^{9028} \sqrt{\boldsymbol{q}_e^\top(t)\boldsymbol{q}_e(t)}dt$ drastically reduces from 5.04×10^{-4} to 1.40×10^{-5} if the NDO is used. Figures 5.6, 5.7, 5.8 and 5.9 show that the NDO proposed in this chapter can effectively estimate external disturbances and improve attitude control accuracy.

5.6.4 Simulation of Fault Diagnosis Scheme

In this simulation, the fault diagnosis results under the proposed scheme, the fault scenario of gimbals is described in Table 5.1. rand(t) is a random number between $[-1, 1]$, which changes every 10 s. The relationship between fault $f_i, i = 1, 2, 3, 4$ and gimbals spinning rate $\dot{\delta}_i, i = 1, 2, 3, 4$ can be seen in (5.27). The actual fault, fault

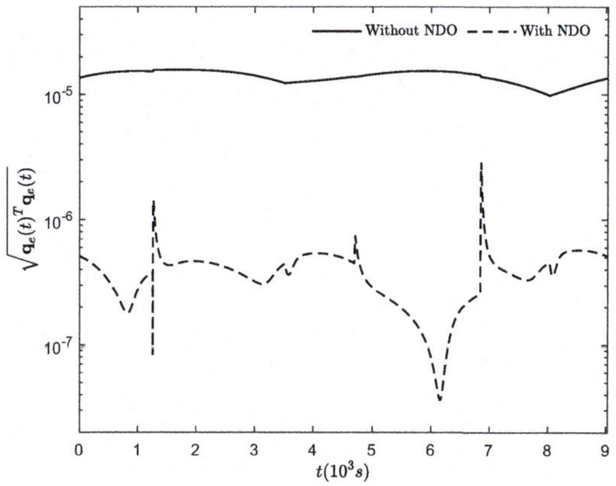

Fig. 5.9 Comparison of $\sqrt{\boldsymbol{q}_e^\top(t)\boldsymbol{q}_e(t)}$ without and with NDO

Table 5.1 SGCMGs fault scenario

Actuator	Fault	
CMG#1	$f_1 = 0.2\sin(3(t-160)) + 0.1\text{rand}(t)$	$160 \le t \le 190$
CMG#2	$f_2 = -0.2\sin(3(t-50)) + 0.1\text{rand}(t)$	$50 \le t \le 80$
CMG#3	$f_3 = -0.2\sin(3(t-100)) + 0.1\text{rand}(t)$	$100 \le t \le 130$
	$f_3 = 0.2 + 0.1\text{rand}(t)$	$150 \le t \le 230$
CMG#4	$f_4 = 0$	

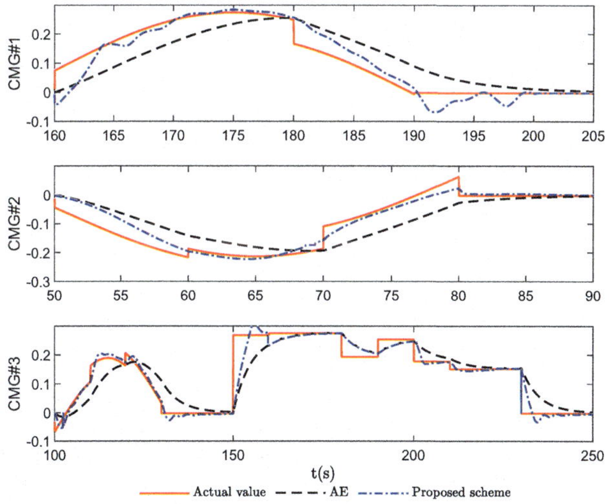

Fig. 5.10 Actual and estimated faults under AE and proposed scheme

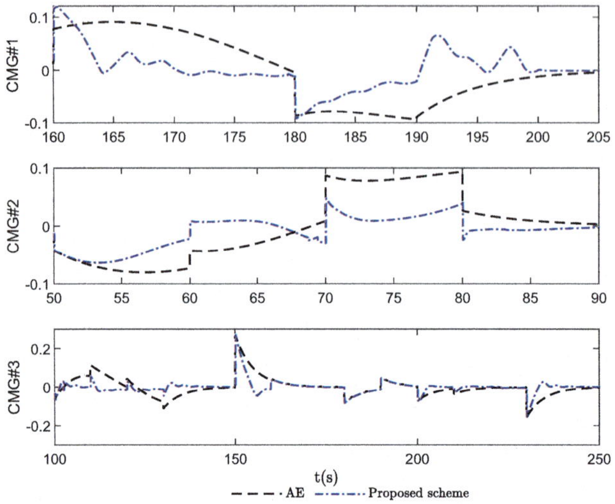

Fig. 5.11 Fault estimation error under AE and proposed scheme

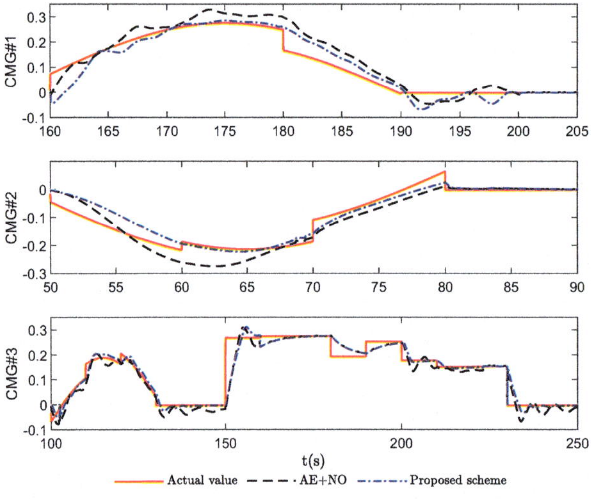

Fig. 5.12 Actual and estimated faults under AE+NO and proposed scheme

estimated by AE and the proposed scheme are shown in Fig. 5.10. The fault estimation error can be seen in Fig. 5.11. As shown in Figs. 5.10 and 5.11, the proposed scheme can improve the accuracy of fault estimation for both sudden and continuous faults.

Define the fault estimation error index $\sum_{i=1}^{4}|f_i - \hat{f}_i|$, and the $\int_0^{300}\sum_{i=1}^{4}|f_i(t) - \hat{f}_i(t)|dt$ for AE and proposed scheme are 9.57 and 4.13, respectively. At the same time, as shown in Figs. 5.12 and 5.13, the proposed NDO can greatly improve the accuracy of fault estimation. Since NDO estimates and compensates the external dis-

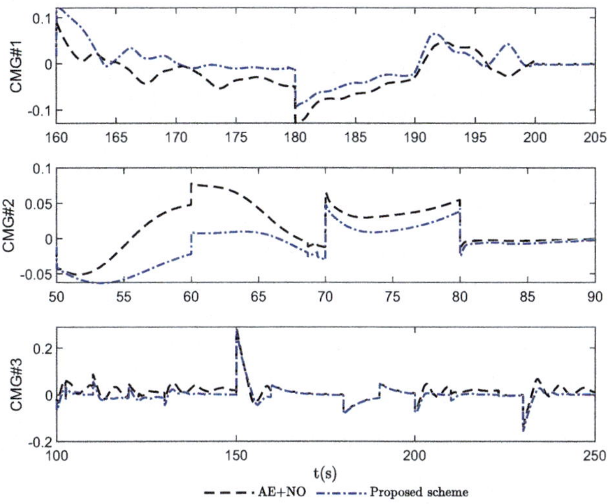

Fig. 5.13 Fault estimation error under AE+NO and proposed scheme

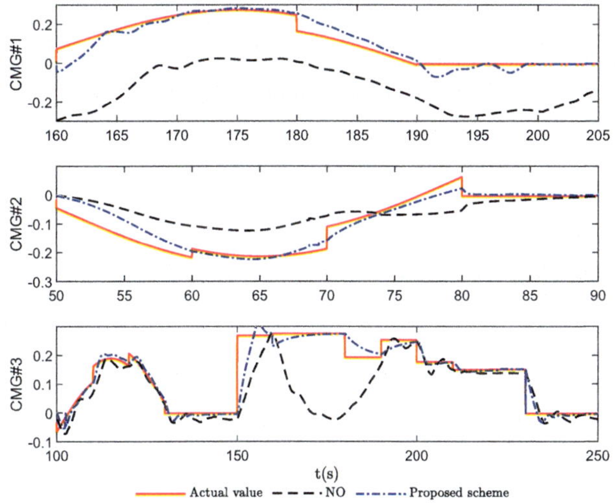

Fig. 5.14 Actual and estimated faults under NO and proposed scheme

turbances, when compared with the fault diagnosis scheme without NDO (AE+NO), the proposed scheme (AE+NO+NDO) performs better. The $\int_0^{300} \sum_{i=1}^{4} |f_i(t) - \hat{f}_i(t)|dt$ for AE+NO and the proposed scheme are 5.95 and 4.13, respectively. The simulation also considers the situation where only NO is used and AE is not applicable (see Fig. 5.14). When multiple actuators occur time-varying faults at the same time, fault isolation cannot be achieved by only using attitude information, so the estimated accuracy is poor. The $\int_0^{300} \sum_{i=1}^{4} |f_i(t) - \hat{f}_i(t)|dt$ for NO is 21.09.

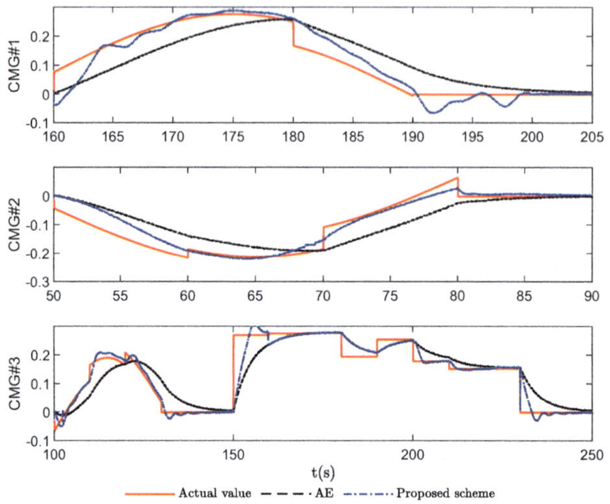

Fig. 5.15 Actual and estimated faults under AE and proposed scheme with noise

The measurement noise of the gimbal angles is considered in Fig. 5.15, and the $\int_0^{300}\sum_{i=1}^4|f_i(t)-\hat{f}_i(t)|dt$ for AE and proposed scheme are 123.94 and 10.61, respectively. The above simulations illustrate that the proposed scheme can achieve higher fault diagnosis accuracy.

5.6.5 *Simulation of Fault-Tolerant Control Scheme*

The following simulation is to prove the efficiency of the adaptive FTC scheme proposed in Sect. 5.5. The fault scenario set in this section is consistent with Sect. 5.6.2 and NDO is applied to all cases for active anti-disturbance. As shown in Fig. 5.16, although the use of sliding mode control (SMC) can ensure the stability of the system, the control accuracy is poor, and the maximum error reaches 0.04 after 100 s. After the use of AE to estimate and compensate the faults, the control accuracy is improved, and the maximum attitude error is about 0.01 after 100s as shown in Fig. 5.17. Using the fault diagnosis scheme proposed in this chapter (see (5.33)), the attitude control error reduces to less than 0.005 after 100 s as shown in Fig. 5.18. Utilize the attitude error index $\sqrt{\boldsymbol{q}_e^\top(t)\boldsymbol{q}_e(t)}$ defined in Sect. 5.6.1, the values of $\int_{100}^{300}\sqrt{\boldsymbol{q}_e^\top(t)\boldsymbol{q}_e(t)}dt$ for SMC, SMC+NO, SMC+AE and SMC+AE+NO are 3.3043, 2.3868, 0.8080 and 0.3282, respectively. The 3D and terminal enlargement diagram for attitude trajectory can be seen in Figs. 5.19 and 5.20, the proposed scheme keeps the spacecraft trajectory within a smaller range due to higher fault diagnosis accuracy.

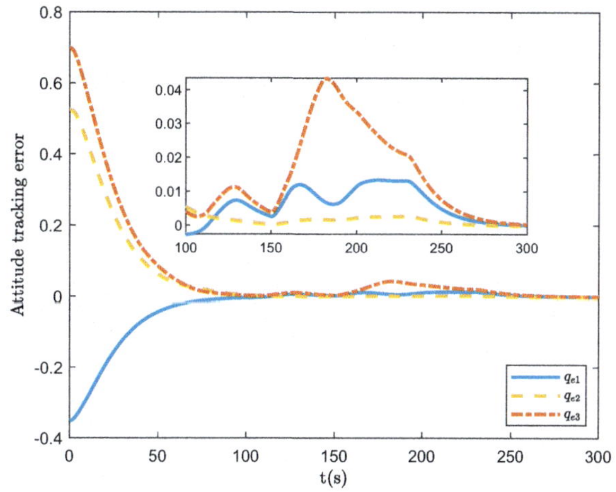

Fig. 5.16 Actual tracking error under SMC

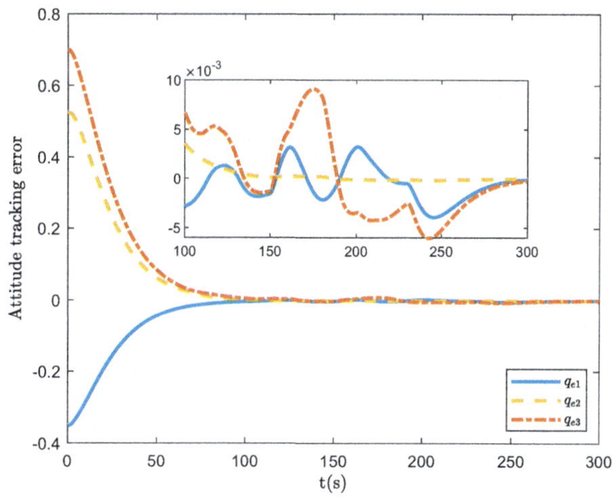

Fig. 5.17 Actual tracking error under SMC+AE

5.7 Summary

In this chapter, the fault diagnosis scheme for SGCMGs with spacecraft in periodic orbit has been studied. The disturbance observation scheme based on neural network proposed in this chapter has the ability of fitting and memory disturbance torques. Using this observer for active anti-interference can improve the attitude control and fault diagnosis performance of the spacecraft. And this scheme takes into account the fact that the fault may occur at any time. Theoretical analysis and simulation proved the effectiveness of the proposed scheme. For spacecraft ACS with redundant actuators, when multiple actuators have time-varying faults at the same time, fault isolation and estimation cannot be achieved only through spacecraft attitude data. In this chapter, several local fault adaptive estimators are used to achieve fault isolation

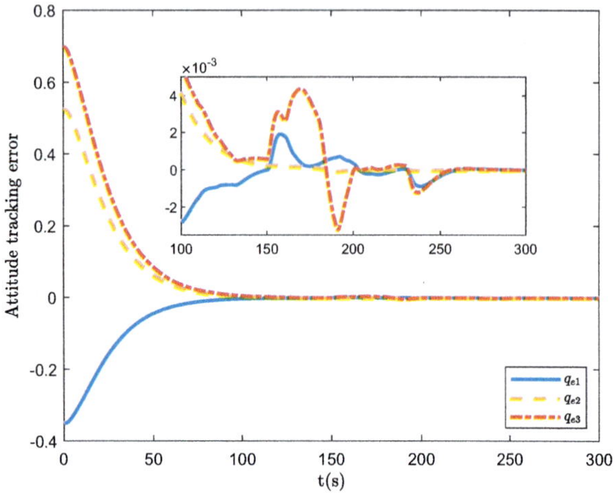

Fig. 5.18 Actual tracking error under proposed scheme

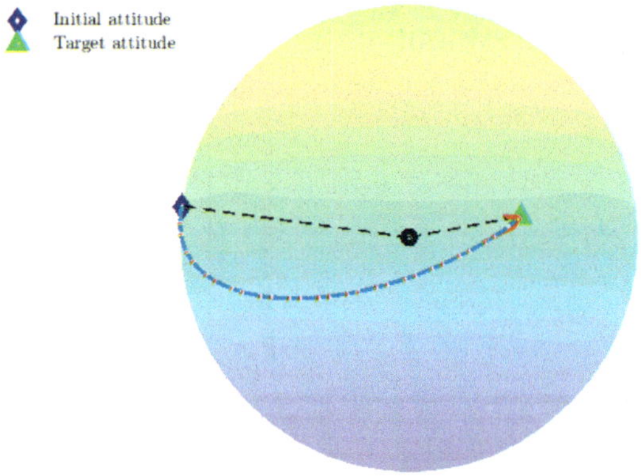

Fig. 5.19 Attitude trajectory of SMC, SMC+NO, SMC+AE and the proposed scheme

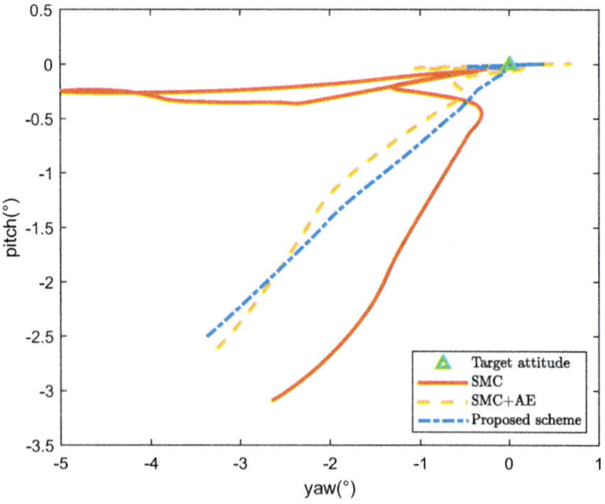

Fig. 5.20 Partial enlarged detail of the attitude trajectory of SMC, SMC+NO, SMC+AE and the proposed scheme

and preliminary fault estimation, and then the neural networks are applied to combine spacecraft attitude data to improve the fault estimation accuracy. Lyapunov method is used to prove the stability of the observer. By using a simple adaptive sliding mode controller for comparative analysis, the FTC scheme proposed in this chapter can effectively improve the ability of spacecraft ACS under SGCMGs fault conditions.

References

1. Cai W, Liao X, Song Y (2008) Indirect robust adaptive fault-tolerant control for attitude tracking of spacecraft. Journal of Guidance, Control, and Dynamics 31(5): 1456–1463
2. Shao X, Hu Q, Shi Y, Jiang B (2018) Fault-tolerant prescribed performance attitude tracking control for spacecraft under input saturation. IEEE Transactions on Control Systems Technology 28(2): 574–582
3. Fan L, Hai H, Zhou K (2020) Robust fault-tolerant attitude control for satellite with multiple uncertainties and actuator faults. Chinese Journal of Aeronautics 33(12): 3380–3394
4. Ma Y, Jiang B, Tao G, Badihi H (2020) Minimum-eigenvalue-based fault-tolerant adaptive dynamic control for spacecraft. Journal of Guidance, Control, and Dynamics 43(9): 1764–1771
5. Qian L, Hao Y, Dong Z, Jiang B (2020) Fault-tolerant control and vibration suppression of flexible spacecraft: An interconnected system approach. Chinese Journal of Aeronautics 33(7): 2014–2023
6. Shen Q, Yue C, Goh CH, Wang D (2018) Active fault-tolerant control system design for spacecraft attitude maneuvers with actuator saturation and faults. IEEE Transactions on Industrial Electronics 66(5): 3763–3772
7. Zhang Y, Jiang J (2008) Bibliographical review on reconfigurable fault-tolerant control systems. Annual Reviews in Control 32(2): 229–252
8. Hu Q, Shao X, Guo L (2017) Adaptive fault-tolerant attitude tracking control of spacecraft with prescribed performance. IEEE/ASME Transactions on Mechatronics 23(1): 331–341
9. Fonod R, Henry D, Charbonnel C, Bornschlegl E, Losa D, Bennani S (2015) Robust fdi for fault-tolerant thrust allocation with application to spacecraft rendezvous. Control Engineering Practice 42: 12–27
10. Shen Q, Yue C, Yu X, Goh CH (2020) Fault modeling, estimation, and fault-tolerant steering logic design for single-gimbal control moment gyro. IEEE Transactions on Control Systems Technology 29(1): 428–435
11. Zhu S, Wang D, Shen Q, Poh EK (2017) Satellite attitude stabilization control with actuator faults. Journal of Guidance, Control, and Dynamics 40(5): 1304–1313
12. Yue C, Shen Q, Cao X, Wang F, Goh CH, Lee TH (2019) Development of a general momentum exchange devices fault model for spacecraft fault-tolerant control system design. arXiv preprint arXiv:1907.06751
13. Farahani HV, Rahimi A (2020) Fault diagnosis of control moment gyroscope using optimized support vector machine. In: 2020 IEEE International Conference on Systems, Man, and Cybernetics (SMC), Toronto, ON, Canada, pp 3111–3116
14. Breiten T, Kunisch K (2021) Neural network based nonlinear observers. Systems & Control Letters 148: 104829
15. Wu Q, Saif M (2005) Neural adaptive observer based fault detection and identification for satellite attitude control systems. In: Proceedings of the American Control Conference, Portland, OR, United states, pp 1054–1059
16. Talebi HA, Khorasani K, Tafazoli S (2008) A recurrent neural-network-based sensor and actuator fault detection and isolation for nonlinear systems with application to the satellite's attitude control subsystem. IEEE Transactions on Neural Networks 20(1): 45–60

17. Talebi HA, Khorasani K (2012) A neural network-based multiplicative actuator fault detection and isolation of nonlinear systems. IEEE Transactions on Control Systems Technology 21(3): 842–851
18. Abbaspour A, Aboutalebi P, Yen KK, Sargolzaei A (2017) Neural adaptive observer-based sensor and actuator fault detection in nonlinear systems: Application in UAV. ISA Transactions 67: 317–329
19. Shen Q, Jiang B, Shi P, Lim CC (2014) Novel neural networks-based fault tolerant control scheme with fault alarm. IEEE Transactions on Cybernetics 44(11): 2190–2201
20. Li Y, Du X, Wan F, Wang X, Yu H (2020) Rotating machinery fault diagnosis based on convolutional neural network and infrared thermal imaging. Chinese Journal of Aeronautics 33(2): 427–438
21. Sun L, Zheng Z (2017) Disturbance-observer-based robust backstepping attitude stabilization of spacecraft under input saturation and measurement uncertainty. IEEE Transactions on Industrial Electronics 64(10): 7994–8002
22. Sun S, Wei X, Zhang H, Karimi HR, Han J (2018) Composite fault-tolerant control with disturbance observer for stochastic systems with multiple disturbances. Journal of the Franklin Institute 355(12): 4897–4915
23. Cheng Y, Wang R, Xu M (2015) A combined model-based and intelligent method for small fault detection and isolation of actuators. IEEE Transactions on Industrial Electronics 63(4): 2403–2413

Chapter 6
Reinforcement Learning-Based Dynamic Control Allocation for Spacecraft Attitude Stabilization

6.1 Introduction

As space missions become more and more complex, higher requirements are placed on the rapid maneuverability of spacecraft. The control moment gyros (CMGs) are more suitable for the rapid maneuvering of large-mass satellites due to the characteristic of moment amplification [1]. In order to ensure the safety and maneuverability of the spacecraft, the CMGs are generally configured redundantly. Since the number of gyroscopes is greater than the spacecraft attitude degrees of freedom, control allocation is required. Moreover, due to the configuration and pointing characteristics of the CMGs, singularity avoidance is required in the process of control allocation [2]. Aiming at the problem of control allocation of redundant actuators, many control allocation algorithms are proposed [2–4].

When the number of actuators is greater than the desired torque direction, there are different control combinations for the actuators to meet the requirements of generating the desired control torques. According to the requirements of control error, energy consumption, torque saturation constraints and other conditions in the actual situation, different optimal control allocation schemes are designed [5–7]. However, most algorithms only regard control allocation as a static problem of searching for the optimal allocation scheme at each time point, while ignoring the dynamic characteristics of the control allocation process. This scheme is difficult to achieve global optimality and requires a large amount of optimization calculations. In contrast, Tjonnas and Johansen [8] proposed an asymptotically optimized dynamic control allocation scheme with search and adaptive capabilities. The authors of [9, 10] established an $\mathcal{H}_\infty$ optimal dynamic allocator by converting the control allocation problem into an $\mathcal{H}_\infty$ optimization problem, and adopted an online reinforcement learning method for the problem that partial differential equations cannot be solved analytically.

Reinforcement learning is an intelligent optimization scheme based on information technology, the main process includes evaluation and optimization [11]. Combining reinforcement learning and neural networks can avoid the curse of dimensionality encountered in solving dynamic programming problems [12]. For the problem

Q. Hu et al., *Intelligent Autonomous Control of Spacecraft with Multiple Constraints*,
https://doi.org/10.1007/978-981-99-0681-9_6

of $\mathcal{H}_\infty$ optimization, Wu and Luo [13] proposed an online simultaneous policy update algorithm based on reinforcement learning, and Modares et al. [10] proposed an Off-Policy integral reinforcement learning algorithm for completely unknown models. In addition, Mu et al. [14] used reinforcement learning to design additional control laws for hypersonic vehicles, which improved the control robustness under external disturbances. Kong et al. [15] designed a robust control law based on reinforcement learning for robots. Dong et al. [16] designed a reinforcement learning control law for spacecraft attitude maneuvers that satisfies state constraints.

This chapter proposed a dynamic control allocation scheme for CMGs based on reinforcement learning. Compared with the scheme in [9], it is suitable for the spacecraft attitude dynamic model and takes into account the avoidance of singularities. Compared with the scheme in [17], the method proposed in this chapter takes into account the dynamic characteristics of control allocation, and reduces the amount of calculation by improving the efficiency of parameter optimization.

The remainder of this chapter is organized as follows. Section 6.2 introduces the problem formulation and the singularity problem of CMGs. The dynamic control allocation scheme are presented in Sect. 6.3. Then, numerical simulations are conducted in Sect. 6.4 to show the effectiveness of the proposed method. Finally, concluding remarks are given in Sect. 6.5.

6.2 Problem Formulation

The attitude dynamics of a fully-actuated spacecraft are given in (2.13) and (2.14). The torques generated by the CMGs are $\boldsymbol{\tau} = -h_0 \boldsymbol{A}_s \dot{\boldsymbol{\delta}} - \boldsymbol{S}(\boldsymbol{\omega})\boldsymbol{h}$ as shown in Sect. 5.2. In this chapter, we consider the pyramidal CMGs (as shown in Fig. 5.3), i.e.,

$$\boldsymbol{A}_s = \begin{bmatrix} -\cos\beta\cos\delta_1 & \sin\delta_2 & \cos\beta\cos\delta_3 & -\sin\delta_4 \\ -\sin\delta_1 & -\cos\beta\cos\delta_2 & \sin\delta_3 & \cos\beta\cos\delta_4 \\ \sin\beta\cos\delta_1 & \sin\beta\cos\delta_2 & \sin\beta\cos\delta_3 & \sin\beta\cos\delta_4 \end{bmatrix}, \tag{6.1}$$

where β is the gyroscope installation angle, and $[\delta_1, \delta_2, \delta_3, \delta_4]$ are the frame positions.

The singular problem of CMGs means under some specific frames angle combination, all the torques generated by the CMGs are in a plane. At this time, no torques perpendicular to the plane can be generated, so it is fall into a singular state. The singular state index is

$$V(\boldsymbol{\delta}) = \frac{1}{\det(\boldsymbol{A}^\top(\boldsymbol{\delta})\boldsymbol{A}(\boldsymbol{\delta}))}. \tag{6.2}$$

The larger the $V(\delta)$ is, the closer it is to the singular state.

The problem to be solved in this chapter is to design the dynamic control allocation law for the spacecraft which uses the CMGs as the actuators of the attitude control system. In the control allocation process, the CMGs must avoid singularity, save energy and do not produce the control torques error.

6.3 Dynamic Control Allocation Scheme

6.3.1 Cost Function

Based on the null-space-based control allocation method in [9], a dynamic control allocation scheme suitable for controlling the moment gyroscope is designed

$$\dot{\boldsymbol{\kappa}}(t) = \boldsymbol{\mu}(t), \tag{6.3}$$

$$\boldsymbol{\delta}(t) = -\frac{\boldsymbol{A}^{\dagger}}{h_0}\boldsymbol{v}_d(t) - \frac{\boldsymbol{A}_{\perp}}{h_0}\boldsymbol{\kappa}(t), \tag{6.4}$$

where the pseudo-inverse of $\boldsymbol{A}$ can be computed as $\boldsymbol{A}^{\dagger} = \boldsymbol{A}^{\top}(\boldsymbol{A}\boldsymbol{A}^{\top})^{-1}$, $\boldsymbol{A}_{\perp} \in \mathbb{R}^{g\times(g-3)}$ represents the basis of the null space of $\boldsymbol{A}$, $\boldsymbol{v}_d$ represents the expected control torque, $\boldsymbol{\kappa} \in \mathbb{R}^{(g-3)}$ is the amount of additional control allocation, and $\boldsymbol{\mu} \in \mathbb{R}^{(g-3)}$ assigns the input volume for the control. Using the null-space-based dynamic control allocation scheme based on formula (6.4), the resulting control torque is $\boldsymbol{v} = -h_0\boldsymbol{A}\dot{\boldsymbol{\delta}} = -h_0\boldsymbol{A}(-\boldsymbol{A}^{\dagger}\boldsymbol{v}_d/h_0 - \boldsymbol{A}_{\perp}\boldsymbol{\omega}/h_0) = \boldsymbol{v}_d$. Therefore, the control error generated by this control allocation scheme is zero, and the control allocation is unidirectionally decoupled from the outer loop control. That is to say, the control allocation does not affect the outer loop control, but the outer loop control will affect the control allocation.

Consider the spacecraft attitude dynamics described by (2.13) and (2.14), and design the sliding mode vector $\boldsymbol{s} = \boldsymbol{\omega} + \beta\boldsymbol{q}_v$, where $\beta > 0$ is a design constant. Taking the time derivative of $\boldsymbol{s}$ along (2.13) and (2.14) yields

$$\dot{\boldsymbol{s}} = -\boldsymbol{J}^{-1}\boldsymbol{S}(\boldsymbol{\omega})\boldsymbol{J}\boldsymbol{\omega} + \boldsymbol{J}^{-1}\boldsymbol{v} + \beta\frac{1}{2}(\boldsymbol{S}(\boldsymbol{q}_v) + q_4\boldsymbol{I}_3)\boldsymbol{\omega}. \tag{6.5}$$

Let $\boldsymbol{H} = -\boldsymbol{J}^{-1}\boldsymbol{S}(\boldsymbol{\omega})\boldsymbol{J}\boldsymbol{\omega} + \beta\frac{1}{2}(\boldsymbol{S}(\boldsymbol{q}_v) + q_4\boldsymbol{I}_3)\boldsymbol{\omega}$, and $\boldsymbol{G} = \boldsymbol{J}^{-1}$. Then, (6.5) can be rewritten as follows:

$$\dot{\boldsymbol{s}} = \boldsymbol{H} + \boldsymbol{G}\boldsymbol{v}, \tag{6.6}$$

where the controlling linearization assumption $\boldsymbol{v} = K\boldsymbol{s}$ is adopted [9].

For the control allocation problem, an augmented variable is set as $\boldsymbol{w} \equiv [\boldsymbol{\kappa}^{\top}, \boldsymbol{v}^{\top}, \boldsymbol{\delta}^{\top}]^{\top}$. In the control distribution system, $\boldsymbol{H}$ is regarded as disturbance, that is, $\boldsymbol{d} \equiv \boldsymbol{H}$. Then the dynamic equation of the augmented system can be represent as

$$\dot{\boldsymbol{\omega}} \equiv \begin{bmatrix} \dot{\boldsymbol{\kappa}} \\ \dot{\boldsymbol{v}} \\ \dot{\boldsymbol{\delta}} \end{bmatrix} \equiv \begin{bmatrix} \boldsymbol{0} & \boldsymbol{0} & \boldsymbol{0} \\ \boldsymbol{0} & \boldsymbol{K}\boldsymbol{G} & \boldsymbol{0} \\ -\boldsymbol{A}_{\perp}/h_0 & -\boldsymbol{A}^{\dagger}/h_0 & \boldsymbol{0} \end{bmatrix} \boldsymbol{\omega} + \begin{bmatrix} \boldsymbol{I} \\ \boldsymbol{0} \\ \boldsymbol{0} \end{bmatrix} \boldsymbol{\mu} + \begin{bmatrix} \boldsymbol{0} \\ \boldsymbol{K} \\ \boldsymbol{0} \end{bmatrix} \boldsymbol{d}. \tag{6.7}$$

Considering energy consumption and singularity avoidance, the cost function can be regarded as:

$$J(\boldsymbol{w}, \boldsymbol{d}, \boldsymbol{\mu}) \equiv \int_t^{\infty} e^{-\alpha(\tau-t)}(\dot{\boldsymbol{\delta}}^{\top} \boldsymbol{Q}_E \dot{\boldsymbol{\delta}} + \boldsymbol{Q}_S V(\boldsymbol{\delta}) + \|\boldsymbol{\mu}\|^2 - \gamma^2 \|\boldsymbol{d}\|^2) d\tau, \tag{6.8}$$

where $\boldsymbol{Q}_E$ stands for energy weight, $\boldsymbol{Q}_S$ represents the singular weight, $V(\cdot)$ is the singular coefficient, and γ is regarded as the interference coefficient. Then, the purpose of control allocation is to design $\boldsymbol{\mu}$ to minimize $J(\boldsymbol{w}, \boldsymbol{d}, \boldsymbol{\mu})$.

Since reinforcement learning is a data-based method, the dynamics based on sliding mode in (6.5) reduces the system variables compared with the traditional spacecraft dynamics model, which is more conducive to data fitting for reinforcement learning. Compared with the method in [9], the augmented system dynamics (6.5) is suitable for the spacecraft attitude dynamics and takes into account the structural characteristics of the CMGs, which provides a basis for singularity avoidance.

6.3.2 *Optimal Manipulation Law Based on Reinforcement Learning*

Formula (6.8) obtains the cost function of the control allocation problem, and then we design the manipulation law $\boldsymbol{\mu}$. Since $\boldsymbol{d}$ is an unknown quantity in the control allocation system, it is necessary to design $\boldsymbol{d}$, which has the greatest influence on the cost function.

The optimal policy and worst disturbance can be defined as

$$\boldsymbol{\mu}^* \equiv \arg\min_{\boldsymbol{\mu}} J(\boldsymbol{w}, \boldsymbol{d}^*, \boldsymbol{\mu}), \ \boldsymbol{d}^* \equiv \arg\max_{\boldsymbol{d}} J(\boldsymbol{w}, \boldsymbol{d}, \boldsymbol{\mu}^*). \tag{6.9}$$

If $\boldsymbol{\mu}^*$ and $\boldsymbol{d}^*$ exist at the same time, they form the saddle point solution of the zero-sum game problem, that is, form the Nash equilibrium. In order to solve (6.9), the Bellman formula is constructed by deriving the cost function (6.8)

$$\begin{aligned} H^{\mu}(\boldsymbol{w}, \boldsymbol{d}, J) \equiv\ & \dot{\boldsymbol{\delta}}^{\top} \boldsymbol{Q}_E \dot{\boldsymbol{\delta}} + \boldsymbol{Q}_S V(\boldsymbol{\delta}) + \|\boldsymbol{\mu}\|^2 - \gamma^2 \|\boldsymbol{d}\|^2 \\ & + \dot{\boldsymbol{w}}^{\top} \nabla_{\boldsymbol{w}} J(\boldsymbol{w}, \boldsymbol{d}, \boldsymbol{\mu}) - \alpha J(\boldsymbol{w}, \boldsymbol{d}, \boldsymbol{\mu}) = 0. \end{aligned} \tag{6.10}$$

With the balance condition in mind, the optimal control and the worst disturbance are expressed as

$$\boldsymbol{\mu}^* = -\frac{1}{2} \boldsymbol{B}_a^{\top} \nabla_{\boldsymbol{w}} J^*(\boldsymbol{w}), \ \boldsymbol{d}^* = \frac{1}{2\gamma^2} \boldsymbol{D}_a^{\top} \nabla_{\boldsymbol{w}} J^*(\boldsymbol{w}), \tag{6.11}$$

where $J^*(\boldsymbol{w})$ is the optimization cost function. Substituting (6.9) into (6.8) gets the Hamilton-Jacobi-Isaacs (HJI) formula

$$H^{\mu^*}(\boldsymbol{w}, \boldsymbol{d}^*, J^*) \equiv \dot{\boldsymbol{\delta}}^{\top} \boldsymbol{Q}_E \dot{\boldsymbol{\delta}} + \boldsymbol{Q}_S V(\boldsymbol{\delta}) + \dot{\boldsymbol{w}}^{\top} \nabla_{\boldsymbol{w}} J^*(\boldsymbol{w}) - \alpha J^*(\boldsymbol{w}) - \frac{1}{4} \nabla_{\boldsymbol{w}}^{\top} J^*(\boldsymbol{w}) \boldsymbol{B}_a \boldsymbol{B}_a^{\top} \nabla_{\boldsymbol{w}} J^*(\boldsymbol{w}) + \frac{1}{4\gamma^2} \nabla_{\boldsymbol{w}}^{\top} J^*(\boldsymbol{w}) \boldsymbol{D}_a \boldsymbol{D}_a^{\top} \nabla_{\boldsymbol{w}} J^*(\boldsymbol{w}) = 0. \tag{6.12}$$

It is noted that $J(\boldsymbol{w})$ in (6.10) is easier to solve than $J^*(\boldsymbol{w})$ in (6.12). However, there are partial differential terms in (6.10), and it is difficult to obtain an analytical solution. To overcome this issue, the reinforcement learning technique is used to obtain the optimal allocation law and the worst disturbance. First, the On-Policy reinforcement learning algorithm for this problem is given. Then, an arbitrary acceptable stability control allocation strategy is set as $\boldsymbol{\mu}_0$. For the control input $\boldsymbol{\mu}_i$ and disturbance strategy $\boldsymbol{d}_i$, (6.13) is used to calculate $J(\boldsymbol{w}, \boldsymbol{d}_i, \boldsymbol{\mu}_i)$.

$$H^{\mu_i}(\boldsymbol{w}, \boldsymbol{d}_i, J_i) \equiv \dot{\boldsymbol{\delta}}^{\top} \boldsymbol{Q}_E \dot{\boldsymbol{\delta}} + \boldsymbol{Q}_S V(\boldsymbol{\delta}) + \|\boldsymbol{\mu}_i\|^2 - \gamma^2 \|\boldsymbol{d}_i\|^2 + \dot{\boldsymbol{w}}^{\top} \nabla_{\boldsymbol{w}} J(\boldsymbol{w}, \boldsymbol{d}_i, \boldsymbol{\mu}_i) - \alpha J(\boldsymbol{w}, \boldsymbol{d}_i, \boldsymbol{\mu}_i) = 0. \tag{6.13}$$

Equation (6.13) is used to update the interference strategy

$$\boldsymbol{d}_{i+1} \equiv \arg\max_{\boldsymbol{d}} J(\boldsymbol{w}, \boldsymbol{d}_i, \boldsymbol{\mu}_i) = \frac{1}{2\gamma^2} \boldsymbol{D}_a^{\top} \nabla_{\boldsymbol{w}} J(\boldsymbol{w}, \boldsymbol{d}_i, \boldsymbol{\mu}_i). \tag{6.14}$$

While (6.14) is used to update the control strategy

$$\boldsymbol{\mu}_{i+1} \equiv \arg\min_{\boldsymbol{\mu}} J(\boldsymbol{w}, \boldsymbol{d}_i, \boldsymbol{\mu}_i) = -\frac{1}{2} \boldsymbol{B}_a^{\top} \nabla_{\boldsymbol{w}} J(\boldsymbol{w}, \boldsymbol{d}_i, \boldsymbol{\mu}_i). \tag{6.15}$$

As pointed out in [13], (6.13) is essentially the same as the Newton iteration, so it converges to the unique solution of the HJI (6.12), then the algorithm is stable. Since the On-Policy algorithm needs to adjust the interference $\boldsymbol{d}$, it does not match the actual situation of the spacecraft. Therefore, an adjustable Off-Policy algorithm reinforcement learning algorithm without interference $\boldsymbol{d}$ is constructed.

Rewrite the system dynamics (6.7) as

$$\dot{\boldsymbol{w}} = \boldsymbol{C}_a \boldsymbol{w} + \boldsymbol{B}_a \boldsymbol{\mu}_i + \boldsymbol{D}_a \boldsymbol{d}_i + \boldsymbol{B}_a(\boldsymbol{\mu} - \boldsymbol{\mu}_i) + \boldsymbol{D}_a(\boldsymbol{d} - \boldsymbol{d}_i). \tag{6.16}$$

According to the kinetic formula, (6.14), and (6.15), $J(\boldsymbol{w}, \boldsymbol{d}_i, \boldsymbol{\mu}_i)$ is derived that

$$\begin{aligned} \nabla_{\boldsymbol{w}} J(\boldsymbol{w}, \boldsymbol{d}_i, \boldsymbol{\mu}_i) &= \nabla_{\boldsymbol{w}}^{\top} J(\boldsymbol{w}, \boldsymbol{d}_i, \boldsymbol{\mu}_i)(\boldsymbol{C}_a \boldsymbol{w} + \boldsymbol{B}_a \boldsymbol{\mu}_i + \boldsymbol{D}_a \boldsymbol{d}_i) \\ &\quad + \nabla_{\boldsymbol{w}}^{\top} J(\boldsymbol{w}, \boldsymbol{d}_i, \boldsymbol{\mu}_i) \boldsymbol{B}_a(\boldsymbol{\mu} - \boldsymbol{\mu}_i) + \nabla_{\boldsymbol{w}}^{\top} J(\boldsymbol{w}, \boldsymbol{d}_i, \boldsymbol{\mu}_i) \boldsymbol{D}_a(\boldsymbol{d} - \boldsymbol{d}_i) \\ &= \alpha J(\boldsymbol{w}, \boldsymbol{d}_i, \boldsymbol{\mu}_i) - \dot{\boldsymbol{\delta}}^{\top} \boldsymbol{Q}_E \dot{\boldsymbol{\delta}} - \boldsymbol{Q}_S V(\boldsymbol{\delta}) - \|\boldsymbol{\mu}^j\|^2 - \gamma^2 \|\boldsymbol{d}^j\|^2 \\ &\quad - 2(\boldsymbol{\mu} - \boldsymbol{\mu}_j)^{\top} \boldsymbol{\mu}_{j+1} + 2\gamma^2 (\boldsymbol{d} - \boldsymbol{d}_j)^{\top}. \end{aligned} \tag{6.17}$$

Then, multiplying both sides of (6.17) by $e^{-\alpha(\tau-t)}$, and integrating the resulting formula, we get

$$\begin{aligned} e^{-\alpha T} J^j(\boldsymbol{w}(t+T)) - J^j(\boldsymbol{w}(t)) = \int_t^{t+T} & e^{-\alpha(\tau-t)}\{-\dot{\boldsymbol{\delta}}^\top \boldsymbol{Q}_E \dot{\boldsymbol{\delta}} - \boldsymbol{Q}_S V(\boldsymbol{\delta}) \\ & - \|\boldsymbol{\mu}^j\|^2 + \gamma^2 \|\boldsymbol{d}^j\|^2 - 2(\boldsymbol{\mu} - \boldsymbol{\mu}^j)^\top \boldsymbol{\mu}^{j+1} \\ & + 2\gamma^2 (\boldsymbol{d} - \boldsymbol{d}^j)^\top \boldsymbol{d}^{j+1}\} d\tau, \end{aligned} \tag{6.18}$$

where $e^{-\alpha(\tau-t)}(\nabla_{\boldsymbol{w}} J(\boldsymbol{w}, \boldsymbol{d}_i, \boldsymbol{\mu}_i) - \alpha J(\boldsymbol{w}, \boldsymbol{d}_i, \boldsymbol{\mu}_i)) = \frac{d}{dt}(e^{-\alpha(\tau-t)} J(\boldsymbol{w}, \boldsymbol{d}_i, \boldsymbol{\mu}_i))$. The reinforcement learning algorithm is summarized in the following:

Algorithm 1 Off-Policy Integral Reinforcement Learning

(0) Initialization: $\boldsymbol{\mu}^0(\boldsymbol{w}), \boldsymbol{d}^0(\boldsymbol{w}), j = 0, i = 0$
(1) Act $\boldsymbol{\mu}, \boldsymbol{d}$ on system (6.7); save data on $[t_i, t_i + T]$; $i \leftarrow i + 1$
(2) After sufficient data is collected, $J^j(\boldsymbol{w}), \boldsymbol{\mu}^{j+1}(\boldsymbol{w}), \boldsymbol{d}^{j+1}(\boldsymbol{w})$ can be obtained by solving formula (6.16); $j \leftarrow j + 1$

The stability proof of the Off-Policy algorithm is given below [10].

Proof The formula (6.18) is derived by T and take the limit to get

$$\begin{aligned} & \lim_{T\to 0} \frac{e^{-\alpha T} J^j(\boldsymbol{w}(t+T)) - J^j(\boldsymbol{w}(t))}{T} \\ & + \lim_{T\to 0} \frac{\int_t^{t+T} e^{-\alpha(\tau-t)}(\dot{\boldsymbol{\delta}}^\top \boldsymbol{Q}_E \dot{\boldsymbol{\delta}} + \boldsymbol{Q}_S V(\boldsymbol{\delta}) + \|\boldsymbol{\mu}^j\|^2 - \gamma^2 \|\boldsymbol{d}^j\|^2) d\tau}{T} \\ & + \lim_{T\to 0} \frac{\int_t^{t+T} e^{-\alpha(\tau-t)}(2(\boldsymbol{\mu} - \boldsymbol{\mu}^j)^\top \boldsymbol{\mu}^{j+1} - 2\gamma^2 (\boldsymbol{d} - \boldsymbol{d}^j)^\top \boldsymbol{d}^{j+1}) d\tau}{T} \\ & = 0. \end{aligned} \tag{6.19}$$

Using L'Hopital's rule, the first term of (6.19) is

$$\begin{aligned} & \lim_{T\to 0} \frac{e^{-\alpha T} J^j(\boldsymbol{w}(t+T)) - J^j(\boldsymbol{w}(t))}{T} \\ & = \lim_{T\to 0} [-\alpha e^{-\alpha T} J^j(\boldsymbol{w}(t+T)) + e^{-\alpha T} \nabla_t J^j(\boldsymbol{w}(t+T))] \\ & = - \alpha J^j(\boldsymbol{w}(t+T)) \\ & \quad + \nabla_{\boldsymbol{w}} J^j(\boldsymbol{w}(t+T))(\boldsymbol{C}_a \boldsymbol{w} + \boldsymbol{B}_a \boldsymbol{\mu}_i + \boldsymbol{D}_a \boldsymbol{d}_i + \boldsymbol{B}_a(\boldsymbol{\mu} - \boldsymbol{\mu}_i) + \boldsymbol{D}_a(\boldsymbol{d} - \boldsymbol{d}_i)). \end{aligned} \tag{6.20}$$

The second and third terms of (6.19) are

$$
\begin{aligned}
&\lim_{T\to 0}\frac{\int_t^{t+T} e^{-\alpha(\tau-t)}(\dot{\delta}^\top Q_E\dot{\delta}+Q_S V(\delta)+\|\mu^j\|^2-\gamma^2\|d^j\|^2)d\tau}{T}\\
&=\dot{\delta}^\top Q_E\dot{\delta}+Q_S V(\delta)+\|\mu^j\|^2-\gamma^2\|d^j\|^2,
\end{aligned}
\tag{6.21}
$$

$$
\begin{aligned}
&\lim_{T\to 0}\frac{\int_t^{t+T} e^{-\alpha(\tau-t)}(2(\mu-\mu^j)^\top\mu^{j+1}-2\gamma^2(d-d^j)^\top d^{j+1})d\tau}{T}\\
&=2(\mu-\mu^j)^\top\mu^{j+1}-2\gamma^2(d-d^j)^\top d^{j+1}.
\end{aligned}
\tag{6.22}
$$

By combining (6.19)–(6.22), one can get

$$
\begin{aligned}
&-\alpha J^j(w(t+T))+\nabla_w J^j(w(t+T))(C_a w+B_a\mu_i+D_a d_i\\
&+B_a(\mu-\mu_i)D_a(d-d_i))+\dot{\delta}^\top Q_E\dot{\delta}+Q_S V(\delta)+\|\mu^j\|^2\\
&-\gamma^2\|d^j\|^2+2(\mu-\mu^j)^\top\mu^{j+1}-2\gamma^2(d-d^j)^\top d^{j+1}=0.
\end{aligned}
\tag{6.23}
$$

By substituting (6.14) and (6.15) into (6.23), Behrman's formula (6.11) can be obtained. Since the On-Policy algorithm and the Off-Policy algorithm have the same Behrman formula and iteration rule, they have the same convergence characteristics. The stability of the Off-Policy algorithm can be known from the stability of the On-Policy algorithm. □

Remark 6.1 The On-Policy algorithm requires the model information to be fully known and the control allocation and disturbance to be adjustable. However, the Off-Policy algorithm needs no model information and only uses historical data, whilst does not have to make the allocation or disturbance adjustable, thus enhancing the engineering applicability.

6.3.3 Parameters Solving Based on Neural Network

To solve (6.16), three neural networks are established to approximate the cost function, control allocation strategy, and disturbance strategy, namely, $J^j(w)$, $\mu^{j+1}(w)$, and $d^{j+1}(w)$.

The data accumulated in the process of control allocation is to train the network. It is assumed that

$$
J^j(w)=W_J^j\sigma_J(w), \tag{6.24}
$$

$$
\mu^j(w)=W_\mu^j\sigma_\mu(w), \tag{6.25}
$$

$$
d^j(w)=W_d^j\sigma_d(w), \tag{6.26}
$$

where the vectors $\sigma_J = [\sigma_{J1}, \cdots, \sigma_{Jl_1}]^\top \in \mathbb{R}^{l_1}$, $\sigma_\mu = [\sigma_{\mu 1}, \cdots, \sigma_{\mu l_2}]^\top \in \mathbb{R}^{l_2}$, and $\sigma_d = [\sigma_{d1}, \cdots, \sigma_{dl_3}]^\top \in \mathbb{R}^{l_3}$ represent the activation functions; $\boldsymbol{W}_J^j \in \mathbb{R}^{1\times l_1}$, $\boldsymbol{W}_\mu^j \in \mathbb{R}^{(n-3)\times l_2}$, $\boldsymbol{W}_d^j \in \mathbb{R}^{3\times l_3}$ represent the networks' weights; and l_1, l_2, $l_3 > 0$ are the numbers of neurons; In the i-th time of Off-Policy integral reinforcement learning, the control allocation, disturbance and cost functions in (6.13) are approximated by (6.23)–(6.26). Then, it follows that

$$\boldsymbol{W}^j \cdot h_i^\top = y_i,\ \boldsymbol{W}^j = [\boldsymbol{W}_J^j, vec(\boldsymbol{W}_\mu^j)^\top, vec(\boldsymbol{W}_d^j)^\top], \tag{6.27}$$

$$h_i = \begin{bmatrix} e^{-\alpha T}\sigma_J(\boldsymbol{w}(t+T) - \sigma_J(\boldsymbol{w}(t))) \\ 2\int_t^{t+T} e^{-\alpha(\tau-t)}\left\{(\boldsymbol{\mu} - \boldsymbol{\mu}^{\boldsymbol{W}_\mu^j}) \otimes \sigma_\mu(\boldsymbol{w})\right\} d\tau \\ -2\gamma^2\int_t^{t+T} e^{-\alpha(\tau-t)}\left\{(\boldsymbol{d} - \boldsymbol{d}^{\boldsymbol{W}_\mu^j}) \otimes \sigma_d(\boldsymbol{w})\right\} d\tau \end{bmatrix}, \tag{6.28}$$

$$y_i = 2\int_t^{t+T} e^{-\alpha(\tau-t)}\left\{-\dot{\boldsymbol{\delta}}^\top \boldsymbol{Q}_E \dot{\boldsymbol{\delta}} - \boldsymbol{Q}_S V(\boldsymbol{\delta}) - \|\boldsymbol{\mu}^j\|^2 + \gamma^2\|\boldsymbol{d}^j\|^2\right\} d\tau, \tag{6.29}$$

where $\otimes$ is Kronecker product, $vec(\cdot)$ is the heap of matrix columns. Data needs to be collected at least $l_1 + (3-m)l_2 + 3l_3$ times to solve the Off-Policy algorithm. Using the least square method, we can get the neural network weights as

$$\boldsymbol{W}^j = (\boldsymbol{H}\boldsymbol{H}^\top)^{-1},\ \boldsymbol{H} \equiv [h_0, \cdots, h_{N-1}],\ \boldsymbol{Y} \equiv [y_0, \ldots, y_{N-1}]^\top. \tag{6.30}$$

6.4 Simulation

The inertia matrix of spacecraft is given by

$$\boldsymbol{J} = \begin{bmatrix} 200 & 0 & 9 \\ 0 & 170 & 0 \\ 9 & 0 & 150 \end{bmatrix} \text{kg} \cdot \text{m}^2. \tag{6.31}$$

The initial attitude of the spacecraft is $\boldsymbol{q}(0) = [-0.5, 0.7, 0.2, 0.469]^\top$, the desired attitude is $\boldsymbol{q}_d = [0, 0, 0, 1]^\top$, and the initial angular velocity is $\boldsymbol{0}$ rad/s. The control parameters are taken as $\beta = 0.1$ and $\boldsymbol{K} = \{-10, -10, -10\}$.

6.4.1 *Simulation of Singularity Problem*

Firstly, the singular configuration problem is illustrated. Singular configuration means that when the CMGs in a specific structure, the control torque cannot be generated in a certain direction. If the moment of inertia of the flywheel is set to

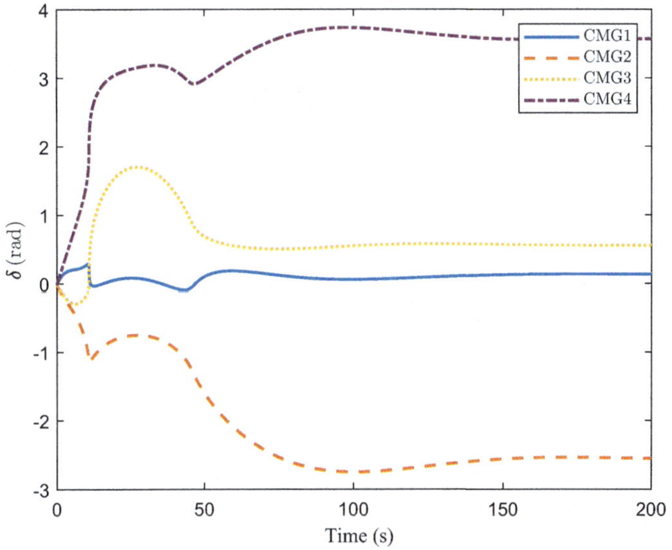

Fig. 6.1 Frame angular positions

5Nm, the pseudo-inverse control allocation may result in singularity in this case. As shown in Figs. 6.1, 6.2 and 6.3, the extreme value of the singularity index (more than 600) appears at about 10 s, indicating that the CMGs fall into a singular state. At this time, the gimbal angular velocity also has an extreme value, which may exceed the usable capability of the gyro gimbal. When the CMGs are trapped in a singular state, the traditional method is to make the gyro escape from the singular state by additional gimbal rotation operations, but if the singularity avoidance is considered in the control allocation process, the CMGs can try to control the singular state and keep in a more appropriate configuration. In practical engineering, the gimbal angular velocity must be limited by the level of physical components. As shown in Figs. 6.4, 6.5 and 6.6, if the angular velocity is limited to 10°/s, the gimbal speed will saturate, which will have an impact on the safe and stable operation of the spacecraft.

6.4.2 Simulation of Dynamic Control Allocation

To compare the existing pseudo inversion control allocation method and the proposed reinforcement learning based control allocation method, the inertia moment of the flywheel in the CMGs is set to 7.5 Nm. Let the activation functions be $\sigma_J = \sigma_\mu = \sigma_d = [\omega, \omega^2, s_1, s_2, s_3, s_1 s_2, s_1 s_3, s_2 s_3, V, V^2]$, where $\boldsymbol{s} = [s_1, s_2, s_3]^\top$ is the sliding mode vector, and V is the singularity index. The data generation time is 20 s, and the integration interval is 0.1 s. The neural network is trained during this period and $\omega = 3^{-0.005t}(\sin(0.025t) + 2\sin(0.05t))$.

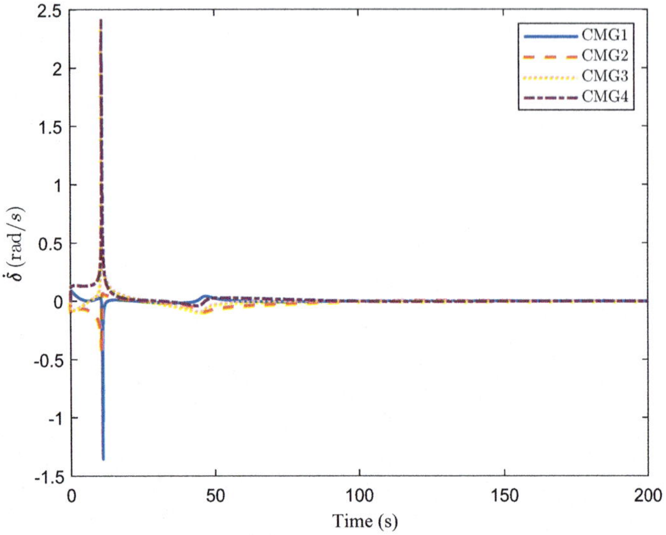

Fig. 6.2 Frame angular velocities

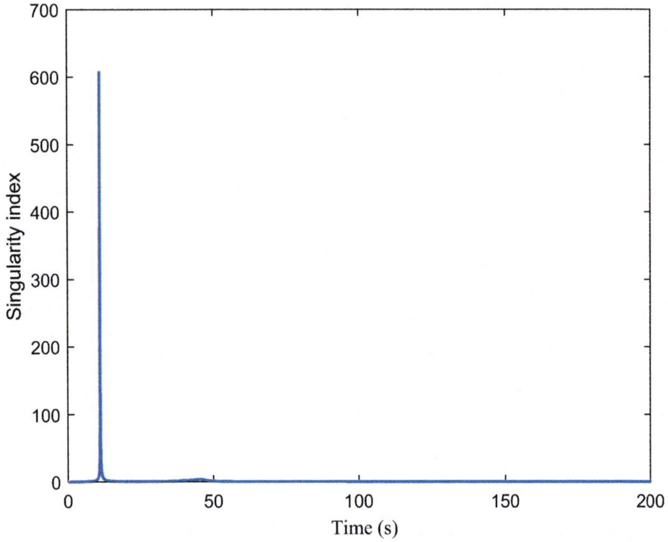

Fig. 6.3 Singularity index

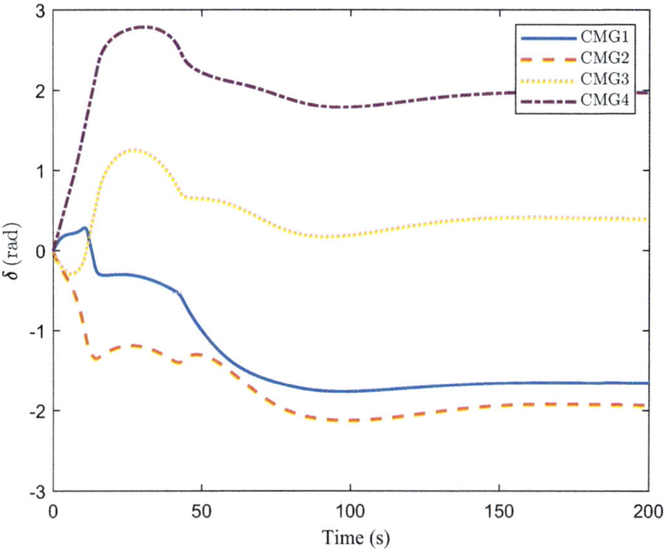

Fig. 6.4 Frame angular positions in the presence of angular velocity limits

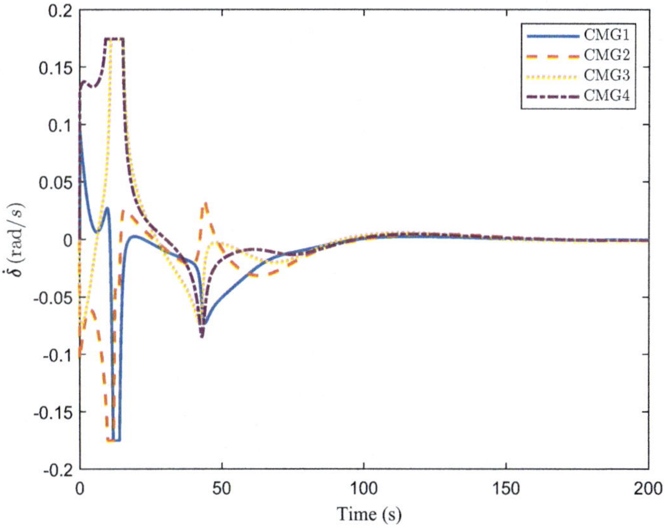

Fig. 6.5 Frame angular velocities in the presence of angular velocity limits

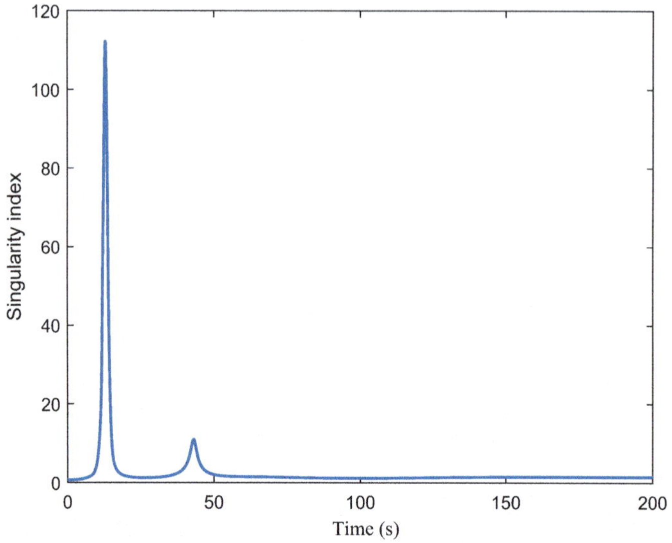

Fig. 6.6 Singularity index in the presence of angular velocity limits

The gimbal angular velocity and position results of pseudo inverse control allocation are shown in Figs. 6.7 and 6.8. The pseudo inverse allocation method is the energy optimal allocation law, but it does not consider singularity avoidance. The results under the proposed reinforcement-learning-based control allocation scheme is shown in Figs. 6.9 and 6.10. By comparing Fig. 6.11, it can be seen that the control allocation scheme based on reinforcement learning has carried out greater manipulation on the CMGs, and the change of gimbal angle is greater, which will inevitably result in greater energy consumption. Singularity index and energy consumption index are shown in Figs. 6.12 and 6.13. The neural network weights are shown in Fig. 6.14, and it is stable after 5 iterations. The comparison of singularity index is shown in Fig. 6.15. The control allocation method based on reinforcement learning takes singularity avoidance into the cost function, so the singularity index is lower, which reduces the risk of losing the control ability of the spacecraft due to the configuration singularity of the CMGs. The cost is that the energy consumption index is higher, but considering that the CMGs does not consume fuel, and the onboard power is sustainable, it is worth to use greater energy consumption to gain a safer control space.

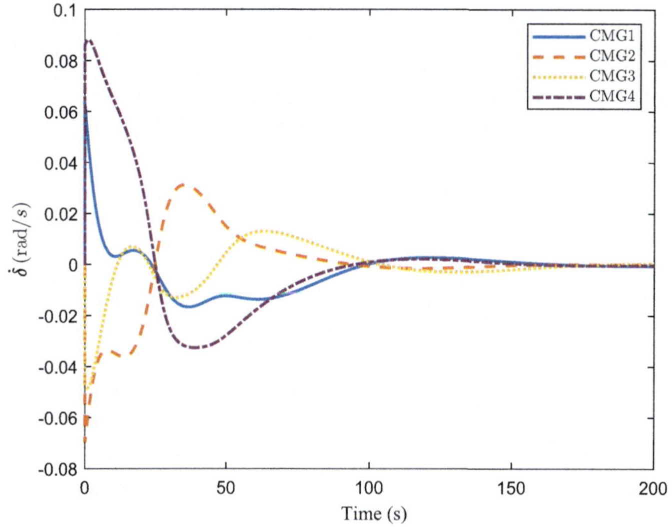

Fig. 6.7 Gimbal angular velocities under pseudo inverse based control allocation

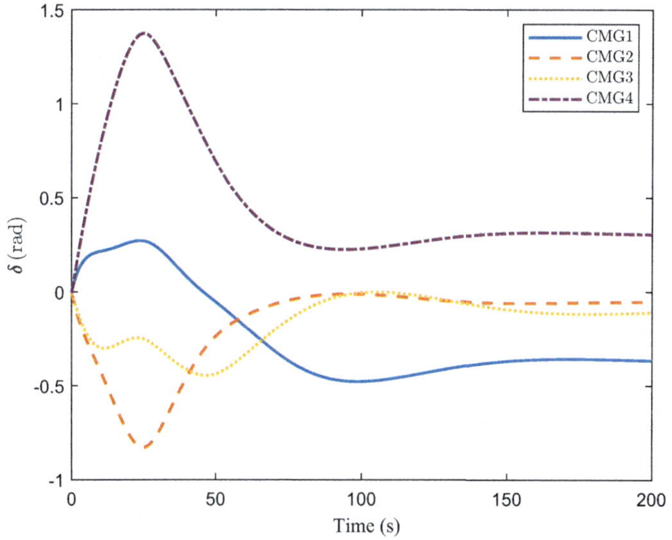

Fig. 6.8 Gimbal angular positions under pseudo inverse based control allocation

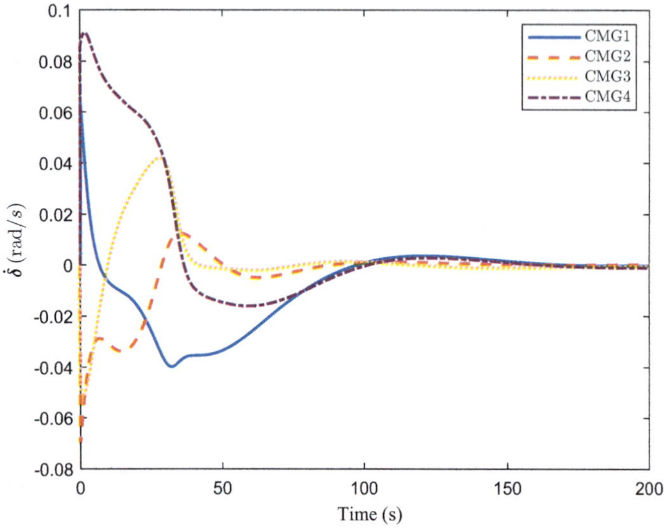

Fig. 6.9 Gimbal angular velocities under reinforcement learning based control allocation

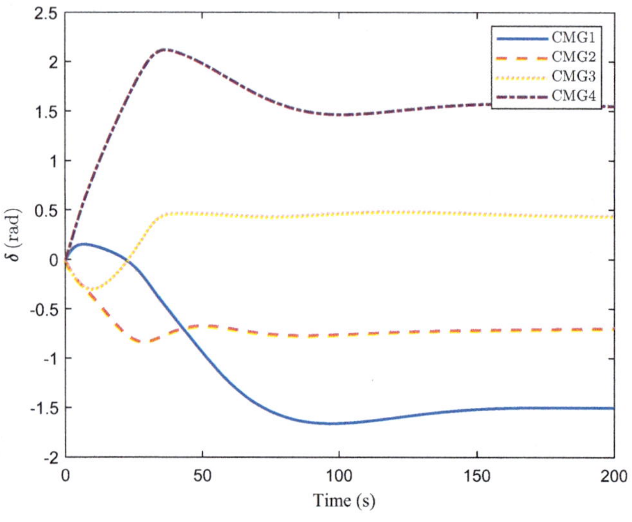

Fig. 6.10 Gimbal angular positions under reinforcement learning based control allocation

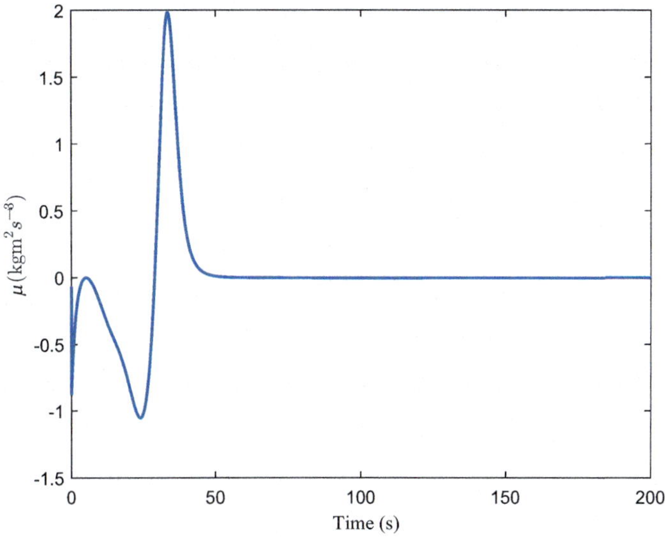

Fig. 6.11 Additional input to control allocation

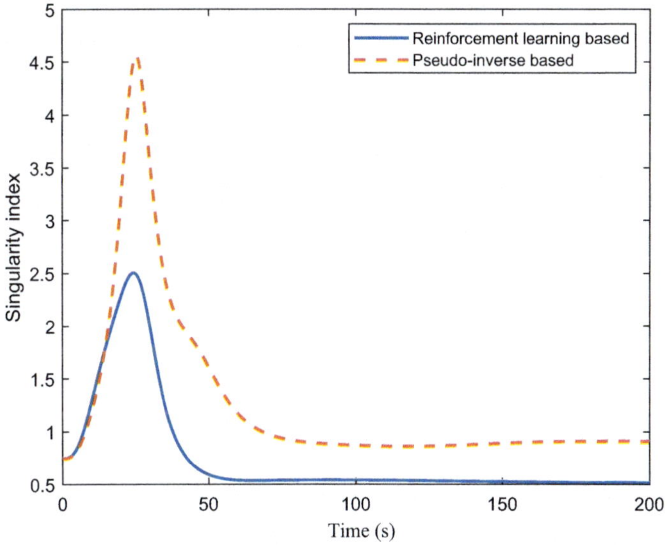

Fig. 6.12 Singularity index

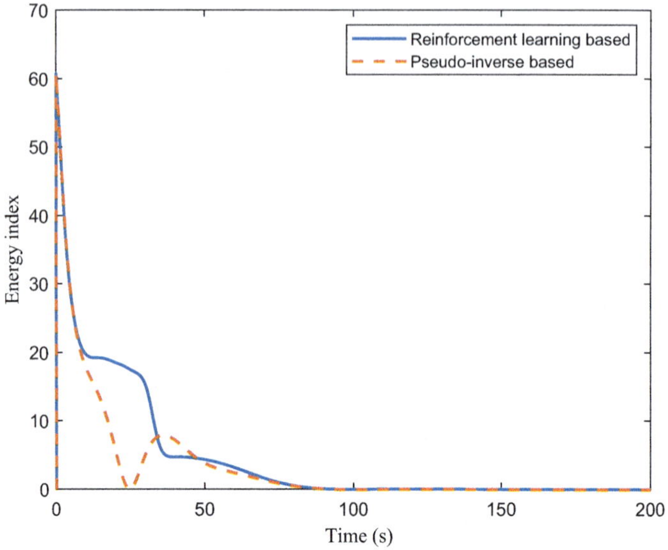

Fig. 6.13 Energy index

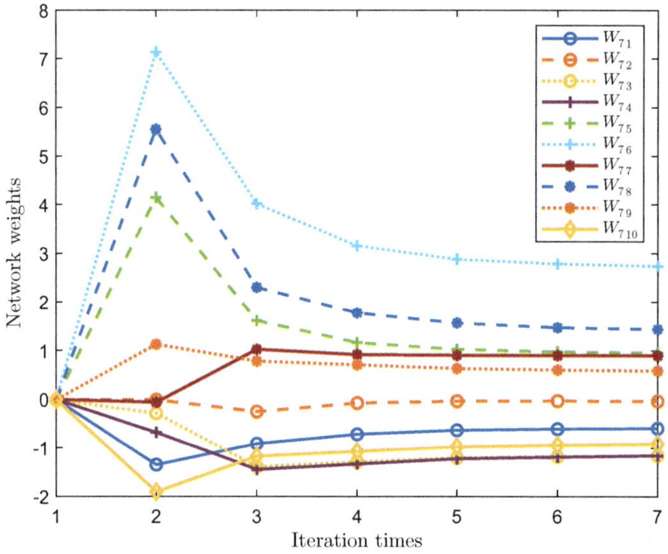

Fig. 6.14 Neural network weights

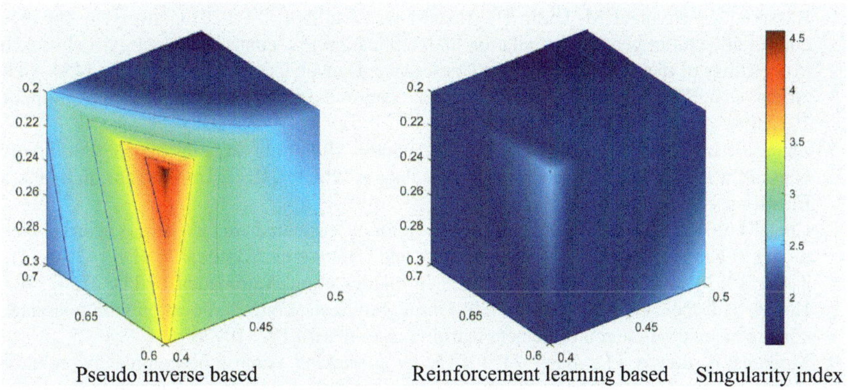

Fig. 6.15 Comparison of singularity index

6.5 Summary

In this chapter, a dynamic control allocation method based on reinforcement learning is proposed to address the singularity avoidance and energy saving problem for attitude stabilization of spacecraft using CMGs. First, a control allocation scheme is designed to decouple the external loop control based on the null-space method, so no control error is introduced into the control allocation problem. Then, the spacecraft attitude dynamics is rewritten using sliding mode variables to reduce system order. The CMGs and attitude dynamics are modeled as an augmented system using the control linearization assumption, and the control allocation is transformed into a dynamic problem. The cost function is constructed and transformed into the Bellman equation. As it is difficult to obtain the analytical solution of the partial differential equation, an integral reinforcement learning algorithm based on Off Policy strategy is designed to estimate the parameters. The algorithm does not require system model and the adjustability of disturbance. The simulation results show the effectiveness of the algorithm proposed in this chapter.

References

1. Hill DE (2016) Dynamics and control of spacecraft using control moment gyros with friction compensation. Journal of Guidance, Control, and Dynamics 39(10): 2406–2418
2. Leve FA (2013) Evaluation of steering algorithm optimality for single-gimbal control moment gyroscopes. IEEE Transactions on Control Systems Technology 22(3): 1130–1134
3. de Vries PS, Van Kampen EJ (2019) Reinforcement learning-based control allocation for the innovative control effectors aircraft. In: AIAA Scitech 2019 Forum, San Diego, CA, United states, p 0144

4. Kassarian E, Rognant M, Evain H, Alazard D, Chauffaut C (2020) Convergent ekf-based control allocation: general formulation and application to a control moment gyro cluster. In: Proceedings of the American Control Conference, Denver, CO, United states, pp 4454–4459
5. Argha A, Su SW, Celler BG (2019) Control allocation-based fault tolerant control. Automatica 103: 408–417
6. Zhang D, Liu G, Zhou H, Zhao W (2018) Adaptive sliding mode fault-tolerant coordination control for four-wheel independently driven electric vehicles. IEEE Transactions on Industrial Electronics 65(11): 9090–9100
7. Chen M (2015) Constrained control allocation for overactuated aircraft using a neurodynamic model. IEEE Transactions on Systems, Man, and Cybernetics: Systems 46(12): 1630–1641
8. Tjønnås J, Johansen TA (2008) Adaptive control allocation. Automatica 44(11): 2754–2765
9. Kolaric P, Lopez VG, Lewis FL (2020) Optimal dynamic control allocation with guaranteed constraints and online reinforcement learning. Automatica 122: 109265
10. Modares H, Lewis FL, Jiang ZP (2015) H_∞ tracking control of completely unknown continuous-time systems via off-policy reinforcement learning. IEEE Transactions on Neural Networks and Learning Systems 26(10): 2550–2562
11. Wang D, He H, Liu D (2017) Adaptive critic nonlinear robust control: A survey. IEEE Transactions on Cybernetics 47(10): 3429–3451
12. Wei C, Luo J, Dai H, Duan G (2018) Learning-based adaptive attitude control of spacecraft formation with guaranteed prescribed performance. IEEE Transactions on Cybernetics 49(11): 4004–4016
13. Wu HN, Luo B (2012) Neural network based online simultaneous policy update algorithm for solving the hji equation in nonlinear H_∞ control. IEEE Transactions on Neural Networks and Learning Systems 23(12): 1884–1895
14. Mu C, Ni Z, Sun C, He H (2016) Air-breathing hypersonic vehicle tracking control based on adaptive dynamic programming. IEEE Transactions on Neural Networks and Learning Systems 28(3): 584–598
15. Kong L, He W, Yang C, Sun C (2020) Robust neurooptimal control for a robot via adaptive dynamic programming. IEEE Transactions on Neural Networks and Learning Systems 32(6): 2584–2594
16. Dong H, Zhao X, Yang H (2020) Reinforcement learning-based approximate optimal control for attitude reorientation under state constraints. IEEE Transactions on Control Systems Technology 29(4): 1664–1673
17. Hu Q, Tan X (2019) Dynamic near-optimal control allocation for spacecraft attitude control using a hybrid configuration of actuators. IEEE Transactions on Aerospace and Electronic Systems 56(2): 1430–1443

Chapter 7
Learning-Based Adaptive Optimal Event-Triggered Control for Spacecraft Formation Flying

7.1 Introduction

The tremendous advances in aerospace technology and optical communication boom the launch/deployment of high-reliable, flexible, and less-cost multi-satellite platform applications, ranging from constellations to spacecraft formation flying (SFF) missions [1–3]. Especially for the SFF missions, such as Earth observation and environment monitoring, a graceful capability of maneuver-maintenance on some specified spatial configurations is an essential prerequisite for mission execution. For instance, TanDEM-X, a radar satellite developed by German Aerospace Center (DLR) [4], has been orbiting Earth, while keeping a pre-appointed formation flying with its "twin" satellite TerraSAR-X, aiming to accurately acquire a global digital elevation model. As a result, designing efficient relative motion coordinated control schemes for the SFF system exhibits significant research value, and has been invoking an ever-growing attention over the past decade.

Up to now, a good deal of relative motion coordinated control schemes for the SFF system have been proposed in literature (e.g., [5–8] and references therein). To list a few, a discrete-time formation tracking protocol was developed in [5] to handle communication delays existing in the leader-follower SFF system. In [6], four different cooperative control protocols were presented for the inner-formation spacecraft system subject to composite external disturbances including solar rational pressure, atmospheric drag, and J2 perturbation. However, aforementioned works [5–8] rely on continuous or periodic communication/computation among multiple spacecraft, leading to heavy computation cost and storage burdens, especially when the number of spacecraft grows or the SFF system approaches to its stable-state phase. In this case, event-triggered control (ETC) method has been reported [9] to handle such issues so that the information transmission or control command update of the SFF system occurs only when certain triggering condition is satisfied [10–12]. Moreover, it should be pointed out that Zeno phenomenon, an infinite number of triggering occurring in a finite time period, is contradictory to the intention of ETC, thus needing to be excluded strictly. For multiple Euler-Lagrange systems with limited

Q. Hu et al., *Intelligent Autonomous Control of Spacecraft with Multiple Constraints*,
https://doi.org/10.1007/978-981-99-0681-9_7

communication rate, an event-based formation containment control method was proposed in [10], where a six-leader-four-follower SFF example was presented to demonstrate the effectiveness of the designed scheme. Later, the triggering mechanism proposed in [10] was extended in [11], and an event-triggered adaptive coordination control approach was developed, guaranteeing the formation-keeping maneuvers of the SFF system in case of parameter uncertainties and external disturbances. Subsequently, a novel event-triggered coordinated control law independent of the neighbors' velocity information was put forward in [12], where both the inter-satellite information transmission and control command update are activated intermittently. It is remarkable, however, that the above-mentioned results [10–12] cannot guarantee the system optimality.

To minimize the energy expenditure or enhance the system performance, the optimal control problem of the SFF system has been formulated in [13–15]. In particular, benefiting from the reinforcement-learning-based actor-critic architecture, the adaptive dynamic programming (ADP) technique without expensive computation cost [16, 17], offers a new point of view on optimal control and has been utilized to stabilize multiple spacecraft. In [18], an optimal impulsive control algorithm was presented in conjunction with an off-line parameter tuning strategy to accomplish the preset-time orbital maneuver in a fuel-optimal manner. In [19] and [20], the prescribed performance control technique was first adopted to characterize the preset transient and steady-state performances of the SFF system. Then, under the actor-critic neural network (NN) structure, the attitude coordinated control schemes were designed in [19, 20] for the SFF system with unknown uncertainties, respectively. It is worth noting that although the works in [18–20] considered the energy cost optimal control problem, the unnecessary resources consumption issue still exists, which restricts the usage of these works in the practical engineering applications to a certain extent.

Motivated by the aforementioned discussions, the first question naturally arises: *can we develop an event-triggered optimal control protocol based on ADP technique for the SFF system?* In this respect, an event-triggered Hamilton-Jacobi-Bellman (HJB) equation was first derived in [21] for the continuous-time nonlinear systems, where the critic and actor NNs are introduced to estimate the cost function and the optimal event-based controller, respectively. Combining a feedforward-based identifier network, an event-driven neuron control scheme was developed in [22]. After that, an event-sampled integral reinforcement learning algorithm was presented in [23] for the partially unknown continuous nonlinear systems without identifying the drift dynamics, and the static and dynamic triggering rules were developed to reduce the controller update frequency, respectively. Similarly, with the aid of the integral reinforcement learning algorithm, the result in [23] was further extended to the case of external disturbances. Besides, the event-triggered optimal control problem under the ADP framework was also sufficiently addressed for the discrete-time nonlinear systems [24, 25]. However, it should be pointed out that two open problems still exist in the aforementioned literature [21–25]. On the one hand, the event-triggered strategy heavily relies on the Lipschitz continuous assumption on the controller (like [21, 22, 25]) or on the closed-loop system (like [21, 23, 24]). In this sense, it is extremely challenging to precisely determine the triggering condition due to the presence of unknown Lipschitz constant. On the other hand, on account of the limitation on the triggering parameter selection, the Hamiltonian under the event-based optimal controller may not be strictly equal to zero, thus it cannot be directly ignored in the Bellman error (see [21, 22]). In light of this, the purpose of this chapter is to find a feasible event-based optimal control scheme by resorting to the ADP approach to achieve the formation tracking control of the leader-follower SFF system. The main innovations are three-fold:

- A Zeno-free event-triggered near optimal tracking control protocol is proposed for the SFF system, which not only ensures the controller being updated intermittently but also preserves the optimal performance of the closed-loop system as far as possible.
- To remove the dependence on unknown Lipschitz constants of the ETC method in [21–25], we propose a new event-triggered mechanism only hinging on the system states and user-defined design parameters.
- By introducing an adaptive projection rule for the weight update of the critic NN, the weight estimate error is shown to be bounded ultimately, which contributes to the rigor of closed-loop stability analysis

The rest of this chapter is stated as follows. In Sect. 7.2, some preliminary knowledge on formation system dynamics is introduced, and the optimal control problem is formulated. The near optimal tracking controller and the event-triggered mechanism are developed in Sect. 7.3, along with the rigorous stability analysis and Zeno-free behavior validation. Section 7.4 provides the numerical simulation results to examine the theoretical scheme. Finally, some conclusive remarks are included in Sect. 7.5.

7.2 Problem Formulation

In line with [5], the desired trajectory of the follower spacecraft is generated based on the force-free Hill's equation, which can be considered being provided by the leader spacecraft. Thus, for the sake of convenience, the desired trajectory is denoted as $\boldsymbol{r}_t$. Further, define $\boldsymbol{r} = \boldsymbol{r}_p - \boldsymbol{r}_t$ as the relative position tracking error. Then, the resulting relative motion error dynamics can be modeled as (2.26). For ease of the control policy development, the Euler-Lagrangian relative position dynamics (2.26) is further transformed as

$$\dot{\boldsymbol{x}} = \boldsymbol{f}(\boldsymbol{x}) + g\boldsymbol{u}, \tag{7.1}$$

where $\boldsymbol{x} = [\boldsymbol{r}^\top, \dot{\boldsymbol{r}}^\top]^\top$, $\boldsymbol{f}(\boldsymbol{x}) = [\dot{\boldsymbol{r}}; -(\boldsymbol{C}_p^*\dot{\boldsymbol{r}} + \boldsymbol{D}_p^*\boldsymbol{r} + \boldsymbol{n}_p^* + m_p\ddot{\boldsymbol{r}}_t)]$, and $g = [\boldsymbol{0}_{3\times3}; 1/\boldsymbol{M}_p^*]$. It is noted that $\boldsymbol{u} = \boldsymbol{f}^{\mathcal{L}}$.

Assumption 7.1 The nonlinear function $\boldsymbol{f}(\boldsymbol{x})$ is Lipschiz continuous on a compact set $\Omega_1 \in \mathbb{R}^6$, satisfying $\|\boldsymbol{f}(\boldsymbol{x})\| \leq L_f\|\boldsymbol{x}\|$, where $L_f > 0$ is the Lipschitz constant.

Given the nonlinear dynamics (7.1), the control objective is to seek an ideal control policy $\boldsymbol{u}^* \in \mathbb{R}^3$ that is capable of rendering $\lim_{t\to\infty} \|\boldsymbol{x}(t)\| \to \mathbf{0}$, while minimizing the following infinite-horizon cost function

$$J(\boldsymbol{x}) = \int_t^{\infty} L(\boldsymbol{x}, \boldsymbol{u}) d\tau, \tag{7.2}$$

where the utility function $L(\boldsymbol{x}, \boldsymbol{u}) = \boldsymbol{x}^\top \boldsymbol{Q}\boldsymbol{x} + \boldsymbol{u}^\top \boldsymbol{R}\boldsymbol{u}$ with the positive-definite symmetric weight matrices $\boldsymbol{Q} \in \mathbb{R}^{6\times 6}$ and $\boldsymbol{R} \in \mathbb{R}^{3\times 3}$.

Define the Hamiltonian as

$$H\left(\boldsymbol{x}, \frac{\partial J(\boldsymbol{x})}{\partial \boldsymbol{x}}, \boldsymbol{u}\right) = \left(\frac{\partial J(\boldsymbol{x})}{\partial \boldsymbol{x}}\right)^\top (\boldsymbol{f}(\boldsymbol{x}) + g\boldsymbol{u}) + \boldsymbol{u}^\top \boldsymbol{R}\boldsymbol{u} + \boldsymbol{x}^\top \boldsymbol{Q}\boldsymbol{x}, \ \forall \boldsymbol{x}, \boldsymbol{u}, \tag{7.3}$$

where $\partial J(\boldsymbol{x})/\partial \boldsymbol{x}$ denotes the gradient of the cost function $J(\boldsymbol{x})$ with respect to $\boldsymbol{x}$. Further, the optimal cost function $J^*(\mathbf{x}) = \min_{\mathbf{u}} \int_t^{\infty} L(\mathbf{x}, \mathbf{u}) d\tau$ and the ideal optimal control policy $\boldsymbol{u}^*$ satisfy the following Hamilton-Jacobi-Bellman (HJB) function

$$\frac{\partial J^*(\boldsymbol{x})}{\partial t} + H\left(\boldsymbol{x}, \frac{\partial J^*(\boldsymbol{x})}{\partial \boldsymbol{x}}, \boldsymbol{u}^*\right) = 0 \Rightarrow H\left(\boldsymbol{x}, \frac{\partial J^*(\boldsymbol{x})}{\partial \boldsymbol{x}}, \boldsymbol{u}^*\right) = 0, \tag{7.4}$$

where $\partial J^*(\boldsymbol{x})/\partial t = 0$. By solving the stationary condition $\partial H(\boldsymbol{x}, \frac{\partial J^*(\boldsymbol{x})}{\partial \boldsymbol{x}}, \boldsymbol{u})/\partial \boldsymbol{u} = 0$, the optimal tracking control policy $\boldsymbol{u}^*$ is given by

$$\boldsymbol{u}^* = \mathrm{argmin}_{\boldsymbol{u}} \, H\left(\boldsymbol{x}, \frac{\partial J^*(\mathbf{x})}{\partial \boldsymbol{x}}, \boldsymbol{u}\right) = -\frac{1}{2}\boldsymbol{R}^{-1}\mathbf{g}^\top \frac{\partial J^*(\boldsymbol{x})}{\partial \boldsymbol{x}}. \tag{7.5}$$

Substituting (7.5) into (7.4), the HJB function is rewritten as

$$\begin{aligned} H\left(\boldsymbol{x}, \frac{\partial J^*(\boldsymbol{x})}{\partial \boldsymbol{x}}, \boldsymbol{u}^*\right) = & \frac{\partial J^*(\boldsymbol{x})}{\partial \boldsymbol{x}}^\top \boldsymbol{f}(\boldsymbol{x}) + \boldsymbol{x}^\top \boldsymbol{Q}\boldsymbol{x} \\ & - \frac{1}{4}\frac{\partial J^*(\boldsymbol{x})}{\partial \boldsymbol{x}}^\top g\boldsymbol{R}^{-1}g^\top \frac{\partial J^*(\boldsymbol{x})}{\partial \boldsymbol{x}}. \end{aligned} \tag{7.6}$$

Remark 7.1 Intuitively, the optimal tracking control problem can be tackled by solving the HJB function (7.6). Nonetheless, it is technically infeasible to seek the analytical solution of this equation directly due to its inherent nonlinear nature. In light of this, the ADP algorithm will be adopted to approximate the optimal tracking control policy. Additionally, to further lower the computation cost of the digital platforms, the control command is generally expected to be updated in a resource-friendly fashion instead of the traditional continuous/fixed update pattern. By means of the ETC method, an appropriate event generating mechanism needs to be developed for the implementation of the near optimal tracking control policy.

7.3 Event-Based Adaptive Optimal Control

This section first designs a traditional continuous near optimal tracking control law with the adaptive projection rule. Then, to reduce the unnecessary control update, an input-state-dependent event-triggered mechanism is developed to generate specific events which are used to trigger the control command execution. Finally, the detailed discussions on the system stability and Zeno-free behavior are provided, respectively.

7.3.1 Continuous Near Optimal Tracking Control Law

Based on the ADP algorithm [16], a three-layer critic NN (i.e., the input, hidden and output layers) is incorporated to reconstruct the continuous optimal cost function $J^*(\boldsymbol{x})$ and its gradient $\frac{\partial J^*(\boldsymbol{x})}{\partial \boldsymbol{x}}$, which are given by

$$J^*(\boldsymbol{x}) = \boldsymbol{W}^\top \boldsymbol{\phi}(\boldsymbol{x}) + \epsilon(\boldsymbol{x}), \tag{7.7}$$

$$\frac{\partial J^*(\boldsymbol{x})}{\partial \boldsymbol{x}} = \nabla \boldsymbol{\phi}^\top \boldsymbol{W} + \nabla \epsilon, \tag{7.8}$$

where $\boldsymbol{W} = [W_1, W_2, \ldots, W_p]^\top \in \mathbb{R}^p$ is the ideal but unknown weight vector, p stands for the neuron number of the hidden layer, $\epsilon(\boldsymbol{x}) \in \mathbb{R}$ represents the reconstruction error of NN, $\boldsymbol{\phi}(\boldsymbol{x}) = [\phi_1(\boldsymbol{x}), \phi_2(\boldsymbol{x}), \ldots, \phi_p(\boldsymbol{x})]^\top \in \mathbb{R}^p$ denotes the nonlinear NN activation function, $\nabla \boldsymbol{\phi} = \frac{\partial \boldsymbol{\phi}(\boldsymbol{x})}{\partial \boldsymbol{x}} \in \mathbb{R}^{p\times 6}$, and $\nabla \epsilon = \frac{\partial \epsilon(\boldsymbol{x})}{\partial \boldsymbol{x}} \in \mathbb{R}^6$. Besides, it has been shown that $\epsilon(\boldsymbol{x})$ and $\nabla\epsilon$ will tend to zero as $p \to \infty$.

Assumption 7.2 The ideal weight vector $\boldsymbol{W}$ is bounded. Here, suppose $\boldsymbol{W} \in \Omega_2$, where $\Omega_2 = \{\boldsymbol{W} \mid \boldsymbol{W}^\top \boldsymbol{W} \le W_m, W_m > 0\}$. Besides, the nonlinear activation function $\boldsymbol{\phi}(\boldsymbol{x})$ and the reconstruction error $\epsilon(\boldsymbol{x})$ are norm-bounded, satisfying $\|\boldsymbol{\phi}(\boldsymbol{x})\| \le \phi_m$ and $\|\epsilon(\boldsymbol{x})\| \le \epsilon_m$, where $\boldsymbol{\phi}_m > 0$ and $\epsilon_m > 0$ are positive constants. Meanwhile, the gradient vectors of $\boldsymbol{\phi}(\boldsymbol{x})$ and $\epsilon(\boldsymbol{x})$ with respect to $\boldsymbol{x}$ are also norm-bounded, i.e., $\|\nabla\boldsymbol{\phi}\| \le \nabla\phi_m$ and $\|\nabla\epsilon\| \le \nabla\epsilon_m$, where $\nabla\phi_m > 0$ and $\nabla\epsilon_m > 0$.

By combining (7.5) and (7.8), the ideal optimal tracking control policy $\boldsymbol{u}^*$ can be directly given by

$$\boldsymbol{u}^* = -\frac{1}{2}\boldsymbol{R}^{-1} g^\top (\nabla\boldsymbol{\phi}^\top \boldsymbol{W} + \nabla\epsilon). \tag{7.9}$$

With the help of the Weierstrass higher-order approximation theorem, the optimal cost function and its gradient with respect to $\boldsymbol{x}$ can be estimated as

$$\hat{J}^*(\boldsymbol{x}) = \hat{\boldsymbol{W}}^\top \boldsymbol{\phi}(\boldsymbol{x}), \quad \frac{\partial \hat{J}^*(\boldsymbol{x})}{\partial \boldsymbol{x}} = \nabla\boldsymbol{\phi}^\top \hat{\boldsymbol{W}}, \tag{7.10}$$

where $\hat{\boldsymbol{W}} \in \mathbb{R}^p$ denotes the estimate value of the ideal weight vector $\boldsymbol{W}$. The weight estimate error is defined as $\tilde{\boldsymbol{W}} = \boldsymbol{W} - \hat{\boldsymbol{W}}$. Accordingly, the continuous near optimal tracking controller $\hat{\boldsymbol{u}}$ can be obtained

$$\hat{\boldsymbol{u}} = -\frac{1}{2}\boldsymbol{R}^{-1}g^{\top}\nabla\boldsymbol{\phi}^{\top}\hat{\boldsymbol{W}}. \tag{7.11}$$

Substituting (7.10) and (7.11) into (7.3), the approximate Hamiltonian is written as

$$\hat{H}(\boldsymbol{x}, \frac{\partial \hat{J}^*(\boldsymbol{x})}{\partial \boldsymbol{x}}, \hat{\boldsymbol{u}}) = \hat{\boldsymbol{W}}^{\top}\nabla\boldsymbol{\phi}\left(\boldsymbol{f}(\boldsymbol{x}) + g\hat{\boldsymbol{u}}\right) + \boldsymbol{x}^{\top}\boldsymbol{Q}\boldsymbol{x} + \hat{\boldsymbol{u}}^{\top}\boldsymbol{R}\hat{\boldsymbol{u}}. \tag{7.12}$$

Then, the Bellman residual error δ_b between the optimal and approximate Hamiltonians is given by

$$\delta_b = \hat{H}(\boldsymbol{x}, \frac{\partial \hat{J}^*(\boldsymbol{x})}{\partial \boldsymbol{x}}, \hat{\boldsymbol{u}}) - H(\boldsymbol{x}, \frac{\partial J^*(\boldsymbol{x})}{\partial \boldsymbol{x}}, \boldsymbol{u}^*). \tag{7.13}$$

To render the estimate weight vector $\hat{\boldsymbol{W}}$ approaching to its ideal true value $\boldsymbol{W}$, a proper weight update law needs to be developed. Similar to [26], the normalized gradient descent algorithm is used here to derive the weight update law, aiming at minimizing the squared Bellman error $E_b = \frac{1}{2}\delta_b^2$. Moreover, in order to prevent $\hat{\boldsymbol{W}}$ drifting arbitrarily away from its ideal true value, a smooth projection rule [27] is employed to modify the traditional gradient-based update law. By doing this, the estimate weight vector $\hat{\boldsymbol{W}}$ will always remain in a convex hypercube $\Omega_3 = \{\boldsymbol{W} \mid \boldsymbol{W}^{\top}\boldsymbol{W} \leq W_m + \kappa,\ \kappa > 0\}$. Specifically, the adaptive weight update law based on the gradient decent algorithm and parameter projection rule is designed as

$$\dot{\hat{\boldsymbol{W}}} = \mathrm{Proj}(\hat{\boldsymbol{W}}, \boldsymbol{\Gamma}) = \begin{cases} \boldsymbol{\Gamma}, & \text{if } \hat{\boldsymbol{W}}^{\top}\hat{\boldsymbol{W}} \leq W_m \text{ or} \\ & \text{if } \hat{\boldsymbol{W}}^{\top}\hat{\boldsymbol{W}} > W_m \text{ and } \hat{\boldsymbol{W}}^{\top}\boldsymbol{\Gamma} \leq 0\ , \\ \boldsymbol{\Gamma} - \frac{(\hat{\boldsymbol{W}}^{\top}\hat{\boldsymbol{W}} - W_m)\hat{\boldsymbol{W}}^{\top}\boldsymbol{\Gamma}}{\kappa\hat{\boldsymbol{W}}^{\top}\hat{\boldsymbol{W}}}\hat{\boldsymbol{W}}, & \text{otherwise} \end{cases} \tag{7.14}$$

with

$$\boldsymbol{\Gamma} = -\alpha\frac{\delta_b\boldsymbol{\theta}}{(\boldsymbol{\theta}^{\top}\boldsymbol{\theta} + 1)^2}, \tag{7.15}$$

where $\alpha > 0$ denotes the adaptive gain, $\boldsymbol{\theta} = \nabla\boldsymbol{\phi}(\boldsymbol{f}(\boldsymbol{x}) + g\hat{\boldsymbol{u}})$, and $1/(\boldsymbol{\theta}^{\top}\boldsymbol{\theta} + 1)^2$ is the normalization term. By applying (7.14), it is straightforward to verify that

$$\hat{\boldsymbol{W}}(0) \in \Omega_2 \Rightarrow \hat{\boldsymbol{W}}(t) \in \Omega_3. \tag{7.16}$$

It is worth noting that the upper bound W_m and initial estimate weight $\hat{\boldsymbol{W}}(0)$ should be selected properly so that the initial condition $\hat{\boldsymbol{W}}(0) \in \Omega_2$ holds true.

Remark 7.2 Recalling the definition of the Bellman residual error δ_b in (7.13), it is easy to reveal that $\delta_b \equiv \hat{H}(\boldsymbol{x}, \frac{\partial \hat{J}^*(x)}{\partial x}, \hat{\boldsymbol{u}})$ since (7.4) always holds true. Hence, δ_b is measurable and can be directly used for calculation of (7.15) in the adaptive weight update law. On the other hand, using (7.6)–(7.8) and (7.11)–(7.13), the Bellman residual error δ_b can be rewritten in an unmeasurable form as

$$\delta_b = -\tilde{\boldsymbol{W}}^\top \boldsymbol{\theta} + \frac{1}{4}\tilde{\boldsymbol{W}}^\top \boldsymbol{F}\tilde{\boldsymbol{W}} + \varepsilon_1, \tag{7.17}$$

where $\boldsymbol{\theta} = \nabla\boldsymbol{\phi}(\boldsymbol{f}(\boldsymbol{x}) + g\boldsymbol{u})$, $\boldsymbol{F} = \nabla\boldsymbol{\phi} g\boldsymbol{R}^{-1}g^\top\nabla\boldsymbol{\phi}^\top$, and $\varepsilon_1 = \frac{1}{4}\nabla\epsilon^\top g\boldsymbol{R}^{-1}g^\top\nabla\epsilon + \frac{1}{2}\nabla\epsilon^\top g\boldsymbol{R}^{-1}g^\top\nabla\phi^\top\boldsymbol{W} - \nabla\epsilon^\top\boldsymbol{f}(\boldsymbol{x})$. Although such an unmeasurable form seems like meaningless for computing (7.15), it is actually essential for establishing the stability of the whole closed-loop system, which will be discussed detailedly later.

Assumption 7.3 The nonlinear function $\boldsymbol{\theta}$ is persistently excited. That is, there exists positive constants t_h, τ_1, and τ_2 such that $\tau_1\boldsymbol{I} \leq \int_t^{t+t_h}\boldsymbol{\theta}(r)\boldsymbol{\theta}(r)^\top \mathrm{d}r \leq \tau_2\boldsymbol{I}$, $\forall t > 0$, where $\boldsymbol{I}$ is the identity matrix with appropriate dimensions.

7.3.2 Event-Triggered Mechanism

Define a monotonically increasing discrete-time series $\{t_k, k = 0, 1, \ldots\}$ as the event triggering instant sequence, where t_k denotes the $(k+1)$-th triggering instant and $t_0 = 0$. The triggering sequence is decided by a specific event-triggered mechanism to be designed later. Under the ETC frame, the event-based near optimal tracking control policy $\hat{\boldsymbol{u}}_k$ is formalized as

$$\hat{\boldsymbol{u}}_k = \begin{cases} \hat{\boldsymbol{u}}|_{t=t_k}, \ t \in [t_k, t_{k+1}) \\ \hat{\boldsymbol{u}}|_{t_{k+1}}, \ t = t_{k+1} \end{cases}. \tag{7.18}$$

It is noticed from (7.18) that the control command $\hat{\boldsymbol{u}}_k$ is updated only at some specific event triggering instants. Otherwise, it will always keep constant until the next event occurs. That indicates that a zero-order hold (ZOH) actuator component here is required to ensure $\hat{\boldsymbol{u}}_k$ executable. By using such an event-based control scheme (7.18), the update number of the controller will be economized significantly.

For the sake of designing the event-triggered mechanism, define an input-based triggering measurement error as

$$\boldsymbol{e}(t) = \hat{\boldsymbol{u}}_k - \hat{\boldsymbol{u}}, \ \forall t \in [t_k, t_{k+1}). \tag{7.19}$$

To proceed, an input-state-dependent event-triggered mechanism is developed

$$t_{k+1} = \inf\left\{t > t_k \middle| \|\boldsymbol{e}(t)\|^2 \geq \mathcal{X}\right\}, \tag{7.20a}$$

with the triggering function $\mathcal{X}$,

$$\mathcal{X} = 2m^2\eta_1\lambda_{\min}(\boldsymbol{Q})\|\boldsymbol{x}\|^2 + 2m^2\eta_2\exp(-\omega t), \tag{7.20b}$$

where $0 < \eta_1 < 1$, $\eta_2 > 0$, and $\omega > 0$ are design constants. The triggering function (7.20b) is designed mainly for ensuring the stability of the whole closed-loop system while reducing unnecessary control command update, especially in the steady-state phase. Under the above event-triggered strategy, a specific event will be generated instantly as long as the triggering condition is satisfied. Meanwhile, the measurement error $\boldsymbol{e}(t)$ will be set to zero immediately once an event is triggered.

Remark 7.3 For the existing works on the approximate optimal control methods with event-triggered mechanism (e.g., see [21, 22, 25]), there is a common assumption that the optimal controller is Lipschitz continuous with respect to the system states or the triggering measurement error. Moreover, the Lipschitz constant pertaining to this assumption is directly used to formulate the event-triggered condition. In general, such an assumption can be trivially satisfied for every function with bounded first-order derivative. Moreover, the Lipschitz constant also can be computed by virtue of the definition or triangle-inequality methods for some known functions with a simple form. Nevertheless, since the optimal controller (7.5) cannot be solved as an explicit form of the system states, it is extremely challenging to numerically determine the Lipschitz constant. Therefore, the control schemes in the above-mentioned studies are infeasible in the practical implementation. On contrary, the Lipschitz continuity assumption on the optimal controller is relaxed tactfully in our proposed scheme. All the gains either in the control policy or the event-triggered mechanism can be determined appropriately by the designer.

Theorem 7.1 Consider the Euler-Lagrangian relative position dynamics (2.26), the adaptive weight update law (7.14), and the event-based near optimal tracking control law (7.18). Under Assumptions 1-3, if the near optima tracking controller (7.18) is executed on the basis of the event-triggered mechanism (7.20a, 7.20b) it can ensure that the relative position and relative velocity errors as well as the network weight estimate error are uniformly ultimately bounded (UUB). Moreover, the Zeno phenomenon is ruled out strictly under the event-triggered mechanism (7.20a, 7.20b).

Proof Please see Sects. 7.3.3 and 7.3.4 for the detailed proof. □

7.3.3 Stability Analysis

To investigate the stability of the whole closed-loop system, choose the Lyapunov function candidate as follows

$$V = J^* + \frac{1}{2\beta}\tilde{\boldsymbol{W}}^\top\tilde{\boldsymbol{W}}, \tag{7.21}$$

where $J^* = J^*(\boldsymbol{x})$, and $\beta > 0$. Although the executive near optimal control policy is piece-wise continuous, the above-constructed Lyapunov function V is still continuous differentiable. Next, taking the time derivative of J^* first and after considering (7.1) and (7.4) in mind, it follows that

$$\begin{aligned} \dot{J}^* =& \frac{\partial J^*(\boldsymbol{x})}{\partial \boldsymbol{x}}^\top \left(\boldsymbol{f}(\boldsymbol{x}) + g\hat{\boldsymbol{u}}_k\right) \\ =& \frac{\partial J^*(\boldsymbol{x})}{\partial \boldsymbol{x}}^\top \left(\boldsymbol{f}(\boldsymbol{x}) + g\boldsymbol{u}^*\right) + \frac{\partial J^*(\boldsymbol{x})}{\partial \boldsymbol{x}}^\top g\left(\hat{\boldsymbol{u}}_k - \boldsymbol{u}^*\right) \\ =& -\boldsymbol{x}^\top \boldsymbol{Q}\boldsymbol{x} - \boldsymbol{u}^{*\top}\boldsymbol{R}\boldsymbol{u}^* + \frac{\partial J^*(\boldsymbol{x})}{\partial \boldsymbol{x}}^\top g\left(\hat{\boldsymbol{u}}_k - \boldsymbol{u}^*\right). \end{aligned} \tag{7.22}$$

Based on the definition of $\boldsymbol{e}(t)$ in (7.19), $\dot{J}^*$ becomes

$$\dot{J}^* = -\boldsymbol{x}^\top \boldsymbol{Q}\boldsymbol{x} - \boldsymbol{u}^{*\top}\boldsymbol{R}\boldsymbol{u}^* + \frac{\partial J^*(\boldsymbol{x})}{\partial \boldsymbol{x}}^\top g\left(\hat{\boldsymbol{u}} - \boldsymbol{u}^* + \boldsymbol{e}(t)\right). \tag{7.23}$$

Besides, given the fact that $\frac{\partial J^*(\boldsymbol{x})}{\partial \boldsymbol{x}}^\top g = -2\boldsymbol{u}^{*\top}\boldsymbol{R}$, one has $\frac{\partial J^*(\boldsymbol{x})}{\partial \boldsymbol{x}}^\top g\left(\hat{\boldsymbol{u}} - \boldsymbol{u}^*\right) = 2\boldsymbol{u}^{*\top}\boldsymbol{R}\boldsymbol{u}^* - 2\boldsymbol{u}^{*\top}\boldsymbol{R}\hat{\boldsymbol{u}}$. Then, combining the definitions of $\boldsymbol{u}^*$ and $\hat{\boldsymbol{u}}$ in (7.9) and (7.11), and using the Young's inequality, the following inequality holds

$$\begin{aligned} \frac{\partial J^*(\boldsymbol{x})}{\partial \boldsymbol{x}}^\top g\left(\hat{\boldsymbol{u}} - \boldsymbol{u}^*\right) \leq& \boldsymbol{u}^{*\top}\boldsymbol{R}\boldsymbol{u}^* + \boldsymbol{u}^{*\top}\boldsymbol{R}\boldsymbol{u}^* - 2\boldsymbol{u}^{*\top}\boldsymbol{R}\hat{\boldsymbol{u}} + \hat{\boldsymbol{u}}^\top\boldsymbol{R}\hat{\boldsymbol{u}} \\ =& \boldsymbol{u}^{*\top}\boldsymbol{R}\boldsymbol{u}^* + \frac{1}{4}\nabla\epsilon^\top g\boldsymbol{R}^{-1}g^\top\nabla\epsilon + \frac{1}{4}\tilde{\boldsymbol{W}}^\top\boldsymbol{F}\tilde{\boldsymbol{W}} \\ &+ \frac{1}{2}\nabla\epsilon^\top g\boldsymbol{R}^{-1}g^\top\nabla\boldsymbol{\phi}^\top\tilde{\boldsymbol{W}} \\ \leq& \boldsymbol{u}^{*\top}\boldsymbol{R}\boldsymbol{u}^* + \frac{1}{4}\nabla\epsilon^\top g\boldsymbol{R}^{-1}g^\top\nabla\epsilon + \frac{1}{4}\tilde{\boldsymbol{W}}^\top\boldsymbol{F}\tilde{\boldsymbol{W}} \\ &+ \frac{1}{4}\nabla\epsilon^\top g\boldsymbol{R}^{-1}g^\top\nabla\epsilon + \frac{1}{4}\tilde{\boldsymbol{W}}^\top\boldsymbol{F}\tilde{\boldsymbol{W}} \\ =& \boldsymbol{u}^{*\top}\boldsymbol{R}\boldsymbol{u}^* + \frac{1}{2}\tilde{\boldsymbol{W}}^\top\boldsymbol{F}\tilde{\boldsymbol{W}} + \varepsilon_2 \end{aligned}, \tag{7.24}$$

where $\varepsilon_2 = \frac{1}{2}\nabla\epsilon^\top g\boldsymbol{R}^{-1}g^\top\nabla\epsilon$. Then, inserting (7.24) into (7.23), the expression of $\dot{J}^*$ is transformed to

$$\begin{aligned} \dot{J}^* \leq& -\boldsymbol{x}^\top\boldsymbol{Q}\boldsymbol{x} + \frac{1}{2}\tilde{\boldsymbol{W}}^\top\boldsymbol{F}\tilde{\boldsymbol{W}} + \varepsilon_2 + \frac{\partial J^*(\boldsymbol{x})}{\partial \boldsymbol{x}}^\top g\boldsymbol{e}(t) \\ \leq& -\boldsymbol{x}^\top\boldsymbol{Q}\boldsymbol{x} + \frac{1}{2}\tilde{\boldsymbol{W}}^\top\boldsymbol{F}\tilde{\boldsymbol{W}} + \varepsilon_2 + \frac{1}{2}\|\nabla\boldsymbol{\phi}^\top\boldsymbol{W} + \nabla\epsilon\|^2 + \frac{1}{2m_p^2}\|\boldsymbol{e}(t)\|^2 \\ \leq& -\boldsymbol{x}^\top\boldsymbol{Q}\boldsymbol{x} + \frac{1}{2}\tilde{\boldsymbol{W}}^\top\boldsymbol{F}\tilde{\boldsymbol{W}} + \varepsilon_2 + \|\nabla\boldsymbol{\phi}^\top\boldsymbol{W}\|^2 + \|\nabla\epsilon\|^2 + \frac{1}{2m_p^2}\|\boldsymbol{e}(t)\|^2 \end{aligned}, \tag{7.25}$$

The following inequalities are formulated [28]

$$\left\| \frac{1}{4} \tilde{\boldsymbol{W}}^{\top} \boldsymbol{F} \tilde{\boldsymbol{W}} + \varepsilon_1 \right\| \leq \varepsilon_3, \tag{7.26}$$

$$0 \leq \left\| \frac{\boldsymbol{y}}{\boldsymbol{y}^{\top}\boldsymbol{y}+1} \right\| < \frac{1}{2}, \tag{7.27}$$

$$0 \leq \| \frac{\boldsymbol{y}^{\top}\boldsymbol{y}}{(\boldsymbol{y}^{\top}\boldsymbol{y}+1)^2} \| \leq \frac{1}{4}, \tag{7.28}$$

where $\varepsilon_3 > 0$ and $\boldsymbol{y} \in \mathbb{R}^n$.

To proceed, let $L = \frac{1}{2\beta} \tilde{\boldsymbol{W}}^{\top} \tilde{\boldsymbol{W}}$. Then, differentiating L over t and combining the adaptive projection rule (7.14) and (7.17) result in

$$\dot{L} \leq -\frac{\alpha}{\beta} \lambda_{\min} \left(\frac{\boldsymbol{\theta}^{\top}\boldsymbol{\theta}}{(\boldsymbol{\theta}^{\top}\boldsymbol{\theta}+1)^2} \right) \tilde{\boldsymbol{W}}^{\top} \tilde{\boldsymbol{W}} + \frac{\alpha}{4\beta} \left(\frac{1}{\gamma_1} \tilde{\boldsymbol{W}}^{\top} \tilde{\boldsymbol{W}} + \gamma_1 \varepsilon_3^2 \right) + \varepsilon_4, \tag{7.29}$$

with

$$\varepsilon_4 = \begin{cases} 0, & \text{if } \hat{\boldsymbol{W}}^{\top} \hat{\boldsymbol{W}} \leq W_m \text{ or} \\ & \text{if } \hat{\boldsymbol{W}}^{\top} \hat{\boldsymbol{W}} > W_m \text{ and } \hat{\boldsymbol{W}}^{\top} \boldsymbol{\Gamma} \leq 0 \\ \frac{(\hat{\boldsymbol{W}}^{\top} \hat{\boldsymbol{W}} - W_m) \hat{\boldsymbol{W}}^{\top} \boldsymbol{\Gamma}}{\kappa \hat{\boldsymbol{W}}^{\top} \hat{\boldsymbol{W}}} \tilde{\boldsymbol{W}} \hat{\boldsymbol{W}} \leq 0, & \text{otherwise} \end{cases}, \tag{7.30}$$

where the Young's inequality is applied, and $\gamma_1 > 0$.

Furthermore, combining (7.25) and (7.29) leads to

$$\begin{aligned} \dot{V} =& \dot{J}^* + \dot{L} \\ \leq& -\boldsymbol{x}^{\top} \boldsymbol{Q} \boldsymbol{x} + \frac{1}{2} \tilde{\boldsymbol{W}}^{\top} \boldsymbol{F} \tilde{\boldsymbol{W}} + \varepsilon_2 + \|\nabla \boldsymbol{\phi}^{\top} \boldsymbol{W}\|^2 + \|\nabla \epsilon\|^2 + \frac{1}{2m_p^2} \|\boldsymbol{e}(t)\|^2 \\ & - \frac{\alpha}{\beta} \lambda_{\min} \left(\frac{\boldsymbol{\theta}^{\top}\boldsymbol{\theta}}{(\boldsymbol{\theta}^{\top}\boldsymbol{\theta}+1)^2} \right) \tilde{\boldsymbol{W}}^{\top} \tilde{\boldsymbol{W}} + \frac{\alpha}{4\beta} \left(\frac{1}{\gamma_1} \tilde{\boldsymbol{W}}^{\top} \tilde{\boldsymbol{W}} + \gamma_1 \varepsilon_3^2 \right) \\ \leq& -\lambda_{\min}(\boldsymbol{Q}) \boldsymbol{x}^{\top} \boldsymbol{x} - k_1 \tilde{\boldsymbol{W}}^{\top} \tilde{\boldsymbol{W}} + \frac{\|\boldsymbol{e}(t)\|^2}{2m_p^2} + \varepsilon_5 \end{aligned}, \tag{7.31}$$

where

$$k_1 = \frac{\alpha}{\beta} \lambda_{\min} \left(\frac{\boldsymbol{\theta}^{\top}\boldsymbol{\theta}}{(\boldsymbol{\theta}^{\top}\boldsymbol{\theta}+1)^2} \right) - \frac{1}{2} \lambda_{\max}(\boldsymbol{F}) - \frac{\alpha}{4\beta\gamma_1}, \tag{7.32a}$$

$$\varepsilon_5 = \varepsilon_2 + \nabla \epsilon_m^2 + \nabla \phi_m^2 W_m + \frac{\alpha}{4\beta} \gamma_1 \varepsilon_3^2. \tag{7.32b}$$

Invoking the event-triggered mechanism in (7.20a, 7.20b), it can be ensured that for all $t \in [t_k, t_{k+1})$, it holds that

$$\|e(t)\|^2 < 2m_p^2\eta_1\lambda_{\min}(Q)\|x\|^2 + 2m_p^2\eta_2\exp(-\omega t). \tag{7.33}$$

By using (7.33), we can further rearrange $\dot{V}$ as

$$\dot{V} \leq -(1-\eta_1)\lambda_{\min}(Q)x^\top x - k_1\tilde{W}^\top\tilde{W} + \varepsilon_{\text{total}}, \tag{7.34}$$

where $\varepsilon_{\text{total}} = \varepsilon_5 + \eta_2\exp(-\mu t) \leq \varepsilon_5 + \eta_2$.

From (7.34), it is straightforward to infer that x and $\tilde{W}$ will converge to the following compact sets

$$\Omega_4 = \left\{x \middle| \|x\| \leq \sqrt{\frac{\varepsilon_{\text{total}}}{(1-\eta_1)\lambda_{\min}(Q)}}\right\}, \tag{7.35a}$$

$$\Omega_5 = \left\{\tilde{W} \middle| \|\tilde{W}\| \leq \sqrt{\frac{\varepsilon_{\text{total}}}{k_1}}\right\}. \tag{7.35b}$$

This is because once x and $\tilde{W}$ go outside of the above compact sets, $\dot{V}$ will become negative. In this case, the system state x and the weight estimate error $\tilde{W}$ will be pulled back to the sets Ω_4 and Ω_5, respectively. Hence, the relative position and velocity tracking errors and the network weight estimate error are UUB.

7.3.4 *Zeno-Free Analysis*

Define $T_k = t_{k+1} - t_k$ $(k = 0, 1, \ldots)$ as the time duration of the $(k+1)$-th triggering event. To exclude the Zeno phenomenon, we will prove that $T_k > 0$, $\forall k$. Recalling the definition of $e(t)$ in (7.19), it can be observed that the time derivative of $\|e(t)\|$ satisfies

$$\|\dot{e}(t)\| \leq \|\dot{\hat{u}}\|, \ t \in [t_k, t_{k+1}). \tag{7.36}$$

Since the smooth parameter projection rule has been used in the adaptive weight update law (7.14), the estimate weight $\hat{W}$ satisfies (7.16), that is, $\hat{W} \in \mathcal{L}_\infty$. Combining Assumption 7.2, the near optimal tracking control law $\hat{u}$ in (7.11) is bounded. Moreover, since x is UUB that has been proved in the above stability analysis part, it can be deduced from (7.12) that $\hat{H}(x, \frac{\partial \hat{J}^*(x)}{\partial x}, \hat{u})$ is bounded, which implies the Bellman residual error $\delta_b \in \mathcal{L}_\infty$. Hence, $\dot{\hat{W}}$ given in (7.14) is bounded. Further, dependent on the fact that R and g are known constant matrix/vector, it is obviously concluded from (7.11) that $\dot{\hat{u}} \in \mathcal{L}_\infty$. Therefore, let u_m be the upper bound of $\|\dot{\hat{u}}\|$, where $u_m > 0$. Then, from (7.36), we can obtain that $\|\dot{e}(t)\| \leq u_m$. Integrating this inequality over $t \in [t_k, t_{k+1})$ yields

$$\|\boldsymbol{e}(t)\| \leq \int_{t_k}^{t_{k+1}} \|\dot{\boldsymbol{e}}(t)\|\mathrm{d}t < u_m T_k. \tag{7.37}$$

Reminding the event-triggered mechanism (7.20a, 7.20b), it follows that

$$\|\boldsymbol{e}(t)\| \leq \sqrt{2m^2\eta_1\lambda_{\min}(\boldsymbol{Q})\|\boldsymbol{x}\|^2 + 2m^2\eta_2\exp(-\omega t)}, \tag{7.38}$$

for $t \in [t_k, t_{k+1})$. Let $T_{min} = \min\{T_k\}$ be the minimum inter-event time duration. Refer to [29], it is noted that

$$u_m T_{\min} \geq \sqrt{2m^2\eta_2\exp(-\omega(t_k + T_{\min}))}. \tag{7.39}$$

During the finite time, the solution of (7.39) is strictly positive. Therefore, no Zeno phenomenon exhibits under the proposed triggering strategy. The proof of Theorem 7.1 is completed.

Remark 7.4 Recalling the definition of T_k and $T_{\min}$, it is easily inferred that $T_{\min} \geq 0$. That is, there are two different cases, i.e., $T_{\min} = 0$ or $T_{\min} > 0$. Obviously, if $T_{\min} > 0$, the proposed event-triggered strategy is Zeno-free. Conversely, Zeno phenomenon will occur if $T_{\min} = 0$. From a theoretical point of view, it is extremely difficult to obtain the explicit lower bound of $T_{\min}$ by solving (7.39). Thus, we will prove $T_{\min} > 0$ by seeking a contradiction. Firstly, suppose that $T_{\min} = 0$ occurs. In this case, the left-hand part of (7.39) is strictly equal to zero. The inequality (7.39) will hold true as long as its right-hand part is also strictly equal to zero. However, considering the form of the right-hand part of (7.39), it will infinitely approach to zero rather than strictly equaling zero when $t_k \to \infty$. Moreover, the practical formation missions usually are accomplished during the finite time, which implies $t_k \to \infty$ may be impossible. Then, it is concluded that (7.39) cannot be satisfied when $T_{\min} = 0$. This is a contradictory with the fact that (7.39) always holds true. Consequently, $T_{\min}$ is strictly positive during the finite time and no Zeno phenomenon occurs in the proposed event-triggered mechanism.

7.4 Numerical Simulations

In this section, numerical simulations are performed for verifying the effectiveness of the proposed event-based near optimal control policy (7.18) and the event-triggered strategy (7.20a, 7.20b). To be specific, assume that the leader spacecraft runs in a standard circular orbit with an angular velocity $n_o = \sqrt{\mu/a_o^3}$. Similar to [14], the desired trajectory of the follower spacecraft is represented in an explicit projected form, that is, $\boldsymbol{r}_t = 0.5r_t \times [\sin(n_o t + \psi), 2\cos(n_o t + \psi), \sin(n_o t + \psi)]^\top$m, where r_t and ψ denote the formation size and the in-plane phase angle, respectively. In this case, the leader and follower spacecraft ($m_p = 1$kg) will eventually keep a constant relative distance on the $Y_{\mathcal{L}} - Z_{\mathcal{L}}$ plane. The detailed parameters on the orbit and

Table 7.1 Relevant orbit parameters and initial system conditions

Parameters	Values
Initial conditions	$\boldsymbol{r}(0) = [100, 1000, 100]^{\top}$ m
	$\dot{\boldsymbol{r}}(0) = [0.55, 0, 1.11]^{\top}$ m/s
Orbital parameters	$a_o = 6.878 \times 10^6$km, $e_o = 0$
	$\psi = 0$ rad/s, $\mu = 3.986 \times 10^{14}$m^3/s^2, $r_t = 1000$ m

the initial states of the formation system are summarized in Table 7.1. Besides, for the purpose of examining the robustness of the designed control method, the external disturbance $\boldsymbol{d} = 10^{-5} \times [-1.2\sin(n_o t), 0.5\cos(2n_o t), \sin(n_o t)]^{\top}$N related to the orbit angular velocity n_o is injected into the relative position dynamics (2.26).

The neuron number of the hidden layer is set as $p = 6$. The nonlinear activation function is chosen to be $\boldsymbol{\phi} = [x_1 x_4, x_2 x_5, x_3 x_6, x_4^2, x_5^2, x_6^2]^{\top}$. Associated with $\boldsymbol{\phi}$, the weight estimate vector $\hat{\boldsymbol{W}}$ is written as $\hat{\boldsymbol{W}} = [\hat{W}_1, \ldots, \hat{W}_6]^{\top}$ and its initial value is selected as $\hat{\boldsymbol{W}}(0) = 0.01 \times [2, 2, 2, 3, 3, 3]^{\top}$. Meanwhile, the initial control policy is constructed by PD-based controller where both the proportional and differential gains are set as 0.1. Besides, considering the inevitable sensor measurement errors, Gaussian white noises, satisfying $E_\rho \sim (0\,\text{m}, 1 \times 10^{-3}\,\text{m}^2)$ and $E_{\dot{\rho}} \sim (0\,\text{m}, 1 \times 10^{-4}\,\text{m}^2/\text{s}^2)$, are injected into the position and velocity vectors. By trial and error, the remaining parameters in the cost function, controller, adaptive weight update law and event-triggered mechanism, are tuned to be $\boldsymbol{R} = 0.1\boldsymbol{I}_3$, $\boldsymbol{Q} = 10^{-4}\boldsymbol{I}_6$, $W_m = 0.1$, $\kappa = 1 \times 10^{-4}$, $\alpha = 2$, $\eta_1 = 0.01$, $\eta_2 = 0.05$, and $\omega = 0.002$, respectively. The whole simulation period is 6000s and the fixed sampling step is 0.1 s.

The simulation results on the control performance and optimal performance are illustrated in Figs. 7.1, 7.2, 7.3, 7.4, 7.5 and 7.6. Specifically, the time responses of the relative position and velocity tracking errors are displayed in Figs. 7.1 and 7.2, respectively. It can be easily observed from these figures that the follower spacecraft will settle probably at the desired position within a complete orbital period ($\approx$ 5677 s). Meanwhile, Fig. 7.3 illustrates a series of "snapshots" of the relative position between the follower and leader spacecraft, where the red arrows represent the motion direction of the follower spacecraft. Form Fig. 7.3, it clearly shows that the follower and leader spacecraft achieve a satisfactory formation maintenance task as required. The evolution of the event-based near optimal control policy is depicted in Fig. 7.4. Obviously, there is only the aperiodic control update execution instead of continuous/fixed-time update. Hence, the on-board computation resources waste is avoided significantly. Subsequently, the evolution of the Bellman residual error is depicted in Fig. 7.5. It is noticed form Fig. 7.5b that after $t \approx 5677$ s, δ_b reaches convergence approximately and its final convergence accuracy is confined within 5×10^{-7}. Recalling the definition of δ_b, it is inferred that the selected NN exhibits

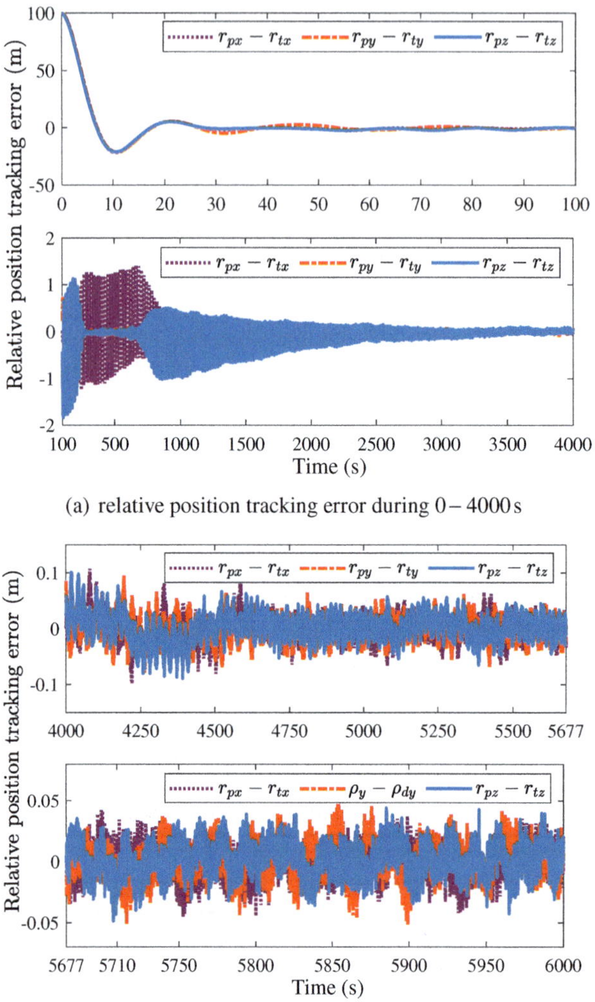

(a) relative position tracking error during 0 – 4000 s

(b) relative position tracking error during 4000 – 6000 s

Fig. 7.1 Time responses of the relative position tracking errors

an excellent approximation performance. Additionally, the evolution of the estimated weight vector $\hat{W}$ is shown in Fig. 7.6, which indicates $\hat{W}$ is UUB.

In addition, to highlight the superiority in terms of relaxing controller update, the traditional time-based sampling method here is considered as a comparison. More specifically, the near optimal tracking control law (7.11) will be actuated in a periodic fixed-time-based sampling method instead of the aperiodic event-triggered one. For more comprehensive comparison, four different sampling periods are chosen, i.e., 0.1 s, 2 s, 4 s, and 5 s. Moreover, for the sake of analysis, we remark the proposed

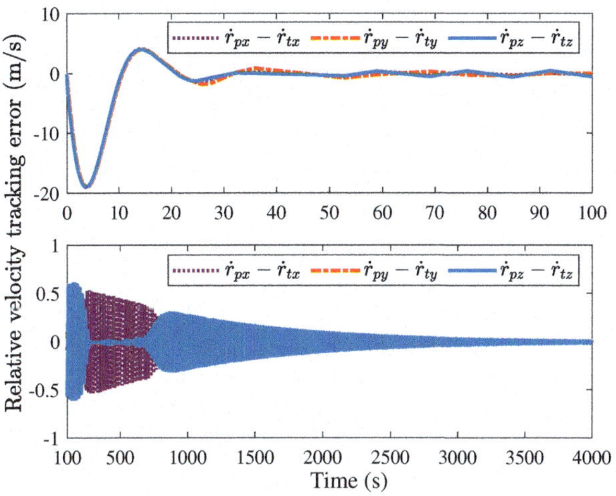

(a) relative velocity tracking error during 0 – 4000 s

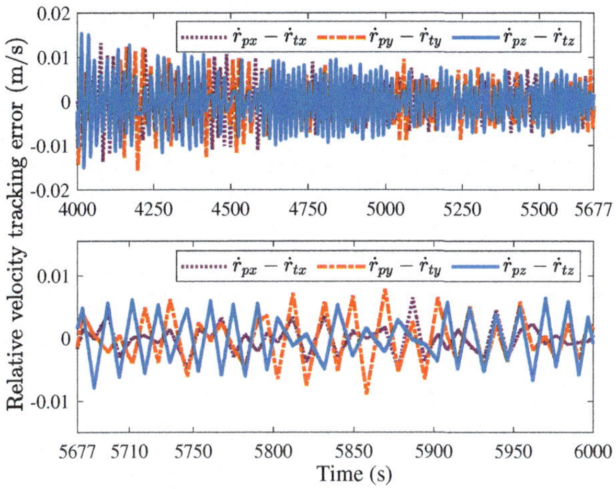

(b) relative velocity tracking error during 4000 – 6000 s

Fig. 7.2 Time responses of the relative velocity tracking errors

event-triggered sampling methods as Case I, and four different time-based periodic sampling methods (i.e., 0.1 s, 2 s, 4 s, and 5 s fixed sampling interval) as Cases II-V, respectively. In addition, to be fair, the simulation conditions, including the initial system states and other parameters associated with the cost function, controller, and adaptive weight update law, keep the same as those in the event-triggered sampling method. In this case, the control performance under Cases I-V is summarized in Table 7.2, where the notation "–" indicates the system states are divergent. Meanwhile, the elaborate condition on the controller update under Cases I-V is reported in Table 7.3, where the maximum, minimum, and average internal triggering time are

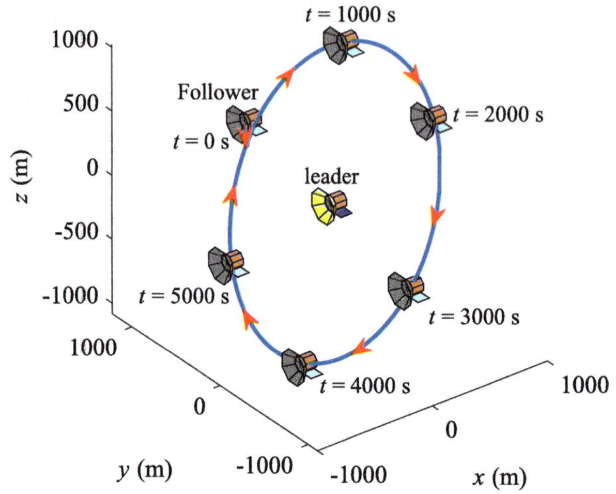

Fig. 7.3 Snapshots of the relative position history

listed in sequence. During the simulation process, it can be found that although the convergence speed in Case II is far faster than that in other cases, the relative position tracking error of the steady-state phase under Cases I-IV is about kept within 0.05m after an orbital period, which is sufficient to ensure the successful implementation of the formation tracking task. In addition, it is obvious that the number of controller update in Case I (717 times) is far less than that in other cases (i.e., 60000, 3000, 1500, and 1200 times, respectively). Theoretically, the update number can be reduced by increasing the fixed sampling period. However, it is obviously observed from Table 7.2 that the resulting system will not longer converge in Case V. That is, although the number of controller update in the time-based sampling method is reduced by increasing the sampling period, the control performance of the whole closed-loop system may not be guaranteed. Consequently, the proposed event-triggered sampling method is not only economized in terms of computational resources consumption, but also is effective for the formation tracking task implementation. Furthermore, the inter-event triggering time interval $T_k = t_{k+1} - t_k$ and the triggering instants are shown in Fig. 7.7. From Table 7.3, it can be obtained that $\min\{T_k\} = 0.2s > 0.1$ s always holds true, which prevents the Zeno phenomenon from occurring successfully.

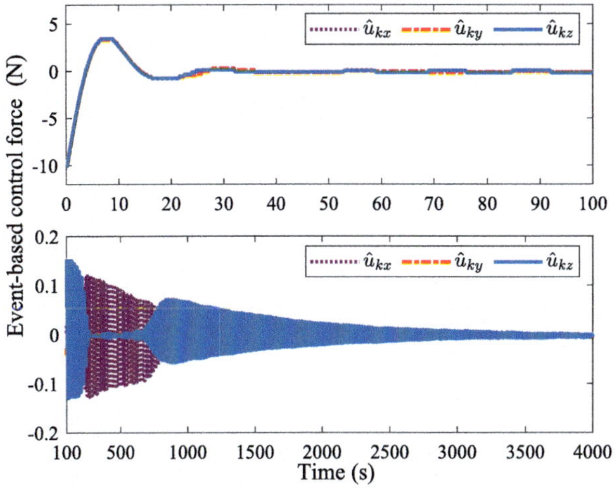

(a) event-based control force during 0 – 4000 s

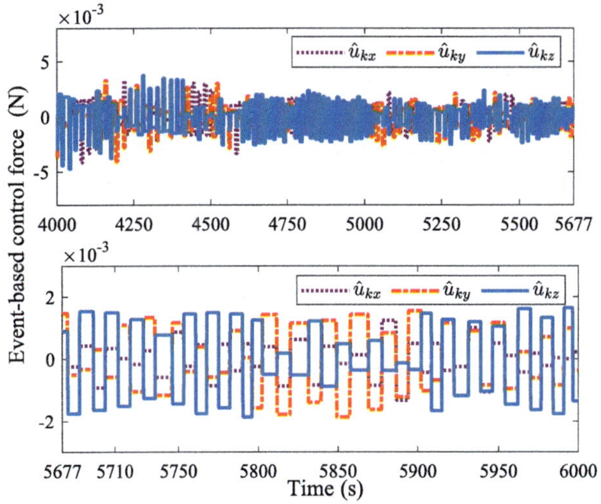

(b) event-based control force during 4000 – 6000 s

Fig. 7.4 Time responses of the event-based control forces

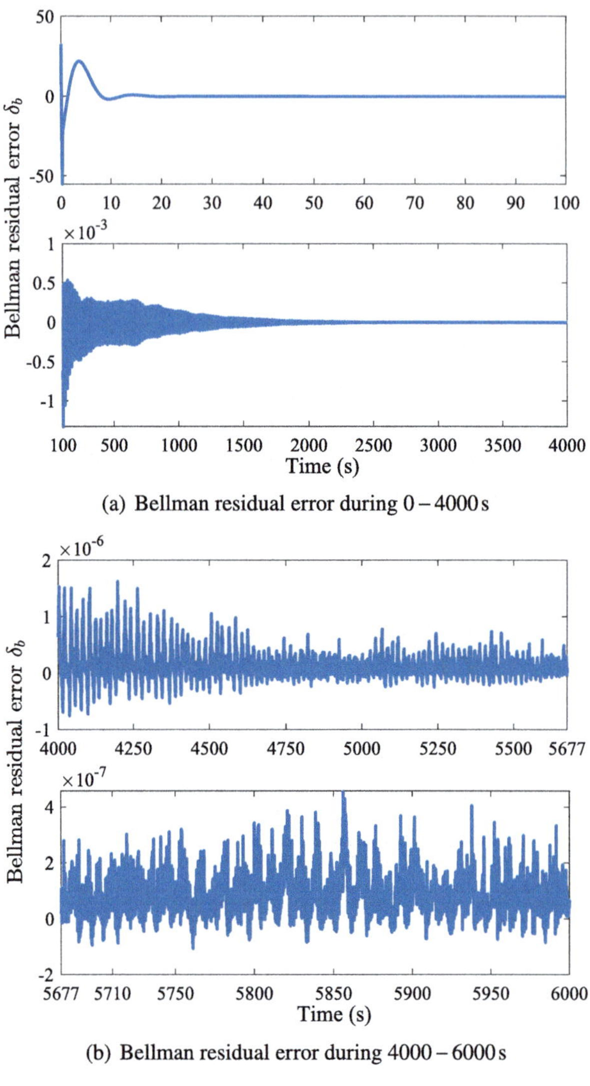

Fig. 7.5 Time response of the Bellman residual error

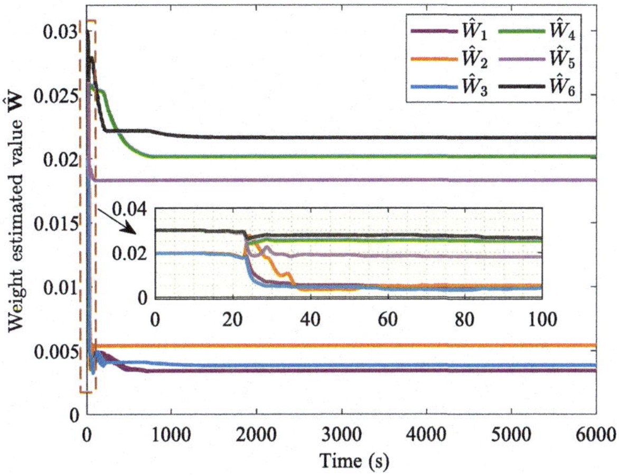

Fig. 7.6 Time responses of the weight estimates $\hat{\mathbf{W}}$

Table 7.2 The steady-state convergence errors under different sampling methods

Steady-state tracking errors	Five different sampling methods				
	Case I	Case II	Case III	Case IV	Case V
Relative position (m)	≤0.05	≤0.05	≤0.05	≤0.05	–
Relative velocity (m/s)	≤0.01	≤0.0005	≤0.01	≤0.02	–

Table 7.3 Controller update analysis of the follower spacecraft

	Case I	Case II	Case III	Case IV	Case V
Total number	717	60000	3000	1500	1200
Max interval (s)	18.5	0.1	2	4	5
Min interval (s)	0.2	0.1	2	4	5
Max frequency (Hz)	5	10	0.5	0.25	0.2
Average interval (s)	7.9	0.1	2	4	5

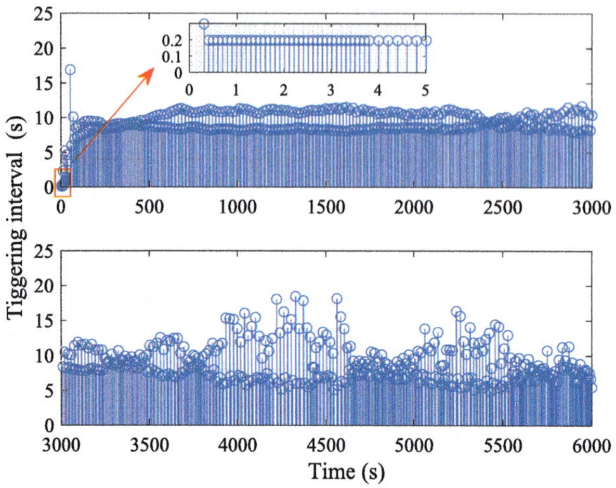

Fig. 7.7 The triggering time and triggering intervals

In summary, the proposed event-based near optimal tracking control policy is capable of achieving the formation control objective with the acceptable accuracy, whilst keeping approximate optimal performance. Meanwhile, such an event-based control scheme is extremely resource-economized and is more feasible in the realistic engineering applications.

7.5 Summary

This chapter has focused on the optimal formation tracking control for the leader-follower spacecraft formation flying system. Based on the ADP technique, a continuous near optimal tracking control scheme was first designed, where the critic-only neural network is established to approximate the optimal controller. Moreover, by resorting to the gradient decent algorithm and the adaptive projection rule, a bounded weight tuning law was derived to update the critic neural network. To further avoid unnecessary resources expenditure, an event-triggered strategy was developed to regulate the update frequency of the controller. By defining an input-based measurement error, the selection of the design parameters is totally independent of the unknown Lipschitz constant, which makes the control scheme easier to be implemented.

References

1. Bandyopadhyay S, Subramanian GP, Foust R, Morgan D, Chung SJ, Hadaegh F (2015) A review of impending small satellite formation flying missions. In: 53rd AIAA Aerospace Sciences Meeting, Kissimmee, FL, United states, pp 1623–1640
2. Guelman M, Kogan A, Kazarian A, Livne A, Orenstein M, Michalik H (2004) Acquisition and pointing control for inter-satellite laser communications. IEEE Transactions on Aerospace and Electronic Systems 40(4): 1239–1248
3. Cui B, Xia Y, Liu K, Wang Y, Zhai DH (2020) Velocity-observer-based distributed finite-time attitude tracking control for multiple uncertain rigid spacecraft. IEEE Transactions on Industrial Informatics 16(4): 2509–2519
4. Di Mauro G, Lawn M, Bevilacqua R (2018) Survey on guidance navigation and control requirements for spacecraft formation-flying missions. Journal of Guidance, Control, and Dynamics 41(3): 581–602
5. Liu X, Kumar KD (2012) Network-based tracking control of spacecraft formation flying with communication delays. IEEE Transactions on Aerospace and Electronic Systems 48(3): 2302–2314
6. Dang Z, Zhang Y (2015) Control design and analysis of an inner-formation flying system. IEEE Transactions on Aerospace and Electronic Systems 51(3): 1621–1634
7. Shouman M, Bando M, Hokamoto S (2019) Output regulation control for satellite formation flying using differential drag. Journal of Guidance, Control, and Dynamics 42(10): 2220–2232
8. Wei C, Wu X, Xiao B, Wu J, Zhang C (2021) Adaptive leader-following performance guaranteed formation control for multiple spacecraft with collision avoidance and connectivity assurance. Aerospace Science and Technology p 107266
9. Zhang XM, Han QL, Zhang BL (2017) An overview and deep investigation on sampled-data-based event-triggered control and filtering for networked systems. IEEE Transactions on Industrial Informatics 13(1): 4–16
10. Chen L, Li C, Xiao B, Guo Y (2019) Formation-containment control of networked Euler-lagrange systems: An event-triggered framework. ISA Transactions 86: 87–97
11. Hu Q, Shi Y, Wang C (2021) Event-based formation coordinated control for multiple spacecraft under communication constraints. IEEE Transactions on Systems, Man, and Cybernetics: Systems 51(5): 3168–3179
12. Hu Q, Shi Y (2020) Event-based coordinated control of spacecraft formation flying under limited communication. Nonlinear Dynamics 99(3): 2139–2159
13. Li J, Chen S, Li C, Wang F (2021) Distributed game strategy for formation flying of multiple spacecraft with disturbance rejection. IEEE Transactions on Aerospace and Electronic Systems 57(1): 119–128
14. Nair RR, Behera L (2018) Robust adaptive gain higher order sliding mode observer based control-constrained nonlinear model predictive control for spacecraft formation flying. IEEE/CAA Journal of Automatica Sinica 5(1): 367–381
15. Broida J, Linares R (2019) Spacecraft rendezvous guidance in cluttered environments via reinforcement learning. In: Advances in the Astronautical Sciences, Maui, HI, United states, pp 1777–1788
16. Lewis FL, Vrabie D (2009) Reinforcement learning and adaptive dynamic programming for feedback control. IEEE Circuits and Systems Magazine 9(3): 32–50
17. Wei Q, Liao Z, Shi G (2021) Generalized actor-critic learning optimal control in smart home energy management. IEEE Transactions on Industrial Informatics 17(10): 6614–6623
18. Heydari A (2021) Optimal impulsive control using adaptive dynamic programming and its application in spacecraft rendezvous. IEEE Transactions on Neural Networks and Learning Systems 32(10): 4544–4552
19. Wei C, Luo J, Dai H, Duan G (2018) Learning-based adaptive attitude control of spacecraft formation with guaranteed prescribed performance. IEEE Transactions on Cybernetics 49(11): 4004–4016

20. Shi XN, Zhou D, Chen X, Zhou ZG (2021) Actor-critic-based predefined-time control for spacecraft attitude formation system with guaranteeing prescribed performance on SO (3). Aerospace Science and Technology 117: 106898
21. Vamvoudakis KG (2014) Event-triggered optimal adaptive control algorithm for continuous-time nonlinear systems. IEEE/CAA Journal of Automatica Sinica 1(3): 282–293
22. Yang X, He H, Liu D (2019) Event-triggered optimal neuro-controller design with reinforcement learning for unknown nonlinear systems. IEEE Transactions on Systems, Man, and Cybernetics: Systems 49(9): 1866–1878
23. Mu C, Wang K, Qiu T (2020) Dynamic event-triggering neural learning control for partially unknown nonlinear systems. IEEE Transactions on Cybernetics
24. Dhar NK, Verma NK, Behera L (2018) Adaptive critic-based event-triggered control for HVAC system. IEEE Transactions on Industrial Informatics 14(1): 178–188
25. Mu C, Wang K, Sun C (2021) Learning control supported by dynamic event communication applying to industrial systems. IEEE Transactions on Industrial Informatics 17(4): 2325–2335
26. Goodwin GC, Mayne DQ (1987) A parameter estimation perspective of continuous time model reference adaptive control. Automatica 23(1): 57–70
27. Thakur D, Srikant S, Akella MR (2015) Adaptive attitude-tracking control of spacecraft with uncertain time-varying inertia parameters. Journal of Guidance, Control, and Dynamics 38(1): 41–52
28. Yang Y, Vamvoudakis KG, Modares H, Yin Y, Wunsch DC (2020) Safe intermittent reinforcement learning with static and dynamic event generators. IEEE Transactions on Neural Networks and Learning Systems 31(12): 5441–5455
29. Liu Q, Ye M, Qin J, Yu C (2019) Event-triggered algorithms for leader–follower consensus of networked Euler–lagrange agents. IEEE Transactions on Systems, Man, and Cybernetics: Systems 49(7): 1435–1447

Chapter 8
Adaptive Prescribed Performance Pose Control of Spacecraft Under Motion Constraints

8.1 Introduction

Spacecraft rendezvous and proximity operations (RPOs), as enabling technologies for several on-orbit missions such as removing space debris [1], on-orbit servicing [2], repairing defunct satellites [3], etc., have gained extensive attention. Different from the cases involving only translation or rotation, the proximity operations require that the pursuer perform the translational and rotational maneuvers with respect to the target simultaneously. This fact necessitates the six-degrees-of-freedom (6-DOF) pose tracking control design and analysis. Some representative solutions to the pose tracking control problem have been reported in the literature. Subbarao and Welsh [4] presented a nonlinear feedback control method for 6-DOF motion synchronization of an active pursuer with a tumbling target. Following a similar framework, Hu et al. [5] further explored the finite-time motion synchronization control problem. Recently, Gui and Ruiter [6] proposed a fault-tolerant pose tracking control scheme through incorporating an adaptive integral SMC and an online control allocator. Note, however, that although fruitful results are now available, the dominant majority only focus on achieving the ultimate goals while ignoring underlying motion and performance constraints in RPOs.

To accomplish the RPOs, the pursuer usually needs to maneuver towards the desired position while adjusting its attitude such that the boresight of its onboard vision sensor points towards the target for relative navigation. This involves two synchronously occurring maneuvers: relative position tracking and boresight pointing adjustment. Regarding the former, proximity safety concern requires the pursuer to maneuver inside a certain approach corridor. Thus, for relative position tracking, a candidate controller, apart from achieving the ultimate control objectives, should also ensure approaching path constraint. Toward this direction, various control methods such as artificial potential field (APF) approach [7–9], model predictive control (MPC) [10, 11], optimal control [12], etc., have been presented. While for the latter, the common practice is to design an attitude controller to track the desired attitude that is extracted according to the boresight pointing requirement [13, 14]. It is noted, however, that such an idea may not satisfy the field-of-view (FOV) constraint that inherently arises owing to the limited FOV of the vision sensor onboard the pursuer,

Q. Hu et al., *Intelligent Autonomous Control of Spacecraft with Multiple Constraints*,
https://doi.org/10.1007/978-981-99-0681-9_8

especially during the transient phase; on the other hand, it may suffer from a singularity problem in the desired attitude extraction. Lee and Mesbahi [15] proposed an APF-based feedback control for spacecraft reorientation under attitude constraints, which can be extended to handle the FOV constraint. The caveat here is that most of the above results can only be used for the translational- or rotational-only maneuver under a simplified 3-DOF motion constraint, and a significant challenge arises for the 6-DOF pose control design when both motion constraints are considered simultaneously. Based upon the dual quaternion algebra, Lee and Mesbahi [16] solved the 6-DOF guidance and control problem for autonomous precision landing within the MPC framework, where the path and FOV constraints, as well as some other constraints were considered. Later, Dong et al. [17] solved a similar problem using the APF method.

Apart from the spatial motion constraints, guaranteeing transient and steady-state tracking performance is also a constraint of great importance for spacecraft RPOs, owing to its direct relationship with the mission-oriented demands on time window, permitted overshoot, and accuracy tolerance. In fact, achieving prescribed performance guarantees is a practical design aspect that is difficult to be embodied in the closed-loop design but is nonetheless crucial to ensure mission success. Fortunately, Bechlioulis and Rovithakis [18] creatively proposed the prescribed performance control (PPC) technique, which could fill this gap. Typically, Hu et al. [19] and Shao et al. [20] investigated the attitude tracking control problem of spacecraft in the PPC framework. However, since exponentially decaying performance functions are commonly used to build the performance envelops (see, e.g., [18–20]), the output tracking errors may only be steered to the predefined residual sets as time tends to infinite, which is undesirable for the time-critical missions. Recently, the appointed-time PPC design problem has been explored in [21–23]. Nonetheless, the simultaneous treatment of both motion and performance constraints would increase significantly the complexity of the pose control design, and to the author's knowledge, no results in this matter have been found in the related literature.

When viewed from a practical implementation perspective, mass and inertia uncertainties caused, for example, by payload variations, rotation of solar arrays, and fuel consumption, may have adverse effects on the response of the closed-loop system. The adaptive control theory paves the way for achieving stable pose tracking in the presence of parameter uncertainties. In [5, 6, 24], adaptive controllers were derived based upon the certainty-equivalence (CE) principle for pose tracking problem in spacecraft RPOs. However, they often render the closed-loop performance poor when compared to the ideal deterministic control case, due to nonsatisfaction of the persistent excitation (PE) condition or slow convergence rates in parameter estimations [25]. As a remedy for this drawback, several non-CE adaptive controllers have recently been presented in [25–27], based upon the immersion and invariance (I&I) approach originally proposed in [28]. However, nearly every one of these existing adaptive control approaches does not take into account the motion and/or performance constraints. Up to now, how to develop a non-CE adaptive pose tracking control scheme for spacecraft RPOs under both kinds of constraints still remains open.

To address the issues discussed above, in this chapter, a novel adaptive PPC framework is presented for spacecraft RPOs with a passive target (with particular attention to a freely tumbling vehicle), under parameter uncertainties as well as motion and performance constraints. By visualizing the motion constraints and the prescribed performance metrics as the pose tracking error bounds, the key idea behind the PPC design is to transform the original constrained tracking error dynamics into an equivalent "state-constrained" one, whose stabilization is shown to be sufficient to address the problems under study. Then, a non-CE adaptive controller is designed using the barrier function in conjunction with the backstepping control, capable of guaranteeing that the transformed errors remain within the specified ranges, despite the presence of parameter uncertainties. As a consequence, the overall control scheme accomplishes the RPOs, whilst complying with the underlying motion and performance constraints. The spatial motion constraints, together with the imposed performance metrics, are artfully converted into the pose tracking error bounds. Given this fact, a non-CE adaptive pose controller is designed, which, by method of the PPC methodology integrating the appointed-time performance functions developed in [23] permits the achievement of proximity operations in a designer-appointed time, whilst complying with the motion and performance constraints, despite the presence of parameter uncertainties. Besides, the underlying singularity problem in the attitude extraction algorithm is avoided through properly choosing the performance bounds for the position tracking errors.

The remainder of the chapter is structured as follows. Section 8.2 presents the control problem formulation and describes the spatial motion constraints. The non-CE adaptive pose tracking controller is derived in Sect. 8.3, along with detailed stability proofs. The simulation results are given in Sect. 8.4, followed by the concluding remarks in Sect. 8.5.

8.2 Problem Formation

Four reference frames are involved in describing the spacecraft relative position and attitude motions for RPOs: the ECI frame $\mathcal{I}$, LVLH frame $\mathcal{L}$, pursuer's body frame $\mathcal{P}$, and target's body frame $\mathcal{T}$. The reader is referred to Sect. 2.3 for more details on the involved references. Assume without loss of generality that the docking ports of the pursuer and target spacecraft locate on the $+X$-axis of $\mathcal{P}$ and $-X$-axis of $\mathcal{T}$, respectively. Since the engagement time for close proximity operations is much shorter than the target's orbital period, we neglect the target orbital perturbations, as stated in Sect. 2.3.

Generally speaking, spacecraft RPOs with a freely tumbling target can be divided into two synchronously occurring maneuvers: relative position tracking and boresight pointing adjustment. As stated in Sect. 8.1, each maneuver suffers from both kinematic and dynamic constraints. For ease of illustration, in the following we separately formulate the control problems for these two maneuvers, along with the detailed constraint descriptions.

Assumption 8.1 The mass and inertia matrix of the pursuer are constant (or slowly varying), but otherwise unknown.

8.2.1 Relative Position Tracking

By resorting to the transport theorem [29], a new relative translational dynamics (2.30) is established in the target's body-fixed frame, which can help facilitate the position control design, when compared with the fully nonlinear Clohessy-Wiltshire (CW) equations commonly used in the related literature (see, for example, [4, 5]), especially under the path constraint. The ultimate control objective for relative position tracking is to steer the pursuer to the desired position on $+X$-axis of $\mathcal{T}$. To this end, a desired relative position vector $\boldsymbol{\rho}_d = [\rho_d, 0, 0]^\top$ resolved in $\mathcal{T}$ is introduced, where $\rho_d > 0$ is chosen by the designer according to the mission demands. The position and velocity tracking errors can thus be specified as $\boldsymbol{\rho}_e = \boldsymbol{\rho} - \boldsymbol{\rho}_d$ and $\boldsymbol{v}_e = \dot{\boldsymbol{\rho}}$ (due to $\dot{\boldsymbol{\rho}}_d = \boldsymbol{0}$). Then, the translational tracking error dynamics are given in (2.52). Since both the pursuer and target are not dimensionless, the position controller, except from achieving the ultimate control objective, should also guarantee that the pursuer maneuvers within a certain approach corridor (a cone-shaped zone around docking axis, in general) for safety concern, as shown in Fig. 8.1. This is the approaching path constraint and, as can be seen from Fig. 8.1 (more precisely, the outer light-cone), the condition that ensures its satisfaction is as follows:

$$\sqrt{\rho_2^2 + \rho_3^2} < \tan\alpha\rho_1.$$

However, it is difficult to embody the above constraint in the tracking control design, owing to its indirect relationship with the position tracking errors ρ_{ei}, $i = 1, 2, 3$. To surmount this difficulty, another cone-shaped zone, i.e., the inner light-cone in Fig. 8.1, is introduced. Intuitively, it is sufficient to guarantee constraint satisfaction if the pursuer keeps maneuvering within the inner light-cone. Given this, the constraint equation for the approaching path constraint becomes

$$\sqrt{\rho_{e2}^2 + \rho_{e3}^2} < \tan\bar{\alpha}\rho_{e1}. \tag{8.1}$$

On the other hand, to satisfy the mission-oriented demands on maneuver completion time, permitted overshoot, and accuracy tolerance, the position controller should additionally guarantee certain transient and steady-state tracking performance.

The control problem for relative position tracking is formulated as follows:

Problem 8.1 Consider the translational tracking error dynamics (2.52) with initial conditions satisfying (8.1). Design properly the position control law $\boldsymbol{f}_c$ such that the following goals are achieved in the presence of mass uncertainty:

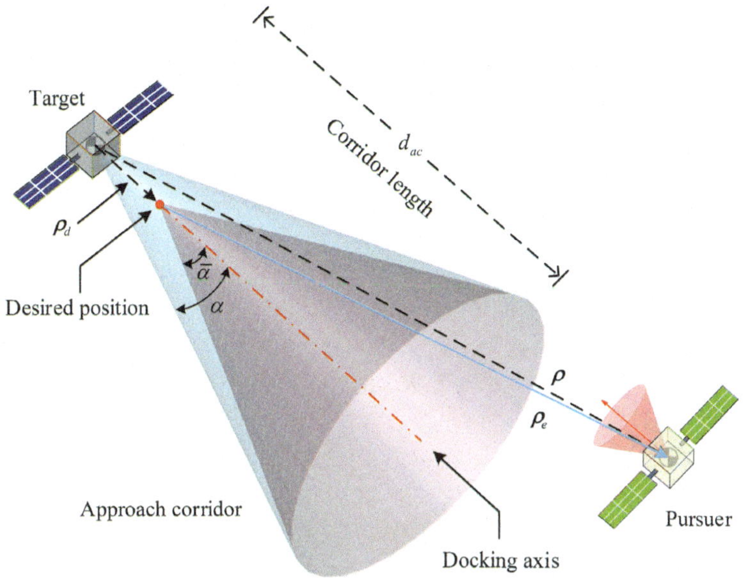

Fig. 8.1 Illustration of the approaching path constraint

- The position tracking errors $\rho_{ei}, i = 1, 2, 3$ converge to preassigned residual sets in a preset time t_f, while exhibiting their maximum overshoots less than given constants;
- The constraint equation (8.1) holds for all $t \geq 0$.

8.2.2 *Boresight Pointing Adjustment*

Assume that the boresight of the vision sensor onboard the pursuer is perfectly aligned with the unit vector $\boldsymbol{x}_{\mathcal{P}} = [1, 0, 0]^\top$ in $\mathcal{P}$, as depicted in Fig. 8.2, where the cone with half-cone angle β represents the FOV of the vision sensor. Intuitively, the ultimate control objective for boresight pointing adjustment is to keep the boresight vector $\boldsymbol{x}_{\mathcal{P}}$ oriented towards the target, which is actually equivalent to achieving $\boldsymbol{x}_{\mathcal{P}} = -\boldsymbol{R}_{\mathcal{PT}}\boldsymbol{\rho}/\|\boldsymbol{\rho}\|$. To this end, a desired LOS frame $\mathcal{D}$ is introduced, whose attitude orientation w.r.t. $\mathcal{I}$ is described as the unit quaternion $\boldsymbol{q}_d = [\boldsymbol{q}_{dv}^\top, q_{d4}]^\top \in \mathbb{R}^4$ such that (2.35) holds. By building upon the work of Roberts and Tayebi [30], a desired attitude quaternion $\boldsymbol{q}_d$ corresponding to a minimum-angle rotation is extracted, whereby the boresight pointing adjustment maneuver can be achieved through attitude tracking. To obtain $\boldsymbol{q}_d$, we first extract the quaternion $\bar{\boldsymbol{q}} = [\bar{\boldsymbol{q}}_v^\top, \bar{q}_4]^\top \in \mathbb{R}^4$ of $\mathcal{D}$ w.r.t. $\mathcal{T}$ through Lemma 2.1 in Sect. 2.4.3.2.

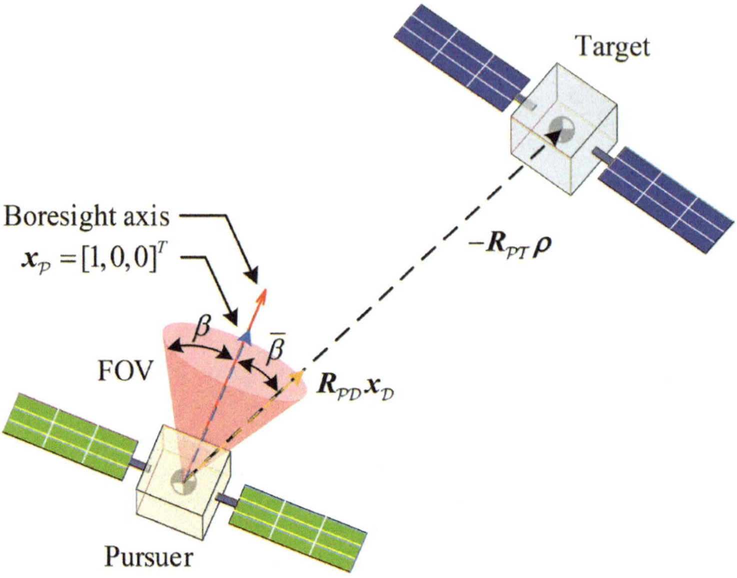

Fig. 8.2 Illustration of the field-of-view constraint

Remark 8.1 The attitude extraction algorithm summarized by Lemma 2.1 will suffer from a singularity problem when $\|\rho\| = 0$ and/or $x = -x_{\mathcal{D}}$. The former condition can be avoided, as long as Problem 8.1 is solved through a properly designed position controller. While the latter one corresponds to ρ being located along $x_{\mathcal{T}}$ in $\mathcal{T}$. To avoid this condition, the position controller is additionally required to guarantee that the inequality $\rho_2^2 + \rho_3^2 \neq 0$ (equivalent to $\rho_{e2}^2 + \rho_{e3}^2 \neq 0$) holds. At first glance, this constraint is incompatible with the ultimate control objective of the relative position tracking. But, from a practical viewpoint, the convergence of ρ_{e2} and ρ_{e3} to nonzero but sufficiently small values satisfying the mission demands on accuracy tolerance, instead of zero, is acceptable.

According to (2.42)–(2.45), the rotational tracking error dynamics (2.46) and (2.47) can be obtained. To facilitate the pose tracking controller design, the tracking error dynamics described by (2.46) and (2.47) are further transformed into the Euler-Lagrange equation (2.48). To ensure the transformation is valid, the following condition must hold (recalling Remark 2.3):

$$\det(q_{e4} \boldsymbol{I}_3 + \boldsymbol{S}(\boldsymbol{q}_{ev})) = q_{e4}(t) \neq 0 \ \ \forall t \geq 0. \tag{8.2}$$

Owing to the intrinsically limited FOV of the vision sensor, the pursuer needs to perform constrained attitude maneuver in order that the target always stays within the FOV of the vision sensor, as shown in Fig. 8.2. This is called here the FOV constraint, whose satisfaction can be determined by:

$$\cos\overline{\beta} = \boldsymbol{x}_{\mathcal{P}} \cdot \frac{-\boldsymbol{R}_{\mathcal{PT}}\boldsymbol{\rho}}{\|\boldsymbol{\rho}\|} > \cos\beta. \tag{8.3}$$

From (8.3), it is evident that the $+X$-axis of $\mathcal{D}$ points towards the CoM of the target, and therefore the left-hand side of (8.3) is equivalent to

$$\cos\overline{\beta} = \boldsymbol{x}_{\mathcal{P}} \cdot \boldsymbol{R}_{\mathcal{PD}}\boldsymbol{x}_{\mathcal{D}} = 1 - 2(q_{e2}^2 + q_{e3}^2), \tag{8.4}$$

where the fact that $\boldsymbol{x}_{\mathcal{D}} = \boldsymbol{x}_{\mathcal{P}} = [1, 0, 0]^\top$ and the definition of $\boldsymbol{R}_{\mathcal{PD}}$ have been used. By combining (8.3) and (8.4), the constraint equation for the FOV constraint can be easily derived that

$$1 - 2(q_{e2}^2 + q_{e3}^2) > \cos\beta. \tag{8.5}$$

Formally, the control problem for boresight pointing adjustment can now be formulated as follows:

Problem 8.2 Consider the rotational tracking error dynamics described by (2.48) with initial conditions satisfying (8.2) and (8.5). Design the attitude control law $\boldsymbol{\tau}_c$ such that the following hold in the presence of inertia uncertainties:

- The attitude tracking errors q_{ei}, $i = 1, 2, 3$ converge to preassigned residual sets in a pre-appointed time t_f^*, while exhibiting maximum overshoots less than given constants.
- The conditions (8.2) and (8.5) hold for all $t \geq 0$.

Remark 8.2 From (2.35), it is clear that the boresight pointing adjustment is directly related to the relative position vector $\boldsymbol{\rho}$, indicating close translation-rotation coupling. Another aspect that needs to be emphasized is that it is a time-consuming and sometimes troublesome task to obtain the analytical expression for $\dot{\boldsymbol{\omega}}_d$. To bypass this barrier, we can let $\boldsymbol{\omega}_d$ pass through a low-pass filter of the form $c\dot{\boldsymbol{z}} = -\boldsymbol{z} + \boldsymbol{\omega}_d$, where $c > 0$ is the filter time constant denoting the bandwidth of the filter. By choosing c sufficient small, $\boldsymbol{z}$ can be viewed as equivalent to $\boldsymbol{\omega}_d$, and at the same time, $\dot{\boldsymbol{z}} \approx \dot{\boldsymbol{\omega}}_d$. Therefore, $\dot{\boldsymbol{z}}$ can be used in lieu of $\dot{\boldsymbol{\omega}}_d$ in the subsequent control design.

Remark 8.3 It should be noted that although the attitude extraction algorithm outlined in (2.37) corresponds to a minimum-angle rotation, it may not ensure satisfaction of the final mating conditions on the lateral axes for either direct docking or robotic capture. For RPOs with a freely tumbling target, the main focus is on real-time target monitoring and accuracy relative navigation, thus the lateral attitude mismatching is still acceptable (to some extent). In reality, the matching condition could be satisfied in the final docking/berthing phase through further attitude maneuvers.

8.3 Problem Solution

In this section, we focus on the design of a pose controller using the PPC methodology integrating appointed-time performance functions, with the aim of providing effective solutions to Problems 8.1 and 8.2 as formulated in Sect. 8.2.

Define $\boldsymbol{x}_1 = [\boldsymbol{\rho}_e^\top, \boldsymbol{q}_{ev}^\top]^\top \in \mathbb{R}^6$ as the pose tracking error and $\boldsymbol{x}_2 = \dot{\boldsymbol{x}}_1$. The integrated tracking error dynamics (2.53) can be rewritten as

$$\begin{cases} \dot{\boldsymbol{x}}_1 = \boldsymbol{x}_2, \\ \boldsymbol{M}\dot{\boldsymbol{x}}_2 + \boldsymbol{C}\boldsymbol{x}_2 + \boldsymbol{G} = \boldsymbol{A}\boldsymbol{D}\boldsymbol{u}, \end{cases} \tag{8.6}$$

which satisfies Properties 2.1 and 2.2 in Sect. 2.4.3.3.

8.3.1 Prescribed Performance

Definition 8.1 ([23]) A C^n (for an integer $n \geq 1$) continuous function $\eta(t) : \mathbb{R}_{\geq 0} \to \mathbb{R}_+$ will be called an appointed-time performance function (ATPF), if there exist a pre-defined time T and a preset constant $\eta_s > 0$ such that the following hold:

- $\eta(t)$ is positive and decreasing over time for $t \in [0, T)$;
- $\eta(t) \equiv \eta_s$ for all $t \geq T$.

To facilitate subsequent design and analysis, the initial value of the relative position vector $\boldsymbol{\rho}$ is required to satisfy

$$0 < \sqrt{\rho_{e2}^2(0) + \rho_{e3}^2(0)} < d_{ac} \tan\alpha, \;\; \rho_{e1}(0) > d_{ac} - \rho_d, \tag{8.7}$$

where $d_{ac} > 0$ is the corridor length (see Fig. 8.1). In practice, such a restriction is very mild, since $\boldsymbol{\rho}$ can always be properly initialized to render the satisfaction of the aforesaid conditions, for example, via a fly-around maneuver. Within this setting, it is sufficient to solve Problem 8.1 and, at the same time, ensure that $\rho_2^2 + \rho_3^2 \neq 0$ (recall Remark 8.1 for the reason), through achieving the following performance metrics imposed on the position tracking errors x_{1i}, $i = 1, 2, 3$:

$$P_{li}(t) < x_{1i}(t) < P_{ui}(t), \;\; i = 1, 2, 3, \tag{8.8}$$

with $P_{ui}(t)$ and $P_{li}(t)$ defined by

$$\begin{cases} P_{u1}(t) = \eta_1(t), & (8.9a) \\ P_{l1}(t) = \sqrt{\eta_2^2(t) + \eta_3^2(t)} \Big/ \tan\overline{\alpha}, & (8.9b) \\ P_{ui}(t) = [k_{si} + (1 - k_{si})k_{oi}]\eta_i(t),\ i = 2, 3, & (8.9c) \\ P_{li}(t) = (k_{si} - 1 - k_{si}k_{oi})\eta_i(t),\ i = 2, 3, & (8.9d) \end{cases}$$

where the design constants k_{si}, $i = 2, 3$ are chosen in a manner that: $k_{si} = 1$ if $x_{1i}(0) > 0$ and $k_{si} = 0$ if $x_{1i}(0) < 0$; $k_{oi} = 0$, $i = 2, 3$ are taken for non-overshoot time responses of $\rho_{e2}(t)$ and $\rho_{e3}(t)$; and $\eta_i(t)$, $i = 1, 2, 3$ are C^2 continuous ATPFs of the following form [23]

$$\eta_i(t) = \begin{cases} a_{i,0} + \sum_{j=1}^{4}(a_{i,j}t^j), & t < t_f \\ \eta_{i,s}, & t \geq t_f \end{cases}, \tag{8.10}$$

where t_f and $\eta_{i,s}$ ($\eta_{i,s} < a_{i,0}$) represent the settling time and steady value of $\eta_i(t)$, respectively, and the design parameters $a_{i,j}$, $j \in \{0, 1, 2, 3, 4\}$ are determined by a group of condition equations given below:

$$\begin{cases} \eta_i(0) &= a_{i,0} > 0, \\ \dot{\eta}_i(0) &= a_{i,1} = 0, \\ \eta_i(t_f) &= a_{i,0} + \sum_{j=1}^{4}(a_{i,j}t_f^j) = \eta_{i,s}, \\ \dot{\eta}_i(t_f) &= a_{i,1} + 2a_{i,2}t_f + 3a_{i,3}t_f^2 + 4a_{i,4}t_f^3 = 0, \\ \ddot{\eta}_i(t_f) &= 2a_{i,2} + 6a_{i,3}t_f + 12a_{i,4}t_f^2 = 0. \end{cases} \tag{8.11}$$

It should be noted that the function defined in (8.10) is only one choice of ATPFs, and there are other choices, as discussed in [23]. The design constants $a_{i,0}$, $i = 1, 2, 3$ need to satisfy the following conditions:

$$a_{i,0} > x_{1i}(0), \text{ and } \sqrt{a_{2,0}^2 + a_{3,0}^2} \leq d_{ac} \tan\alpha, \tag{8.12}$$

which, together with the fact that $\rho_{e1}(0) > d_{ac} - \rho_d$, guarantee that (8.8) holds at $t = 0$. From (8.8)–(8.11), it is evident that t_f specifies the upper bound on the settling time of $x_{1i}(t)$, while $\eta_{i,s}$ specifies the maximum allowable size of $x_{1i}(t)$ at the steady state and can be set to a sufficiently small value reflecting the resolution of the navigation sensors, thus establishing practical convergence of $x_{1i}(t)$ to zero; moreover, no overshoot is permitted by $x_{1i}(t)$. The caveat is that, for the validity of (8.8), the condition $\eta_{1,s} > \sqrt{\eta_{2,s}^2 + \eta_{3,s}^2} \Big/ \tan\overline{\alpha}$ should hold.

Provided that (8.8) always holds, it can be derived from (8.8) and (8.9b) that

$$\sqrt{\rho_{e2}^2(t) + \rho_{e3}^2(t)} < \sqrt{\eta_2^2(t) + \eta_3^2(t)} < \tan\overline{\alpha}\rho_{e1}(t),$$

which satisfies the constraint Eq. (8.1) for the approaching path constraint; moreover, from (8.9c) and (8.9d), together with the selection of $k_{oi} = 0$, $i = 2, 3$, it easily follows that $\rho_{e2}^2(t) + \rho_{e3}^2(t) \neq 0\ \forall t \geq 0$. Thus, Problem 1 can be addressed and the constraint $\rho_2^2(t) + \rho_3^2(t) \neq 0$, $\forall t \geq 0$ can be guaranteed, through guaranteeing that (8.8) holds for all time.

To address Problem 8.2, we similarly impose the performance bounds on the attitude tracking errors x_{1i}, $i = 4, 5, 6$:

$$P_{li}(t) < x_{1i}(t) < P_{ui}(t),\ i = 4, 5, 6, \tag{8.13}$$

where $P_{ui}(t)$ and $P_{li}(t)$ have the same forms as that in (8.9c) and (8.9d). Specifically, k_{si} is chosen in an analogous way as that for k_{s2} and k_{s3} (both cases of $k_{si} = 1$ and $k_{si} = 0$ can be used for zero initial condition), $0 \leq k_{oi} \leq 1$ is a constant used to specify the maximum level of overshoot in x_{1i} (notice that in case $x_{1i}(0) = 0$, k_{oi} cannot be chosen equal to zero), and $\eta_i(t)$ is also a C^2 continuous ATPF and has the same form as that given in (8.10), but with t_f^* instead of t_f. In general, t_f^* should be chosen much smaller than t_f, in order to accomplish the boresight pointing adjustment within a short period of time. Of particular note is that the design constants $a_{i,0}$, $i = 4, 5, 6$ should be chosen such that the following hold:

$$\begin{cases} a_{i,0} > x_{1i}(0), & (8.14a) \\ a_{4,0}^2 + a_{5,0}^2 + a_{6,0}^2 \leq 1, & (8.14b) \\ a_{5,0}^2 + a_{6,0}^2 \leq (1 - \cos\beta)/2, & (8.14c) \end{cases}$$

where (8.14a) ensures that (8.13) holds at $t = 0$, while (8.14b) and (8.14c) are introduced to pave the way towards guaranteeing the satisfaction of the constraints in (8.2) and (8.5).

Assume momentarily that (8.13) holds for all time. Then, from (8.13) and (8.14b), we deduce that $\|\boldsymbol{q}_{ev}\|^2 < a_{4,0}^2 + a_{5,0}^2 + a_{6,0}^2 \leq 1$. As is well known, the unit quaternion conforms to the normalization constraint $\|\boldsymbol{q}_e\| = 1$, then it follows that $q_{e4}^2 = 1 - \|\boldsymbol{q}_{ev}\|^2 > 0$. It is thus concluded that $q_{e4}(t) \neq 0$ for all $t \geq 0$. Next, inspecting (8.13) and (8.14c) finds that

$$q_{e2}^2(t) + q_{e3}^2(t) < a_{5,0}^2 + a_{6,0}^2 \leq (1 - \cos\beta)/2,$$

whereby it can be further inferred that

$$1 - 2(q_{e2}^2(t) + q_{e3}^2(t)) > \cos\beta,$$

which complies with (8.5) for the FOV constraint. With the above in mind, Problem 8.2 can be solved by guaranteeing that (8.13) holds for all time.

8.3.2 Non-CE Adaptive Pose Control

To incorporate the prescribed performance metrics described by (8.8) and (8.13) into the closed-loop design, we here introduce an error transformation [19]

$$\varepsilon_i(t) = \frac{2x_{1i}(t) - (P_{ui}(t) + P_{li}(t))}{P_{ui}(t) - P_{li}(t)}, \tag{8.15}$$

where $\varepsilon_i(t)$ $(i = 1, 2, \ldots, 6)$ denote the transformed errors. As claimed in [19], the pose tracking errors $x_{1i}(t)$, $i = 1, 2, \ldots, 6$ will evolve strictly along their respective performance bounds, provided that $|\varepsilon_i(t)| < 1$, $i = 1, 2, \ldots, 6$. Hereafter, addressing Problems 8.1 and 8.2 boils down to guaranteeing that $|\varepsilon_i(t)| < 1$, $i = 1, 2, \ldots, 6$ are never transgressed.

From now on, for notational conciseness, we drop the time argument t except where this omission can cause confusion. Taking the time derivative of ε_i leads to

$$\dot{\varepsilon}_i = \frac{2\dot{x}_{1i} - \xi_i}{P_{ui} - P_{li}},$$

with $\xi_i = \dot{P}_{ui} + \dot{P}_{li} + \varepsilon_i(\dot{P}_{ui} - \dot{P}_{li})$. Define $\boldsymbol{\varepsilon} = [\varepsilon_1, \varepsilon_2, \ldots, \varepsilon_6]^\top$, $\boldsymbol{\xi} = [\xi_1, \xi_2, \ldots, \xi_6]^\top$, and $\boldsymbol{\vartheta P} = \mathrm{diag}_{i \in \{1,2,\ldots,6\}}[1/(P_{ui} - P_{li})]$. The dynamic equation for the transformed errors can then be compactly expressed as

$$\dot{\boldsymbol{\varepsilon}} = \boldsymbol{\vartheta P}(2\boldsymbol{x}_2 - \boldsymbol{\xi}). \tag{8.16}$$

Theoretically speaking, it is a challenging task to develop a non-CE adaptive controller using the I&I adaptive approach to address the problem under study, owing to the existence of a non-cascaded term $\boldsymbol{\xi}$ in (8.16). To circumvent this difficulty, the controller design is carried out following the backstepping approach. Define an error vector $\boldsymbol{s} = \boldsymbol{x}_2 - \boldsymbol{x}_c$, where $\boldsymbol{x}_c$ is the virtual control law to be designed at a later stage. The detailed design procedure is outlined as follows:

Step 1: To respect the prescribed performance bounds, we consider a logarithmic barrier function, given by

$$V_1 = \frac{k_0}{2}\log\left(\prod_{i=1}^{6} \frac{1}{1 - \varepsilon_i^2}\right), \tag{8.17}$$

where $0 < k_0 \leq 1$ is a constant weighting parameter. It can be verified that V_1 in (8.17) is positive definite and C^1 continuous in the set $\mathbb{D}_\varepsilon = \{\boldsymbol{\varepsilon} \in \mathbb{R}^6 : |\varepsilon_i| < 1,\ i = 1, 2, \ldots, 6\}$, and thus a valid Lyapunov function candidate. Then, taking the time derivative of V_1 along (8.16) would lead to

$$\dot{V}_1 = k_0 \boldsymbol{\varepsilon}^\top \boldsymbol{\vartheta N}[2(\boldsymbol{s} + \boldsymbol{x}_c) - \boldsymbol{\xi}], \tag{8.18}$$

where $\boldsymbol{\vartheta N} = \mathrm{diag}_{i \in \{1,2,\ldots,6\}}[1/(1 - \varepsilon_i^2)]\boldsymbol{\vartheta P}$.

Choose the virtual control law $\boldsymbol{x}_c$ as

$$\boldsymbol{x}_c = 0.5(\boldsymbol{\xi} - k_1 \vartheta \boldsymbol{N}^{-1} \boldsymbol{\varepsilon}), \tag{8.19}$$

where $k_1 > 0$ is a control gain. Then, using (8.19) in (8.18) yields

$$\dot{V}_1 = -k_0 k_1 \|\boldsymbol{\varepsilon}\|^2 + \boldsymbol{y}^\top \boldsymbol{s}, \tag{8.20}$$

where $\boldsymbol{y} = 2k_0 \vartheta \boldsymbol{N} \boldsymbol{\varepsilon}$ is defined just for notational brevity.

Step 2: This step is devoted to designing a non-CE adaptive control law $\boldsymbol{u}$ to nullify $\boldsymbol{s}$. Unlike a CE adaptive controller, the parameter estimates of which come only from the update law, a function that satisfies a certain partial differential equation (PDE) is combined with the states of the update law to form the estimates. This requires solving a PDE. However, note that there usually exist no solutions to such a PDE for multi-input nonlinear systems. This is the so-called "integrability obstacle" that arises in the I&I adaptive control design. This obstacle is overcome in [25] through introducing tactfully low-pass filters for the state and regressor matrix. In this step, we employ the method in [25] to circumvent the integrability obstacle.

Taking the time derivative of $\boldsymbol{s}$ and using (8.6) render

$$\boldsymbol{M}\dot{\boldsymbol{s}} = -\boldsymbol{C}\boldsymbol{s} - \underbrace{(\boldsymbol{M}\dot{\boldsymbol{x}}_c + \boldsymbol{C}\boldsymbol{x}_c + \boldsymbol{G})}_{\boldsymbol{H}} + \boldsymbol{A}\boldsymbol{D}\boldsymbol{u}. \tag{8.21}$$

To deal with the mass and inertia uncertainties, we perform the following affine parameterization:

$$\begin{aligned} \boldsymbol{W}\boldsymbol{\theta}^* =& - \boldsymbol{C}\boldsymbol{s} - \boldsymbol{H} + \boldsymbol{M}(c_1 \dot{\boldsymbol{y}} + c_f c_1 \boldsymbol{y} + c_2 \boldsymbol{s}) \\ &+ \dot{\boldsymbol{M}}(c_1 \boldsymbol{y} - c_1 \boldsymbol{s}_f + \boldsymbol{s}), \end{aligned} \tag{8.22}$$

where $\boldsymbol{\theta}^* = [m_p, J_{p,11}, J_{p,22}, J_{p,33}, J_{p,12}, J_{p,23}, J_{p,13}]^\top$ with $J_{p,ij}$ the independent (ij)-th elements of $\boldsymbol{J}_p$, $\boldsymbol{W} \in \mathbb{R}^{6\times 7}$ is the regressor matrix whose detailed expression can be derived following [5], $c_1, c_2 > 0$ are any design constants, $c_f = c_1 + c_2$, and $\boldsymbol{s}_f$ is obtained from a low-pass filter of the form:

$$\dot{\boldsymbol{s}}_f = -c_f \boldsymbol{s}_f + \boldsymbol{s}, \tag{8.23}$$

whose initial condition is given by

$$\boldsymbol{s}_f(0) = (\boldsymbol{s}(0) + c_1 \boldsymbol{y}(0))/c_1. \tag{8.24}$$

By simple algebraic manipulations, (8.21) is rewritten as

$$\begin{aligned} \boldsymbol{M}\dot{\boldsymbol{s}} =& \boldsymbol{W}\boldsymbol{\theta}^* + \boldsymbol{A}\boldsymbol{D}\boldsymbol{u} - \boldsymbol{M}(c_1 \dot{\boldsymbol{y}} + c_f c_1 \boldsymbol{y} + c_2 \boldsymbol{s}) \\ &- \dot{\boldsymbol{M}}(\dot{\boldsymbol{s}}_f + c_1 \boldsymbol{y} + c_2 \boldsymbol{s}_f). \end{aligned} \tag{8.25}$$

Further, a filtered regressor matrix is generated by

$$\dot{\boldsymbol{W}}_f = -c_f \boldsymbol{W}_f + \boldsymbol{W}, \ \ \boldsymbol{W}_f(0) = \boldsymbol{0}. \tag{8.26}$$

For purpose of stability analysis, let us consider another linear filter involving the control input signal defined by

$$\dot{\boldsymbol{u}}_f = -c_f \boldsymbol{s}_f + \boldsymbol{A}\boldsymbol{D}\boldsymbol{u}. \tag{8.27}$$

Differentiating both sides of the filtered dynamics expressed in (8.23), followed by the use of (8.25)–(8.27), it follows that

$$\dot{\boldsymbol{\delta}} = -c_f \boldsymbol{\delta}, \ \boldsymbol{\delta} = \boldsymbol{M}(\dot{\boldsymbol{s}}_f + c_1 \boldsymbol{y} + c_2 \boldsymbol{s}_f) - \boldsymbol{W}_f \boldsymbol{\theta}^* - \boldsymbol{u}_f, \tag{8.28}$$

for which the solution can be calculated by

$$\dot{\boldsymbol{s}}_f = -c_1 \boldsymbol{y} - c_2 \boldsymbol{s}_f + \boldsymbol{M}^{-1}(\boldsymbol{W}_f \boldsymbol{\theta}^* + \boldsymbol{u}_f + \boldsymbol{\delta}(0) e^{-c_f t}). \tag{8.29}$$

Design the filtered input signal as

$$\boldsymbol{u}_f = -\boldsymbol{W}_f(\hat{\boldsymbol{\theta}} + \boldsymbol{\varphi}), \tag{8.30}$$

with $\hat{\boldsymbol{\theta}}$ and $\boldsymbol{\varphi}$ determined by

$$\dot{\hat{\boldsymbol{\theta}}} = \gamma \boldsymbol{W}_f^{\top}[(c_f + c_2)\boldsymbol{s}_f + c_1 \boldsymbol{y}] - \gamma \boldsymbol{W}^{\top} \boldsymbol{s}_f, \tag{8.31}$$

$$\boldsymbol{\varphi} = \gamma \boldsymbol{W}_f^{\top} \boldsymbol{s}_f, \tag{8.32}$$

where $\gamma > 0$ is a design constant. In this manner the composite term $\hat{\boldsymbol{\theta}} + \boldsymbol{\varphi}$ actually acts as an estimation of the unknown vector $\boldsymbol{\theta}^*$. We therefore define the estimation error as $\boldsymbol{z} = \hat{\boldsymbol{\theta}} + \boldsymbol{\varphi} - \boldsymbol{\theta}^*$. Since $\boldsymbol{W}_f(0) = \boldsymbol{0}$ (see (8.26)), we have $\boldsymbol{u}_f(0) = \boldsymbol{0}$ from (8.30). In addition, recalling (8.23), (8.24) and the fact that $c_f = c_1 + c_2$, it follows that $\dot{\boldsymbol{s}}_f(0) + c_1 \boldsymbol{y}(0) + c_2 \boldsymbol{s}_f(0) = \boldsymbol{0}$. Then, from (8.28), it can be easily checked that $\boldsymbol{\delta}(0) = \boldsymbol{0}$. As a consequence, $\boldsymbol{\delta}(t) = \boldsymbol{\delta}(0) e^{-c_f t} = \boldsymbol{0}$ holds for all $t \geq 0$. In view of this, (8.29) further reduces to

$$\dot{\boldsymbol{s}}_f = -c_1 \boldsymbol{y} - c_2 \boldsymbol{s}_f - \boldsymbol{M}^{-1} \boldsymbol{W}_f \boldsymbol{z}. \tag{8.33}$$

By (8.26), (8.31) and (8.32), the estimation error dynamics is derived that

$$\dot{\boldsymbol{z}} = -\gamma \boldsymbol{W}_f^{\top} \boldsymbol{M}^{-1} \boldsymbol{W}_f \boldsymbol{z}. \tag{8.34}$$

Choose the overall Lyapunov function candidate as

$$V_2 = V_1 + \frac{1}{2} s_f^\top s_f + \frac{\psi}{2\lambda_{\min}(M)} z^\top z, \tag{8.35}$$

where $\psi > 9/(4\gamma \min\{c_1, c_2\})$. Now taking the derivative of V_2 w.r.t. time along (8.33) and (8.34), and recalling (8.20) and (8.23), we have the following:

$$\begin{aligned}
\dot{V}_2 =& - k_0 k_1 \|\varepsilon\|^2 + y^\top(-c_1 y - M^{-1} W_f z) + s_f^\top(-c_2 s_f \\
& - M^{-1} W_f z) - \frac{\gamma\psi}{\lambda_{\min}(M)} z^\top W_f^\top M^{-1} M M^{-1} W_f z \\
\leq & - k_0 k_1 \|\varepsilon\|^2 - c_1 \|y\|^2 - c_2 \|s_f\|^2 - y^\top M^{-1} W_f z \\
& - s_f^\top M^{-1} W_f z - \gamma\psi \|M^{-1} W_f z\|^2 \\
=& - k_0 k_1 \|\varepsilon\|^2 - \frac{2c_1}{3}\|y\|^2 - \frac{2c_2}{3}\|s_f\|^2 - \frac{\gamma\psi}{3}\|M^{-1} W_f z\|^2 \\
& - \frac{c_1}{3}(\|y\|^2 + \frac{3}{c_1} y^\top M^{-1} W_f z + \frac{\gamma\psi}{c_1}\|M^{-1} W_f z\|^2) \\
& - \frac{c_2}{3}(\|s_f\|^2 + \frac{3}{c_2} s_f^\top M^{-1} W_f z + \frac{\gamma\psi}{c_2}\|M^{-1} W_f z\|^2) \\
\leq & - k_0 k_1 \|\varepsilon\|^2 - \frac{2c_1}{3}\|y\|^2 - \frac{2c_2}{3}\|s_f\|^2 - \frac{\gamma\psi}{3}\|M^{-1} W_f z\|^2.
\end{aligned} \tag{8.36}$$

Theorem 8.1 *Consider the integrated tracking error dynamics described by* (8.6) *under Assumption 8.1. Given initial conditions satisfying* (8.2), (8.5), *and* (8.7), *design properly the ATPFs* η_i, $i = 1, 2, \ldots, 6$ *satisfying* (8.12) *and* (8.14), *and choose corresponding design constants* k_{si}, $i = 2, 3, \ldots, 6$ *and* k_{oi}, $i = 4, 5, 6$, *such that* $\varepsilon(0) \in \mathbb{D}_\varepsilon$. *Then, the controller given below*

$$u = (AD)^{-1}\{-W(\hat{\theta} + \varphi) - \gamma W_f W_f^\top[c_1(y - s_f) + s]\}, \tag{8.37}$$

where $\hat{\theta}$ *and* φ *are determined, respectively, by* (8.31) *and* (8.32), *implemented in conjunction with the filters* (8.23) *and* (8.26), *can guarantee that the following results hold, despite the presence of mass and inertia uncertainties:*

- *The pose tracking errors* x_{1i}, $i = 1, 2, \ldots, 6$ *would evolve strictly along the preset performance envelops, thus providing an effective solution to Problems 8.1 and 8.2; moreover, the condition* $\rho_2^2(t) + \rho_3^2(t) \neq 0$ *holds for all* $t \geq 0$, *thereby ensuring nonsingularity in the desired attitude extraction.*
- *The transformed error* $\varepsilon(t)$, *the velocity error* $v_e(t)$, *and the angular velocity error* $\omega_e(t)$ *converge asymptotically to zero, that is,* $\lim_{t\to\infty}[\varepsilon(t), v_e(t), \omega_e(t)] = \mathbf{0}$.

Proof The proof proceeds in three bullet points below:

(1) As a consequence of (8.36), $\dot{V}_2$ is negative-semidefinite, indicating the boundedness of V_2 and hence V_1. From the fact that $V_1(t) \in \mathcal{L}_\infty$ and $\varepsilon(0) \in \mathbb{D}_\varepsilon$, we con-

clude that $\varepsilon(t) \in \mathbb{D}_\varepsilon$ for all $t \geq 0$. Thus, the pose tracking errors x_{1i}, $i = 1, 2, \ldots, 6$ evolve strictly along their respective prescribed performance envelops, in the sense that Problems 8.1 and 8.2 are successfully solved, as stated in Sect. 8.3.1. As no overshoot is permitted by the position tracking errors, invoking (8.8) further demonstrates that the condition $\rho_2^2(t) + \rho_3^2(t) \neq 0$ (for singularity avoidance as stated in Remark 8.1) holds for all $t \geq 0$.

(2) As V_2 is upper bounded by $V_2(0)$ and lower bounded by 0, we known that $\int_0^\infty \dot{V}_2(t)\,dt$ exists and is finite, which in turn suggests ε, $\boldsymbol{y}$, $\boldsymbol{s}_f$, and $\boldsymbol{M}^{-1}\boldsymbol{W}_f\boldsymbol{z} \in \mathcal{L}_2 \cap \mathcal{L}_\infty$. This also implies $\dot{\boldsymbol{s}}_f$, $\boldsymbol{s} \in \mathcal{L}_\infty$ from (8.23) and (8.33). Further, from the definition of $\boldsymbol{s}$ aided by the boundedness of $\boldsymbol{x}_c$, we have $\boldsymbol{x}_2 \in \mathcal{L}_\infty$ and thus $\boldsymbol{v}$, $\dot{\boldsymbol{q}}_{ev}$, $\dot{\boldsymbol{x}}_c$, and $\dot{\boldsymbol{y}} \in \mathcal{L}_\infty$. By invoking Lemma 2.2 and noting the fact that $|q_{e4}(t)| > \varrho > 0$ with ϱ a constant (recall our previous analysis for Remark 2.3), it can be shown from $\boldsymbol{\omega}_e = \boldsymbol{P}\dot{\boldsymbol{q}}_{ev}$ that $\boldsymbol{\omega}_e \in \mathcal{L}_\infty$. Consequently, from (2.46), (8.22), (8.26), and (8.34), we further claim that $\boldsymbol{W}$, $\boldsymbol{W}_f$, $\dot{\boldsymbol{W}}_f$, $\boldsymbol{M}^{-1}$, $\dot{\boldsymbol{M}}^{-1}$, $\boldsymbol{z}$, and $\dot{\boldsymbol{z}} \in \mathcal{L}_\infty$.

Based on the above arguments, we conclude that ε, $\boldsymbol{s}_f$, and $\boldsymbol{M}^{-1}\boldsymbol{W}_f\boldsymbol{z}$ are square integrable; moreover from their derivative given by

$$\begin{cases} \dot{\varepsilon} = \vartheta \boldsymbol{P}(2\boldsymbol{x}_2 - \boldsymbol{\xi}), \\ \dot{\boldsymbol{s}}_f = -c_1\boldsymbol{y} - c_2\boldsymbol{s}_f - \boldsymbol{M}^{-1}\boldsymbol{W}_f\boldsymbol{z}, \\ \dfrac{d}{dt}(\boldsymbol{M}^{-1}\boldsymbol{W}_f\boldsymbol{z}) = \dot{\boldsymbol{M}}^{-1}\boldsymbol{W}_f\boldsymbol{z} + \boldsymbol{M}^{-1}\dot{\boldsymbol{W}}_f\boldsymbol{z} + \boldsymbol{M}^{-1}\boldsymbol{W}_f\dot{\boldsymbol{z}}, \end{cases}$$

it can be inferred that $\dot{\varepsilon}$, $\dot{\boldsymbol{s}}_f$, and $\frac{d}{dt}(\boldsymbol{M}^{-1}\boldsymbol{W}_f\boldsymbol{z})$ are uniformly continuous. Then, applying Barbalat's Lemma concludes

$$\lim_{t \to \infty} [\varepsilon(t), \boldsymbol{s}_f(t), \boldsymbol{M}^{-1}\boldsymbol{W}_f\boldsymbol{z}(t)] = \boldsymbol{0},$$

whereby we have $\lim_{t\to\infty} \boldsymbol{y}(t) = \boldsymbol{0}$ (recalling the definition of $\boldsymbol{y} = 2k_0\vartheta \boldsymbol{N}\varepsilon$), and from (8.33), it therefore follows that $\lim_{t\to\infty} \dot{\boldsymbol{s}}_f(t) = \boldsymbol{0}$. Based on the latter result and from (8.23), we can deduce that $\lim_{t\to\infty} \boldsymbol{s}(t) = \boldsymbol{0}$. As can be seen from (8.9) and (8.10), $\lim_{t\to\infty}[\dot{P}_{ui}(t), \dot{P}_{li}(t)] = \boldsymbol{0}$ $(i = 1, 2, \ldots, 6)$, whereby we can derive from (8.15) and (8.19) the asymptotic convergence of $\boldsymbol{x}_c(t)$. Finally, by the definition of $\boldsymbol{s}$, we have $\lim_{t\to\infty}[\boldsymbol{x}_2(t), \boldsymbol{v}_e(t), \boldsymbol{\omega}_e(t)] = \boldsymbol{0}$.

(3) Our last concern is to recover the actual control input $\boldsymbol{u}$ from the filtered input signal $\boldsymbol{u}_f$ defined by (8.30). To this end, we perform the following algebraic manipulations:

$$\begin{aligned} \boldsymbol{u} &= (\boldsymbol{A}\boldsymbol{D})^{-1}(\dot{\boldsymbol{u}}_f + c_f\boldsymbol{u}_f) \\ &= (\boldsymbol{A}\boldsymbol{D})^{-1}[-\dot{\boldsymbol{W}}_f(\hat{\boldsymbol{\theta}} + \boldsymbol{\varphi}) - \boldsymbol{W}_f(\dot{\hat{\boldsymbol{\theta}}} + \dot{\boldsymbol{\varphi}}) + c_f\boldsymbol{u}_f] \\ &= (\boldsymbol{A}\boldsymbol{D})^{-1}[-\boldsymbol{W}(\hat{\boldsymbol{\theta}} + \boldsymbol{\varphi}) - \boldsymbol{W}_f(\dot{\hat{\boldsymbol{\theta}}} + \dot{\boldsymbol{\varphi}})]. \end{aligned} \tag{8.38}$$

Using (8.31) and (8.32) in (8.38) and noting (8.23) and (8.26) directly lead to the expression (8.37), thus completing the proof. □

Remark 8.4 It has been concluded from the proof of Theorem 8.1 that

$$\lim_{t\to\infty} \boldsymbol{M}^{-1}\boldsymbol{W}_f z(t) = \mathbf{0},$$

indicating the establishment of an attracting manifold $\mathcal{S}$ defined by

$$\mathcal{S} = \{z \in \mathbb{R}^7 | \boldsymbol{M}^{-1}\boldsymbol{W}_f z = \mathbf{0}\}.$$

All closed-loop trajectories would ultimately end up inside $\mathcal{S}$, whereby the closed-loop dynamics reduce to

$$\dot{\varepsilon} = \vartheta \boldsymbol{P}(2\boldsymbol{x}_2 - \boldsymbol{\xi}), \ \dot{\boldsymbol{s}}_f = -c_1\boldsymbol{y} - c_2\boldsymbol{s}_f.$$

This is a salient feature of the proposed non-CE-based adaptive pose tracking controller, which permits the recovery of the ideal closed-loop performance gained in the absence of parameter uncertainties. Furthermore, unlike the existing CE-based adaptive controllers (e.g., see [5, 6]), the estimation error dynamics in (8.34) is independent of system tracking errors and linear only w.r.t. the estimation error z. Hence, if $z(t) = \mathbf{0}$ occurs at any instance of time t^*, then the adaptation stops thereafter and the adaptive parameters thereby stay locked at their true values. Note that the above two features can hardly be obtained in the CE-based adaptive controllers, even though they are applicable to solving the problems under study.

Remark 8.5 It is important to underscore that $\lim_{t\to\infty} \boldsymbol{M}^{-1}\boldsymbol{W}_f z(t) = \mathbf{0}$ does not necessarily mean $\lim_{t\to\infty} z(t) = \mathbf{0}$. In reality, in this chapter we cannot establish the convergence of the adaptive estimates to their corresponding true values from (8.34), unless the filtered regressor matrix $\boldsymbol{W}_f$ satisfies the restrictive PE condition [31]. However, this PE condition is rarely met in practical applications. But nonetheless, since establishing the attractive manifold $\mathcal{S} = \{z \in \mathbb{R}^7 | \boldsymbol{M}^{-1}\boldsymbol{W}_f z = \mathbf{0}\}$ is sufficient to recover the ideal closed-loop performance, as stated in Remark 8.4, the main focus is therefore on showing $\lim_{t\to\infty} \boldsymbol{M}^{-1}\boldsymbol{W}_f z(t) = \mathbf{0}$, without requiring the PE condition on $\boldsymbol{W}_f$.

Remark 8.6 The APF-based control method (see, indicatively, [17]), as a most common method for the treatment of motion constraints, could only achieve asymptotic pose tracking, and usually its control gains need to be finely tuned toward yielding a desired tracking performance (without any a priori guarantees). In comparison to the APF-based method, the proposed control scheme, apart from guaranteeing the motion constraints (through visualizing them as error bounds), is also capable of achieving prescribed performance (which can be specified *a priori* by the designer according to the mission requirements) without resorting to a judicious parameters selection, and thus is more preferable in practical engineering.

8.4 Numerical Simulations

In this section, numerical simulations are carried out to illustrate the effectiveness of the proposed control scheme. Assume that the freely tumbling target orbits the Earth in a Molniya orbit with initial orbital elements listed in Table 8.1. The attitude evolution of the target is described by (2.31) and (2.32) with the inertia matrix $\boldsymbol{J}_t = \text{diag}[22, 20, 23]\,\text{kg}\cdot\text{m}^2$ and the initial conditions: $\boldsymbol{q}_t(0) = [0, 0, 0, 1]^\top$ and $\boldsymbol{\omega}_t(0) = [0.01, -0.01, 0.01]^\top$ rad/s. Moreover, the gravity-gradient torque is considered as the external disturbances, given by $\boldsymbol{\tau}_{td} = 3\mu(\boldsymbol{S}(\boldsymbol{\rho}_t)\boldsymbol{J}_t\boldsymbol{\rho}_t)/\|\boldsymbol{\rho}_t\|^5$.

The nominal mass and inertia of the pursuer are as follows:

$$m_p = 200\,\text{kg}, \text{ and } \boldsymbol{J}_p = \begin{bmatrix} 55 & 0.3 & 0.5 \\ 0.3 & 65 & 0.2 \\ 0.5 & 0.2 & 58 \end{bmatrix} \text{kg}\cdot\text{m}^2.$$

The thruster distribution matrix $\boldsymbol{D}$ is given by (2.55) with $d_x = d_y = d_z = 1$ m, each pair of thrusters can provide a bidirectional thrust with a fixed magnitude of 10 N. Initially, the relative position and velocity of the pursuer w.r.t. the target are $\boldsymbol{\rho}(0) = [230, -35, 30]^\top$ m and $\dot{\boldsymbol{\rho}}(0) = [-0.1, 0.5, -0.3]^\top$ m/s, while the attitude and angular velocity of the pursuer are set as $\boldsymbol{q}_p(0) = [0.1176, 0.0665, -0.9852, 0.1044]^\top$ and $\boldsymbol{\omega}_p(0) = [-0.01, 0.01, 0.02]^\top$ rad/s. The approach corridor is defined by $d_{ac} = 200$ m and $\alpha = 15°$, and the FOV of the vision sensor is $\beta = 30°$. The desired relative position is $\boldsymbol{\rho}_d = [15, 0, 0]^\top$ m. The initial configuration actually satisfies both the approaching path and FOV constraints.

8.4.1 Nominal Simulation Campaign

In this subsection, we present a nominal simulation example to demonstrate the theoretical validity of the proposed non-CE adaptive pose tracking controller. In this case, the mass and inertia of the pursuer and the thrust distribution matrix are assumed to maintain their nominal values throughout the whole period of proximity operations,

Table 8.1 Initial orbital elements

Orbital elements	Values	Units
Semimajor axis	26628	km
Eccentricity	0.7417	–
Inclination	63.4	deg
RAAN	0	deg
Argument of perigee	–90	deg
True anomaly	0	deg

Table 8.2 Baseline conditions for mission success

Conditions	Position errors		Attitude errors
	x-axis	Lateral axis	Three axis
Precision tolerance	15 mm	2 mm	10 arcmin
Maneuver completion time (s)	600	600	200

Table 8.3 Performance function parameters

i	$a_{i,0}$	$\eta_{i,s}$	k_{si}	k_{oi}	t_f/t_f^*
1	230	0.015	–	–	600
2	40	0.002	0	0	600
3	35	0.002	1	0	600
4	0.8	0.0014[a]	1	0	200
5	0.2	0.0014[a]	0	0	200
6	0.15	0.0014[a]	0	0	200

[a] The value is derived from the transformation between the quaternion $\boldsymbol{q}_e$ and the 3-1-2 Euler angles

and the effects of the thruster modulation and external disturbances are ignored. Ideal mission-oriented demands on accuracy tolerance and maneuver completion time are tabulated in Table 8.2. To address Problems 8.1 and 8.2, at the same time, meet the mission demands, we impose *a priori* performance envelops described by (8.8) and (8.13) on the pose tracking errors x_{1i}, $i = 1, 2, \ldots, 6$, through choosing the design constants listed in Table 8.3. The control gains and the other design parameters are obtained by trial and error: $k_0 = 0.01$, $k_1 = 0.01$, $c_1 = 0.01$, $c_2 = 3$, and $\gamma = 10$. The initial value of $\hat{\boldsymbol{\theta}}$ is taken as $[210, 60, 70, 65, 1, 1, 1]^\top$.

The responses of the transformed errors are plotted in Fig. 8.3, from which it is clearly seen that $\varepsilon(t)$ always remains with the set $\mathbb{D}_\varepsilon$. This observation suggests the satisfaction of prescribed performance, as will be witnessed later. Figure 8.4a and b depict the evolution of the position and attitude tracking errors along the performance envelops imposed by (8.8) and (8.13). For reference, we also mark in these two figures the corresponding settling times. As can be seen, $x_{1i}(t)$ $(i = 1, 2, \ldots, 6)$ indeed evolves strictly inside the performance envelop which is upper and lower bounded by $P_{ui}(t)$ and $P_{li}(t)$, and thus Problems 8.1 and 8.2 are successfully addressed and, at the same time, $\rho_2^2(t) + \rho_3^2(t) \neq 0$ is strictly guaranteed.

The time responses of the velocity and angular velocity errors are depicted in Fig. 8.5, from which the asymptotic convergence of both $\boldsymbol{v}_e$ and $\boldsymbol{\omega}_e$ is observed, as theoretical analysis predicted. The time response of $\boldsymbol{M}^{-1}\boldsymbol{W}_f\boldsymbol{z}$ is plotted in Fig. 8.6. It is clear that $\boldsymbol{M}^{-1}\boldsymbol{W}_f\boldsymbol{z}$ converges asymptotically to zero, indicating the establishment of the attracting manifold $\mathcal{S}$, which is consistent with the discussion in Remark 8.4. Since thruster modulation is not considered in this scenario, only the time responses of the desired driving forces and torques are provided, as plotted in Fig. 8.7. Finally, to intuitively illustrate the effectiveness of the proposed control scheme, the 3-D motion

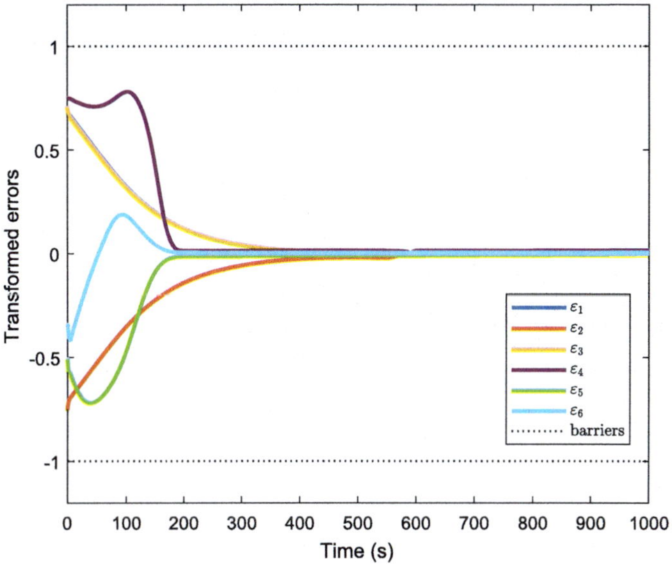

Fig. 8.3 Time responses of the transformed errors

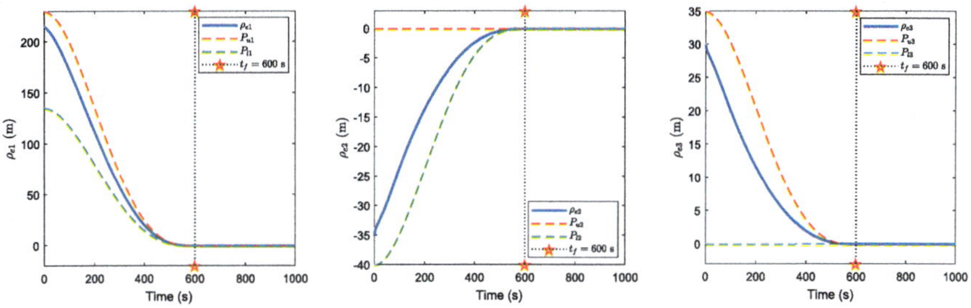

(a) position tracking errors

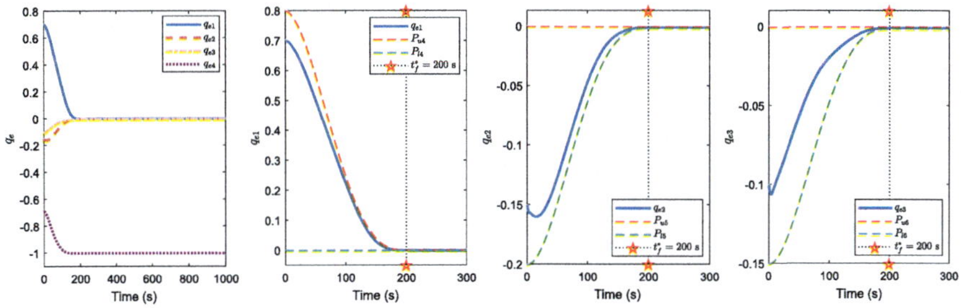

(b) attitude tracking errors

Fig. 8.4 Evolution of the pose tracking errors along the performance envelops

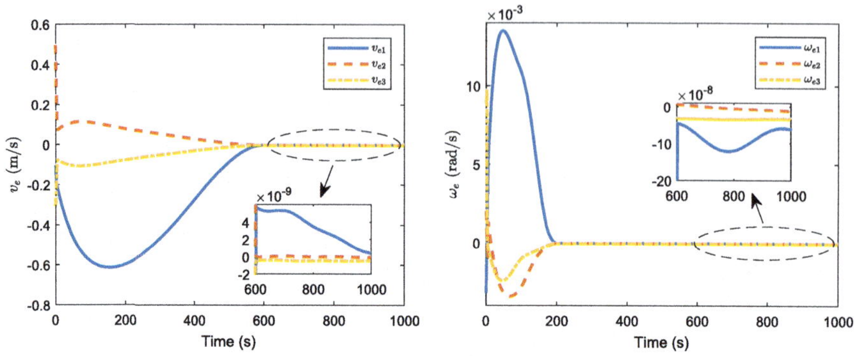

(a) velocity errors (b) angular velocity errors

Fig. 8.5 Time responses of the velocity and angular velocity errors

Fig. 8.6 Time response of $\boldsymbol{M}^{-1}\boldsymbol{W}_f z$

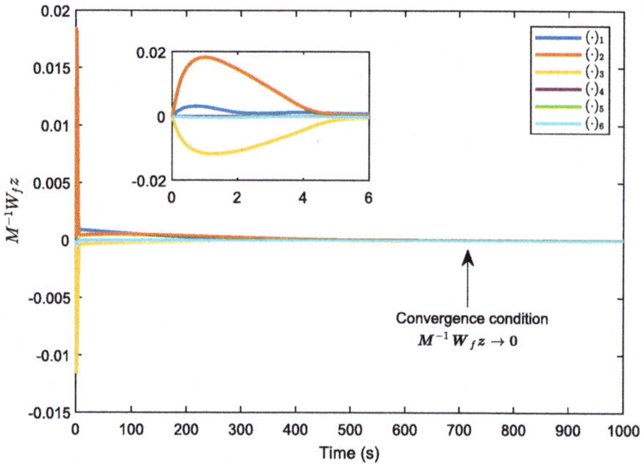

trajectory and partial attitude snapshots of the pursuer w.r.t. the target observed in $\mathcal{T}$ are shown in Fig. 8.8, where both the pursuer and the target are portrayed as the cubes with solar panels. As can be seen, the pursuer ultimately reaches the desired position along the approach corridor, while the target always lies within the FOV of the pursuer's vision sensor. In particular, the boresight of the vision sensor keeps pointing towards the target after preset $t_f^* = 200\,\text{s}$. As a consequence, the proximity operations are successfully accomplished.

8.4.2 *Practical Simulation Campaign*

This scenario aims to demonstrate the utility of the proposed method in a realistic environment. Several practical aspects including thruster modulation and installation deviation, external disturbances, and mass and inertia changes are considered. The

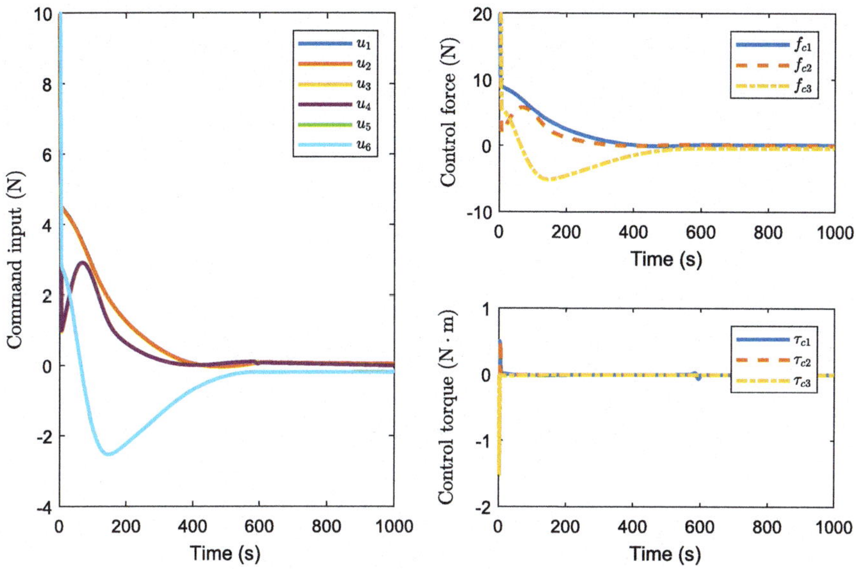

Fig. 8.7 Time responses of the command input and driving forces and torques

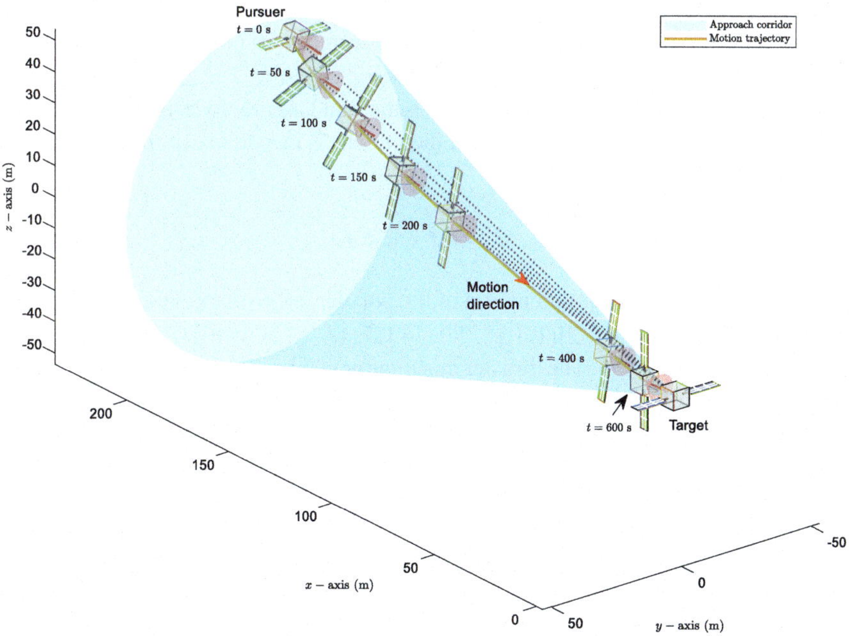

Fig. 8.8 3-D motion trajectory of the pursuer w.r.t. the target observed in $\mathcal{T}$

disturbance forces and torques exerting on the pursuer are modeled as

$$\boldsymbol{f}_d = \boldsymbol{f}_d^{J_2} + \boldsymbol{f}_d^o \text{ (N) and } \boldsymbol{\tau}_d = \boldsymbol{\tau}_d^g + \boldsymbol{\tau}_d^o \text{ (N} \cdot \text{m)},$$

where $\boldsymbol{f}_d^{J_2}$ and $\boldsymbol{\tau}_d^g$ denote the perturbing acceleration due to the Earth's oblateness and the gravity-gradient torque, and their respective expressions are given by

$$\boldsymbol{f}_d^{J_2} = \boldsymbol{R}_{\mathcal{TN}} \boldsymbol{a}_{J_2}^{\mathcal{I}},$$

$$\boldsymbol{a}_{J_2}^{\mathcal{I}} = -\frac{3\mu J_2 R_e^2}{2\|[\boldsymbol{\rho}_p]_{\mathcal{I}}\|^5} \left(\boldsymbol{U} - 5 \left(\frac{z_p}{\|[\boldsymbol{\rho}_p]_{\mathcal{I}}\|} \right)^2 \boldsymbol{I}_3 \right) [\boldsymbol{\rho}_p]_{\mathcal{I}},$$

$$\boldsymbol{\tau}_d^g = 3\mu \frac{\boldsymbol{S}([\boldsymbol{\rho}_p]_{\mathcal{P}}) \boldsymbol{J}_p [\boldsymbol{\rho}_p]_{\mathcal{P}}}{\|[\boldsymbol{\rho}_p]_{\mathcal{P}}\|^5},$$

where $J_2 = 0.0010826267$, $R_e = 6378.137\,\text{km}$ is the mean equatorial radius of the Earth, $\boldsymbol{U} = \text{diag}[1, 1, 3]$, and z_p is the third component of $[\boldsymbol{\rho}_p]_{\mathcal{I}}$. Similar to [24], in the simulations, we do not explicitly consider other external forces and torques arising from, for example, atmospheric drag, solar radiation, and third bodies; instead, we only consider constant $\boldsymbol{f}_d^o$ and $\boldsymbol{\tau}_d^o$ that capture all neglected external forces and torques. Here, we simply choose $\boldsymbol{f}_d^o = [0.001, 0.001, 0.001]^\top$ N and $\boldsymbol{\tau}_d^o = [0.001, 0.001, 0.001]^\top$ Nm.

The mass variation induced by fuel consumption is modeled as [32]: $\dot{m}_p = -\sum_{i=1}^{6} |u_i|/(I_{sp} g)$ with initial value of 200 kg, where $I_{sp} = 250\,\text{s}$ is the specific impulse of the thrusters, and $g = \mu/\|\boldsymbol{\rho}_p\|^2$ is the standard gravity. Due to, for instance, fuel consumption and payload motion, the pursuer's inertia matrix is uncertain and here takes the following form [33]:

$$\boldsymbol{J}_p = \boldsymbol{J}_p^* + \Delta \boldsymbol{J},$$

where $\boldsymbol{J}_p^*$ denotes the nominal inertia of the pursuer, as given previously, and $\Delta \boldsymbol{J}$ is given by $\Delta \boldsymbol{J} = \text{diag}[4, 3, 2](1 + e^{-0.01t}) - 1.5\chi(t - 30)$ with $\chi(\cdot)$ defined as $\chi(\cdot < 0) = 0$ and $\chi(\cdot \geq 0) = 1$. The thruster misalignment is modeled by 5% multiplicative uncertainty on the commands u_i, $i = 1, 2, \ldots, 6$. The PWPF modulators are applied for pulse modulation, and the corresponding parameters are chosen as $K_m = 0.8$, $T_m = 0.1$, $\delta_{on} = 0.45$, and $\delta_{off} = 0.15$. As it is well known that the use of the PWPF modulator inevitably results in degradation to the steady-state tracking accuracy, the requirement for precision tolerance has to be relaxed from a practical viewpoint. Here, the precision tolerances required are 0.3 m on x-axis (0.05 m on lateral axis) for the position tracking errors, and 60 arcmin for the attitude tracking errors. In the light of this, the design parameters $\eta_{i,s}$, $i = 1, 2, \ldots, 6$ are retaken as $\eta_{1,s} = 0.3$, $\eta_{i,s} = 0.05$ $(i = 2, 3)$ and $\eta_{i,s} = 8 \times 10^{-3}$ $(i = 4, 5, 6)$, while keeping other parameters unchanged.

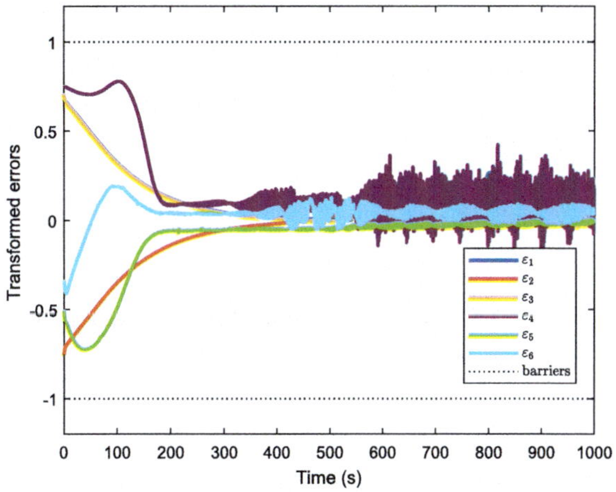

Fig. 8.9 Time responses of the transformed errors

The time responses of the transformed errors are shown in Fig. 8.9. As can be clearly seen, ε is strictly confined within the set $\mathbb{D}_\varepsilon$ throughout the simulation period, despite the adverse effects of some practical factors. This suggests that the pose tracking errors would be forced to evolve along their respective performance envelops. In Fig. 8.10a and b, we depict the time responses of the pose tracking errors. It can be seen that the prescribed performance specifications imposed on x_{1i}, $i = 1, 2, \ldots, 6$ are guaranteed, thus solving Problems 8.1 and 8.2, whilst guaranteeing that the constraint $\rho_2^2(t) + \rho_3^2(t) \neq 0$, $\forall t \geq 0$ hold. An intuitive illustration showing the above result is given in Fig. 8.11, wherein the motion trajectory of the pursuer w.r.t. the target in $\mathcal{T}$ on $x_\mathcal{T} - y_\mathcal{T}$ plane is depicted. A careful inspection of Fig. 8.11 reveals that the proximity operations are achieved without violating both motion constraints, but relatively large steady-state errors are observed. In fact, such relatively rough results are still acceptable owing to the inevitable performance degradation caused by those practical factors.

8.4.3 *Monte Carlo Simulation Campaign*

A 500-run Monte Carlo simulation is further carried out for different initial conditions and control parameters to provide a comprehensive insight into the effectiveness of the proposed control method. To this end, the initial conditions and control parameters are randomly selected in the ranges listed in Table 8.4 for each simulation cycle, wherein we likewise simulate a realistic case, as done in Sect. 8.4.2. For simplicity, the initial values, steady-state values and settling times of the ATPFs are uniformly chosen for all simulation cycles, as given in Table 8.5, while $k_{oi} = 0$, $i = 2, 3, \ldots, 6$ are taken for no overshoots. Note that, in the figure plotted latter, every single point corresponds to one simulation cycle.

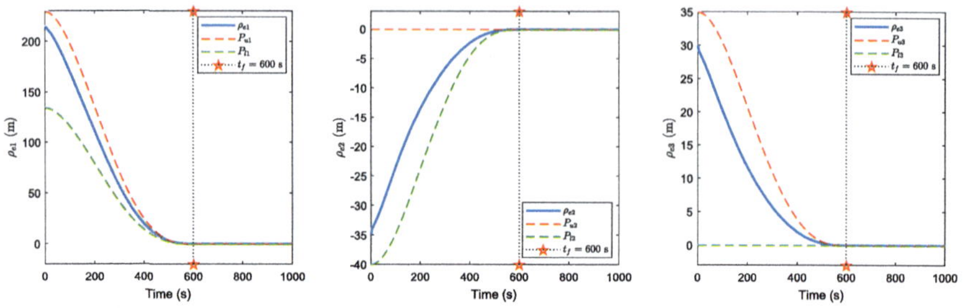

(a) position tracking errors

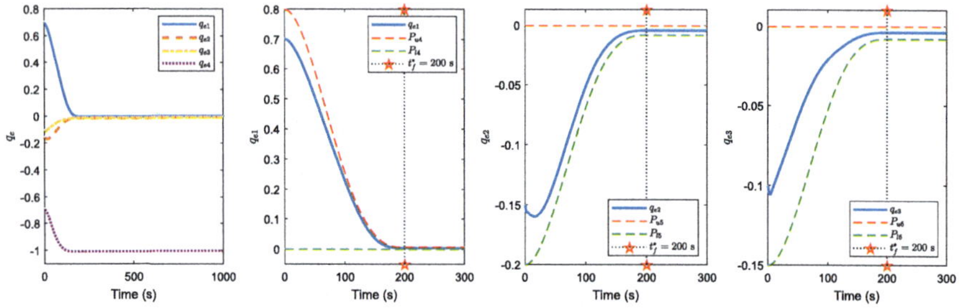

(b) attitude tracking errors

Fig. 8.10 Evolution of the pose tracking errors along the performance envelops

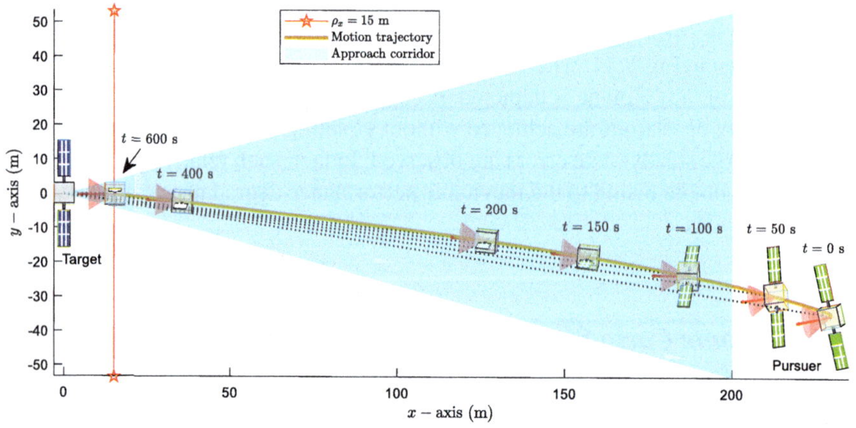

Fig. 8.11 Motion trajectory of the pursuer w.r.t. the target in $x_\mathcal{T} - y_\mathcal{T}$ plane

Table 8.4 Randomized variables and their respective ranges

Descriptions	Variables	Ranges
Control parameters	k_0, k_1, c_1	$(0.001, 0.02)$
	c_2, γ	$(0.1, 10), (1, 20)$
Initial conditions	$\rho(0)$, m	$(200, 245) \times (1, 35) \times (1, 35)$
	$\boldsymbol{q}_p(0)$	s.t. $\boldsymbol{q}_{ev} \in (0.01, 0.75) \times (0.01, 0.17) \times (0.01, 0.17)$
	$\boldsymbol{v}(0)$, m/s	$(-0.1, 0.1) \times (-0.1, 0.1) \times (-0.1, 0.1)$
	$\boldsymbol{\omega}_p(0)$, rad/s	$(-0.01, 0.01) \times (-0.01, 0.01) \times (-0.01, 0.01)$

Table 8.5 Performance function parameters

i	$a_{i,0}$	$\eta_{i,s}$	t_f/t_f^*
1	250	0.3	600
2	37.5	0.05	600
3	37.5	0.05	600
4	0.8	0.008	200
5	0.18	0.008	200
6	0.18	0.008	200

For the purpose of illustration, we here introduce two new vectors $\boldsymbol{\varepsilon}_t = [\varepsilon_1, \varepsilon_2, \varepsilon_3]^\top$ and $\boldsymbol{\varepsilon}_r = [\varepsilon_4, \varepsilon_5, \varepsilon_6]^\top$, and denote by $\max(\|\boldsymbol{\varepsilon}_j\|_\infty)$ $(j \in \{t, r\})$ the maximum value of $\|\boldsymbol{\varepsilon}_j\|_\infty$ in one run. It is straightforward to check that if $\max(\|\boldsymbol{\varepsilon}_t\|_\infty) < 1$ and $\max(\|\boldsymbol{\varepsilon}_r\|_\infty) < 1$ hold simultaneously, then $\boldsymbol{\varepsilon}(t)$ will stay within the set $\mathbb{D}_\varepsilon$ all the time, which, in turn, indicates the satisfaction of the prescribed performance for the pose tracking errors and thus the success of the proximity operations. In view of this, the distributions of $\max(\|\boldsymbol{\varepsilon}_t\|_\infty)$ and $\max(\|\boldsymbol{\varepsilon}_r\|_\infty)$ are shown in Fig. 8.12, where the colorbar specifies the cycle index. As can be seen, in the overwhelming majority of the simulation runs, $\boldsymbol{\varepsilon}(t) \in \mathbb{D}_\varepsilon$ $\forall t \geq 0$ is guaranteed, indicating that Problems 8.1 and 8.2 are solved and that $\rho_2^2(t) + \rho_3^2(t) \neq 0$ holds.

Further analyzing the recorded dates reveals that there are 35 simulation cycles that fail to guarantee prescribed performance bounds for the pose tracking errors and hence render a possible failure of the proximity mission. By checking their simulation conditions, it is found that the bad results are mainly attributed to inappropriate control parameters selection. To shed light on this point, we emphatically consider two cases that correspond to the points P1 and P2 marked in Fig. 8.12. Concerning P1, the control parameters are taken as $k_0 = 0.0034$, $k_1 = 0.02$, $c_1 = 0.0043$, $c_2 = 0.4227$, and $\gamma = 11.6628$. We find that c_2 is so small that the proposed controller cannot render a fast convergence of the filtered state $\boldsymbol{s}_f$ and hence the error vector $\boldsymbol{s}$, which, in turn, indicates that $\boldsymbol{x}_2$ cannot provide adequate damping in time for (some of) the pose tracking errors $x_{1i}, i = 1, 2, \ldots, 6$ to prevent them from approaching the margins of their respective performance bounds (they are observed to have the tendency during certain phases). Further, due to thrust limit (i.e., $u_{\max} = 10\,\text{N}$),

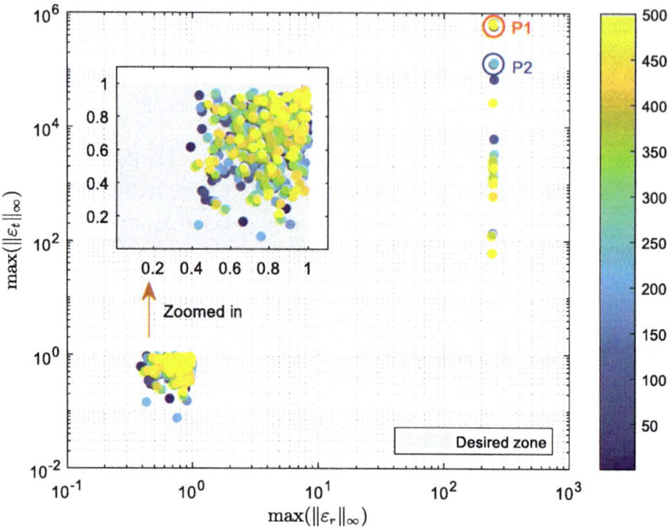

Fig. 8.12 Distributions of the maximum values of $\|\varepsilon_t\|_\infty$ and $\|\varepsilon_r\|_\infty$

it is practically impossible to prevent x_{1i}, $i = 1, 2, \ldots, 6$ from traversing the prescribed performance bounds. While for P2, the control parameters are $k_0 = 0.0173$, $k_1 = 0.0195$, $c_1 = 0.0118$, $c_2 = 9.9688$, and $\gamma = 11.5173$. Although c_2 in this case is large enough, k_0 is chosen relatively large such that the command input signals are large due to the existence of $\boldsymbol{y}$ ($\boldsymbol{y} = 2k_0\vartheta \boldsymbol{N}\varepsilon$) in $\boldsymbol{u}$ (see (8.37)). Likewise, due to thrust limits, the thrusters cannot provide sufficient actuation for the pursuer to regulate the pose tracking errors within the prescribed performance envelops. Overall, $k_0 \in [0.005, 0.01]$, $k_1, c_1 \in [0.001, 0.02]$, $c_2 \in [3, 10]$, and $\gamma \in [5, 20]$ might be reasonable selection; of course, the final selection should also hinge upon the initial conditions.

8.5 Summary

In this chapter, a novel adaptive pose tracking control scheme has been presented for spacecraft proximity operations with a freely tumbling target, capable of accomplishing the proximity operations in a preset time with overshoot and accuracy tolerance less predefined levels, whilst ensuring that both path and FOV constraints are satisfied, despite the presence of mass and inertia uncertainties. The core idea behind this solution is to transform the desired performance specifications and motion constraints as the pose tracking error bounds, and then design a non-CE adaptive controller using the PPC approach integrating a class of ATPFs to ensure that the pose tracking errors remain within the required ranges, even in the presence of parameter uncertainties.

Another aspect is that through properly choosing the performance envelops for the position tracking errors, the underlying singularity in the attitude extraction algorithm can be strictly avoided. Finally, simulation results illustrated the effectiveness of the overall control scheme.

References

1. Shan M, Guo J, Gill E (2016) Review and comparison of active space debris capturing and removal methods. Progress in Aerospace Sciences 80: 18–32
2. Stoll E, Letschnik J, Walter U, Artigas J, Kremer P, Preusche C, Hirzinger G (2009) On-orbit servicing. IEEE Robotics & Automation Magazine 16(4): 29–33
3. Pelton JN (2019) On-orbit servicing, active debris removal and repurposing of defunct spacecraft. In: Space 2.0, Springer, pp 87–101
4. Subbarao K, Welsh SJ (2008) Nonlinear control of motion synchronization for satellite proximity operations. Journal of Guidance, Control, and Dynamics 31(5): 1284–1294
5. Hu Q, Shao X, Chen WH (2018) Robust fault-tolerant tracking control for spacecraft proximity operations using time-varying sliding mode. IEEE Transactions on Aerospace and Electronic Systems 54(1): 2–17
6. Gui H, de Ruiter AH (2019) Adaptive fault-tolerant spacecraft pose tracking with control allocation. IEEE Transactions on Control Systems Technology 27(2): 479–494
7. Lopez I, McInnes CR (1995) Autonomous rendezvous using artificial potential function guidance. Journal of Guidance, Control, and Dynamics 18(2): 237–241
8. Dong H, Hu Q, Akella MR (2017) Safety control for spacecraft autonomous rendezvous and docking under motion constraints. Journal of Guidance, Control, and Dynamics 40(7): 1680–1692
9. Zappulla R, Park H, Virgili-Llop J, Romano M (2018) Real-time autonomous spacecraft proximity maneuvers and docking using an adaptive artificial potential field approach. IEEE Transactions on Control Systems Technology (99): 1–8
10. Di Cairano S, Park H, Kolmanovsky I (2012) Model predictive control approach for guidance of spacecraft rendezvous and proximity maneuvering. International Journal of Robust and Nonlinear Control 22(12): 1398–1427
11. Zagaris C, Park H, Virgili-Llop J, Zappulla R, Romano M, Kolmanovsky I (2018) Model predictive control of spacecraft relative motion with convexified keep-out-zone constraints. Journal of Guidance, Control, and Dynamics 41(9): 2054–2062
12. Epenoy R (2011) Fuel optimization for continuous-thrust orbital rendezvous with collision avoidance constraint. Journal of Guidance, Control, and Dynamics 34(2): 493–503
13. Zhu Z, Yan Y (2014) Space-based line-of-sight tracking control of geo target using nonsingular terminal sliding mode. Advances in Space Research 54(6): 1064–1076
14. Huang Y, Jia Y (2018) Adaptive fixed-time six-dof tracking control for noncooperative spacecraft fly-around mission. IEEE Transactions on Control Systems Technology (99): 1–9
15. Lee U, Mesbahi M (2014) Feedback control for spacecraft reorientation under attitude constraints via convex potentials. IEEE Transactions on Aerospace and Electronic Systems 50(4): 2578–2592
16. Lee U, Mesbahi M (2016) Constrained autonomous precision landing via dual quaternions and model predictive control. Journal of Guidance, Control, and Dynamics 40(2): 292–308
17. Dong H, Hu Q, Akella MR (2017) Dual-quaternion-based spacecraft autonomous rendezvous and docking under six-degree-of-freedom motion constraints. Journal of Guidance, Control, and Dynamics 41(5): 1150–1162
18. Bechlioulis CP, Rovithakis GA (2008) Robust adaptive control of feedback linearizable mimo nonlinear systems with prescribed performance. IEEE Transactions on Automatic Control 53(9): 2090–2099

19. Hu Q, Shao X, Guo L (2017) Adaptive fault-tolerant attitude tracking control of spacecraft with prescribed performance. IEEE/ASME Transactions on Mechatronics 23(1): 331–341
20. Shao X, Hu Q, Shi Y, Jiang B (2018) Fault-tolerant prescribed performance attitude tracking control for spacecraft under input saturation. IEEE Transactions on Control Systems Technology 28(2): 574–582
21. Wei C, Luo J, Yin Z, Yuan J (2018) Leader-following consensus of second-order multi-agent systems with arbitrarily appointed-time prescribed performance. IET Control Theory & Applications 12(16): 2276–2286
22. Yin Z, Luo J, Wei C (2019) Quasi fixed-time fault-tolerant control for nonlinear mechanical systems with enhanced performance. Applied Mathematics and Computation 352: 157–173
23. Liu M, Shao X, Ma G (2019) Appointed-time fault-tolerant attitude tracking control of spacecraft with double-level guaranteed performance bounds. Aerospace Science and Technology 92: 337–346
24. Filipe N, Tsiotras P (2014) Adaptive position and attitude-tracking controller for satellite proximity operations using dual quaternions. Journal of Guidance, Control, and Dynamics 38(4): 566–577
25. Seo D, Akella MR (2008) High-performance spacecraft adaptive attitude-tracking control through attracting-manifold design. Journal of Guidance, Control, and Dynamics 31(4): 884–891
26. Seo D (2015) Fast adaptive pose tracking control for satellites via dual quaternion upon non-certainty equivalence principle. Acta Astronautica 115: 32–39
27. Sun L, Huo W, Jiao Z (2017) Robust nonlinear adaptive relative pose control for cooperative spacecraft during rendezvous and proximity operations. IEEE Transactions on Control Systems Technology 25(5): 1840–1847
28. Astolfi A, Ortega R (2003) Immersion and invariance: A new tool for stabilization and adaptive control of nonlinear systems. IEEE Transactions on Automatic control 48(4): 590–606
29. Junkins JL, Schaub H (2009) Analytical mechanics of space systems. American Institute of Aeronautics and Astronautics, Reston, Virginia
30. Roberts A, Tayebi A (2011) Adaptive position tracking of vtol uavs. IEEE Transactions on Robotics 27(1): 129–142
31. Boyd S, Sastry SS (1986) Necessary and sufficient conditions for parameter convergence in adaptive control. Automatica 22(6): 629–639
32. Capello E, Punta E, Dabbene F, Guglieri G, Tempo R (2017) Sliding-mode control strategies for rendezvous and docking maneuvers. Journal of Guidance, Control, and Dynamics 40(6): 1481–1487
33. Cai W, Liao X, Song Y (2008) Indirect robust adaptive fault-tolerant control for attitude tracking of spacecraft. Journal of Guidance, Control, and Dynamics 31(5): 1456–1463

Chapter 9
I&I Adaptive Pose Control of Spacecraft Under Kinematic and Dynamic Constraints

9.1 Introduction

Over the past two decades, autonomous rendezvous and proximity operations (RPOs) of spacecraft has elicited widespread attention due to its important roles in many current and near-future space missions, such as on-orbit servicing, active debris removal, and repairing of defunct satellites, etc. [1]. To data, some international space organizations, like the NASA of the United States and the NASDA of Japan, have implemented RPOs demonstration and verification in orbit [2]. In April 2017, China also conducted RPOs tests of Tianzhou-1 cargo resupply spacecraft with Tiangong-2 space lab. Safe and high-precision proximity operations are very critical for the success of an assigned RPOs mission. During proximity operations, the pursuer is usually required to perform simultaneously the relative translational and rotational motions. This necessitates the design of six-degrees-of-freedom (6-DoF) pose controllers. In this respect, some representative results have been reported in the literature [3–5] (to name a few). However, the majority of these existing works focus on achieving the ultimate control goals while ignoring some underlying constraints (as discussed later), and thus may result in RPOs mission failure.

In general, spacecraft proximity operations involve two synchronously occurring maneuvers, i.e., relative position tracking and boresight pointing adjustment. For the former, the pursuer should maneuver towards the desired anchoring point along the approach corridor to ensure proximity safety. This imposes a path constraint upon the pose control design. Several methods such as artificial potential function (APF) based method [6, 7], trajectory planning method [8], model predictive control (MPC) [9], etc., have been developed to make sure compliance with path constraint. While for the latter, since the pursuer is usually equipped with a vision sensor (e.g., optical camera) for real-time target inspection and accurate relative navigation, the sensor's boresight should always point towards the target to generate the image patches. To this aim, the common practice is to develop an attitude controller for the pursuer to track the desired attitude that is extracted on the basis of the boresight

Q. Hu et al., *Intelligent Autonomous Control of Spacecraft with Multiple Constraints*,
https://doi.org/10.1007/978-981-99-0681-9_9

pointing requirement [10]. However, such a method may lead to loss of the target in the sensor's field-of-view (FOV) during a certain time period, which, in turn, may give rise to system instability. In practice, the pursuer should perform constrained attitude maneuvers to ensure that the target is always in sight. This is called here the FOV constraint. The APF-based control methods have been proposed in [11–13] to deal with such a constraint or its complement. In addition, the path planning based methods [14, 15] provide an alternative way to ensure satisfaction of attitude constraints. It is noteworthy that all the foregoing methods can only be used to deal with either path or FOV constraint. The study of the 6-DoF pose control problem under both constraints (hereafter termed kinematic constraints) has attracted less attention from both academia and aerospace sectors. In [16], the authors proposed a prescribed performance control strategy to deal with kinematic constraints. Using the dual quaternion representation, Lee and Mesbahi [17] provided a MPC framework for the guidance and control of autonomous precision landing under kinematic constraints and some other constraints; while Dong et al. presented an APF-based control scheme in [18] and its adaptive extension in [19] for spacecraft proximity operations to handle kinematic constraints.

In proximity operations, another practically important issue that deserves special attention is velocity and angular velocity constraints (hereafter termed dynamic constraints), which arise due to two aspects: (1) for proximity safety concern, the relative velocity between the pursuer and the target should not exceed a maximum value; (2) for overlapping image patches obtained from the vision sensor, the relative angular velocity should be constrained below a certain maximum value dependent on the sensor specifications [20]. The pose control design should take dynamic constraints into consideration. But, in this direction, only a very few results have been reported in the literature. Li et al. [21] developed MPC methods for spacecraft proximity operations to ensure compliance with velocity constraint and some other safety and physical requirements. In [22], a Linear Quadratic Regulator (known as LQR) controller was designed to account for velocity constraint. There are nowadays several solutions for angular velocity constraint, such as backstepping control [23], APF-based control [24], successive convex optimisation [20], etc. It is noted, however, that nearly every one of the foregoing references can only accommodate either one of the two dynamic constraints. In [25], an adaptive backstepping control method integrating barrier functions (a special type of APF) was put forth for spacecraft proximity operations, which can ensure that both dynamic constraints are satisfied.

The simultaneous treatment of both kinematic and dynamic constraints poses significant challenges for the pose controller design. At present, potential solutions to this problem include the optimization-based methods (e.g., MPC and LQR) and the APF-based methods. The former is computationally expensive, while the latter based on Lyapunov theory can deliver analytic control laws that are easy for onboard implementation. In view of this, in this chapter we will explore the use of the APF-based method to deal with both kinematic and dynamic constraints. However, one important caveat here is that the mass and inertia uncertainties caused by, for example, onboard payload motion, fuel consumptions, and rotation of solar arrays, would increase the difficulty in deriving a pose controller based on the APFs for spacecraft proximity operations. In reality, the vast majority of the APF-based control methods require exact knowledge of the mass and inertia parameters of the pursuer, and therefore are not applicable to our case. Adaptive control, as an effective way to tackle parameter uncertainties, has been widely applied to spacecraft proximity operations (see [3–5] and references therein), but nonetheless, it is a non-trivial task to develop an adaptive pose controller in the context of both kinematic and dynamic constraints. The technical barriers lie in two aspects: (1) some adaptive control strategies can hardly be extended to tackle state constraints (e.g., model reference adaptive control (MRAC) and its extensions including $\mathcal{L}_1$ adaptive control and simple adaptive control); (2) some other adaptive solutions obey the classical certainty equivalence (CE) principle, and require a "realizability" condition, which does not hold in the Lyapunov sense when considering dynamic constraints. Even though the adaptive control scheme in [25] can effectively circumvent the realizability condition by using both element-wise and norm-wise adaptive estimation techniques, it has a certain degree of conservativeness due to the utilization of norm-wise adaptive estimations [26]; moreover, it cannot recover the deterministic-case of closed-loop performance (no parameter uncertainties), unless the persistent excitation (PE) condition is met such that the adaptive estimates converge to their true values.

Motivated by the above discussion, a novel adaptive control scheme based on APFs is proposed in this chapter. The core idea behind this strategy is to employ the immersion and invariance (I&I) adaptive control philosophy, which is originally proposed by Astolfi and Ortega in [27], to circumvent the realizability problem, yielding naturally a non-certainty-equivalence (non-CE) adaptive controller. Moreover, the dynamic scaling technique is applied in the I&I adaptive control design to overcome the integrability obstacle that arises owing to unsolvable partial differential equations (PDEs). By Lyapunov stability theory, it is shown that the proposed control method enables asymptotic convergence of the position- and velocity-level tracking errors, while complying with both kinematic and dynamic constraints, despite the presence of parameter uncertainties. As a result, the proximity operations are successfully accomplished. The main contributions of the chapter are summarized below:

- A transformed relative translational dynamics described in the target's body-fixed frame and a line-of-sight (LOS) frame are introduced to facilitate the pose control design for spacecraft proximity operations under kinematic constraints. Within this setting, four APFs free of stubborn local minima problem are designed to ensure satisfaction of both kinematic and dynamic constraints.
- A novel dynamically scaled I&I adaptive control scheme is proposed to circumvent the realizability condition and to ensure compliance with both kinematic and dynamic constraints with the help of tactfully constructed APFs. Moreover, the proposed algorithm preserves all the key beneficial features of the I&I adaptive control methodology and can, therefore, achieve high-performance pose tracking in the presence of parameter uncertainties.

The remainder of the chapter is structured as follows. Sect. 9.2 designs two APFs to handle kinematic and dynamic constraints, and formulates the control problems for spacecraft constrained RPOs. The I&I adaptive pose controller is derived in Sect. 9.3. The simulation results are given in Sect. 9.4. Finally, this chapter is wrapped up by concluding remarks in Sect. 9.5.

9.2 Problem Formulation

Similar to Chap. 8, the spacecraft RPOs with a freely tumbling target are also divided into two synchronously occurring maneuvers: relative position tracking and boresight pointing adjustment. For ease of illustration, in the following we separately formulate the control problems for the two maneuvers, along with detailed constraint descriptions and APF designs. To proceed, we make the following reasonable assumptions:

Assumption 9.1 The mass and inertia matrix of the pursuer are constant (or slowly varying), but otherwise unknown.

Assumption 9.2 The angular velocity vector of the target ω_t is bounded, and its time derivatives up to order two, i.e., $\dot{\omega}_t$ and $\ddot{\omega}_t$, are continuous and bounded.

Remark 9.1 In close proximity operations, the pursuer's mass and inertia properties may change due to fuel consumptions, payload motion, or rotation of solar arrays, inevitably rendering the mass m_p and the inertia matrix $\boldsymbol{J}_p$ time-varying. But, their variation rates are very slow, and thus Assumption 9.1 is practically reasonable. In addition, Assumption 9.2 is common for spacecraft RPOs with a tumbling target.

9.2.1 *Relative Position Tracking*

The final goal for relative position tracking is to steer the pursuer to the desired anchoring point lying on the $-X$-axis of $\mathcal{T}$, i.e., $\boldsymbol{\rho}_d = [\rho_d, 0, 0]^\top$ ($\rho_d < 0$), as shown in Fig. 9.1. To this aim, the position and velocity tracking errors can be defined as $\boldsymbol{\rho}_e = \boldsymbol{\rho} - \boldsymbol{\rho}_d$ and $\boldsymbol{v}_e = \boldsymbol{v}$ (due to $\dot{\boldsymbol{\rho}}_d = \mathbf{0}$). In addition, considering that the actuators are fixed in the pursuer's body frame $\mathcal{P}$, the control force $\boldsymbol{f}$ in $\mathcal{T}$ should be rewritten as $\boldsymbol{f} = \boldsymbol{R}_{\mathcal{PT}}^\top \boldsymbol{f}_c$, where $\boldsymbol{f}_c$ denotes the control force in $\mathcal{P}$, and $\boldsymbol{R}_{\mathcal{PT}} = \boldsymbol{R}_{\mathcal{PI}} \boldsymbol{R}_{\mathcal{TI}}^\top$ with $\boldsymbol{R}_{\mathcal{PI}}$ and $\boldsymbol{R}_{\mathcal{TI}}$ being calculated by (2.12) in terms of $\boldsymbol{q}_p$ and $\boldsymbol{q}_t$, respectively. To facilitate the design, the translational tracking error dynamics (2.52) is rewritten as

$$\dot{\boldsymbol{\rho}}_e = \boldsymbol{v}_e, \tag{9.1}$$

$$\boldsymbol{M}_p \dot{\boldsymbol{v}}_e + \boldsymbol{C}_p \boldsymbol{v}_e + \boldsymbol{G}_p = \boldsymbol{R}_{\mathcal{PT}}^\top \boldsymbol{f}_c. \tag{9.2}$$

For safety concern, the position controller to be developed, except from achieving the final goal, it should also ensure that the pursuer is within a certain approach

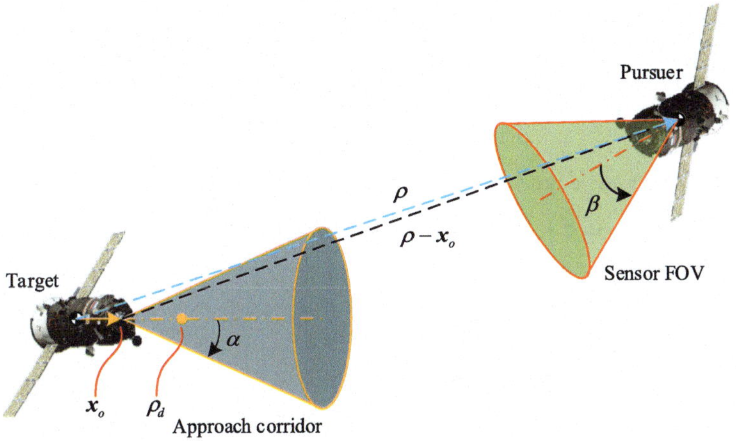

Fig. 9.1 Illustration of kinematic constraints

corridor, as shown in Fig. 9.1. This is called here the path constraint. In the following, an APF free of local minima will be designed to handle this constraint. Although the following contents pertain to the path constraint are almost the same as that in [7], they are still given to make the book self-contained.

The path constraint boundary can be given as:

$$h_t(\boldsymbol{\rho}) = (\boldsymbol{\rho} - \boldsymbol{x}_o)^\top \boldsymbol{W}_t(\boldsymbol{\rho} - \boldsymbol{x}_o), \tag{9.3}$$

with

$$\boldsymbol{W}_t = \mathrm{diag}\{1, -\cot^2(\alpha), -\cot^2(\alpha)\},$$

where $\boldsymbol{x}_o = [a, 0, 0]^\top$ and $\alpha > 0$ represent, respectively, the vertex and half-aperture of the approach corridor. As such, the permissible zone for relative position tracking is given by the set $\mathbb{K}_\rho = \{\boldsymbol{\rho} \in \mathbb{R}^3 : h_t(\boldsymbol{\rho}) > 0\}$.

To ensure compliance with the path constraint, i.e., $\boldsymbol{\rho} \in \mathbb{K}_\rho$, we consider the following APF $V_p : \mathbb{K}_\rho \to \mathbb{R}$ [7]:

$$V_p = \underbrace{k_{a1}\boldsymbol{\rho}_e^\top \boldsymbol{\rho}_e}_{\text{attractive potential}} + \underbrace{k_{r1}\boldsymbol{\rho}_e^\top \boldsymbol{\rho}_e / h_t(\boldsymbol{\rho})}_{\text{repulsive potential}}, \tag{9.4}$$

where $k_{a1} > 0$ and $k_{r1} > 0$ are the weighting parameters. A closer inspection of (9.4) reveals that $V_p \to \infty$ when $h_t(\boldsymbol{\rho}) = 0$. Thus, given initial conditions such that $h_t(\boldsymbol{\rho}(0)) > 0$, if the controller $\boldsymbol{f}_c$ can ensure that $V_p \in \mathcal{L}_\infty$, the path constraint can then be satisfied.

Taking the time derivative of V_p yields

$$\dot{V}_p = (\nabla_\rho V_p)^\top \boldsymbol{v}_e = \boldsymbol{v}_e^\top \boldsymbol{s}_t, \tag{9.5}$$

where the gradient $\nabla_\rho V_p$ can be calculated as

$$\nabla_\rho V_p = 2k_{a1}\boldsymbol{\rho}_e + 2k_{r1}\frac{h_t(\boldsymbol{\rho})\boldsymbol{\rho}_e - \|\boldsymbol{\rho}_e\|^2 \boldsymbol{W}_t(\boldsymbol{\rho}-\boldsymbol{x}_o)}{h_t^2(\boldsymbol{\rho})},$$

where $\boldsymbol{s}_t = \nabla_\rho V_p$ is defined just for notational conciseness.

Lemma 9.1 *The APF V_p satisfies the following properties:*

- $V_p > 0$ *for all* $\boldsymbol{\rho} \in \mathbb{K}_\rho \backslash \{\boldsymbol{\rho}_d\}$;
- V_p *has a global minimum* $V_p = 0$ *at* $\boldsymbol{\rho} = \boldsymbol{\rho}_d$;
- $\nabla_\rho V_p = \boldsymbol{0}$ *holds only when* $\boldsymbol{\rho} = \boldsymbol{\rho}_d$ *in the set* $\mathbb{K}_\rho$.

Proof The proof is relegated Appendix A. □

Furthermore, for safety concern again, the relative velocity $\boldsymbol{v}$ of the pursuer with respect to (w.r.t.) the target cannot exceed beyond a certain maximum value. This is called here the velocity constraint and poses another requirement for the position controller design. Since $\boldsymbol{v} = \boldsymbol{v}_e$, the permissible set for velocity constraint can be represented as $\mathbb{D}_{v_e} = \{\boldsymbol{v}_e \in \mathbb{R}^3 : |v_{ei}| < v_{e,\max}, i = 1, 2, 3\}$, where $v_{e,\max} > 0$ represents the common magnitude limit on $\boldsymbol{v}_e$. To ensure that the relative velocity vector always remains within the set $\mathbb{D}_{v_e}$, a APF $V_{v_e} : \mathbb{D}_{v_e} \to \mathbb{R}$ is constructed:

$$V_{v_e} = \frac{1}{2}\log\left(\prod_{i=1}^{3}\frac{v_{e,\max}^2}{v_{e,\max}^2 - v_{ei}^2}\right). \tag{9.6}$$

Notably, V_{v_e} has a global minimum at $\boldsymbol{v}_e = \boldsymbol{0}$ and is free of local minima problem; moreover, if $|v_{ei}| = v_{e,\max}$, $V_{v_e} \to \infty$. Thus, if the position controller can ensure that $V_{v_e}(t) \in \mathcal{L}_\infty\ \forall t \geq 0$, then the velocity constraint can be satisfied.

Evaluating the time derivative of V_{v_e} along (9.2) gives

$$\dot{V}_{v_e} = \boldsymbol{v}_e^\top \boldsymbol{N}_{v_e}\dot{\boldsymbol{v}}_e = \boldsymbol{v}_e^\top \boldsymbol{N}_{v_e} m_p^{-1}(-\boldsymbol{C}\boldsymbol{v}_e - \boldsymbol{G} + \boldsymbol{R}_{\mathcal{PT}}^\top \boldsymbol{f}_c), \tag{9.7}$$

where $\boldsymbol{N}_{v_e} = \text{diag}[N_{v_e,i}]$ with $N_{v_e,i} = 1/(v_{e,\max}^2 - v_{ei}^2)$.

Now we formulate the control problem for relative position tracking as follows:

Problem 9.1 Consider the translational tracking error dynamics given by (9.1) and (9.2). Design properly the position control law $\boldsymbol{f}_c$ in the context of the APFs (9.4) and (9.6) to render $\lim_{t\to\infty}[\boldsymbol{\rho}_e(t), \boldsymbol{v}_e(t)] = \boldsymbol{0}$, while guaranteeing $\boldsymbol{\rho}(t) \in \mathbb{K}_\rho$ and $\boldsymbol{v}_e(t) \in \mathbb{D}_{v_e}$ for all $t \geq 0$, despite mass uncertainty.

9.2.2 Boresight Pointing Adjustment

Assume that the boresight axis of the vision sensor onboard the pursuer is aligned with the $+X$-axis of $\mathcal{P}$. The final control goal for boresight pointing adjustment is to keep the boresight vector $\boldsymbol{x}_{\mathcal{P}}$ oriented towards the target, that is, to achieve $\boldsymbol{x}_{\mathcal{P}} = \boldsymbol{R}_{\mathcal{P}\mathcal{T}}\boldsymbol{x}_{\rho}$, where $\boldsymbol{x}_{\rho} = -\boldsymbol{\rho}/\|\boldsymbol{\rho}\|$. To achieve this, we construct an LOS frame $\mathcal{D}$, whose attitude orientation w.r.t. $\mathcal{I}$ is described as the unit quaternion $\boldsymbol{q}_d = [\boldsymbol{q}_{dv}^{\top}, q_{d4}]^{\top} \in \mathbb{R}^4$, such that $\boldsymbol{R}_{\mathcal{D}\mathcal{T}}\boldsymbol{x}_{\rho} = \boldsymbol{x}_{\mathcal{D}}$ holds, where $\boldsymbol{x}_{\mathcal{D}}$ is the X-axis of $\mathcal{D}$. To obtain $\boldsymbol{q}_d$, we first extract the quaternion $\bar{\boldsymbol{q}} = [\bar{\boldsymbol{q}}_v^{\top}, \bar{q}_4]^{\top} \in \mathbb{R}^4$ of $\mathcal{F}$ w.r.t. $\mathcal{T}$ through Lemma 2.1 in Sect. 2.4.3.2 of Chap. 2.

Remark 9.2 A singularity problem will occur in Lemma 2.1 when $\|\boldsymbol{\rho}\| = 0$ or $\boldsymbol{x}_{\rho} = -\boldsymbol{x}_{\mathcal{D}}$. With closer inspection, it can be claimed that these two conditions do not hold when the path constraint is satisfied. Specifically, from Fig. 9.1, onc can intuitively observe that $\|\boldsymbol{\rho}\| \neq 0$ if the pursuer keeps maneuvering within the approach corridor. On the other hand, if $\boldsymbol{x}_{\rho} = -\boldsymbol{\rho}/\|\boldsymbol{\rho}\| = -\boldsymbol{x}_{\mathcal{D}}$ holds, it then follows that $\boldsymbol{\rho} = [\rho_1, 0, 0]^{\top}$ with $\rho_1 > 0$. Obviously, $\boldsymbol{\rho}$ locates on the $+X$-axis of the frame $\mathcal{T}$, and hence lies outside the approach corridor, indicating that $\boldsymbol{x}_{\rho} \neq -\boldsymbol{x}_{\mathcal{D}}$ in the approach corridor. Thus, we conclude that the singularity problem will not occur, as long as Problem 9.1 is solved.

Now, the rotational tracking error dynamics (2.46) and (2.47) are rewritten as

$$\dot{\boldsymbol{q}}_e = \frac{1}{2}\boldsymbol{Q}(\boldsymbol{q}_e)\boldsymbol{\omega}_e = \frac{1}{2}(\boldsymbol{q}_e \odot \vec{\boldsymbol{\omega}}_e), \tag{9.8}$$

$$\boldsymbol{J}_p\dot{\boldsymbol{\omega}}_e = -\boldsymbol{S}(\boldsymbol{\omega}_p)\boldsymbol{J}_p\boldsymbol{\omega}_p + \boldsymbol{J}_p(\boldsymbol{S}(\boldsymbol{\omega}_e)\boldsymbol{\Omega} - \bar{\boldsymbol{\Omega}}) + \boldsymbol{\tau}_c, \tag{9.9}$$

where $\vec{\boldsymbol{\omega}}_e = [\boldsymbol{\omega}_e^{\top}, 0]^{\top} \in \mathbb{R}^4$ is an extended version of $\boldsymbol{\omega}_e$. It should be noted that in this chapter, the pursuer's inertia matrix $\boldsymbol{J}_p$ is assumed to be diagonal, that is, $\boldsymbol{J}_p = \mathrm{diag}(J_{p,11}, J_{p,22}, J_{p,33})$.

Since the FOV of the vision sensor is limited, the pursuer needs to perform constrained attitude maneuvers to ensure that the target is always in sight, as shown in Fig. 9.1. This is the FOV constraint, whose constraint equation is of the form:

$$\boldsymbol{x}_{\mathcal{D}}\boldsymbol{R}_{\mathcal{P}\mathcal{D}}^{\top}\boldsymbol{x}_{\mathcal{P}} > \cos\beta. \tag{9.10}$$

Using $\boldsymbol{R}_{\mathcal{P}\mathcal{D}}$ in (9.10) and after some algebra, the FOV constraint can be expressed as the following quadratic inequality:

$$h_r(\boldsymbol{q}_e) = \boldsymbol{q}_e^{\top}\boldsymbol{W}_r\boldsymbol{q}_e > 0, \tag{9.11}$$

where $\boldsymbol{W}_r$ is a symmetric matrix given below:

$$\boldsymbol{W}_r = \begin{bmatrix} 2\boldsymbol{x}_{\mathcal{D}}\boldsymbol{x}_{\mathcal{P}}^{\top} - (\boldsymbol{x}_{\mathcal{D}}^{\top}\boldsymbol{x}_{\mathcal{P}} + \cos\beta)\boldsymbol{I}_3 & \boldsymbol{x}_{\mathcal{P}} \times \boldsymbol{x}_{\mathcal{D}} \\ (\boldsymbol{x}_{\mathcal{P}} \times \boldsymbol{x}_{\mathcal{D}})^{\top} & \boldsymbol{x}_{\mathcal{D}}^{\top}\boldsymbol{x}_{\mathcal{P}} - \cos\beta \end{bmatrix}.$$

With additional consideration of the unwinding phenomenon, the permissible set for the attitude tracking error $\boldsymbol{q}_e$ is represented as $\mathbb{K}_{q_e} = \{\boldsymbol{q}_e \in \mathbb{S}^3 : h_r(\boldsymbol{q}_e) > 0, q_{e4} \neq 0\}$.

To ensure compliance with the FOV constraint, i.e., $\boldsymbol{q}_e \in \mathbb{K}_{q_e}$, we consider the following APF $V_f : \mathbb{K}_{q_e} \to \mathbb{R}$:

$$V_f = \underbrace{-k_{a2}\log(q_{e4}^2)}_{\text{attractive potential}} \underbrace{-k_{r2}\log(q_{e4}^2)/h_r(\boldsymbol{q}_e)}_{\text{repulsive potential}}, \tag{9.12}$$

where $k_{a2} > 0$ and $k_{r2} > 0$ are the weighting parameters. It is not difficult to check from (9.12) that $V_f \to \infty$ when $h_r(\boldsymbol{q}_e) = 0$. Thus, given initial conditions such that $h_r(\boldsymbol{q}_e(0)) > 0$, if the attitude controller $\boldsymbol{\tau}_c$ can ensure that $V_f \in \mathcal{L}_\infty$, the FOV constraint can then be satisfied.

Evaluating the time derivative of V_f along (9.8) leads to

$$\dot{V}_f = \frac{1}{2}(\nabla_{\boldsymbol{q}_e} V_f)^\top (\boldsymbol{q}_e \odot \boldsymbol{\omega}_e), \tag{9.13}$$

where the gradient $\nabla_{\boldsymbol{q}_e} V_f$ can be calculated by

$$\nabla_{\boldsymbol{q}_e} V_f = -2k_{a2}\frac{\boldsymbol{q}_I}{q_{e4}} - 2k_{r2}\frac{(\boldsymbol{q}_I/q_{e4})h_r(\boldsymbol{q}_e) - \log(q_{e4}^2)\boldsymbol{W}_r\boldsymbol{q}_e}{h_r^2(\boldsymbol{q}_e)},$$

with $\boldsymbol{q}_I = [0, 0, 0, 1]^\top$. Further, using (2.7) in (9.13) yields

$$\dot{V}_f = -\frac{1}{2}\boldsymbol{\omega}_e^\top \mathrm{Vec}[(\nabla_{\boldsymbol{q}_e} V_f)^* \odot \boldsymbol{q}_e] = \boldsymbol{\omega}_e^\top \boldsymbol{s}_r, \tag{9.14}$$

where $\mathrm{Vec}[\cdot]$ is the 3×1 vector part of the argument, while $\boldsymbol{s}_r = -0.5\mathrm{Vec}[(\nabla_{\boldsymbol{q}_e} V_f)^* \odot \boldsymbol{q}_e]$ is defined for brevity.

Lemma 9.2 *The APF in* (9.12) *has the following properties:*

- $V_f > 0$ *for all* $\boldsymbol{q}_e \in \mathbb{K}_{q_e} \backslash \{\pm \boldsymbol{q}_I\}$;
- V_f *has two global minima* $V_f = 0$ *at* $\boldsymbol{q}_e = \pm \boldsymbol{q}_I$;
- $\boldsymbol{s}_r = \boldsymbol{0}$ *holds only when* $\boldsymbol{q}_e = \pm \boldsymbol{q}_I$ *in the set* $\mathbb{K}_{q_e}$.

Proof The detailed proof is given in Appendix B. □

Furthermore, for overlapping image patches or raster scans, the relative angular velocity of $\mathcal{P}$ w.r.t. $\mathcal{T}$, i.e., $\boldsymbol{\omega}_{\mathcal{TP}}^{\mathcal{P}}$, needs to be constrained below a certain maximum allowable value imposed by the sensor specifications [20]. This is called here the angular velocity constraint. Note that the angular velocity tracking error can be rewritten as $\boldsymbol{\omega}_e = \boldsymbol{\omega}_{\mathcal{TP}}^{\mathcal{P}} - \boldsymbol{R}_{\mathcal{PD}}\bar{\boldsymbol{\omega}}$, and that $\bar{\boldsymbol{\omega}}$ is bounded in the context of limited relative velocity, the angular velocity constraint can thus be indirectly formulated as the hard constraints on $\boldsymbol{\omega}_e$. At this point, the permissible set for this type of constraint is defined as $\mathbb{D}_{\omega_e} = \{\boldsymbol{\omega}_e \in \mathbb{R}^3 : |\omega_{ei}| < \omega_{e,\max},\ i = 1, 2, 3\}$, where $\omega_{e,\max} > 0$ is the common magnitude limit on $\boldsymbol{\omega}_e$.

To ensure that $\boldsymbol{\omega}_e$ always remains within the set $\mathbb{D}_{\omega_e}$, a APF $V_{\omega_e} : \mathbb{D}_{\omega_e} \to \mathbb{R}$ is introduced:

$$V_{\omega_e} = \frac{1}{2}\log\left(\prod_{i=1}^{3}\frac{\omega_{e,\max}^2}{\omega_{e,\max}^2 - \omega_{ei}^2}\right). \tag{9.15}$$

Evidently, V_{ω_e} has a global minimum at $\boldsymbol{\omega}_e = 0$ and is free of local minima problem; moreover, if $|\omega_i| = \omega_{e,\max}$, $V_{\omega_e} \to \infty$. Hence, if the attitude controller can ensure that $V_{\omega_e}(t) \in \mathcal{L}_\infty\ \forall t \geq 0$, then the angular velocity constraint can be met.

Taking the time derivative of V_{v_e} along (9.9) leads to

$$\dot{V}_{\omega_e} = \boldsymbol{\omega}_e^\top \boldsymbol{N}_{\omega_e}\boldsymbol{J}_p^{-1}[-\boldsymbol{S}(\boldsymbol{\omega}_p)\boldsymbol{J}_p\boldsymbol{\omega}_p + \boldsymbol{J}_p(\boldsymbol{S}(\boldsymbol{\omega}_e)\boldsymbol{\Omega} - \bar{\boldsymbol{\Omega}}) + \boldsymbol{\tau}_c], \tag{9.16}$$

where $\boldsymbol{N}_{\omega_e} = \mathrm{diag}[N_{\omega_e,i}]$ with $N_{\omega_e,i} = 1/(\omega_{e,\max}^2 - \omega_{ei}^2)$.

Formally, the control problem for boresight pointing adjustment can now be formulated as follows:

Problem 9.2 Consider the rotational tracking error dynamics given by (9.8) and (9.9). Design properly the attitude control law $\boldsymbol{\tau}_c$ in the context of the APFs (9.12) and (9.15) to render $\lim_{t\to\infty}[\boldsymbol{q}_e(t), \boldsymbol{\omega}_e(t)] = \boldsymbol{0}$, while guaranteeing $\boldsymbol{q}_e(t) \in \mathbb{K}_{q_e}$ and $\boldsymbol{\omega}_e(t) \in \mathbb{D}_{\omega_e}$ for all $t \geq 0$, despite inertia uncertainties.

9.2.3 Challenges

Theoretically speaking, it is a rather difficult task to design an adaptive pose controller using the classical adaptive control methodologies to address Problems 9.1 and 9.2. To provide more insights on the theoretical difficulty, in the following we revisit several kinds of typical adaptive control methods.

- *The MRAC method* [28] *and its modified versions including the* $\mathcal{L}_1$ *adaptive control* [29], *and the simple adaptive control* [30]: In such kind of methods, an ideal reference model is introduced, and the differences between the real plant output and the reference model output are used to predict and cancel the parameter uncertainties. Note, however, that they have a relatively fixed design architecture with low flexibility, and can hardly be extended to handle kinematic and dynamic constraints due to the involvement of a reference model to be tracked. Currently, it is still unclear how to design an adaptive controller in the MRAC framework to handle kinematic and dynamic constraints, as well as parameter uncertainties.
- *The CE-based adaptive control methods such as the direct adaptive control* [3–5], *the composite adaptive control* [31], *and the composite learning control* [32]: Although this kind of methods can be augmented with the APFs to deal with kinematic and dynamic constraints, the adaptive law designs certainly require the "realizability" condition, which does not hold in our case. To demonstrate this fact, let us here take the position control loop as an example, and postulate a typical Lyapunov function candidate $V = V_p + V_{v_e} + \tilde{m}_p^2/2\gamma$, where $\gamma > 0$, and $\tilde{m}_p = \hat{m}_p - m_p$ is the parameter estimation error. In the CE framework, the

control law is designed as $\boldsymbol{f}_c = -\boldsymbol{R}_{\mathcal{PT}}(\boldsymbol{\Phi}\hat{m}_p + k\boldsymbol{N}_{\upsilon_e}\boldsymbol{v}_e)$, where $\boldsymbol{\Phi}$ is a known regressor matrix satisfying $\boldsymbol{\Phi}m_p = -\boldsymbol{C}\boldsymbol{v}_e - \boldsymbol{G} + m_p\boldsymbol{N}_{\upsilon_e}^{-1}\boldsymbol{s}_t$, and $k > 0$. In view of this, we have $\dot{V} = -\boldsymbol{v}_e^\top\boldsymbol{N}_{\upsilon_e}m_p^{-1}(k\boldsymbol{N}_{\upsilon_e}\boldsymbol{v}_e + \boldsymbol{\Phi}\tilde{m}_p) + \tilde{m}_p\dot{\hat{m}}_p/\gamma$. It is clearly, it is impossible to derive an adaptive law to cancel the $\tilde{m}_p$-dependent term in $\dot{V}$, since $(\partial V/\partial \boldsymbol{v}_e)m_p^{-1} = \boldsymbol{v}_e^\top\boldsymbol{N}_{\upsilon_e}m_p^{-1}$ is directly related to the unknown m_p, hence unrealizable. Based on the above argument, we know that the afore-mentioned CE-based adaptive control methods are inapplicable to our case.

- *The indirect nonregressor-based adaptive control method in* [25]: This method effectively circumvents the realizability condition and achieves full-state constrained pose control for spacecraft proximity operations, through using both the element-wise and norm-wise adaptive estimation techniques. But, it is noted that this method is essentially a robust adaptive control method, and hence has a certain level of conservativeness; moreover, it cannot recover the ideal-case closed-loop performance (no effect of parameter uncertainties), in the absence of PE condition.

Motivated by the above discussion, it is necessary to explore a new route towards tailoring feasible and promising adaptive solutions to Problems 9.1 and 9.2 with less conservatism.

9.3 Adaptive Controller Design

In this section, a dynamic scaling based I&I adaptive control scheme is proposed to overcome the theoretical difficulty as discussed in Sect. 9.2.3, yielding non-CE adaptive controllers that can effectively solve Problems 9.1 and 9.2. The block diagram of the resulted closed-loop system is shown in Fig. 9.2.

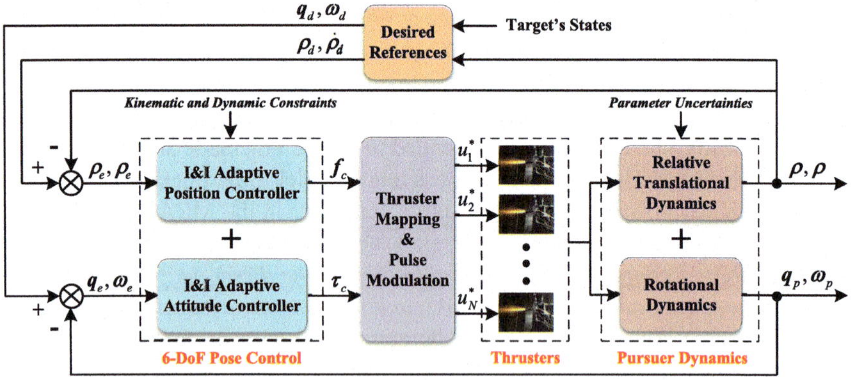

Fig. 9.2 Block diagram of the closed-loop system

A thrusters-only propulsion system is used in this work for 6-DoF pose control. The thruster configuration is detailed in Sect. 2.4.3.4. Then, the control force $\boldsymbol{f}_c$ and the control torque $\boldsymbol{\tau}_c$ satisfy the following thruster mapping:

$$[\boldsymbol{f}_c^\top, \boldsymbol{\tau}_c^\top]^\top = \boldsymbol{A}\boldsymbol{u}, \tag{9.17}$$

where $\boldsymbol{A} \in \mathbb{R}^{6\times N}$ (with $N > 0$ the number of thrusters) is the thruster distribution matrix, and $\boldsymbol{u} = [u_1, u_2, \ldots, u_N]^\top \in \mathbb{R}^N$ is the input command providing the desired force outputs for N thrusters. We emphasize here that, due to thrusters' on-off nature, it is additionally required to transform the continuous commands u_i $(i = 1, 2, \ldots, N)$ into the pulsed signals u_i^* via a pulse modulation mechanism, in practical implementation of the proposed control algorithm, as shown in Fig. 9.2.

9.3.1 I&I Adaptive Position Controller

To solve Problem 9.1, we rewrite (9.2) as

$$\dot{\boldsymbol{v}}_e = -\boldsymbol{N}_{v_e}^{-1}\boldsymbol{s}_t - k_{d1}(t)\boldsymbol{N}_{v_e}\boldsymbol{v}_e + m_p^{-1}(\boldsymbol{\Phi}_t(\cdot)m_p + \boldsymbol{R}_{\mathcal{PT}}^\top \boldsymbol{f}_c), \tag{9.18}$$

where $k_{d1}(t) > 0$ (independent of $\boldsymbol{v}_e$) is a time-varying gain to be determined, and $\boldsymbol{\Phi}_t(\cdot) \in \mathbb{R}^3$ is the regressor matrix such that $\boldsymbol{\Phi}_t(\cdot)m_p = -\boldsymbol{C}_p\boldsymbol{v}_e - \boldsymbol{G}_p + m_p(\boldsymbol{N}_{v_e}^{-1}\boldsymbol{s}_t + k_{d1}(t)\boldsymbol{N}_{v_e}\boldsymbol{v}_e)$. For subsequent analysis, $\boldsymbol{\Phi}_t(\cdot)$ is decomposed into two parts: $\boldsymbol{\Phi}_{t1}(\cdot) = -\bar{\boldsymbol{G}}_p + \boldsymbol{N}_{v_e}^{-1}\boldsymbol{s}_t + k_{d1}(t)\boldsymbol{N}_{v_e}\boldsymbol{v}_e$, and $\boldsymbol{\Phi}_{t2}(\cdot) = -\bar{\boldsymbol{C}}_p\boldsymbol{v}_e$, where $\bar{\boldsymbol{G}}_p = \boldsymbol{G}_p/m_p$ and $\bar{\boldsymbol{C}}_p = \boldsymbol{C}_p/m_p$ are known matrices. With regard to $\boldsymbol{\Phi}_{t1}(\cdot)$, since, for all $i, j \in \{1, 2, 3\}$, $\partial\phi_{t1,i}/\partial v_{ej} = \partial\phi_{t1,j}/\partial v_{ei}$, where $\phi_{t1,i}$ and $\phi_{t1,j}$ are the i-th and j-th elements of $\boldsymbol{\Phi}_{t1}^\top(\cdot)$, respectively, there certainly exists $\mu_{t1} \in \mathbb{R}$ (but not unique) satisfying the following solvable PDE:

$$\partial\mu_{t1}/\partial\boldsymbol{v}_e = \boldsymbol{\Phi}_{t1}^\top(\cdot). \tag{9.19}$$

Direct deduction from (9.19) delivers a valid solution

$$\mu_{t1} = \mu_{t1,1} + \mu_{t1,2}, \tag{9.20}$$

with

$$\mu_{t1,1} = -\bar{\boldsymbol{G}}^\top \boldsymbol{v}_e,$$

$$\mu_{t1,2} = \sum_{i=1}^{3}\left[s_{ti}v_{ei}\left(v_{e,\max}^2 - \frac{1}{3}v_{ei}^2\right) - \frac{k_{d1}(t)}{2}\log(\boldsymbol{N}_{v_e,i}^{-1})\right].$$

However, unlike $\boldsymbol{\Phi}_{t1}(\cdot)$, there exists no $\mu_{t2} \in \mathbb{R}$ such that the PDE $\partial\mu_{t2}/\partial\boldsymbol{v}_e = \boldsymbol{\Phi}_{t2}^\top(\cdot)$ holds, due to the skew-symmetric matrix $\bar{\boldsymbol{C}}_p$ rendering $\partial\phi_{t2,i}/\partial v_{ej} \neq \partial\phi_{t2,j}/$

∂v_{ei}, for all $i, j \in \{1, 2, 3\}$ other than $i = j$. This is the so-called "integrability obstacle" arising in the I&I adaptive design. To overcome this obstruction, we deliberately introduce a vector $\boldsymbol{\Psi}_t(\cdot) \in \mathbb{R}^3$ to construct a solvable PDE through satisfying the following relationship

$$\frac{\partial \phi_{t2,i}}{\partial v_{ej}} + \frac{\partial \psi_{ti}}{\partial v_{ej}} = \frac{\partial \phi_{t2,j}}{\partial v_{ei}} + \frac{\partial \psi_{tj}}{\partial v_{ei}}, \tag{9.21}$$

for all $i, j \in \{1, 2, 3\}$, where ψ_{ti} and ψ_{tj} stand for the i-th and j-th columns of $\boldsymbol{\Psi}_t^\top(\cdot)$. A simple choice of $\boldsymbol{\Psi}_t(\cdot)$ that suffices (9.21) is $\boldsymbol{\Psi}_t(\cdot) = -\boldsymbol{\Phi}_{t2}(\cdot)$. With this in mind, the following PDE is solvable

$$\partial \mu_{t2}/\partial \boldsymbol{v}_e = \boldsymbol{\Phi}_{t2}^\top(\cdot) + \boldsymbol{\Psi}_t^\top(\cdot), \tag{9.22}$$

and a valid solution is $\mu_{t2} = 0$.

For notational brevity, $\mu_t = \mu_{t1} + \mu_{t2}$ is further defined, and in the rest of this subsection, function arguments are dropped whenever no confusion can occur.

Design the adaptive position control law as

$$\boldsymbol{f}_c = -\boldsymbol{R}_{\mathcal{PT}} \boldsymbol{\Phi}_t (\hat{m}_p + \beta_t), \tag{9.23}$$

with $\hat{m}_p$ and β_t determined by

$$\dot{\hat{m}}_p = \gamma_1 [-\dot{\bar{\mu}}_t + (\boldsymbol{\Phi}_t + \tilde{\boldsymbol{\Psi}}_t)^\top (\boldsymbol{N}_{v_e}^{-1} \boldsymbol{s}_t + k_{d1}(t) \boldsymbol{N}_{v_e} \boldsymbol{v}_e) + \dot{\hat{\boldsymbol{\Psi}}}_t^\top \boldsymbol{v}_e], \tag{9.24}$$

$$\beta_t = \gamma_1 (\mu_t - \hat{\boldsymbol{\Psi}}_t^\top \boldsymbol{v}_e), \tag{9.25}$$

wherein $\gamma_1 > 0$ is a scalar constant, $\dot{\bar{\mu}}_t = \dot{\mu}_t - (\partial \mu_t / \partial \boldsymbol{v}_e) \dot{\boldsymbol{v}}_e$, $\tilde{\boldsymbol{\Psi}}_t = \boldsymbol{\Psi}_t - \hat{\boldsymbol{\Psi}}_t$ with $\hat{\boldsymbol{\Psi}}_t$ defined by replacing $\boldsymbol{v}_e$ with $\hat{\boldsymbol{v}}_e$ in $\boldsymbol{\Psi}_t$, $\hat{\boldsymbol{v}}_e$ is the filtered velocity constructed by

$$\dot{\hat{\boldsymbol{v}}}_e = -\boldsymbol{N}_{v_e}^{-1} \boldsymbol{s}_t - k_{d1}(t) \boldsymbol{N}_{v_e} \boldsymbol{v}_e + k_{f1}(t) \tilde{\boldsymbol{v}}_e, \tag{9.26}$$

where $k_{f1}(t)$ is also a time-varying gain to be determined, and $\tilde{\boldsymbol{v}}_e = \boldsymbol{v}_e - \hat{\boldsymbol{v}}_e$. The composite term $(\hat{m}_p + \beta_t) \in \mathbb{R}$ actually acts as the estimate of m_p. Thus, the estimation error is defined as $\tilde{m}_p = \hat{m}_p + \beta_t - m_p$.

Once the control law (9.23) is plugged in, (9.18) reduces to

$$\dot{\boldsymbol{v}}_e = -\boldsymbol{N}_{v_e}^{-1} \boldsymbol{s}_t - k_{d1}(t) \boldsymbol{N}_{v_e} \boldsymbol{v}_e - m_p^{-1} \boldsymbol{\Phi}_t \tilde{m}_p, \tag{9.27}$$

In addition, with (9.24)–(9.27) in mind, the time derivative of $\tilde{m}_p$ can be derived as

$$\dot{\tilde{m}}_p = -\gamma_1 (\boldsymbol{\Phi}_t + \tilde{\boldsymbol{\Psi}}_t)^\top m_p^{-1} \boldsymbol{\Phi}_t \tilde{m}_p. \tag{9.28}$$

Here, $\tilde{\boldsymbol{\Psi}}_t$ acts like a "perturbation" to the adaptive parameter estimation. To deal with it, a dynamic scaling factor $r_t(t) \in \mathbb{R}$ satisfying $r_t(t) \geq 1$ for all $t \geq 0$ is introduced to form the scaled estimation error $z_t \in \mathbb{R}$ as follows [33]:

$$z_t = \frac{e^{\frac{1}{2}(\frac{1}{m_{p,\min}^2}+1)}}{\sqrt{m_{p,\min}}} \cdot \frac{\tilde{m}_p}{e^{\frac{\sqrt{\log r_t+1}}{m_{p,\min}}}}, \tag{9.29}$$

where the unknown scalar $m_{p,\min} > 0$ is the minimum of m_p (note that $m_{p,\min}$ is introduced just for the subsequent stability analysis and will not be used in the control implementation), and r_t is determined by

$$\dot{r}_t = \gamma_1 r_t \sqrt{\log r_t + 1} \|\tilde{\boldsymbol{\Psi}}_t\|^2, \ r_t(0) = 1. \tag{9.30}$$

It is noteworthy that if the scaled estimation error z_t is defined as $z_t = \tilde{m}_p / r_t$ with r_t determined by $\dot{r}_t = \gamma_1 \Gamma_1 \|\tilde{\boldsymbol{\Psi}}_t\|^2 r_t$ ($\Gamma_1 > 0$), as similarly done in [34, 35], then the parameter Γ_1 should be chosen such that $\Gamma_1 > 1/(2m_{p,\min})$ holds. This implies that the I&I adaptive control design requires knowledge of the lower bound of m_p. But, in the present manner, such a problem is effectively eliminated by introducing the modified dynamic scaling factor r_t given by (9.30). Then, differentiating z_t w.r.t. time along (9.28) and (9.30) gives

$$\dot{z}_t = -\gamma_1 (\boldsymbol{\Phi}_t + \tilde{\boldsymbol{\Psi}}_t)^\top m_p^{-1} \boldsymbol{\Phi}_t z_t - \frac{\gamma_1}{2m_{p,\min}} \|\tilde{\boldsymbol{\Psi}}_t\|^2 z_t. \tag{9.31}$$

Theorem 9.1 *Consider the translational tracking error dynamics given by* (9.1) *and* (9.2) *with the mass* m_p *being unknown. If the time-varying gains are chosen as:* $k_{d1}(t) = \kappa_1 r_t$ *and* $k_{f1}(t) = \kappa_1 r_t + (c_1/2)\|\bar{\boldsymbol{C}}_p\|^2$*, where* κ_1, $c_1 > 0$ *are design constants, then, for the initial conditions satisfying* $\boldsymbol{\rho}(0) \in \mathbb{K}_\rho$ *and* $\boldsymbol{v}_e(0) \in \mathbb{K}_{v_e}$*, the control law in* (9.23) *can guarantee that:*

- V_p *and* V_{v_e} *remain bounded for all* $t \geq 0$*, indicating that both path and velocity constraints are satisfied.*
- *The scaled estimation error dynamics* (9.31) *has a globally stable equilibrium at* $z_t = 0$.
- *The vector* $m_p^{-1}\boldsymbol{\Phi}_t z_t$*, as well as the position and velocity tracking errors, converge asymptotically to zero, that is,* $\lim_{t\to\infty}[m_p^{-1}\boldsymbol{\Phi}_t(t)z_t(t), \boldsymbol{\rho}_e(t), \boldsymbol{v}_e(t)] = \boldsymbol{0}$.

Proof Consider the Lyapunov-like function:

$$V_1 = V_p + V_{v_e} + \frac{1}{2}\tilde{\boldsymbol{v}}_e^\top \tilde{\boldsymbol{v}}_e + \frac{\eta_1}{2\gamma_1} z_t^2 + \frac{c_1}{\gamma_1}\sqrt{\log r_t + 1}, \tag{9.32}$$

where $\eta_1 = (2/\kappa_1) + \varrho$ with ϱ a positive constant is chosen just for stability analysis. Now, taking the time derivative of V_1 along (9.5), (9.7), (9.26), (9.27), (9.30) and (9.31) leads to

$$\begin{aligned}\dot{V}_1 =& -k_{d1}(t)\|\boldsymbol{N}_{v_e}\boldsymbol{v}_e\|^2 - \boldsymbol{v}_e^\top \boldsymbol{N}_{v_e} m_p^{-1}\boldsymbol{\Phi}_t\tilde{m}_p - k_{f1}(t)\|\tilde{\boldsymbol{v}}_e\|^2\\ &- \tilde{\boldsymbol{v}}_e^\top m_p^{-1}\boldsymbol{\Phi}_t\tilde{m}_p - \eta_1 z_t(\boldsymbol{\Phi}_t + \tilde{\boldsymbol{\Psi}}_t)^\top m_p^{-1}\boldsymbol{\Phi}_t z_t\\ &- \frac{\eta_1}{2m_{p,\min}}\|\tilde{\boldsymbol{\Psi}}_t\|^2 z_t^2 + \frac{c_1}{2}\|\tilde{\boldsymbol{\Psi}}_t\|^2.\end{aligned} \tag{9.33}$$

By Young's inequality, one can easily deduce that

$$e^{\frac{\sqrt{\log r_t+1}}{m_{p,\min}}} \le e^{\left(\frac{\log r_t+1}{2}+\frac{1}{2m_{p,\min}^2}\right)} = e^{\frac{1}{2}\left(\frac{1}{m_{p,\min}^2}+1\right)}\sqrt{r_t},$$

$$\sqrt{m_{p,\min} r_t}|\boldsymbol{v}_e^\top \boldsymbol{N}_{v_e} m_p^{-1}\boldsymbol{\Phi}_t z_t| \le \frac{\kappa_1 r_t}{2}\|\boldsymbol{N}_{v_e}\boldsymbol{v}_e\|^2 + \frac{m_{p,\min}}{2\kappa_1}\|m_p^{-1}\boldsymbol{\Phi}_t z_t\|^2,$$

$$\sqrt{m_{p,\min} r_t}|\tilde{\boldsymbol{v}}_e^\top m_p^{-1}\boldsymbol{\Phi}_t z_t| \le \frac{\kappa_1 r_t}{2}\|\tilde{\boldsymbol{v}}_e\|^2 + \frac{m_{p,\min}}{2\kappa_1}\|m_p^{-1}\boldsymbol{\Phi}_t z_t\|^2,$$

$$\eta_1|z_t\tilde{\boldsymbol{\Psi}}_t^\top m_p^{-1}\boldsymbol{\Phi}_t z_t| \le \frac{\eta_1}{2m_{p,\min}}\|\tilde{\boldsymbol{\Psi}}_t\|^2 z_t^2 + \frac{\eta_1 m_{p,\min}}{2}\|m_p^{-1}\boldsymbol{\Phi}_t z_t\|^2.$$

With the above inequalities in mind, and noting the fact that $\tilde{\boldsymbol{\Psi}}_t = \bar{\boldsymbol{C}}_p\tilde{\boldsymbol{v}}_e$ and $r_t(t) \ge 1$ $\forall t \ge 0$, we further have

$$\dot{V}_1 \le -\frac{\kappa_1}{2}(\|\boldsymbol{N}_{v_e}\boldsymbol{v}_e\|^2 + \|\tilde{\boldsymbol{v}}_e\|^2) - \frac{m_{p,\min}\varrho}{2}\|m_p^{-1}\boldsymbol{\Phi}_t z_t\|^2. \tag{9.34}$$

The following observations are now in order:

(1) Inspecting (9.34) reveals that $\dot{V}_1 \le 0$ and accordingly $V_1(t)$ and thus $V_p(t)$ and $V_{v_e}(t)$ are uniformly bounded for all $t \ge 0$. Since $\boldsymbol{\rho}(0) \in \mathbb{K}_\rho$ and $\boldsymbol{v}_e(0) \in \mathbb{K}_{v_e}$, the boundedness of V_p and V_{v_e} implies that $\boldsymbol{\rho}(t) \in \mathbb{K}_\rho$ and $\boldsymbol{v}_e(t) \in \mathbb{K}_{v_e}$ for all time. Thus, both path and velocity constraints are satisfied.

(2) Consider the positive-definite function $V_t(z_t) = \frac{1}{2}z_t^2$, whose time derivative along (9.31) satisfies

$$\begin{aligned}\dot{V}_t =& -\gamma_1(\boldsymbol{\Phi}_t + \tilde{\boldsymbol{\Psi}}_t)^\top m_p^{-1}\boldsymbol{\Phi}_t z_t^2 - \frac{\gamma_1}{2m_{p,\min}}\|\tilde{\boldsymbol{\Psi}}_t\|^2 z_t^2\\ \le& -\gamma_1 m_{p,\min}\|m_p^{-1}\boldsymbol{\Phi}_t z_t\|^2 + \frac{\gamma_1}{2m_{p,\min}}\|\tilde{\boldsymbol{\Psi}}_t\|^2 z_t^2\\ &+ \frac{\gamma_1 m_{p,\min}}{2}\|m_p^{-1}\boldsymbol{\Phi}_t z_t\|^2 - \frac{\gamma_1}{2m_{p,\min}}\|\tilde{\boldsymbol{\Psi}}_t\|^2 z_t^2\\ \le& -\frac{\gamma_1 m_{p,\min}}{2}\|m_p^{-1}\boldsymbol{\Phi}_t z_t\|^2 \le 0.\end{aligned} \tag{9.35}$$

Then, based on the Lyapunov stability theorem, we conclude that the scaled parameter estimation error dynamics (9.31) has a globally stable equilibrium at $z_t = 0$.

(3) As $\dot{V}_1 \le 0$, $V_1(t)$ is upper bounded by $V_1(0)$, from which we deduce that z_t and $r_t \in \mathcal{L}_\infty$, and that $\int_0^\infty \dot{V}_1(t)\,dt$ exists and is finite. From the latter, we have $\boldsymbol{N}_{v_e}\boldsymbol{v}_e$,

$\tilde{\boldsymbol{v}}_e$ and $m_p^{-1}\boldsymbol{\Phi}_t z_t \in \mathcal{L}_2 \cap \mathcal{L}_\infty$. Then, from the definitions of $k_{d1}(t)$ and $k_{f1}(t)$, it is evident that $k_{d1}(t)$ and $k_{f1}(t) \in \mathcal{L}_\infty$. Further, invoking (1) obtains that $\boldsymbol{s}_t$, $\boldsymbol{v}_e$, $\boldsymbol{N}_{v_e}$, and $\boldsymbol{N}_{v_e}^{-1} \in \mathcal{L}_\infty$, which, together with Assumption 9.1, lead to $\boldsymbol{\Phi}_t$ and $\tilde{\boldsymbol{\Psi}}_t \in \mathcal{L}_\infty$. As a result, it follows from (9.26)–(9.28) and (9.31) that $\dot{\boldsymbol{v}}_e$, $\dot{\tilde{\boldsymbol{v}}}_e$, $\dot{\tilde{m}}_p$ and $\dot{z}_t \in \mathcal{L}_\infty$. With $\boldsymbol{v}_e$ and $\dot{\boldsymbol{v}}_e$ bounded, we can claim that $\dot{\boldsymbol{s}}_t$, $\dot{\boldsymbol{N}}_{v_e}$, and $\dot{\boldsymbol{N}}_{v_e}^{-1} \in \mathcal{L}_\infty$. Furthermore, from (9.30), it can be easily deduced that $\dot{k}_{d1}(t)$ and hence $\dot{\boldsymbol{\Phi}}_t \in \mathcal{L}_\infty$.

As per the above discussion, it can be concluded that the vectors $m_p^{-1}\boldsymbol{\Phi}_t z_t$, $\boldsymbol{N}_{v_e}\boldsymbol{v}_e$, and $\tilde{\boldsymbol{v}}_e$ are square integrable. In addition, from the fact that $\dot{\tilde{\boldsymbol{v}}}_e \in \mathcal{L}_\infty$, and that $\frac{d}{dt}(m_p^{-1}\boldsymbol{\Phi}_t z_t) = m_p^{-1}(\dot{\boldsymbol{\Phi}}_t z_t + \boldsymbol{\Phi}_t \dot{z}_t)$ and $\frac{d}{dt}(\boldsymbol{N}_{v_e}\boldsymbol{v}_e) = \dot{\boldsymbol{N}}_{v_e}\boldsymbol{v}_e + \boldsymbol{N}_{v_e}\dot{\boldsymbol{v}}_e$, we can infer that $m_p^{-1}\boldsymbol{\Phi}_t z_t$, $\boldsymbol{N}_{v_e}\boldsymbol{v}_e$, and $\tilde{\boldsymbol{v}}_e$ are uniformly continuous. Then, applying Barbalat's Lemma establishes

$$\lim_{t\to\infty} [m_p^{-1}\boldsymbol{\Phi}_t(t) z_t(t), \boldsymbol{N}_{v_e}(t)\boldsymbol{v}_e(t), \tilde{\boldsymbol{v}}_e(t)] = \boldsymbol{0}.$$

Note that $\lim_{t\to\infty} \boldsymbol{N}_{v_e}(t)\boldsymbol{v}_e(t) = \boldsymbol{0}$ is, in essence, equivalent to $\lim_{t\to\infty} \boldsymbol{v}_e(t) = \boldsymbol{0}$.

Next, we show the asymptotic convergence of $\boldsymbol{\rho}_e$. Actually, with the aid of the above analyses, it is an easy task to establish the boundedness of $\ddot{\boldsymbol{v}}_e$, that is, the uniformly continuous of $\dot{\boldsymbol{v}}_e$. Then, together with the convergence of $\boldsymbol{v}_e$ to the origin, it can be claimed from Barbalat's Lemma that $\lim_{t\to\infty} \dot{\boldsymbol{v}}_e(t) = \boldsymbol{0}$. From (9.27), it follows that $\lim_{t\to\infty} \boldsymbol{s}_t(t) = \boldsymbol{0}$. Further, recalling Lemma 9.1, it can be concluded that $\lim_{t\to\infty} \boldsymbol{\rho}_e(t) = \boldsymbol{0}$. □

9.3.2 I&I Adaptive Attitude Controller

To deal with inertia uncertainties, we here introduce a linear operator $\boldsymbol{L}(\cdot) : \mathbb{R}^3 \to \mathbb{R}^{3\times 3}$ such that $\boldsymbol{J}_p\boldsymbol{x} = \boldsymbol{L}(\boldsymbol{x})\boldsymbol{\theta}$ for any $\boldsymbol{x} \in \mathbb{R}^3$, where $\boldsymbol{\theta} = [J_{p,11}, J_{p,22}, J_{p,33}]^\top$ with $J_{p,ii}$, $i = 1, 2, 3$ the diagonal elements of $\boldsymbol{J}_p$. Then, to solve Problem 9.2, (9.9) is rewritten as

$$\dot{\boldsymbol{\omega}}_e = -\boldsymbol{N}_{\omega_e}^{-1}\boldsymbol{s}_r - k_{d2}(t)\boldsymbol{N}_{\omega_e}\boldsymbol{\omega}_e + \boldsymbol{J}_p^{-1}(\boldsymbol{\Phi}_r(\cdot)\boldsymbol{\theta} + \boldsymbol{\tau}_c), \tag{9.36}$$

where $k_{d2}(t) > 0$ (independent of $\boldsymbol{\omega}_e$), similar to $k_{d1}(t)$, is also a time-varying gain to be determined, and $\boldsymbol{\Phi}_r(\cdot) \in \mathbb{R}^{3\times 3}$ is the regressor matrix such that $\boldsymbol{\Phi}_r(\cdot)\boldsymbol{\theta} = -\boldsymbol{S}(\boldsymbol{\omega}_p)\boldsymbol{J}_p\boldsymbol{\omega}_p + \boldsymbol{J}_p(\boldsymbol{S}(\boldsymbol{\omega}_e)\boldsymbol{\Omega} - \bar{\boldsymbol{\Omega}}) + \boldsymbol{J}_p(\boldsymbol{N}_{\omega_e}^{-1}\boldsymbol{s}_r + k_{d2}(t)\boldsymbol{N}_{\omega_e}\boldsymbol{\omega}_e)$. For subsequent analysis, $\boldsymbol{\Phi}_r(\cdot)$ is decomposed into two parts: $\boldsymbol{\Phi}_{r1}(\cdot) = -\boldsymbol{S}(\boldsymbol{\Omega})\boldsymbol{L}(\boldsymbol{\Omega}) - \boldsymbol{L}(\bar{\boldsymbol{\Omega}}) + \boldsymbol{L}(\boldsymbol{N}_{\omega_e}^{-1}\boldsymbol{s}_r) + k_{d2}(t)\boldsymbol{L}(\boldsymbol{N}_{\omega_e}\boldsymbol{\omega}_e)$, and $\boldsymbol{\Phi}_{r2}(\cdot) = -\boldsymbol{S}(\boldsymbol{\omega}_e)\boldsymbol{L}(\boldsymbol{\omega}_e + \boldsymbol{\Omega}) - \boldsymbol{S}(\boldsymbol{\Omega})\boldsymbol{L}(\boldsymbol{\omega}_e) + \boldsymbol{L}(\boldsymbol{S}(\boldsymbol{\omega}_e)\boldsymbol{\Omega})$.

Since, for all $i, j \in \{1, 2, 3\}$, $\partial\phi_{r1,i}/\partial\omega_{ej} = \partial\phi_{r1,j}/\partial\omega_{ei}$, where $\phi_{r1,i}$ and $\phi_{r1,j}$ are the i-th and j-th columns of $\boldsymbol{\Phi}_{r1}^\top(\cdot)$, respectively, there exists $\boldsymbol{\mu}_{r1} \in \mathbb{R}^3$ (but not unique) satisfying the following PDE:

$$\partial \boldsymbol{\mu}_{r1}/\partial \boldsymbol{\omega}_e = \boldsymbol{\Phi}_{r1}^{\top}(\cdot). \tag{9.37}$$

It is easy to find a valid solution to (9.37), given by

$$\boldsymbol{\mu}_{r1} = \boldsymbol{\mu}_{r1,1} + \boldsymbol{\mu}_{r1,2}, \tag{9.38}$$

with

$$\boldsymbol{\mu}_{r1,1} = (\boldsymbol{L}^{\top}(\boldsymbol{\Omega})\boldsymbol{S}(\boldsymbol{\Omega}) - \boldsymbol{L}^{\top}(\bar{\boldsymbol{\Omega}}))\boldsymbol{\omega}_e,$$

$$\boldsymbol{\mu}_{r1,2} = \begin{bmatrix} s_{r1}\omega_{e,\max}^2\omega_{e1} - \frac{1}{3}s_{r1}\omega_{e1}^3 - \frac{k_{d2}(t)}{2}\log(N_{\omega_e,1}^{-1}) \\ s_{r2}\omega_{e,\max}^2\omega_{e2} - \frac{1}{3}s_{r2}\omega_{e2}^3 - \frac{k_{d2}(t)}{2}\log(N_{\omega_e,2}^{-1}) \\ s_{r3}\omega_{e,\max}^2\omega_{e3} - \frac{1}{3}s_{r3}\omega_{e3}^3 - \frac{k_{d2}(t)}{2}\log(N_{\omega_e,3}^{-1}) \end{bmatrix}.$$

Unfortunately, there exists no $\boldsymbol{\mu}_{r2} \in \mathbb{R}^3$ satisfying the PDE $\partial \boldsymbol{\mu}_{r2}/\partial \boldsymbol{\omega}_e = \boldsymbol{\Phi}_{r2}^{\top}(\cdot)$, since $\partial \phi_{r2,i}/\partial \omega_{ej} \neq \partial \phi_{r2,j}/\partial \omega_{ei}$, for all $i, j \in \{1, 2, 3\}$ other than $i = j$. As done in Sect. 9.3.1, a vector $\boldsymbol{\Psi}_r(\cdot) \in \mathbb{R}^{3\times 3}$ is also introduced here to render

$$\frac{\partial \phi_{r2,i}}{\partial \omega_{ej}} + \frac{\partial \psi_{ri}}{\partial \omega_{ej}} = \frac{\partial \phi_{r2,j}}{\partial \omega_{ei}} + \frac{\partial \psi_{rj}}{\partial \omega_{ei}}, \tag{9.39}$$

for all $i, j \in \{1, 2, 3\}$, where ψ_{ri} and ψ_{rj} stand for the i-th and j-th columns of $\boldsymbol{\Psi}_r^{\top}(\cdot)$. A simple choice of $\boldsymbol{\Psi}_r(\cdot)$ that suffices (9.39) is $\boldsymbol{\Psi}_r(\cdot) = -\boldsymbol{\Phi}_{r2}(\cdot)$. Given this, it is not difficult to check that the following PDE is solvable

$$\partial \boldsymbol{\mu}_{r2}/\partial \boldsymbol{\omega}_e = \boldsymbol{\Phi}_{r2}^{\top}(\cdot) + \boldsymbol{\Psi}_r^{\top}(\cdot), \tag{9.40}$$

and a direct solution is $\boldsymbol{\mu}_{r2} = \boldsymbol{0}$. Define $\boldsymbol{\mu}_r = \boldsymbol{\mu}_{r1} + \boldsymbol{\mu}_{r2}$.

Design the adaptive attitude control law as

$$\boldsymbol{\tau}_c = -\boldsymbol{\Phi}_r(\hat{\boldsymbol{\theta}} + \boldsymbol{\beta}_r), \tag{9.41}$$

with $\hat{\boldsymbol{\theta}}$ and $\boldsymbol{\beta}_r$ determined by

$$\dot{\hat{\boldsymbol{\theta}}} = \gamma_2[-\dot{\bar{\boldsymbol{\mu}}}_r + (\boldsymbol{\Phi}_r + \tilde{\boldsymbol{\Psi}}_r)^{\top}(\boldsymbol{N}_{\omega_e}^{-1}\boldsymbol{s}_r + k_{d2}(t)\boldsymbol{N}_{\omega_e}\boldsymbol{\omega}_e) + \dot{\hat{\boldsymbol{\Psi}}}_r^{\top}\boldsymbol{\omega}_e], \tag{9.42}$$

$$\boldsymbol{\beta}_r = \gamma_2(\boldsymbol{\mu}_r - \hat{\boldsymbol{\Psi}}_r^{\top}\boldsymbol{\omega}_e), \tag{9.43}$$

where $\gamma_2 > 0$ is a scalar constant, $\dot{\bar{\boldsymbol{\mu}}}_r = \dot{\boldsymbol{\mu}}_r - (\partial \boldsymbol{\mu}_r/\partial \boldsymbol{\omega}_e)\dot{\boldsymbol{\omega}}_e$, $\tilde{\boldsymbol{\Psi}}_r = \boldsymbol{\Psi}_r - \hat{\boldsymbol{\Psi}}_r$ with $\hat{\boldsymbol{\Psi}}_r$ defined by replacing $\boldsymbol{\omega}_e$ with $\hat{\boldsymbol{\omega}}_e$ in $\boldsymbol{\Psi}_r$, and $\hat{\boldsymbol{\omega}}_e$ is the filtered angular velocity constructed in the following fashion:

$$\dot{\hat{\boldsymbol{\omega}}}_e = -\boldsymbol{N}_{\omega_e}^{-1}\boldsymbol{s}_r - k_{d2}(t)\boldsymbol{N}_{\omega_e}\boldsymbol{\omega}_e + k_{f2}(t)\tilde{\boldsymbol{\omega}}_e, \tag{9.44}$$

where $k_{f2}(t)$ is a time-varying gain to be determined, and $\tilde{\boldsymbol{\omega}}_e = \boldsymbol{\omega}_e - \hat{\boldsymbol{\omega}}_e$. After some straightforward algebra, we can rewrite $\tilde{\boldsymbol{\Psi}}_r$ as

$$\tilde{\boldsymbol{\Psi}}_r = [\boldsymbol{H}(\boldsymbol{I}_3 \otimes \tilde{\boldsymbol{\omega}}_e)]^\top, \tag{9.45}$$

where "$\otimes$" represents the Kronecker product and $\boldsymbol{H} \in \mathbb{R}^{3\times 9}$ is defined as

$$\boldsymbol{H} = [\boldsymbol{H}_1 \ \boldsymbol{H}_2 \ \boldsymbol{H}_3],$$

with the block matrices $\boldsymbol{H}_i \in \mathbb{R}^{3\times 3}$, $i = 1, 2, 3$ given by:

$$\boldsymbol{H}_1 = \begin{bmatrix} 0 & -\Omega_3 & \Omega_2 \\ 0 & -(\hat{\omega}_{e3} + \Omega_3) & -(\omega_{e2} + \Omega_2) \\ 0 & \omega_{e3} + \Omega_3 & \hat{\omega}_{e2} + \Omega_2 \end{bmatrix},$$

$$\boldsymbol{H}_2 = \begin{bmatrix} \hat{\omega}_{e3} + \Omega_3 & 0 & \omega_{e1} + \Omega_1 \\ \Omega_3 & 0 & -\Omega_1 \\ -(\omega_{e3} + \Omega_3) & 0 & -(\hat{\omega}_{e1} + \Omega_1) \end{bmatrix},$$

$$\boldsymbol{H}_3 = \begin{bmatrix} -(\hat{\omega}_{e2} + \Omega_2) & -(\omega_{e1} + \Omega_1) & 0 \\ \omega_{e2} + \Omega_2 & \hat{\omega}_{e1} + \Omega_1 & 0 \\ -\Omega_2 & \Omega_1 & 0 \end{bmatrix}.$$

Define the adaptive estimation error as $\tilde{\boldsymbol{\theta}} = \hat{\boldsymbol{\theta}} + \boldsymbol{\beta}_r - \boldsymbol{\theta}$. Then, inserting the control law (9.41) into (9.36) leads to

$$\dot{\boldsymbol{\omega}}_e = -\boldsymbol{N}_{\omega_e}^{-1}\boldsymbol{s}_r - k_{d2}(t)\boldsymbol{N}_{\omega_e}\boldsymbol{\omega}_e - \boldsymbol{J}_p^{-1}\boldsymbol{\Phi}_r\tilde{\boldsymbol{\theta}}. \tag{9.46}$$

In addition, with (9.37), (9.40), (9.42) and (9.43) in mind, the time derivative of $\tilde{\boldsymbol{\theta}}$ can be derived as

$$\dot{\tilde{\boldsymbol{\theta}}} = -\gamma_2(\boldsymbol{\Phi}_r + \tilde{\boldsymbol{\Psi}}_r)^\top \boldsymbol{J}_p^{-1}\boldsymbol{\Phi}_r\tilde{\boldsymbol{\theta}}. \tag{9.47}$$

To counter the effect of $\tilde{\boldsymbol{\Psi}}_r$, a dynamic scaling factor $r_r(t) \in \mathbb{R}$ satisfying $r_r(t) \geq 1$ for all $t \geq 0$ is introduced to form the scaled estimation error $z_r \in \mathbb{R}^3$ [33]

$$z_r = \frac{e^{\frac{1}{2}\left(\frac{1}{J_{p,\min}^2}+1\right)}}{\sqrt{J_{p,\min}}} \cdot \frac{\tilde{\boldsymbol{\theta}}}{e^{\frac{\sqrt{\log r_r+1}}{J_{p,\min}}}}, \tag{9.48}$$

where $J_{p,\min} > 0$ denotes the minimum eigenvalue of $\boldsymbol{J}_p$ (note that $J_{p,\min}$ will not be used in the control implementation), and r_r is determined by

$$\dot{r}_r = \gamma_2 r_r \sqrt{\log r_r + 1}\|\tilde{\boldsymbol{\Psi}}_r\|^2, \; r_r(0) = 1. \tag{9.49}$$

Taking the time derivative of z_r along (9.47) and (9.49) outputs

$$\dot{z}_r = -\gamma_2(\boldsymbol{\Phi}_r + \tilde{\boldsymbol{\Psi}}_r)^\top \boldsymbol{J}_p^{-1}\boldsymbol{\Phi}_r z_r - \frac{\gamma_2}{2J_{p,\min}}\|\tilde{\boldsymbol{\Psi}}_r\|^2 z_r. \tag{9.50}$$

Theorem 9.2 *Consider the rotational tracking error dynamics described by* (9.8) *and* (9.9) *with the inertia matrix* $\boldsymbol{J}_p$ *unknown. If the time-varying gains are chosen as:* $k_{d2}(t) = \kappa_2 r_r$ *and* $k_{f2}(t) = \kappa_2 r_r + (c_2/2)\|\boldsymbol{H}\|^2$*, where* κ_2, $c_2 > 0$ *are constants, then, for the initial conditions that satisfy* $\boldsymbol{q}_e(0) \in \mathbb{K}_{q_e}$ *and* $\boldsymbol{\omega}_e(0) \in \mathbb{K}_{\omega_e}$*, the control law in* (9.41) *can guarantee that:*

- V_f *and* V_{ω_e} *remain bounded for all* $t \geq 0$*, indicating that both FOV and angular velocity constraints are satisfied.*
- *The scaled estimation error dynamics* (9.50) *has a globally stable equilibrium at* $z_r = \mathbf{0}$.
- $\boldsymbol{J}_p^{-1}\boldsymbol{\Phi}_r z_r$*, as well as the attitude and angular velocity tracking errors, converge asymptotically to zero, that is,* $\lim_{t\to\infty}[\boldsymbol{J}_p^{-1}\boldsymbol{\Phi}_r(t)z_r(t), \boldsymbol{q}_{ev}(t), \boldsymbol{\omega}_e(t)] = \mathbf{0}$.

Proof Consider the following Lyapunov-like function:

$$V_2 = V_f + V_{\omega_e} + \frac{1}{2}\tilde{\boldsymbol{\omega}}_e^\top \tilde{\boldsymbol{\omega}}_e + \frac{\eta_2}{2\gamma_2} z_r^\top z_r + \frac{c_2}{\gamma_2}\sqrt{\log r_r + 1}, \tag{9.51}$$

where $\eta_2 = (2/\kappa_2) + \varrho$ is chosen just for stability analysis. Now, taking the time derivative of V_2 along (9.14), (9.16), (9.44), (9.46), (9.49), and (9.50) leads to

$$\begin{aligned}\dot{V}_2 = &- k_{d2}(t)\|\boldsymbol{N}_{\omega_e}\boldsymbol{\omega}_e\|^2 - \boldsymbol{\omega}_e^\top \boldsymbol{N}_{\omega_e}\boldsymbol{J}_p^{-1}\boldsymbol{\Phi}_r\tilde{\boldsymbol{\theta}} - k_{f2}(t)\|\tilde{\boldsymbol{\omega}}_e\|^2 \\ &- \tilde{\boldsymbol{\omega}}_e^\top \boldsymbol{J}_p^{-1}\boldsymbol{\Phi}_r\tilde{\boldsymbol{\theta}} - \eta_2 z_r \boldsymbol{J}_p^{-1}(\boldsymbol{\Phi}_r + \tilde{\boldsymbol{\Psi}}_r)^\top \boldsymbol{\Phi}_r z_r \\ &- \frac{\eta_2}{2J_{p,\min}}\|\tilde{\boldsymbol{\Psi}}_r\|^2\|z_r\|^2 + \frac{c_2}{2}\|\tilde{\boldsymbol{\Psi}}_r\|^2.\end{aligned} \tag{9.52}$$

Following a similar procedure for deriving (9.34), and noticing (9.45) and the fact that $\|\tilde{\boldsymbol{\Psi}}_r\| = \|\tilde{\boldsymbol{\Psi}}_r^\top\|$ and $\|\boldsymbol{I}_3 \otimes \tilde{\boldsymbol{\omega}}_e\| = \|\tilde{\boldsymbol{\omega}}_e\|$, (9.52) becomes

$$\dot{V}_2 \leq -\frac{\kappa_2}{2}(\|\boldsymbol{N}_{\omega_e}\boldsymbol{\omega}_e\|^2 + \|\tilde{\boldsymbol{\omega}}_e\|^2) - \frac{J_{p,\min}\varrho}{2}\|\boldsymbol{J}_p^{-1}\boldsymbol{\Phi}_r z_r\|^2. \tag{9.53}$$

Considering the positive-definite function $V_r(z_r) = \frac{1}{2}z_r^\top z_r$, we conclude that the z_r-dynamics (9.50) has a globally stable equilibrium at zero, similar to z_t-dynamics (9.31). In addition, from (9.53) and previous analyses, it can be inferred that both FOV and angular velocity constraints are satisfied, and that $\lim_{t\to\infty}[\boldsymbol{J}_p^{-1}\boldsymbol{\Phi}_r(t)z_r(t), \boldsymbol{s}_r(t), \boldsymbol{\omega}_e(t), \tilde{\boldsymbol{\omega}}_e(t)] = \mathbf{0}$. Then, invoking Lemma 9.2, we can further deduce from

$\lim_{t\to\infty} \boldsymbol{s}_r(t) = \mathbf{0}$ that $\lim_{t\to\infty} \boldsymbol{q}_{ev}(t) = \mathbf{0}$. It is noted that the detailed analysis for the above conclusion is almost the same as that in Theorem 9.1 and thus is omitted to save space. □

9.3.3 Discussion

The following observations are now in order.

(1) To assist the reader in understanding the design process and the practical implementation of the proposed I&I adaptive control algorithm, a flowchart of the control design procedure is provided in Fig. 9.3, where T is the simulation duration.

(2) The theoretical result of this chapter is partly motivated by the dynamically scaled I&I adaptive control scheme [34], but nonetheless preserves obvious distinctions in three-fold. First, to accommodate both kinematic and dynamic constraints, the gradient-related terms are tactfully incorporated into the target

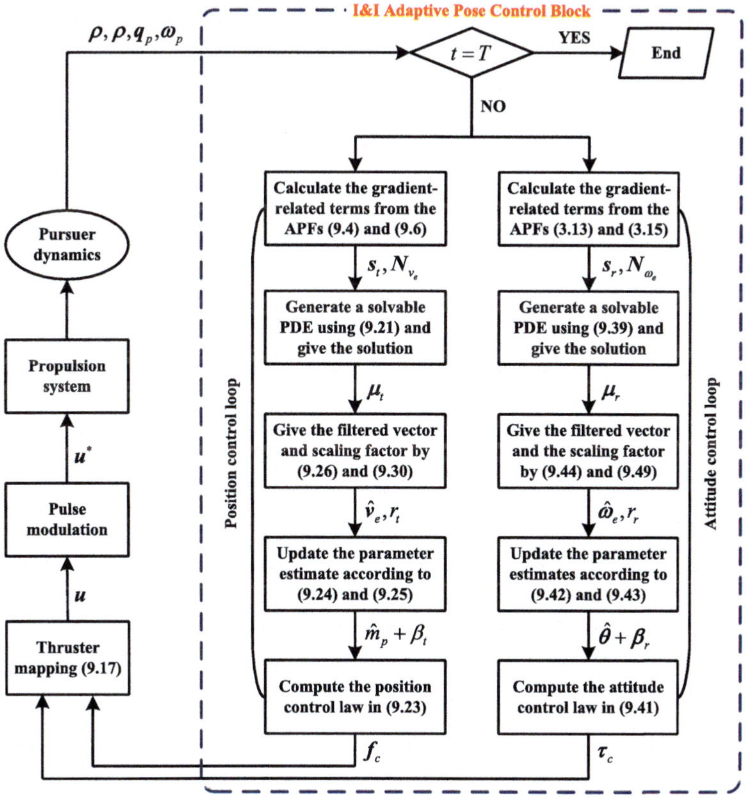

Fig. 9.3 Flow chart of the control design procedure

dynamics and velocity-level state filters. Second, two dynamic gains, i.e., $k_{di}(t)$ and $k_{fi}(t)$, $i \in \{1, 2\}$, are introduced in I&I adaptive control design to avoid the use of partial Lyapunov strictification to show asymptotic convergence of the position-level tracking errors, thus simplifying the closed-loop stability analysis. Third, a special type of dynamic scaling factor (see (9.30) and (9.49)), which is originally developed in [33], is used to eliminate the dependence of the I&I adaptive control methods in [34, 35] on the lower bound of the unknown parameters.

(3) Compared with the classical CE-based adaptive methods, the adaptive algorithm developed, except from bypassing the realizability condition, it also has two superior attributes. On the one hand, two judiciously designed functions β_t and $\boldsymbol{\beta}_r$ are combined with $\hat{m}_p$ and $\hat{\boldsymbol{\theta}}$, respectively, to form the parameter estimates. This makes the estimation error dynamics (9.28) and (9.47) directly related to the estimation errors instead of closed-loop tracking errors, which, to some extent, behaves like the indirect adaptive control [31, 32]. In this manner, it can be intuitively concluded that, if the parameter estimation error vector is equal to zero at any instant of time t^*, the parameter estimates will stay locked at their true values thereafter. On the other hand, introducing the nonlinear functions β_t and $\boldsymbol{\beta}_r$ in adaptive estimation helps us obtain the convergence condition $\lim_{t\to\infty}[m_p^{-1}\boldsymbol{\Phi}_t(t)z_t(t), \boldsymbol{J}_p^{-1}\boldsymbol{\Phi}_r(t)z_r(t)] = \mathbf{0}$. Since both r_t and r_r are bounded, we can further deduce from the above result that $\lim_{t\to\infty}[m_p^{-1}\boldsymbol{\Phi}_t(t)\tilde{m}_p(t), \boldsymbol{J}_p^{-1}\boldsymbol{\Phi}_r(t)\tilde{\boldsymbol{\theta}}(t)] = \mathbf{0}$. This result directly contributes to establishing an attracting manifold $\mathcal{S}$ defined by

$$\mathcal{S} = \{(\tilde{m}_p, \tilde{\boldsymbol{\theta}}) \in \mathbb{R} \times \mathbb{R}^6 \mid \boldsymbol{\Phi}_t\tilde{m}_p = \mathbf{0},\ \boldsymbol{\Phi}_r\tilde{\boldsymbol{\theta}} = \mathbf{0}\}, \tag{9.54}$$

in the sense that all the closed-loop trajectories ultimately end up inside $\mathcal{S}$. Consequently, the ideal closed-loop performance (no effect of parameter uncertainties) can be recovered without requiring convergence of the parameter estimates to their true values—being replaced, instead, by the presence of attractive manifold $\mathcal{S}$—and, hence, independent of the PE condition.

It is stressed that the establishment of $\mathcal{S}$ does not necessarily mean $\lim_{t\to\infty}[\tilde{m}_p, \tilde{\boldsymbol{\theta}}] = \mathbf{0}$. From a theoretical perspective, we cannot prove the convergence of the adaptive estimates to their true values from (9.28) and (9.47), unless the regressor matrix $\boldsymbol{\Phi}_t$ and $\boldsymbol{\Phi}_r$ satisfy the PE condition [36]. However, this condition does not hold in our case, since $\boldsymbol{\rho}_d$ is a set-point.

(4) From (9.41)–(9.43), it is found that the analytical expressions for $\boldsymbol{u}_r$, $\dot{\boldsymbol{u}}_r$ and $\dot{\boldsymbol{\Psi}}_r$ are necessitated by the implementation of the attitude control law $\boldsymbol{\tau}_c$. However, these terms are related to $\dot{\boldsymbol{\omega}}_f$ and $\ddot{\boldsymbol{\omega}}_f$, which are very difficult or even impossible to be analytically calculated. To overcome this difficulty, a third-order sliding mode differentiator is introduced [37]:

$$\begin{cases} \dot{z}_0 = \boldsymbol{h}_0,\ \boldsymbol{h}_0 = -\lambda_0\|z_0 - \boldsymbol{\omega}_f\|^{3/4}\mathrm{sgn}(z_0 - \boldsymbol{\omega}_f) + z_1, \\ \dot{z}_1 = \boldsymbol{h}_1,\ \boldsymbol{h}_1 = -\lambda_1\|z_1 - \boldsymbol{h}_0\|^{2/3}\mathrm{sgn}(z_1 - \boldsymbol{h}_0) + z_2, \\ \dot{z}_2 = \boldsymbol{h}_2,\ \boldsymbol{h}_2 = -\lambda_2\|z_2 - \boldsymbol{h}_1\|^{1/2}\mathrm{sgn}(z_2 - \boldsymbol{h}_1) + z_3, \\ \dot{z}_3 = -\lambda_3\mathrm{sgn}(z_3 - \boldsymbol{h}_2). \end{cases}$$

where $\lambda_i > 0, i = 0, 1, 2, 3$ are design gains, $z_i \in \mathbb{R}^3, i = 0, 1, 2, 3$ are the states of the differentiator, and for a vector $\boldsymbol{x} = [x_1, x_2, x_3]^\top$, $\text{sgn}(\boldsymbol{x}) = [\text{sign}(x_1), \text{sign}(x_2), \text{sign}(x_3)]^\top$ is defined with sign(·) the standard sign function. According to Theorem 5 in [37], it can be concluded that, in the absence of input noises, $z_0 = \boldsymbol{\omega}_d(t)$ and $z_i = \boldsymbol{h}_{i-1} = \boldsymbol{\omega}_d^{(i)}(t)$, $i = 1, 2, 3$ hold after a finite time. With this in mind, $\dot{\boldsymbol{\omega}}_d$ and $\ddot{\boldsymbol{\omega}}_d$ can be replaced with z_1 and z_2, respectively, for practical implementation of $\boldsymbol{\tau}_c$.

(5) Note that in Sect. 9.3, the position and attitude controllers are derived separately. This is just for ease of problem formulations and control designs, and the actual control synthesis is, in essence, a coupled pose control law derivation. The couplings mainly lie in three-fold: (i) the thrust force vector $\boldsymbol{f} = \boldsymbol{R}_{\mathcal{PT}}^\top \boldsymbol{f}_c$ for relative translational motion is related to the attitude of the pursuer, which implies that the relative translational motion is affected by the relative rotational motion; (ii) the extraction of the LOS frame $\mathcal{D}$ requires knowledge of the relative position $\boldsymbol{\rho}$ (recall Lemma 2.1 in Chap. 2), indicating that the relative rotational motion is also affected by the relative translational motion; and (iii) a thrusters-only actuation system (see Fig. 2.4 in Sect. 2.4.3.4) will be adopted for the pose control, which further induces coupling between the relative translational and rotational motions. We emphasize here that although using the dual-quaternion (or the Lie group SE(3)) representation to derive a pose controller in a compact way has become increasingly popular, their algebraic rules are somewhat complicated and abstruse, and moreover, the position and attitude controllers under them have the same structure, which, to some extent, limits the design freedom and flexibility. Furthermore, when using the APF-based method to deal with kinematic and dynamic constraints, it is very difficult to design proper APFs and analyze the local minima problem in the dual-quaternion or SE(3) framework. In contrast, in this chapter, the idea of separate design provides great freedoms and flexibilities in developing the pose controller, while involving only simple and lucid algebraic operations.

9.4 Numerical Simulations

In this section, numerical simulations are presented to show the effectiveness of the proposed adaptive control scheme.

9.4.1 Baseline Simulation Configuration

It is assumed that the target spacecraft orbits in a Molniya orbit, whose initial orbital elements are listed in Table 9.1. The tumbling nature of the target is simulated through running the attitude dynamics as described by (2.31) and (2.32), with the inertia matrix $\boldsymbol{J}_t = \text{diag}[55, 65, 58]\ \text{kg} \cdot \text{m}^2$ and the initial conditions: $\boldsymbol{q}_t(0) = [0, 0, 0, 1]^\top$ and $\boldsymbol{\omega}_t(0) = [0.01, -0.01, 0.01]^\top$ rad/s. The nominal mass and inertia of the pursuer are $m_p = 100$ kg and $\boldsymbol{J}_p = \text{diag}[22, 20, 23]\ \text{kg} \cdot \text{m}^2$. The relative position and

Table 9.1 Initial orbital elements

Orbital elements	Values	Units
Semimajor axis	26628	km
Eccentricity	0.7417	–
Inclination	63.4	deg
RAAN	0	deg
Argument of perigee	–90	deg
True anomaly	0	deg

velocity of the pursuer w.r.t. the target are initially set to $\boldsymbol{\rho}(0) = [-150, -25, 30]^\top$ m and $\boldsymbol{v}(0) = [0.1, 0.2, -0.3]^\top$ m/s, while the inertial attitude and angular velocity of the pursuer are initially set to $\boldsymbol{q}_p(0) = [0.6968, -0.1593, -0.2237, -0.6626]^\top$ and $\boldsymbol{\omega}_p(0) = [-0.01, 0.01, 0.02]^\top$ rad/s. The desired anchoring point is chosen as $\boldsymbol{\rho}_d = [-15, 0, 0]^\top$ m. Furthermore, the geometrical parameters of the approach corridor and sensor's FOV are $\boldsymbol{x}_o = [-1, 0, 0]^\top$, $\alpha = 15°$, and $\beta = 30°$. The relative velocity $\boldsymbol{v}_e$ and relative angular velocity $\boldsymbol{\omega}_{\mathcal{TP}}^{\mathcal{P}}$ are required to remain within 0.5 m/s and 0.06 rad/s, respectively. Roughly, when considering the relative velocity constraint, the angular velocity $\bar{\boldsymbol{\omega}}$ satisfies $|\bar{\omega}_i| < 0.005$ rad/s. Thus, imposing limit $\omega_{e,\max} = 0.05$ rad/s on $\boldsymbol{\omega}_e$ is sufficient to guarantee that $|\omega_{\mathcal{TP},i}^{\mathcal{P}}| < 0.06$ rad/s (recalling the fact that $\boldsymbol{\omega}_e = \boldsymbol{\omega}_{\mathcal{TP}}^{\mathcal{P}} - \boldsymbol{R}_{\mathcal{PD}}\bar{\boldsymbol{\omega}}$). As such, in the following simulations, we set $v_{e,\max} = 0.5$ m/s and $\omega_{e,\max} = 0.05$ rad/s. Within the above setting, the initial conditions satisfy both kinematic and dynamic constraints. The control parameters are obtained by trial and error; specifically, for the position controller, $k_{a1} = 0.3$, $k_{r1} = 0.01$, $\kappa_1 = 0.25$, $\gamma_1 = 0.05$, and $c_1 = 50$, whereas for the attitude controller, $k_{a2} = 15$, $k_{r2} = 8$, $\kappa_2 = 0.01$, $\gamma_2 = 0.5$, and $c_2 = 50$. The initial values of $\hat{m}_p$ and $\hat{\boldsymbol{\theta}}$ are taken as 80 and $[10, 10, 10]^\top$, respectively. The initial conditions of the velocity-level state filters are selected such that $\tilde{\boldsymbol{v}}_e(0) = \boldsymbol{0}$ and $\tilde{\boldsymbol{\omega}}_e(0) = \boldsymbol{0}$. In addition, the differentiator's parameters are taken as $\lambda_0 = 12$, $\lambda_1 = 8$, $\lambda_2 = 5$, and $\lambda_3 = 0.1$.

A thrusters-only actuation system consisting of 12 thrusters is equipped for the pose control. The configuration is depicted in Fig. 2.4, where $d_x = 1$ m, $d_y = 1$ m, and $d_z = 1$ m denote the moment arms of the thrusters w.r.t. the center of mass (CoM) of the pursuer. Further arrange the thrusters in 6 thruster pairs, i.e., $\{T_i, L_i\}$, $i = 1, 2, \ldots, 6$, and each thruster pair can provide bidirectional thrust with a fixed magnitude of 10 N.

9.4.2 *Ideal Simulation Scenario*

An ideal simulation scenario is considered in this subsection to show the theoretical validity of the proposed I&I adaptive control scheme. In this case, the mass properties of the pursuer remain unchanged, and the effects of thruster modulation and mis-

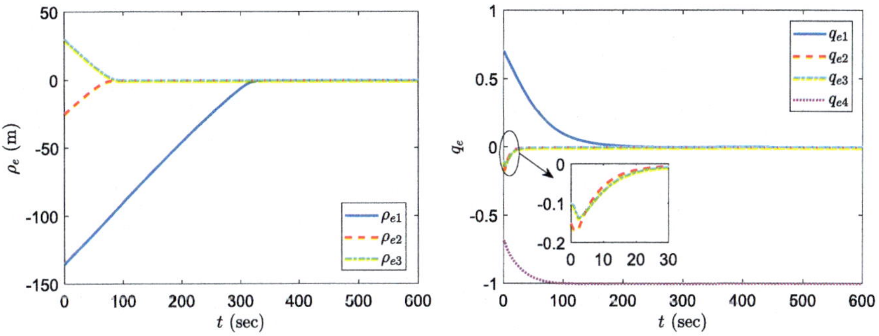

(a) position errors (b) attitude errors

Fig. 9.4 Time histories of position and attitude tracking errors

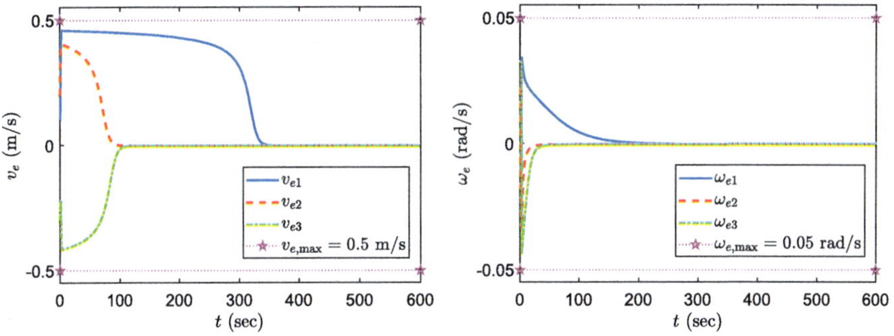

(a) velocity errors (b) angular velocity errors

Fig. 9.5 Time histories of velocity and angular velocity tracking errors

alignment as well as external disturbances are ignored. The closed-loop responses under the proposed control method are shown in Figs. 9.4, 9.5, 9.6 and 9.7. As can be seen, the ultimate control goals are achieved with good transient performance. Notice that the convergence time of the position error in x-channel is much longer than those in y- and z-channel, as observed in Fig. 9.4a. This behavior is actually caused by the velocity constraint (see Fig. 9.5a); interestingly, it renders a straight-line approaching path along the docking axis after 100 s, which is preferable for close-range proximity operations. From Fig. 9.5, it is evident that the dynamic constraints are satisfied, and moreover, a high degree of position and attitude stability is obtained. As thruster modulation is not involved in this case, only the time histories of the continuous control signals are plotted in Fig. 9.6. Besides, the time histories (just first 10 s) of the parameter estimation errors are plotted in Fig. 9.7. From the left-side subfigure of Fig. 9.7, it is obvious that the estimation errors converge rapidly to their steady-state values rather than zero. In theory, for the I&I adaptive control,

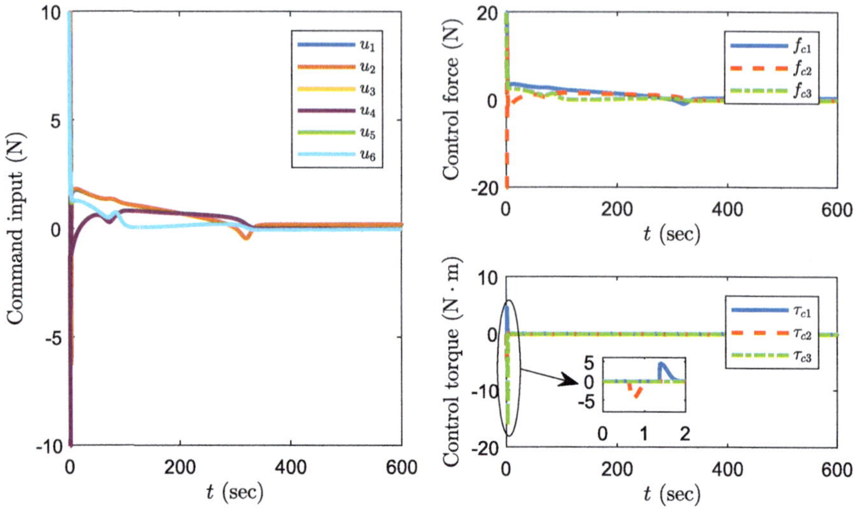

Fig. 9.6 Time histories of control forces and torques

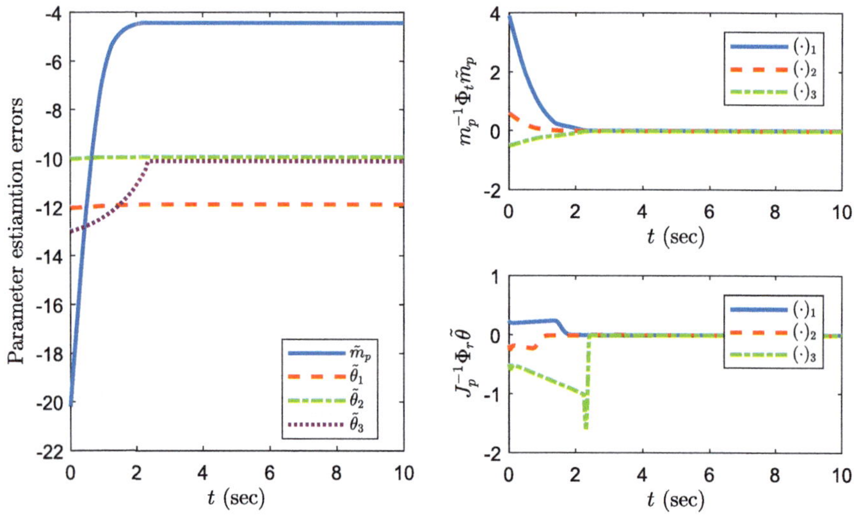

Fig. 9.7 Time histories of parameter estimation errors

the parameter estimations cannot converge to their true values due to the lack of PE condition. Nonetheless, the establishment of attracting manifold $\mathcal{S}$ (see the right-side subfigures of Fig. 9.7) recovers the closed-loop dynamics to the ideal case (no effect of parameter uncertainties).

For comparison, the passivity-based PD+ controller reported in [38] is also simulated. The control laws are given by

$$\boldsymbol{f}_c = \boldsymbol{R}_{\mathcal{PT}}(\boldsymbol{G}_p - k_{p1}\boldsymbol{\rho}_e - k_{d1}\boldsymbol{v}_e),$$

$$\boldsymbol{\tau}_c = \boldsymbol{S}(\boldsymbol{\omega}_p)\boldsymbol{J}_p\boldsymbol{\omega}_p - \boldsymbol{J}_p(\boldsymbol{S}(\boldsymbol{\omega}_e)\boldsymbol{\Omega} - \bar{\boldsymbol{\Omega}}) - k_{p2}\boldsymbol{q}_{ev} - k_{d2}\boldsymbol{\omega}_e,$$

where the control gains are chosen as $k_{p1} = 2$, $k_{d1} = 26$, $k_{p2} = 0.75$, and $k_{d2} = 5$. It is noted that the PD+ controller is dependent on the unknown mass and inertia parameters, and thus cannot be practically implemented. With this in mind, m_p and $\boldsymbol{\theta}$ in the PD+ controller are set to $\hat{m}_p(0)$ and $\hat{\boldsymbol{\theta}}(0)$, which is actually a guess of the unknown parameters. The comparison results are given in Figs. 9.8, 9.9, 9.10 and 9.11. More specifically, to validate whether both path and FOV constraints are satisfied, a polar coordinate is used in Fig. 9.8, where the polar radius is the geodesic distance $\mathrm{d}_\alpha = \arccos(\frac{<\boldsymbol{\rho}-\boldsymbol{x}_o, -\boldsymbol{x}_\mathcal{T}>}{\|\boldsymbol{\rho}-\boldsymbol{x}_o\|})$, while the polar angle is the geodesic distance $\mathrm{d}_\beta = \arccos(\frac{<-\boldsymbol{R}_{\mathcal{PT}}\boldsymbol{\rho}, \boldsymbol{x}_\mathcal{P}>}{\|\boldsymbol{R}_{\mathcal{PT}}\boldsymbol{\rho}\|})$. If $\mathrm{d}_\alpha(t) < 15°$ and $\mathrm{d}_\beta(t) < 30°$ hold for all time, then kinematic constraints are met. From Fig. 9.8, it can be seen that the proposed controller complies strictly with both kinematic constraints during the entire proximity process, but the PD+ controller fails. Furthermore, the three-dimensional (3-D) motion trajectories, together with partial attitude snapshots, of the pursuer w.r.t. the target in $\mathcal{T}$ are depicted in Figs. 9.9 and 9.10 to intuitively illustrate constraint satisfaction. In these two figures, both the pursuer and the target are portrayed as a cube with two solar panels, and the body axes are fixed on the pursuer to show its attitude orientation. Evidently, under the proposed I&I adaptive controller, the pursuer ultimately arrives at the desired anchoring point along the approach corridor, and moreover, in Fig. 9.9, a straight-line approaching path is indeed witnessed after 100 s, as we discussed early. In addition, the boresight of the vision sensor remains pointing towards the target after a short period of time, and in particular, the target is always in sight. As a consequence, the proposed I&I adaptive controller successfully accomplishes the proximity operations mission, whilst complying with both path and FOV constraints. However, even though the PD+ controller can also enable the pursuer to arrive at the anchoring point with the vision sensor's boresight pointing towards the target, two kinds of kinematic constraints are transgressed, as clearly seen

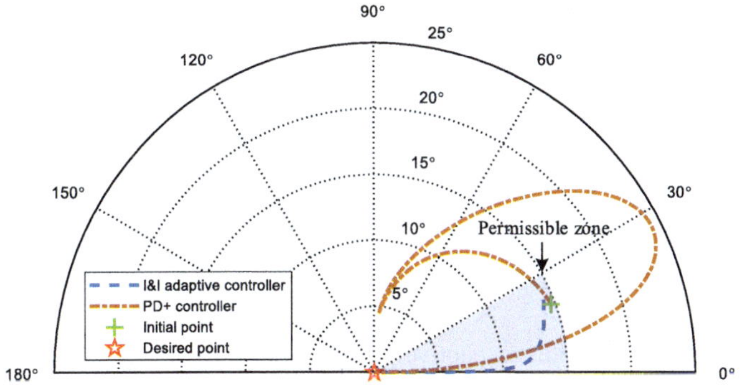

Fig. 9.8 Validation of kinematic constraints in polar coordinate

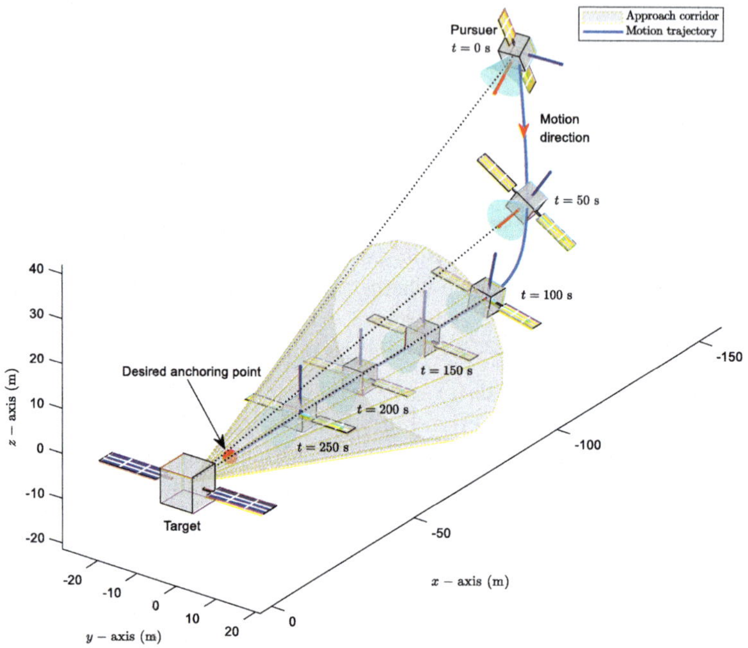

Fig. 9.9 3-D motion trajectory in $\mathcal{T}$ under the proposed controller

in Fig. 9.10. Finally, we depict the time histories of $\|\boldsymbol{v}_e\|_\infty$ and $\|\boldsymbol{\omega}_e\|_\infty$ in Fig. 9.11, from which it is evident that, under the PD+ controller, both the velocity and angular velocity errors exceed the maximum allowable values, thus violating the dynamic constraints.

9.4.3 Practical Simulation Scenario

In this subsection, a practical simulation scenario is considered to testify the effectiveness of the proposed control method in a (close to) realistic environment. Several practical aspects including thruster modulation, installation deviation, parameter variations, and external disturbances are taken into account in the simulation. The disturbance forces and torques exerting on the pursuer in $\mathcal{P}$ are modeled as

$$\boldsymbol{f}_d = \boldsymbol{f}_d^{J_2} + \boldsymbol{f}_d^{o}\,(\mathrm{N}),\ \ \boldsymbol{\tau}_d = \boldsymbol{\tau}_d^{g} + \boldsymbol{\tau}_d^{o}\,(\mathrm{N}\cdot\mathrm{m}),$$

where $\boldsymbol{f}_d^{J_2}$ and $\boldsymbol{\tau}_d^{g}$ denote, respectively, the J_2 perturbing force and the gravity-gradient torque, given by

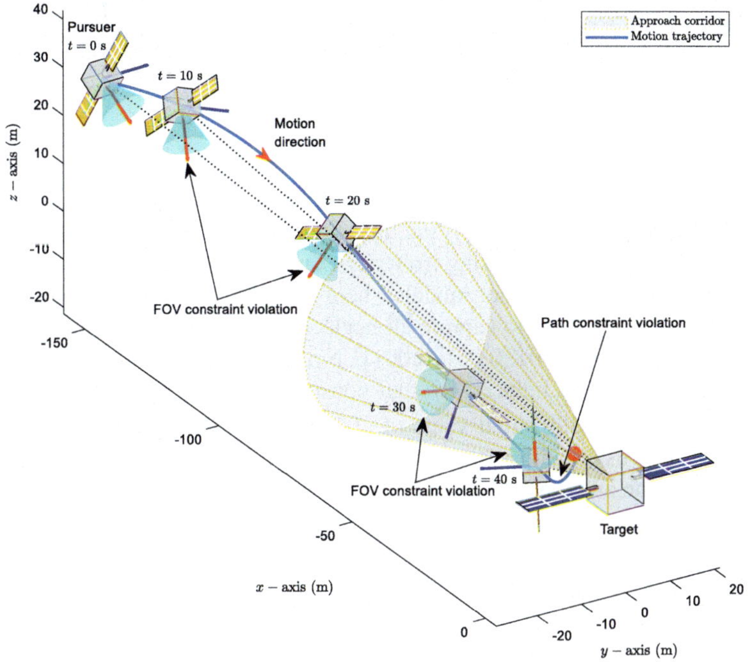

Fig. 9.10 3-D motion trajectory in $\mathcal{T}$ under the PD+ controller

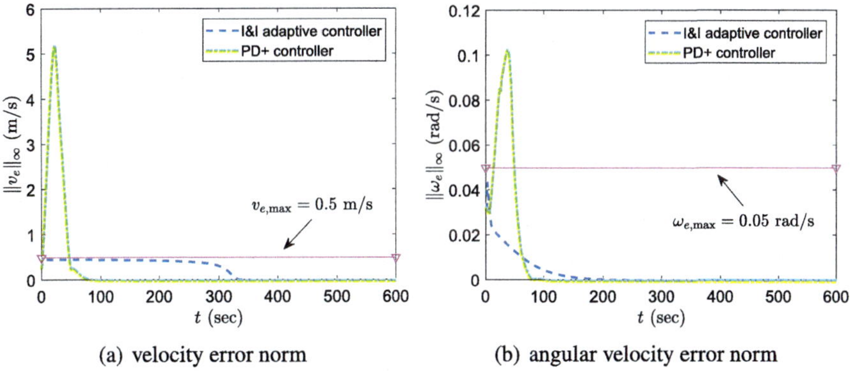

Fig. 9.11 Time histories of $\|\boldsymbol{v}_e\|_\infty$ and $\|\boldsymbol{\omega}_e\|_\infty$

$$f_d^{J_2} = -R_{\mathcal{T}\mathcal{I}}\left(\frac{3\mu J_2 R_e^2}{2\|\rho_p^{\mathcal{I}}\|^5}\left(U - 5\left(\frac{z_{\mathcal{P}}}{\|\rho_p^{\mathcal{I}}\|}\right)^2 I_3\right)\rho_p^{\mathcal{I}}\right),$$

$$\tau_d^g = 3\mu\frac{S(\rho_p^{\mathcal{P}})J_t\rho_p^{\mathcal{P}}}{\|\rho_p^{\mathcal{P}}\|^5},$$

where $J_2 = 0.0010826267$, $R_e = 6378.137$ km is the mean equatorial radius of the Earth, $U = \text{diag}[1, 1, 3]$, and $z_{\mathcal{P}}$ is the third component of $\rho_p^{\mathcal{P}}$. Similar to [3], other external forces and torques arising from, for example, atmospheric drag, solar radiation, and third bodies are not modeled here in detail, but instead, only constant f_d^o and τ_d^o are considered to capture all neglected external forces and torques. We simply choose $f_d^o = [0.001, 0.001, 0.001]^\top$ N and $\tau_d^o = [0.001, 0.001, 0.001]^\top$ N · m. The mass variation induced by fuel consumption is modeled as [39]: $\dot{m}_p = -\sum_{i=1}^{6}|u_i|/(I_{sp}g)$ with initial value of 100 kg, where $I_{sp} = 250$ s is the specific impulse of the thrusters, and $g = \mu/\|\rho_p\|^2$ is the standard gravity. Due to, for instance, fuel consumption and payload motion, the pursuer's inertia matrix is uncertain and here takes the following form [26]:

$$J_p = J_p^* + \Delta J,$$

where J_p^* denotes the nominal inertia of the pursuer, as given in Sect. 9.4.1, and ΔJ is described by $\Delta J = \text{diag}[3, 2, 1](1 + e^{-0.01t}) - 1.5\chi(t - 30)$ with $\chi(\cdot)$ defined as $\chi(\cdot < 0) = 0$ and $\chi(\cdot \geq 0) = 1$. Following the line of [40], the thruster misalignment is simply modeled by 5% multiplicative uncertainty on the thrusts $u_i^*, i = 1, 2, \ldots, 6$. In addition, the pulse-width pulse-frequency (aliased as PWPF) modulator is used in the subsequent simulation. The gain, time constant, switch-on value and switch-off value of the PWPF are set to 0.8, 0.1, 0.45 and 0.15, respectively.

The simulation results are shown in Figs. 9.12, 9.13, 9.14 and 9.15. From Figs. 9.12 and 9.13, it can be observed that, in the practical scenario, the control performance of the proposed controller, especially the steady-state accuracy, degrades dramatically. Specifically, the steady-state accuracies of the position and attitude tracking errors reduce from 1.5×10^{-3} m and 1.5×10^{-4} to 0.1 m and 0.01, respectively (for certain missions, such tolerances are acceptable). But nonetheless, both kinematic and dynamic constraints are still satisfied, as will be witnessed in Fig. 9.14. In addition, as depicted in Fig. 9.13, high-frequent chattering obviously occurs in the velocity and angular velocity tracking errors. This is actually caused by the on-off nature of thruster modulation. The motion trajectory with partial attitude snapshots on $x_{\mathcal{T}} - y_{\mathcal{T}}$ plane is provided in Fig. 9.14 for illustration of kinematic constraints. Inspecting this figure reveals that the proximity operations are still achieved without violating both path and FOV constraints, despite relatively large steady-state errors. The impulsed control signals are plotted in Fig. 9.15.

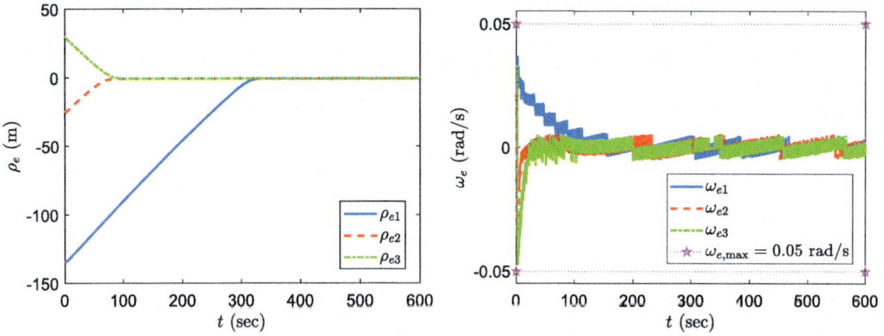

(a) position errors (b) attitude errors

Fig. 9.12 Time histories of position and attitude tracking errors

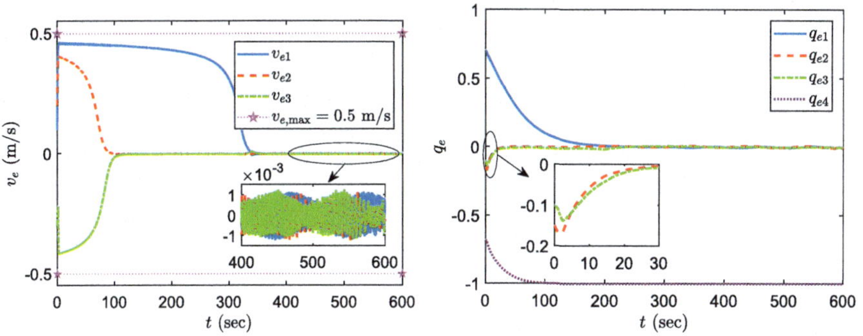

(a) velocity errors (b) angular velocity errors

Fig. 9.13 Time histories of velocity and angular velocity tracking errors

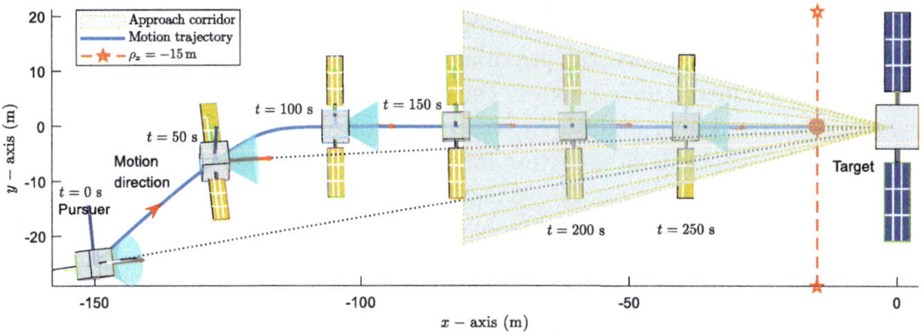

Fig. 9.14 3-D motion trajectory in $\mathcal{T}$ under the proposed controller

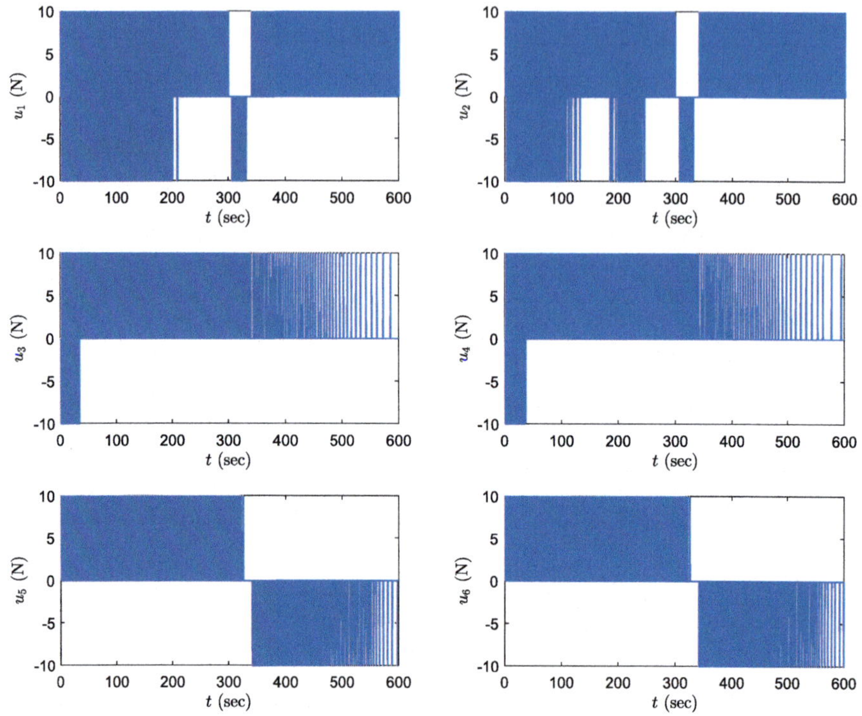

Fig. 9.15 Time histories of impulsed control forces

9.5 Summary

In this chapter, a novel adaptive pose control scheme based on APFs has been presented for constrained spacecraft proximity operations with a freely tumbling target, under mass and inertia uncertainties. The core idea behind this solution is to adopt the I&I adaptive design philosophy to circumvent the realizability condition, and meanwhile, to establish an attracting manifold, whereby the deterministic case of closed-loop performance can be asymptotically recovered. A class of APFs that were proven free of local minima problem were tactfully constructed to deal with both kinematic and dynamic constraints. Furthermore, a special type of dynamic scaling factor, together with a state filter, were designed to overcome the integrability obstacle that arises in the I&I adaptive control design. Based on Lyapunov stability theory, it was shown that the proposed control method can achieve the proximity operations successfully and comply with both kinematic and dynamic constraints, despite parameter uncertainties. Simulation results supported the theoretical findings. Future works would focus on improving the adaptive estimation mechanism such that the parameter estimates could converge to their true values without PE condition.

Appendixes

Appendix 1: Proof of Lemma 9.1

The proofs of the first two properties are straightforward and thus are omitted. Here, we emphatically illustrate the third property. From (9.5) and the fact that $h_t(\boldsymbol{\rho}) > 0$ in the set $\mathbb{K}_\rho$, we can deduce from $\nabla_{\boldsymbol{\rho}} V_p = \mathbf{0}$ that

$$h_t(\boldsymbol{\rho})(k_{r1} + k_{a1}h_t(\boldsymbol{\rho}))\boldsymbol{\rho}_e = k_{r1}\|\boldsymbol{\rho}_e\|^2 \boldsymbol{W}_t(\boldsymbol{\rho} - \boldsymbol{x}_o), \tag{9.55}$$

in the set $\mathbb{K}_\rho$. It is not difficult to check that $\boldsymbol{\rho} = \boldsymbol{\rho}_d$ (i.e., $\boldsymbol{\rho}_e = \mathbf{0}$) is one of possible solutions of (9.55). In the following, we further show that $\boldsymbol{\rho}_e = \mathbf{0}$ is the uniquely reasonable solution. First, we examine other possible solutions. Define $\boldsymbol{\rho} = [\rho_x, \rho_y, \rho_z]^\top$, then (9.55) becomes

$$h_t(\boldsymbol{\rho})(k_{r1} + k_{a1}h_t(\boldsymbol{\rho})) \begin{bmatrix} \rho_x - \rho_d \\ \rho_y \\ \rho_z \end{bmatrix} = k_{r1}\|\boldsymbol{\rho}_e\|^2 \begin{bmatrix} \rho_x - a \\ -\cot^2(\alpha)\rho_y \\ -\cot^2(\alpha)\rho_z \end{bmatrix}. \tag{9.56}$$

As k_{r1} and k_{a1} are positive constants, it can be concluded from (9.56) that $\rho_y = \rho_z = 0$; in other terms, the possible solutions (except for $\boldsymbol{\rho} = \boldsymbol{\rho}_d$), if existed, certainly locate at the X-axis of the frame $\mathcal{T}$.

Using $\boldsymbol{\rho} = [\rho_x, 0, 0]^\top$ ($\rho_x \neq \rho_d$) in (9.56), we have

$$K_{a1}(\rho_x - a)^3 = k_{r1}(a - \rho_d), \tag{9.57}$$

in the set $\mathbb{K}_\rho$. Solving (9.57) yields $\rho_x = \sqrt[3]{(k_{r1}/k_{a1})(a - \rho_d)} + a < a$, as $a < \rho_d$. Obviously, $\boldsymbol{\rho} = [\sqrt[3]{(k_{r1}/k_{a1})(a - \rho_d)} + a, 0, 0]^\top$ is beyond the cone-shaped docking corridor, and thus an invalid solution. Thus, $\boldsymbol{\rho} = \boldsymbol{\rho}_d$ is the uniquely reasonable condition for $\nabla_{\boldsymbol{\rho}} V_p = \mathbf{0}$, which completes the proof.

Appendix 2: Proof of Lemma 9.2

Since the first two properties are straightforward, we therefore emphatically illustrate the last property.

Solving $\mathrm{Vec}[\nabla V_f^* \odot \boldsymbol{q}] = \mathbf{0}$ directly yields

$$\left(\frac{k_{a2}}{k_{r2}} + \frac{1}{h_r^2(\boldsymbol{q}_e)}\right)\boldsymbol{q}_{ev} = \frac{q_{e4}\log(q_{e4}^2)}{h_r^2(\boldsymbol{q}_e)}\mathrm{Vec}[(\boldsymbol{W}_r\boldsymbol{q}_e)^* \odot \boldsymbol{q}_e], \tag{9.58}$$

in the set $\mathbb{K}_{q_e}$. Apparently, the global minima $\boldsymbol{q}_e = \pm\boldsymbol{q}_I$ are the solutions of (9.58). In what follows, we further demonstrate that they are the unique solutions in the set $\mathbb{K}_{q_e}$. To this end, we consider the case in which $\boldsymbol{q}_e \in \mathbb{K}_{q_e} \backslash \{\pm\boldsymbol{q}_I\}$. By simple algebra operations, and recalling the fact that $\boldsymbol{x}_{\mathcal{P}} = \boldsymbol{x}_{\mathcal{D}} = [1, 0, 0]^\top$, we have

$$\text{Vec}[(\boldsymbol{W}_r\boldsymbol{q}_e)^* \odot \boldsymbol{q}_e] = 2\begin{bmatrix} 0 \\ q_{e2}q_{e4} + q_{e1}q_{e3} \\ q_{e3}q_{e4} - q_{e1}q_{e2} \end{bmatrix}. \tag{9.59}$$

From (9.58) and (9.59), it can be concluded that $\boldsymbol{q}_{ev}$ is aligned with $\text{Vec}[(\boldsymbol{W}_r\boldsymbol{q}_e)^* \odot \boldsymbol{q}_e]$, indicating that $q_{e1} = 0$. Then, inserting $q_{e1} = 0$ into (9.59), we further have $\text{Vec}[(\boldsymbol{W}_r\boldsymbol{q}_e)^* \odot \boldsymbol{q}_e] = 2q_{e4} \times [0, q_{e2}, q_{e3}]^\top$. In view of this, (9.59) reduces to

$$\left(\frac{k_{a2}}{k_{r2}} + \frac{1}{h_r^2(\boldsymbol{q}_e)}\right)\begin{bmatrix} 0 \\ q_{e2} \\ q_{e3} \end{bmatrix} = \frac{2q_{e4}^2 \log(q_{e4}^2)}{h_r^2(\boldsymbol{q}_e)}\begin{bmatrix} 0 \\ q_{e2} \\ q_{e3} \end{bmatrix}. \tag{9.60}$$

It is not difficult to check from (9.60) that the scalar coefficients of its left- and right-hand sides have opposite signs. Therefore, (9.58) does not have other solutions in the set $\mathbb{K}_{q_e}$ in addition to global minima $\boldsymbol{q}_e = \pm\boldsymbol{q}_I$. This completes the proof.

References

1. Pelton JN (2019) On-orbit servicing, active debris removal and repurposing of defunct spacecraft. In: Space 2.0, Springer, pp 87–101
2. Kawano I, Mokuno M, Kasai T, Suzuki T (2001) Result of autonomous rendezvous docking experiment of engineering test satellite-vii. Journal of Spacecraft and Rockets 38(1): 105–111
3. Filipe N, Tsiotras P (2014) Adaptive position and attitude-tracking controller for satellite proximity operations using dual quaternions. Journal of Guidance, Control, and Dynamics 38(4): 566–577
4. Sun L, Huo W (2015) 6-dof integrated adaptive backstepping control for spacecraft proximity operations. IEEE Transactions on Aerospace and Electronic Systems 51(3): 2433–2443
5. Hu Q, Shao X, Chen WH (2018) Robust fault-tolerant tracking control for spacecraft proximity operations using time-varying sliding mode. IEEE Transactions on Aerospace and Electronic Systems 54(1): 2–17
6. Zappulla R, Park H, Virgili-Llop J, Romano M (2018) Real-time autonomous spacecraft proximity maneuvers and docking using an adaptive artificial potential field approach. IEEE Transactions on Control Systems Technology (99): 1–8
7. Shao X, Hu Q (2019) Adaptive control for autonomous spacecraft rendezvous with approaching path constraint. In: Chinese Control Conference, Guangzhou, China, pp 8188–8193
8. Lu P, Liu X (2013) Autonomous trajectory planning for rendezvous and proximity operations by conic optimization. Journal of Guidance, Control, and Dynamics 36(2): 375–389
9. Zagaris C, Park H, Virgili-Llop J, Zappulla R, Romano M, Kolmanovsky I (2018) Model predictive control of spacecraft relative motion with convexified keep-out-zone constraints. Journal of Guidance, Control, and Dynamics 41(9): 2054–2062
10. Zhu Z, Yan Y (2014) Space-based line-of-sight tracking control of geo target using nonsingular terminal sliding mode. Advances in Space Research 54(6): 1064–1076

11. Lee U, Mesbahi M (2014) Feedback control for spacecraft reorientation under attitude constraints via convex potentials. IEEE Transactions on Aerospace and Electronic Systems 50(4): 2578–2592
12. Shen Q, Yue C, Goh CH (2017) Velocity-free attitude reorientation of a flexible spacecraft with attitude constraints. Journal of Guidance, Control, and Dynamics 40(5): 1293–1299
13. Hu Q, Chi B, Akella MR (2019) Anti-unwinding attitude control of spacecraft with forbidden pointing constraints. Journal of Guidance, Control, and Dynamics 42(4): 822–835
14. Biggs JD, Colley L (2016) Geometric attitude motion planning for spacecraft with pointing and actuator constraints. Journal of Guidance, Control, and Dynamics 39(7): 1672–1677
15. Tan X, Berkane S, Dimarogonas DV (2020) Constrained attitude maneuvers on SO(3): Rotation space sampling, planning and low-level control. Automatica 112: 108659
16. Shao X, Hu Q, Shi Y (2021) Adaptive pose control for spacecraft proximity operations with prescribed performance under spatial motion constraints. IEEE Transactions on Control Systems Technology 29(4): 1405–1419
17. Lee U, Mesbahi M (2016) Constrained autonomous precision landing via dual quaternions and model predictive control. Journal of Guidance, Control, and Dynamics 40(2): 292–308
18. Dong H, Hu Q, Akella MR (2017) Dual-quaternion-based spacecraft autonomous rendezvous and docking under six-degree-of-freedom motion constraints. Journal of Guidance, Control, and Dynamics 41(5): 1150–1162
19. Dong H, Hu Q, Liu Y, Akella MR (2019) Adaptive pose tracking control for spacecraft proximity operations under motion constraints. Journal of Guidance, Control, and Dynamics 42(10): 2258–2271
20. Hazra S (2019) Autonomous guidance for asteroid descent using successive convex optimisation: Dual quaternion approach. PhD thesis, Delft University of Technology
21. Li Q, Yuan J, Zhang B, Gao C (2017) Model predictive control for autonomous rendezvous and docking with a tumbling target. Aerospace Science and Technology 69: 700–711
22. Chan N, Mitra S (2017) Verifying safety of an autonomous spacecraft rendezvous mission. arXiv preprint arXiv:1703.06930
23. Shen Q, Yue C, Goh CH, Wu B, Wang D (2018) Rigid-body attitude stabilization with attitude and angular rate constraints. Automatica 90: 157–163
24. Hu Q, Chi B, Akella MR (2019) Reduced attitude control for boresight alignment with dynamic pointing constraints. IEEE/ASME Transactions on Mechatronics 24(6): 2942–2952
25. Sun L, Huo W, Jiao Z (2016) Adaptive backstepping control of spacecraft rendezvous and proximity operations with input saturation and full-state constraint. IEEE Transactions on Industrial Electronics 64(1): 480–492
26. Cai W, Liao X, Song Y (2008) Indirect robust adaptive fault-tolerant control for attitude tracking of spacecraft. Journal of Guidance, Control, and Dynamics 31(5): 1456–1463
27. Astolfi A, Ortega R (2003) Immersion and invariance: A new tool for stabilization and adaptive control of nonlinear systems. IEEE Transactions on Automatic control 48(4): 590–606
28. Singla P, Subbarao K, Hughes D, Junkins JL (2003) Structured model reference adaptive control for vision based spacecraft rendezvous and docking. Advances in the Astronautical Sciences 114: 55–74
29. Zuo Z, Ru P (2014) Augmented $\mathcal{L}_1$ adaptive tracking control of quad-rotor unmanned aircrafts. IEEE Transactions on Aerospace and Electronic Systems 50(4): 3090–3101
30. Ulrich S, Saenz-Otero A, Barkana I (2016) Passivity-based adaptive control of robotic spacecraft for proximity operations under uncertainties. Journal of Guidance, Control, and Dynamics 39(6): 1444–1453
31. Slotine JJE, Li W (1989) Composite adaptive control of robot manipulators. Automatica 25(4): 509–519
32. Pan Y, Yu H (2018) Composite learning robot control with guaranteed parameter convergence. Automatica 89: 398–406
33. Wen H, Yue X, Yuan J (2018) Dynamic scaling–based noncertainty-equivalent adaptive spacecraft attitude tracking control. Journal of Aerospace Engineering 31(2): 04017098

34. Yang S, Akella MR, Mazenc F (2017) Dynamically scaled immersion and invariance adaptive control for euler–lagrange mechanical systems. Journal of Guidance, Control, and Dynamics 40(11): 2844–2856
35. Karagiannis D, Sassano M, Astolfi A (2009) Dynamic scaling and observer design with application to adaptive control. Automatica 45(12): 2883–2889
36. Boyd S, Sastry SS (1986) Necessary and sufficient conditions for parameter convergence in adaptive control. Automatica 22(6): 629–639
37. Levant A (2003) Higher-order sliding modes, differentiation and output-feedback control. International Journal of Control 76(9-10): 924–941
38. Kristiansen R, Nicklasson PJ, Gravdahl JT (2008) Spacecraft coordination control in 6DOF: Integrator backstepping vs passivity-based control. Automatica 44(11): 2896–2901
39. Capello E, Punta E, Dabbene F, Guglieri G, Tempo R (2017) Sliding-mode control strategies for rendezvous and docking maneuvers. Journal of Guidance, Control, and Dynamics 40(6): 1481–1487
40. Fonod R, Henry D, Charbonnel C, Bornschlegl E, Losa D, Bennani S (2015) Robust fdi for fault-tolerant thrust allocation with application to spacecraft rendezvous. Control Engineering Practice 42: 12–27

Chapter 10
Composite Learning Pose Control of Spacecraft with Guaranteed Parameter Convergence

10.1 Introduction

The primary goal of adaptive control is to recover the control performance under parameter uncertainties as close as possible to the ideal parameter-deterministic case. To achieve this, the adaptive law should not only ensure that the parameter estimates converge to their true values, but also account for good control response and tracking performance. Zhang and Duan [1] proposed an adaptive integrated finite-time control scheme for a rigid spacecraft, in order to deal with external disturbance, unknown mass property, and thruster misalignment. Aiming at spacecraft RPOs, a series of adaptive position-attitude coupled control schemes were developed in the backstepping design framework [2–5]. Based on the dual-quaternion formalism, two adaptive pose tracking control methods were presented in [6, 7]. By constructing novel sliding mode vectors and adaptive laws, several kinds of adaptive fixed-time control schemes were proposed in [8–10], which provide a new idea for the design of adaptive fixed-time pose controllers under mass and inertia uncertainties. Considering the spacecraft rendezvous and docking problem, Singla et al. [11] proposed an output-feedback MRAC method, and analyzed the impact of measurement noise on closed-loop stability and control performance. However, the adaptive laws in the aforementioned works are derived only using the tracking errors. Composite adaptive control combines the advantages of direct adaptive control and indirect adaptive control [12]. By using both tracking errors and parameter estimation errors to design adaptive laws, the convergence performance of output-tracking and parameter estimation errors as well as the robustness of the closed-loop system can be effectively improved without causing explicit oscillation of parameter estimation. However, it should be noted that the traditional CE-based composite adaptive control can ensure that the parameter estimates converge to their true values, only if the reference trajectory satisfies the PE condition [13, 14]. Although the non-CE I&I adaptive control can improve the closed-loop performance, the PE condition is also required to ensure parameter convergence. The main reason lies in the rank deficiency of information matrix. In fact, PE is a restrictive, and the reference signal usually does not satisfy

Q. Hu et al., *Intelligent Autonomous Control of Spacecraft with Multiple Constraints*,
https://doi.org/10.1007/978-981-99-0681-9_10

this condition in practice. Moreover, the PE condition is difficult to be monitored in real-time, since it is related to the future state.

In order to relax the requirement of PE condition for parameter convergence, Chowdhary and Johnson [15] proposed an adaptive control approach based on concurrent learning (CL), wherein both historical and current data are used to design adaptive laws. Benefiting from the stored historical data, the condition for parameter convergence is significantly relaxed, and the parameter convergence can be guaranteed if the information matrix comprised of historical data is full rank. This condition can be established if the regressor matrix satisfies a interval excitation (IE) condition, which is strictly weaker than PE and can be monitored online. Valverde and Tsiotras [16] successfully applied this method to the integrated pose control of spacecraft. By using the dual-quaternion description, a CL-based adaptive law was designed to accurately estimate the mass and inertia parameters of spacecraft. It should be noted that the practical implementation of the CL-based adaptive controller needs real-time data selection and storage. In order to store the historical data as rich as possible, it is necessary to judiciously select an effective recording scheme that can adequately store linearly independent data. However, this will greatly increase the system design complexity. In addition, the design of CL-based adaptive law requires state derivatives to construct parameter estimation errors, but in general, the required state derivatives are unmeasured signals (such as the relative acceleration and angular acceleration between the pursuer and target spacecraft). Towards this end, a fixed point smoother was used in [15, 17] to estimate the state derivative. However, a prominent problem of this method is the sensitiveness to measurement error and noise. Parikh et al. [18] used the integral of state derivative, control input, and regressor matrix to construct a CL-based adaptive law, thus avoiding the estimation of state derivative. Cho et al. [19] proposed a composite learning adaptive control scheme, which can achieve parameter convergence under the IE condition. By using the filtered state and regressor matrix to design the adaptive law, the parameter estimation error is constructed without using the unmeasurable state derivative. In addition, the integral of filtered regressor matrix is used for data storage, thus avoiding the complex online data selection. Pan and Yu [20] proposed a composite learning adaptive method which is similar to [19] and successfully applied it to the robot control problems. Although the aforementioned CL-based adaptive control methods can ensure that parameter estimates converge to their true values if the regressor matrix satisfies a strictly weak IE condition, a common problem is that: the dynamics of parameter estimation errors are coupled, which makes these methods difficult to balance the convergence rate of each parameter, and the convergence speeds are highly dependent on the excitation strength of relevant signals. Generally speaking, the users need to tune the adaptive gains by trial and error, in order to achieve good performance of parameter convergence.

Motivated by the above discussions, this chapter focuses on the design of an adaptive spacecraft pose control scheme that can synchronously enhance parameter convergence and tracking performance. A novel composite learning pose controller is proposed to ensure that the output-tracking and parameter estimation errors converge exponentially to zero under a strictly weak IE condition. With this design, the pursuer spacecraft can quickly accomplish the proximity operations with high precision. Firstly, an adaptive control law is designed in the framework of certainty equivalence, and the filtered system dynamics are established so that the construction of parameter estimation errors does not require unmeasured state derivatives. Then, the traditional composite adaptive law is given, and by using the CL technique in conjunction with the DREM procedure, a composite learning law is further designed to relax the excitation requirement for parameter convergence. Stability analysis shows that the proposed composite learning control method can ensure that the pose tracking errors and parameter estimation errors converge exponentially to zero under the IE condition. In particular, thanks to the DREM procedure and some special designs, the error dynamics of parameter estimation are independent of each other, whilst the convergent rate of is independent of the excitation strength. This property not only improves the convergence performance of parameter estimation, but also makes the selection of adaptive gain simple and clear. Finally, simulations verify the effectiveness of proposed control scheme.

The remainder of the chapter is structured as follows. Section 10.2 introduces the preliminary results for the gradient descent estimator and DREM procedure. A composite learning pose tracking control scheme is proposed in Sect. 10.3, along with the rigorous stability analysis. Numerical simulations are carried out in Sect. 10.4. Finally, Sect. 10.5 concludes this chapter.

10.2 Preliminaries

The reader is referred to Sect. 3.4 for the definitions of IE and PE conditions.

10.2.1 Gradient Descent Estimator

In typical application scenarios of adaptive control, the relationship between unknown parameters and measured data is usually affinely linear, which is instrumental for the design of parameter estimator. The idea of parameter estimation is to design an online estimator for unknown parameters by using measurement data and regressor matrix. Based on some estimation algorithms, such as gradient descent method, least square method, etc., the estimation error can satisfy specific optimization criteria. As the basis of adaptive law design, this chapter will briefly introduce the design and analysis of a gradient descent-based parameter estimator (referred to the gradient estimator). It should be noted that since gradient estimators are landmark achievements in the field of parameter identification and adaptive control, the following contents about gradient estimators can be found in most classical textbooks on parameter identification and adaptive control, like the book *Adaptive control: stability, convergence and robustness* by Sastry and Bodson [21].

Consider the following linear regression equation (LRE)

$$y(t) = \mathbf{v}^{\top}(t)\phi, \tag{10.1}$$

where $y(t) \in \mathbb{R}$ is the system output, $\mathbf{v}(t) \in \mathbb{R}^m$ is a bounded and differentiable regression vector, and $\phi \in \mathbb{R}^m$ is an unknown parameter vector. Define $\hat{\phi}(t) \in \mathbb{R}^m$ as the estimation of ϕ. Then, define a gradient estimator for LRE (10.1):

$$\dot{\hat{\phi}}(t) = \boldsymbol{\Gamma}\mathbf{v}(t)[\mathbf{v}^{\top}(t)\hat{\phi}(t) - y(t)], \tag{10.2}$$

where $\boldsymbol{\Gamma} > 0$ is the gain matrix. Let $\tilde{\phi}(t) = \hat{\phi}(t) - \phi$ be the parameter estimation error. Substituting (10.1) into (10.2), the parameter estimation error dynamics is obtained

$$\dot{\tilde{\phi}}(t) = -\boldsymbol{\Gamma}\mathbf{v}(t)\mathbf{v}^{\top}(t)\tilde{\phi}(t). \tag{10.3}$$

Consider the Lyapunov function candidate

$$V(\tilde{\phi}) = \frac{1}{2}\tilde{\phi}^{\top}(t)\boldsymbol{\Gamma}^{-1}\tilde{\phi}(t). \tag{10.4}$$

Taking the time derivative of $V(\tilde{\phi})$ and substituting (10.3), the following conclusions are arrived at:

- The norm of parameter estimation error $\tilde{\phi}$ is monotonous and non-increasing, i.e.,

$$\|\tilde{\phi}(t_b)\| \leq \|\tilde{\phi}(t_a)\| \quad \forall t_b > t_a \geq 0. \tag{10.5}$$

- If the regressor $\mathbf{v} \in$ PE, then parameter estimation error $\tilde{\phi}$ converges to zero.

10.2.2 Dynamic Regressor Extension and Mixing

To relax the requirement of parameter convergence on the PE condition, Aranovskiy and Ortega [22, 23] proposed a new parameter estimation procedure, namely dynamic regressor extension and mixing (DREM). Compared with the traditional parameter estimator, the DREM procedure has the following advantages.

- A new parameter convergence condition different from the PE condition is provided, that is, the determinant of the extended regressor satisfies the non-square integrability condition, which is usually weaker than the PE condition;
- It can ensure that every element in the parameter estimation error is monotonously convergent. Compared with the norm convergence, this method not only improves the transient performance of parameter estimation, but also eliminates the potential oscillation and peak value problems in the transient process;

- Each element of parameter estimation is adjusted by a separate scalar gain, which does not affect the transient response of other elements. This property makes the parameter selection more simple and clear.

The core idea of DREM is to generate a set of m scalar LREs that share the same scalar regressor to estimate each unknown parameter independently, and it mainly includes two basic steps:

Step 1: Dynamic regressor extension. In this step, one should introduce a linear, single-input m-output operator $\mathcal{H}$ with stable bounded input and bounded output (BIBO) property, and define the vector $\boldsymbol{Y} \in \mathbb{R}^m$ and matrix $\mathbf{V} \in \mathbb{R}^{m\times m}$

$$\boldsymbol{Y} := \mathcal{H}[y],\ \mathbf{V} := \mathcal{H}[\mathbf{v}^\top]. \tag{10.6}$$

It is noted that the choice of $\mathcal{H}$ is not unique. For example, in [22] a first-order linear time-invariant (LTI) filter is adopted to carry out the regressor extension. We use the LTI filter as an example to illustrate the dynamic regressor extension step. Consider the operator $\mathcal{H}$ of the following filter form:

$$\mathcal{H}(s) = \left[\frac{1}{c_1 s + 1}, \frac{1}{c_2 s + 1}, ..., \frac{1}{c_{m-1} s + 1}\right], \tag{10.7}$$

where $c_i > 0, i = 1, 2, ..., m - 1$ is the filter constants, and s is the Laplace operator. Applying the above operator to LRE (10.1), a set of $m - 1$ scalar LREs are obtained

$$y_{f_i}(t) = \mathbf{v}_{f_i}^\top(t)\phi,\ i = 1, 2, ..., m - 1, \tag{10.8}$$

where y_{f_i} and $\mathbf{v}_{f_i}$ are filtered signals of y and $\mathbf{v}$, respectively. In view of the linearity of the operator $\mathcal{H}$ and its BIBO stability, the original LRE (10.1) can be stacked with $m - 1$ filtered LREs. Then, an extended LRE is given by

$$\boldsymbol{Y}(t) = \mathbf{V}(t)\phi. \tag{10.9}$$

Specifically, $\boldsymbol{Y}$ and $\mathbf{V}$ are of the forms

$$\boldsymbol{Y} = \begin{bmatrix} y \\ y_{f_1} \\ \vdots \\ y_{f_{m-1}} \end{bmatrix},\ \mathbf{V} = \begin{bmatrix} \mathbf{v}^\top \\ \mathbf{v}_{f_1}^\top \\ \vdots \\ \mathbf{v}_{f_{m-1}}^\top \end{bmatrix}. \tag{10.10}$$

Step 2: Mixing. The main purpose of this step is to obtain a set of scalar equations using the extended LRE in (10.9). Note that for any $m \times m$ square matrix $\mathbf{V}$ (possibly singular), $\mathrm{adj}(\mathbf{V})\mathbf{V} = \det(\mathbf{V})\boldsymbol{I}_m$ always holds, where $\mathrm{adj}(\cdot)$ represents the adjoint matrix of a square matrix, and $\det(\cdot)$ represents the determinant of a square matrix. Multiplying both sides of (10.9) by $\mathrm{adj}(\mathbf{V})$, we obtain

$$\mathcal{Y}(t) = \sigma(t)\phi, \tag{10.11}$$

where $\mathcal{Y} = \text{adj}(\mathbf{V})\boldsymbol{Y}$, and $\sigma = \det(\mathbf{V})$. Then, unfolding the above equation yields

$$\mathcal{Y}_i(t) = \sigma(t)\phi_i, \; i = 1, 2, ..., m, \tag{10.12}$$

where $\mathcal{Y}_i$ represents the ith element of the vector $\mathcal{Y}$.

In the DREM framework, design gradient estimators for $\phi_i, i = 1, 2, ..., m$ as

$$\dot{\hat{\phi}}_i(t) = \Gamma_i\sigma(t)[\sigma(t)\hat{\phi}_i(t) - \mathcal{Y}_i(t)]. \tag{10.13}$$

The corresponding parameter estimation error dynamics are

$$\dot{\tilde{\phi}}_i(t) = -\Gamma_i\sigma^2(t)\tilde{\phi}_i(t). \tag{10.14}$$

Consider the following Lyapunov function candidate:

$$V_i(\tilde{\phi}_i) = \frac{1}{2\Gamma_i}\tilde{\phi}_i^2(t). \tag{10.15}$$

Taking the time derivative of $V_i(\tilde{\phi}_i)$ and recalling Eq. (10.14) lead to

$$\dot{V}_i(\tilde{\phi}_i) = -\sigma^2(t)\tilde{\phi}_i^2(t). \tag{10.16}$$

Based on the above analysis, we arrive at the following conclusions:

- Each element $\tilde{\phi}_i$ ($i = 1, 2, ..., m$) of the parameter estimation error vector $\tilde{\phi}$ is monotonically non-increasing, i.e.,

$$|\tilde{\phi}_i(t_b)| \leq |\tilde{\phi}_i(t_a)|, \; \forall t_b > t_a \geq 0. \tag{10.17}$$

- If the determinant $\sigma \notin \mathcal{L}_2$, then $\tilde{\phi}_i$ converges asymptotically to zero;
- If the determinant $\sigma \in$ PE, then $\tilde{\phi}_i$ converges exponentially to zero;
- The adjustment of Γ_i only affects the transient response of $\hat{\phi}_i$.

Remark 10.1 The second conclusion shows that the condition of parameter convergence under the DREM framework is $\sigma \notin \mathcal{L}_2$. As stated in [22], $\sigma \in$ PE is the sufficient (but not necessary) condition for $\sigma \notin \mathcal{L}_2$, that is, $\sigma \in \text{PE} \Rightarrow \sigma \notin \mathcal{L}_2$, but the reverse is not necessarily true. A typical example is $\sigma = 1/\sqrt{1+t}$, which satisfies neither the square integrability nor the PE condition. It is clear that the non-square integrability of σ is weaker than the PE condition. However, this condition can only ensure the asymptotic convergence of parameter estimation errors, rather than exponential convergence. In addition, in practical application, the operator $\mathcal{H}$ should be carefully selected to enhance the non-square integrability of σ.

10.3 Composite Learning Pose Control

In this section, we shall propose a composite learning pose control scheme for spacecraft RPOs, which can synchronously enhance parameter convergence and tracking performance under a strictly weak IE condition. Firstly, an adaptive control law is designed based on the CE principle. Then, considering that the composite adaptive law generally requires the construction of parameter estimation errors, a filtered system dynamics are established according to the regressor filtering method, which avoids the use of unmeasured state derivatives. Next, a traditional composite adaptive law is given, and on this basis, the CL idea and DREM procedure are further used to design the composite learning law. Finally, the closed-loop system stability as well as the exponential convergence of pose tracking and parameter estimation errors under the IE condition are rigorously analyzed.

Recalling the integrated relative position-attitude coupled dynamics in (2.53) and the thruster configuration in (2.54), we have

$$\boldsymbol{M}\ddot{\boldsymbol{e}} + \boldsymbol{C}\dot{\boldsymbol{e}} + \boldsymbol{G} = \boldsymbol{A}\boldsymbol{D}\boldsymbol{u}_c. \tag{10.18}$$

To facilitate the controller design, a filtered tracking error is defined

$$\boldsymbol{s} = \dot{\boldsymbol{e}} + k\boldsymbol{e}, \tag{10.19}$$

where $k > 0$ is a design constant. Evaluating the time derivative of $\boldsymbol{s}$ along the open-loop system dynamics (10.18) leads to

$$\boldsymbol{M}\dot{\boldsymbol{s}} = -\boldsymbol{C}\boldsymbol{s} + \underbrace{k\boldsymbol{M}\dot{\boldsymbol{e}} + k\boldsymbol{C}\boldsymbol{e} - \boldsymbol{G}}_{\boldsymbol{H}} + \boldsymbol{A}\boldsymbol{D}\boldsymbol{u}_c. \tag{10.20}$$

The affine linear parameterization of $\boldsymbol{H}$ is given by

$$\boldsymbol{H} = \boldsymbol{\Phi}\boldsymbol{\theta} = \text{blkdiag}\{\boldsymbol{\Phi}_p, \boldsymbol{\Phi}_r\}[m_p, \boldsymbol{\vartheta}^\top]^\top, \tag{10.21}$$

where $\boldsymbol{\vartheta} = [J_{p,11}, J_{p,22}, J_{p,33}, J_{p,12}, J_{p,23}, J_{p,13}]^\top$, whereas $\boldsymbol{\Phi}_p$ and $\boldsymbol{\Phi}_r$ satisfy

$$\boldsymbol{\Phi}_p m_p = k\boldsymbol{M}_p\dot{\boldsymbol{\rho}}_e + k\boldsymbol{C}_p\boldsymbol{\rho}_e - \boldsymbol{G}_p, \ \boldsymbol{\Phi}_r\boldsymbol{\vartheta} = k\boldsymbol{M}_r\dot{\boldsymbol{q}}_{ev} + k\boldsymbol{C}_r\boldsymbol{q}_{ev} - \boldsymbol{G}_r, \tag{10.22}$$

where $\boldsymbol{\Phi}_p = k\dot{\boldsymbol{\rho}}_e + k\bar{\boldsymbol{C}}_p\boldsymbol{\rho}_e - \bar{\boldsymbol{G}}_p$ with $\bar{\boldsymbol{C}}_p = \boldsymbol{C}_p/m_p$ and $\bar{\boldsymbol{G}}_p = \boldsymbol{G}_p/m_p$, whereas $\boldsymbol{\Phi}_r = k\boldsymbol{P}^\top(\boldsymbol{L}(\boldsymbol{P}\dot{\boldsymbol{q}}_{ev}) + \boldsymbol{L}(\dot{\boldsymbol{P}}\boldsymbol{q}_{ev}) + \boldsymbol{S}(\boldsymbol{P}\boldsymbol{q}_{ev})\boldsymbol{L}(\boldsymbol{P}\dot{\boldsymbol{q}}_{ev})) - \boldsymbol{P}^\top[\boldsymbol{S}(\boldsymbol{\omega}_e)\boldsymbol{L}(\boldsymbol{R}_{\mathcal{PD}}\boldsymbol{\omega}_d) + \boldsymbol{S}(\boldsymbol{R}_{\mathcal{PD}}\boldsymbol{\omega}_d)\boldsymbol{L}(\boldsymbol{\omega}_p) - \boldsymbol{L}(\boldsymbol{S}(\boldsymbol{\omega}_e)\boldsymbol{R}_{\mathcal{PD}}\boldsymbol{\omega}_d - \boldsymbol{R}_{\mathcal{PD}}\dot{\boldsymbol{\omega}}_d)]$.

As per the open-loop dynamics (10.20), the control law is designed as

$$\boldsymbol{u}_c = -(\boldsymbol{A}\boldsymbol{D})^{-1}(\boldsymbol{K}_c\boldsymbol{s} + \boldsymbol{\Phi}\hat{\boldsymbol{\theta}}), \tag{10.23}$$

where $\boldsymbol{K}_c \in \mathbb{R}^{6\times 6}$ is a positive definite diagonal matrix, $\hat{\boldsymbol{\theta}}$ is the estimate of unknown parameter vector $\boldsymbol{\theta}$. Define the estimation error as $\tilde{\boldsymbol{\theta}} = \hat{\boldsymbol{\theta}} - \boldsymbol{\theta}$. Then, substituting the control law $\boldsymbol{u}_c$ into (10.20) gets

$$\boldsymbol{M}\dot{\boldsymbol{s}} = -\boldsymbol{C}\boldsymbol{s} - \boldsymbol{K}_c\boldsymbol{s} - \boldsymbol{\Phi}\tilde{\boldsymbol{\theta}}. \tag{10.24}$$

10.3.1 Filtered System Dynamics

The composite learning law is composed of both the tracking error and parameter estimation error. In general, the construction of parameter estimation error involves the unmeasured state derivatives, like $\dot{\boldsymbol{s}}$. Although it is possible to estimate the unmeasured state derivatives by using a fixed-point smoother or state differentiator, such methods are sensitive to measurement noise and make it impossible to conduct rigorous closed-loop stability analysis. Inspired by [19], we use the regressor filtering method introduced in [24] to establish a filtered system dynamics, which enables the design of composite learning law to use only easily obtained signals. To this end, the affine linear parameterization is conducted

$$\boldsymbol{W}\boldsymbol{\theta} = -\boldsymbol{C}\boldsymbol{s} + \boldsymbol{H} + \dot{\boldsymbol{M}}\dot{\boldsymbol{s}}_f, \tag{10.25}$$

where $\boldsymbol{W} \in \mathbb{R}^{6\times 7}$ is a known regressor matrix, whose specific form is omitted here, and $\boldsymbol{s}_f$ is obtained by a stable low-pass filter:

$$\dot{\boldsymbol{s}}_f = -c\boldsymbol{s}_f + \boldsymbol{s},\ \ \boldsymbol{s}_f(0) = \boldsymbol{s}(0)/c, \tag{10.26}$$

where $c > 0$ is the filter time constant.

In view of (10.25), (10.20) can be rewritten as

$$\boldsymbol{M}\dot{\boldsymbol{s}} = \boldsymbol{W}\boldsymbol{\theta} + \boldsymbol{A}\boldsymbol{D}\boldsymbol{u}_c - \dot{\boldsymbol{M}}\dot{\boldsymbol{s}}_f. \tag{10.27}$$

Next, stable filters are introduced for the regressor matrix $\boldsymbol{W}$ and control input $\boldsymbol{u}_c$

$$\dot{\boldsymbol{W}}_f = -c\boldsymbol{W}_f + \boldsymbol{W},\ \ \boldsymbol{W}_f(0) = \boldsymbol{0}, \tag{10.28}$$

$$\dot{\boldsymbol{u}}_f = -c\boldsymbol{u}_f + \boldsymbol{A}\boldsymbol{D}\boldsymbol{u}_c,\ \ \boldsymbol{u}_f(0) = \boldsymbol{0}. \tag{10.29}$$

Taking the time derivative of $\dot{\boldsymbol{s}}_f$ in (10.26) and using (10.27)–(10.29), we obtain

$$\begin{aligned}\ddot{\boldsymbol{s}}_f &= -c\dot{\boldsymbol{s}}_f + \boldsymbol{M}^{-1}(\boldsymbol{W}\boldsymbol{\theta} + \boldsymbol{A}\boldsymbol{D}\boldsymbol{u}_c - \dot{\boldsymbol{M}}\dot{\boldsymbol{s}}_f)\\ &= -c\dot{\boldsymbol{s}}_f + \boldsymbol{M}^{-1}[(\dot{\boldsymbol{W}}_f + c\boldsymbol{W}_f)\boldsymbol{\theta} + (\dot{\boldsymbol{u}}_f + c\boldsymbol{u}_f) - \dot{\boldsymbol{M}}\dot{\boldsymbol{s}}_f].\end{aligned} \tag{10.30}$$

Rearranging the above equation gives

$$\dot{\boldsymbol{\delta}} = -c\boldsymbol{\delta},\ \boldsymbol{\delta} = \boldsymbol{M}\dot{\boldsymbol{s}}_f - \boldsymbol{W}_f\boldsymbol{\theta} - \boldsymbol{u}_f. \tag{10.31}$$

According to the filters' initial values, it can be easily checked that $\boldsymbol{\delta}(0) = \mathbf{0}$. Solving (10.31) yields $\boldsymbol{\delta}(t) = \boldsymbol{\delta}(0)e^{-ct} \equiv \mathbf{0},\ \forall t \geq 0$. Then, from (10.31), it follows that

$$\boldsymbol{u}_f = \boldsymbol{M}\dot{\boldsymbol{s}}_f - \boldsymbol{W}_f\boldsymbol{\theta} = \boldsymbol{W}_a\boldsymbol{\theta}, \tag{10.32}$$

where $\boldsymbol{W}_a$ is a new regressor matrix that can be easily obtained. The prediction of the filtered input $\boldsymbol{u}_f$ is

$$\hat{\boldsymbol{u}}_f = \boldsymbol{W}_a\hat{\boldsymbol{\theta}}. \tag{10.33}$$

As per (10.32)–(10.33), a prediction error is constructed

$$\boldsymbol{\epsilon} = \hat{\boldsymbol{u}}_f - \boldsymbol{u}_f = \boldsymbol{W}_a\tilde{\boldsymbol{\theta}}. \tag{10.34}$$

Note that although the prediction error $\boldsymbol{\epsilon}$ contains the unknown parameter estimation error $\tilde{\boldsymbol{\theta}}$, it can be calculated from the easily accessible signals $\boldsymbol{u}_f$ and $\hat{\boldsymbol{u}}_f$, and thus can be directly used to design the composite learning law.

Assumption 10.1 There exist constants $t_s \geq 0$ and $T, \alpha > 0$ such that $\boldsymbol{W}_a^\top$ is of (t_s, T, α)-IE, that is,

$$\int_{t_s}^{t_s+T} \boldsymbol{W}_a^\top(\tau)\boldsymbol{W}_a(\tau)d\tau \geq \alpha\boldsymbol{I}_7. \tag{10.35}$$

10.3.2 Traditional Composite Adaptive Law

Generally speaking, in direct adaptive control, the parameter updating law is driven only by the filtered error $\boldsymbol{s}$. While in indirect adaptive control, the parameter updating law is driven only by the parameter estimation error $\tilde{\boldsymbol{\theta}}$. Both of these two adaptive laws have obvious shortcomings. Specifically, the direct adaptive law cannot guarantee parameter convergence, while the indirect adaptive law cannot improve the tracking performance. In order to synchronously improve parameter convergence and tracking performance, Slotine and Li [12] introduced the concept of composite adaptive control and used both the filtered error $\boldsymbol{s}$ and parameter estimation error $\tilde{\boldsymbol{\theta}}$ to design the adaptive law.

According to [12], the following composite adaptive law is designed

$$\dot{\hat{\boldsymbol{\theta}}} = \boldsymbol{\Upsilon}(t)(\boldsymbol{\Phi}^\top\boldsymbol{s} - \gamma\boldsymbol{W}_a^\top\boldsymbol{\epsilon}), \tag{10.36}$$

where $\gamma > 1$ is the weighting factor for parameter estimation, and $\boldsymbol{\Upsilon}(t) \in \mathbb{R}^{7\times 7}$ is a positive definite time-varying gain matrix. Here, the bounded gain forgetting (BGF) [12] method is chosen to update $\boldsymbol{\Upsilon}(t)$

$$\mathrm{d}\boldsymbol{\Upsilon}^{-1}(t)/\mathrm{d}t = -\lambda(t)\boldsymbol{\Upsilon}^{-1}(t) + \boldsymbol{W}_a^\top \boldsymbol{W}_a,\ \boldsymbol{\Upsilon}(0) = \boldsymbol{\Upsilon}_0, \tag{10.37}$$

where $\lambda(t) > 0$ is the forgetting factor, which is given by

$$\lambda(t) = \lambda_0(1 - \|\boldsymbol{\Upsilon}(t)\|/k_0), \tag{10.38}$$

where $\lambda_0 > 0$ and $k_0 \geq \|\boldsymbol{\Upsilon}_0\|$ specify upper bounds for $\lambda(t)$ and $\|\boldsymbol{\Upsilon}(t)\|$, respectively. One can infer from (10.37)-(10.38) that $\lambda(t) \geq 0$ and $\boldsymbol{\Upsilon}(t) \leq k_0\mathbf{I}_{7\times 7}$ hold for all $t \geq 0$.

Theorem 10.1 *Consider the integrated relative position-attitude coupled dynamics* (10.18). *Implementing the adaptive control law* (10.23) *and the accompanying adaptive law* (10.36) *can guarantee the following, despite the presence of mass and inertia uncertainties:*

(1) All of the closed-loop signals are bounded, and the pose tracking error $\boldsymbol{e}$ *and prediction error* $\boldsymbol{\epsilon}$ *converge asymptotically to zero;*
(2) If the regressor matrix $\boldsymbol{W}_a^\top \in$ *PE, then the pose tracking error* $\boldsymbol{e}$ *and parameter estimation error* $\tilde{\boldsymbol{\theta}}$ *converge exponentially to zero.*

Proof Consider the following Lyapunov function candidate

$$V(\boldsymbol{s}, \tilde{\boldsymbol{\theta}}) = \frac{1}{2}\boldsymbol{s}^\top \boldsymbol{M}\boldsymbol{s} + \frac{1}{2}\tilde{\boldsymbol{\theta}}^\top \boldsymbol{\Upsilon}^{-1}\tilde{\boldsymbol{\theta}}. \tag{10.39}$$

Based on the definition of $V(\boldsymbol{s}, \tilde{\boldsymbol{\theta}})$, it can be seen that $V(\mathbf{0}, \mathbf{0}) = 0$, and $V(\boldsymbol{s}, \tilde{\boldsymbol{\theta}}) > 0$ always hold for any $(\boldsymbol{s}, \tilde{\boldsymbol{\theta}}) \neq (\mathbf{0}, \mathbf{0})$. Define $\varsigma = [\boldsymbol{s}^\top, \tilde{\boldsymbol{\theta}}^\top]^\top \in \mathbb{R}^{13}$. Then, $V(\boldsymbol{s}, \tilde{\boldsymbol{\theta}})$ has the following upper and lower bounds,

$$\frac{1}{2}\min\{\lambda_{\min}(\boldsymbol{M}), \lambda_{\min}(\boldsymbol{\Upsilon}^{-1})\}\|\varsigma\|^2 \leq V(\boldsymbol{s}, \tilde{\boldsymbol{\theta}}) \leq \frac{1}{2}\max\{\lambda_{\max}(\boldsymbol{M}), \lambda_{\max}(\boldsymbol{\Upsilon}^{-1})\}\|\varsigma\|^2. \tag{10.40}$$

Substituting the tracking error dynamics (10.24) and (10.36)–(10.37) into the time derivative of $V(\boldsymbol{s}, \tilde{\boldsymbol{\theta}})$, whilst considering that $\dot{\boldsymbol{M}} - 2\boldsymbol{C}$ is a skew symmetric matrix, we can deduce that

$$\begin{aligned}
\dot{V}(\boldsymbol{s}, \tilde{\boldsymbol{\theta}}) &= \boldsymbol{s}^\top \boldsymbol{M}\dot{\boldsymbol{s}} + \frac{1}{2}\boldsymbol{s}^\top \dot{\boldsymbol{M}}\boldsymbol{s} + \tilde{\boldsymbol{\theta}}^\top \boldsymbol{\Upsilon}^{-1}\dot{\hat{\boldsymbol{\theta}}} + \frac{1}{2}\tilde{\boldsymbol{\theta}}^\top \dot{\boldsymbol{\Upsilon}}^{-1}\hat{\boldsymbol{\theta}} \\
&= -\boldsymbol{s}^\top \boldsymbol{K}_c\boldsymbol{s} - \left(\gamma - \frac{1}{2}\right)\boldsymbol{\epsilon}^\top\boldsymbol{\epsilon} - \frac{\lambda(t)}{2}\tilde{\boldsymbol{\theta}}^\top \boldsymbol{\Upsilon}^{-1}\tilde{\boldsymbol{\theta}} \\
&\leq -\lambda_{\min}(\boldsymbol{K}_c)\|\boldsymbol{s}\|^2 - \frac{1}{2}\|\boldsymbol{\epsilon}\|^2 - \frac{\lambda(t)}{2}\tilde{\boldsymbol{\theta}}^\top \boldsymbol{\Upsilon}^{-1}\tilde{\boldsymbol{\theta}},
\end{aligned} \tag{10.41}$$

which indicates that $V(\boldsymbol{s}(t), \tilde{\boldsymbol{\theta}}(t)) \leq V(\boldsymbol{s}(0), \tilde{\boldsymbol{\theta}}(0))$. Thus, according to the definition of $V(\boldsymbol{s}, \tilde{\boldsymbol{\theta}})$, it follows that $\boldsymbol{s}, \tilde{\boldsymbol{\theta}} \in \mathcal{L}_\infty$. From the boundedness of $\boldsymbol{s}$, we can conclude that $\dot{\boldsymbol{e}}, \boldsymbol{e} \in \mathcal{L}_\infty$. In addition, since $\tilde{\boldsymbol{\theta}} \in \mathcal{L}_\infty$, it is clear that $\hat{\boldsymbol{\theta}} \in \mathcal{L}_\infty$. Furthermore, we can easily prove the boundedness of $\boldsymbol{v}_e$ and $\boldsymbol{\omega}_e$ from $\dot{\boldsymbol{e}} \in \mathcal{L}_\infty$. By reviewing the definition of $\boldsymbol{\Phi}$, we can conclude that $\boldsymbol{\Phi} \in \mathcal{L}_\infty$. From Eq. (10.23), it is clear that $\boldsymbol{u}_c$ is also bounded. Therefore, all closed-loop signals are bounded. Next, the asymptotic convergence of $\boldsymbol{s}$ and $\boldsymbol{\epsilon}$ is analyzed. According to (10.41), one has $\boldsymbol{s}, \boldsymbol{\epsilon} \in \mathcal{L}_2 \cap \mathcal{L}_\infty$, showing that $\boldsymbol{s}$ and $\boldsymbol{\epsilon}$ are square integrable. Furthermore, based on their derivatives and the boundedness of closed-loop signals, it can be shown that $\boldsymbol{s}$ and $\boldsymbol{\epsilon}$ are uniformly continuous. Applying the Barbalat's lemma yields $\lim_{t\to\infty}[\boldsymbol{s}(t), \boldsymbol{\epsilon}(t)] = \mathbf{0}$. Since $\boldsymbol{s} = \dot{\boldsymbol{e}} + k\boldsymbol{e}$, $\lim_{t\to\infty} \boldsymbol{s}(t) = \mathbf{0}$ also suggests $\lim_{t\to\infty}[\dot{\boldsymbol{e}}(t), \boldsymbol{e}(t)] = \mathbf{0}$.

If $\boldsymbol{W}_a^\top \in$ PE, then $\exists \lambda_1 > 0$ such that $\lambda(t) \geq \lambda_1$ for all $t \geq 0$. In view of this and noting $\boldsymbol{\Upsilon}(t) \leq k_0 \mathbf{I}_{7\times 7}$, it can be easily shown that

$$\lambda(t)\tilde{\boldsymbol{\theta}}^\top \boldsymbol{\Upsilon}^{-1}\tilde{\boldsymbol{\theta}} \geq (\lambda_1/k_0)\tilde{\boldsymbol{\theta}}^\top\tilde{\boldsymbol{\theta}}. \tag{10.42}$$

Using the above inequality in (10.41) gets

$$\begin{aligned}\dot{V}(\boldsymbol{s}, \tilde{\boldsymbol{\theta}}) \leq & -\lambda_{\min}(\boldsymbol{K}_c)\|\boldsymbol{s}\|^2 - (\lambda_1/2k_0)\|\tilde{\boldsymbol{\theta}}\|^2 \\ \leq & -\frac{1}{2}\min\{2\lambda_{\min}(\boldsymbol{K}_c), \lambda_1/k_0\}\|\boldsymbol{\varsigma}\|^2.\end{aligned} \tag{10.43}$$

Using (10.40) and (10.43), it can be concluded that

$$\dot{V}(\boldsymbol{s}, \tilde{\boldsymbol{\theta}}) \leq -aV(\boldsymbol{s}, \tilde{\boldsymbol{\theta}}), \tag{10.44}$$

where $a := \frac{\min\{2\lambda_{\min}(\boldsymbol{K}_c), \lambda_1/k_0\}}{\max\{\lambda_{\max}(\boldsymbol{M}), \lambda_{\max}(\boldsymbol{\Upsilon}^{-1})\}}$. According to the comparison lemma, $V(\boldsymbol{s}(t), \tilde{\boldsymbol{\theta}}(t))$ converges exponentially to zero and hence the equilibrium point $(\boldsymbol{s}, \tilde{\boldsymbol{\theta}}) \equiv \mathbf{0}$ is exponentially stable. This completes the proof. □

Remark 10.2 Note that the information matrix $\boldsymbol{W}_a^\top \boldsymbol{W}_a$ is a 7×7 square matrix. Recalling the important properties of matrix rank, it is known that for any $\boldsymbol{A}$ and $\boldsymbol{B}$ with matching dimensional, $\text{rank}(\boldsymbol{AB}) \leq \min\{\text{rank}(\boldsymbol{A}), \text{rank}(\boldsymbol{B})\}$ always hold. Using this property, $\text{rank}(\boldsymbol{W}_a^\top \boldsymbol{W}_a) \leq 6$ can be directly derived, which indicates that the information matrix $\boldsymbol{W}_a^\top \boldsymbol{W}_a$ is only semi-positive definite. The rank deficiency of information matrix is the main reason for the requirement of PE condition on parameter convergence. However, the PE condition is very restrictive and rarely met in practical applications. As such, most of the traditional composite adaptive control cannot ensure that parameter estimates converge to their true values.

10.3.3 Composite Learning Law

As stated in Remark 10.2, due to the rank deficiency of information matrix $\boldsymbol{W}_a^\top \boldsymbol{W}_a$, the traditional composite adaptive law can ensure parameter convergence only when $\boldsymbol{W}_a^\top$ is of PE. To relax this requirement, one of the most intuitive ideas is to construct a new information matrix, so that it can become a full-rank matrix after a strictly weak IE condition. Therefore, this subsection aims to design a novel composite learning law by combing the CL and DREM techniques. The learning law not only ensures parameter convergence under an IE condition, but also inherits the advantages of DREM and has enhanced parameter estimation performance.

First, a set of independent scalar LREs are derived using the DREM procedure.

Step 1: dynamic regressor extension. The Kreisselmeier regressor extension method, originally developed in [25], is used to extend the LRE (10.32). Multiplying both sides of (10.32) by $\boldsymbol{W}_a^\top$ gives

$$\boldsymbol{W}_a^\top \boldsymbol{u}_f = \boldsymbol{W}_a^\top \boldsymbol{W}_a \boldsymbol{\theta}. \tag{10.45}$$

Consider the operator $\mathcal{H}(s) = 1/(s+a)$, where s represents the Laplace operator. Applying $\mathcal{H}$ to both sides of (10.45), whilst defining $\boldsymbol{\Omega} := \mathcal{H}[\boldsymbol{W}_a^\top \boldsymbol{W}_a] \in \mathbb{R}^{7\times 7}$ and $\boldsymbol{N} := \mathcal{H}[\boldsymbol{W}_a^\top \boldsymbol{u}_f] \in \mathbb{R}^7$, we have

$$\dot{\boldsymbol{\Omega}} = -a\boldsymbol{\Omega} + \boldsymbol{W}_a^\top \boldsymbol{W}_a, \ \boldsymbol{\Omega}(0) = \boldsymbol{0}, \tag{10.46}$$

$$\dot{\boldsymbol{N}} = -a\boldsymbol{N} + \boldsymbol{W}_a^\top \boldsymbol{u}_f, \ \boldsymbol{N}(0) = \boldsymbol{0}. \tag{10.47}$$

Solving the above two equations gets

$$\boldsymbol{\Omega}(t) = \int_0^t e^{-a(t-\tau)} \boldsymbol{W}_a^\top(\tau) \boldsymbol{W}_a(\tau) d\tau, \tag{10.48}$$

$$\boldsymbol{N}(t) = \int_0^t e^{-a(t-\tau)} \boldsymbol{W}_a^\top(\tau) \boldsymbol{u}_f(\tau) d\tau. \tag{10.49}$$

According to Eqs. (10.32), (10.48) and (10.49), it is clear that,

$$\boldsymbol{N}(t) = \boldsymbol{\Omega}(t)\boldsymbol{\theta} \tag{10.50}$$

It is noted that, when compared with traditional dynamic regressor extension, the Kreisselmeier regressor extension only requires two filters for $\boldsymbol{W}_a^\top \boldsymbol{W}_a$ and $\boldsymbol{W}_a^\top \boldsymbol{u}_f$, so the design is simpler. In addition, Kreisselmeier regressor extension also realizes the storage of historical data by using the forward integration, which is similar to the data storage of CL-based adaptive control method in [15, 26, 27]. However, forward integration is simpler and does not require additional design of data selection mechanism, so it is more conducive for practical application. With Kreisselmeier regressor extension, if $W_a^\top$ satisfies a weak IE condition such that historical data is stored rich enough, then the new information matrix $\boldsymbol{\Omega}$ becomes full-rank, thus relaxing the condition for parameter convergence.

Lemma 10.1 *If $W_a^\top$ satisfies Assumption 10.1, then $\boldsymbol{\Omega}(t) > 0$ hold for any $t \geq t_e$.*

Proof Given some constant vector $\boldsymbol{x} \in \mathbb{R}^7$, consider the following quadratic form with respect to the matrix $\boldsymbol{\Omega}(t)$:

$$\boldsymbol{x}^\top \boldsymbol{\Omega}(t)\boldsymbol{x} = \int_0^t (\sqrt{\boldsymbol{O}(t,\tau)}\|\boldsymbol{W}_a\boldsymbol{x}\|)^2 d\tau, \tag{10.51}$$

where $0 < \boldsymbol{O}(t,\tau) = e^{-a(t-\tau)} \leq 1$. As dictated by Assumption 10.1, if $\boldsymbol{W}_a^\top \in$ IE, then it follows that $\int_{t_s}^{t_e} \boldsymbol{W}_a^\top(\tau)\boldsymbol{W}_a(\tau)d\tau > 0$ $(t_e = t_s + T)$, indicating that $\int_0^t \boldsymbol{W}_a^\top(\tau)\boldsymbol{W}_a(\tau)d\tau > 0$ holds for any $t \geq t_e$. This implies that $\boldsymbol{W}_a$ will not always on an affine hyperplane on $[0, t]$ $(t \geq t_e)$ [19], so there is no $\boldsymbol{x} \neq \boldsymbol{0}$ such that $\boldsymbol{W}_a\boldsymbol{x} \equiv \boldsymbol{0}$ for all $t \in [0, t]$ $(t \geq t_e)$. Further, considering that $\boldsymbol{x}$ can be any vector and $\boldsymbol{O}(t,\tau) > 0$, we can directly conclude from (10.51) that $\boldsymbol{x}^\top\boldsymbol{\Omega}(t)\boldsymbol{x} > 0, \forall t \geq t_e$, that is, $\boldsymbol{\Omega}(t) > 0$, $\forall t \geq t_e$. This completes the proof. □

Step 2: Mixing. Multiplying both sides of (10.50) by adj$(\boldsymbol{\Omega})$, and defining $\boldsymbol{\Psi} :=$ adj$(\boldsymbol{\Omega})\boldsymbol{N}$ and $\Delta :=$ det$(\boldsymbol{\Omega})$, we have

$$\boldsymbol{\Psi}(t) = \Delta(t)\boldsymbol{\theta}. \tag{10.52}$$

In the framework of DREM, composite learning law can generally be designed as $\dot{\hat{\boldsymbol{\theta}}} = \boldsymbol{\Upsilon}[\boldsymbol{\Phi}^\top \boldsymbol{s} - \boldsymbol{\Gamma}\Delta(\Delta\hat{\boldsymbol{\theta}} - \boldsymbol{\Psi})]$. Note that the value of information matrix $\boldsymbol{\Omega}$ at time t is the weighted accumulation of all data on time interval $[0, t]$. However, since (10.46) has an exponential forgetting nature, if $\boldsymbol{W}_a^\top$ only satisfies the IE condition, the information matrix will decay to zero after the excitation disappears, which will result in $\Delta \to 0$. This greatly affect the parameter convergence performance. Therefore, in order to obtain a good and consistent parameter estimation performance under the IE condition, information on the whole time interval should not be fully used to generate $\boldsymbol{\Omega}$ and Δ. Towards this end, we set a time t_e to stop updating the information matrix $\boldsymbol{\Omega}$ and the auxiliary matrix $\boldsymbol{N}$, and use $\boldsymbol{\Omega}(t_e)$ and $\boldsymbol{N}(t_e)$ to calculate $\boldsymbol{\Delta}$ and $\boldsymbol{\Psi}$, that is

$$t_e := \min\left\{\underset{\tau\in(0,t]}{\arg}\ \Delta(\tau) \geq \Delta_{thr}\right\}, \tag{10.53}$$

$$\Delta_e := \Delta(t_e),\ \ \boldsymbol{\Psi}_e := \boldsymbol{\Psi}(t_e), \tag{10.54}$$

where $\Delta_{thr} > 0$ is a user-defined threshold. According to (10.52) and (10.54), we get

$$\boldsymbol{\Psi}_e = \Delta_e\boldsymbol{\theta}. \tag{10.55}$$

In fact, Eq. (10.55) gives a set of scalar LREs, and each one is decoupled from the others

$$\psi_{ei} = \Delta_e \theta_i, \ i = 1, 2, ..., 7, \tag{10.56}$$

where ψ_{ei} represents the ith element of the vector $\boldsymbol{\Psi}_e$. According to Lemma 10.1, for all $t \geq t_e$, $\boldsymbol{\Omega}(t)$ is a positive definite matrix and all of its eigenvalues are greater than zero. Since Δ_e equals to the product of all eigenvalues of $\boldsymbol{\Omega}(t_e)$, we have $\Delta_e > 0$.

Another problem is that if the excitation level of the regressor matrix $\boldsymbol{W}_a$ is very low, the eigenvalues of information matrix $\boldsymbol{\Omega}$ at t_e and the determinant (the product of all eigenvalues) will be very small, which may lead to $\Delta_e^2 \ll 1$, and greatly reduce the rate of parameter convergence. To overcome the aforementioned drawbacks, a composite learning law is designed as follows:

$$\dot{\hat{\boldsymbol{\theta}}} = \boldsymbol{\Upsilon}(\boldsymbol{\Phi}^\top \boldsymbol{s} - \boldsymbol{\Gamma}\boldsymbol{\Theta}), \tag{10.57}$$

where $\boldsymbol{\Upsilon}, \boldsymbol{\Gamma} > 0$ are diagonal matrices, and the prediction-error-related term $\boldsymbol{\Theta}$ is given by

$$\boldsymbol{\Theta}(t) = \begin{cases} \boldsymbol{0}, & t < t_e \\ \Delta_e^{-1}(\Delta_e \hat{\boldsymbol{\theta}} - \boldsymbol{\Psi}_e), & t \geq t_e \end{cases}. \tag{10.58}$$

Remark 10.3 In fact, the parameter estimator (10.57) can be numerically implemented without requiring the real-time computation of the adjoint matrix adj($\boldsymbol{\Omega}$). According to the method proposed in [25], the element ψ_{ei} in (10.56) can be directly calculated by using the well-known Cramer's rule

$$\psi_{ei} = \det(\boldsymbol{\Omega}_{N,i}(t_e)), \ i = 1, 2, ..., 7, \tag{10.59}$$

where $\boldsymbol{\Omega}_{N,i}(t_e)$ is obtained by replacing the ith column of the matrix $\boldsymbol{\Omega}(t_e)$ with the vector $\boldsymbol{N}(t_e)$.

Theorem 10.2 *Consider the integrated relative position-attitude coupled dynamics* (10.18) *and Assumption 10.1. The control law* (10.23) *and the composite learning law* (10.57) *can achieve the following, despite the mass and inertia uncertainties:*

(1) All closed-loop signals are bounded, and the pose tracking error $\boldsymbol{e}$ *asymptotically converges to zero on* $t \in [0, \infty)$*;*
(2) If Assumption 10.1 holds, the pose tracking error $\boldsymbol{e}$ *and parameter estimation error* $\tilde{\boldsymbol{\theta}}$ *exponentially converge to zero on* $t \in [t_e, \infty)$.
(3) The norms of the filtered error $\boldsymbol{s}$ *and parameter estimation error* $\tilde{\boldsymbol{\theta}}$ *satisfy*

$$\|\boldsymbol{s}\| \leq \begin{cases} \sqrt{\dfrac{2V(0)}{\lambda_{\min}(\boldsymbol{M})}}, & t < t_e \\ \sqrt{\dfrac{2V(0)e^{-a(t-t_e)}}{\lambda_{\min}(\boldsymbol{M})}}, & t \geq t_e \end{cases}, \tag{10.60}$$

$$\|\tilde{\boldsymbol{\theta}}\| \leq \begin{cases} \sqrt{\dfrac{2V(0)}{\lambda_{\min}(\boldsymbol{\Upsilon}^{-1})}}, & t < t_e \\ \sqrt{\dfrac{2V(0)e^{-a(t-t_e)}}{\lambda_{\min}(\boldsymbol{\Upsilon}^{-1})}}, & t \geq t_e \end{cases}. \tag{10.61}$$

Proof Consider a Lyapunov candidate similar to (10.39) (the only difference is that $\boldsymbol{\Upsilon}$ is a constant matrix). Substituting the control law (10.23) and the composite learning law (10.57) into the time derivative of $V(\boldsymbol{s}, \tilde{\boldsymbol{\theta}})$ and noting $\tilde{\boldsymbol{\theta}}^{\top}\boldsymbol{\Gamma\Theta} \geq 0$, it follows that

$$\dot{V}(\boldsymbol{s}, \tilde{\boldsymbol{\theta}}) = -\boldsymbol{s}^{\top}\boldsymbol{K}_c\boldsymbol{s} - \tilde{\boldsymbol{\theta}}^{\top}\boldsymbol{\Gamma\Theta} \leq -\lambda_{\min}(\boldsymbol{K}_c)\|\boldsymbol{s}\|^2. \tag{10.62}$$

Following the proof process of Theorem 10.1, it is shown from (10.62) that all closed-loop signals are bounded in the whole time interval, and the pose tracking error $\boldsymbol{e}$ converges asymptotically to zero.

Next, under the Assumption 10.1, the exponential stability of the closed-loop system on time interval $[t_e, \infty)$ is further analyzed. Substituting (10.55) into (10.58) yields

$$\boldsymbol{\Theta} = \Delta_e^{-1}(\Delta_e\hat{\boldsymbol{\theta}} - \boldsymbol{\Psi}_e) = \tilde{\boldsymbol{\theta}},\ \forall t \geq t_e. \tag{10.63}$$

Applying (10.63) to $\dot{V}(\boldsymbol{s}, \tilde{\boldsymbol{\theta}})$ and after some evaluation, one gets

$$\dot{V}(\boldsymbol{s}, \tilde{\boldsymbol{\theta}}) \leq -\frac{1}{2}\min\{2\lambda_{\min}(\boldsymbol{K}_c), 2\lambda_{\min}(\boldsymbol{\Gamma})\}\|\varsigma\|^2,\ \forall t \geq t_e. \tag{10.64}$$

According to (10.40) and (10.63), it follows that

$$\dot{V}(\boldsymbol{s}, \tilde{\boldsymbol{\theta}}) \leq -aV(\boldsymbol{s}, \tilde{\boldsymbol{\theta}})\ \forall t \geq t_e, \tag{10.65}$$

where $a := \frac{\min\{2\lambda_{\min}(\boldsymbol{K}_c), \lambda_{\min}(\boldsymbol{\Gamma})\}}{\max\{\lambda_{\max}(\boldsymbol{M}), \lambda_{\max}(\boldsymbol{\Upsilon}^{-1})\}}$. According to the comparison lemma, $V(\boldsymbol{s}(t), \tilde{\boldsymbol{\theta}}(t))$ exponentially converges to zero, and the equilibrium point $(\boldsymbol{s}, \tilde{\boldsymbol{\theta}}) \equiv \boldsymbol{0}$ is exponentially stable. Further, it can be inferred that $\boldsymbol{e}$ and $\dot{\boldsymbol{e}}$ are also exponentially stable.

In general, it is difficult to predict the exact time t_e when the information matrix $\boldsymbol{\Omega}$ becomes full rank. In order to quantitatively analyze the transient performance of closed-loop system on $t \in [0, t_e]$ and $t \in [t_e, \infty)$, the norm bounds of the filtered tracking error $\boldsymbol{s}$ and parameter estimation error $\tilde{\boldsymbol{\theta}}$ are given later. Let us written $V(\boldsymbol{s}(t), \tilde{\boldsymbol{\theta}}(t))$ as $V(t)$ for the notation brevity. According to the definition of $V(t)$, for all $t \geq 0$, one has

$$\frac{1}{2}\lambda_{\min}(\boldsymbol{M})\|\boldsymbol{s}\|^2 \leq \frac{1}{2}\boldsymbol{s}^{\top}\boldsymbol{M}\boldsymbol{s} < V(t), \tag{10.66}$$

$$\frac{1}{2}\lambda_{\min}(\boldsymbol{\Upsilon}^{-1})\|\tilde{\boldsymbol{\theta}}\|^2 \leq \frac{1}{2}\tilde{\boldsymbol{\theta}}^{\top}\boldsymbol{\Upsilon}^{-1}\tilde{\boldsymbol{\theta}} < V(t). \tag{10.67}$$

From the above two equations, it can be deduced that

$$\|s\| \leq \sqrt{\frac{2V(t)}{\lambda_{\min}(\boldsymbol{M})}}, \ \|\tilde{\theta}\| \leq \sqrt{\frac{2V(t)}{\lambda_{\min}(\boldsymbol{\Upsilon}^{-1})}}. \tag{10.68}$$

To proceed, two cases are considered:

Case 1: $t \in [0, t_e]$. As can be seen from (10.62), $\dot{V}(t) \leq 0$ always holds whether the regressor matrix $\boldsymbol{\Omega}$ is full-rank or not, indicating that $V(t_e) \leq V(t) \leq V(0)$. Hence, the results of $t < t_e$ in (10.60) and (10.61) can be directly obtained from (10.68).

Case 2: $t \in [t_e, \infty)$. In this case, we have $\dot{V}(t) \leq -aV(t)$. According to the comparison lemma, it follows that

$$V(t) \leq V(t_e)e^{-a(t-t_e)} \leq V(0)e^{-a(t-t_e)} \tag{10.69}$$

Applying (10.69)–(10.68), the results of (10.60) and (10.61) for $t \geq t_e$ can be directly obtained. In fact, it can be seen from (10.60) and (10.61) that the filtered tracking error and parameter estimation error are bounded on $t \in [0, t_e]$, and they converge exponentially once the information matrix becomes full-rank. This completes the proof. □

Remark 10.4 For $t \geq t_e$, the parameter estimation error dynamics can be decomposed into the following forms:

$$\dot{\tilde{\theta}}_i = \Upsilon_i[\boldsymbol{\Phi}^\top s]_i - \Upsilon_i \Gamma_i \tilde{\theta}_i, \ i = 1, 2, ..., 7, \tag{10.70}$$

where $[\boldsymbol{\Phi}^\top s]_i$ represents the ith element of the vector $\boldsymbol{\Phi}^\top s$. Obviously, the parameter estimation error dynamics is independent of each other, so the adjustment of adaptive gains Υ_i and Γ_i will not affect the transient response of other parameter estimation error. Moreover, the convergence rate of parameter estimation error does not depend on the signal excitation strength. These two properties not only help to improve the transient performance of parameter estimation, but also make the selection of adaptive gains simple and clear, without resorting to the trial and error process. Compared with the traditional composite adaptive law [12] and the CL-based adaptive law [15–20, 26, 27], the above properties are unique outstanding advantages of the composite learning law proposed in this chapter.

10.4 Numerical Simulations

In this section, a typical simulation is performed for the integrated position-attitude coupled dynamics (10.18), so as to verify the effectiveness of the composite learning control method summarized in Theorem 10.2. The target spacecraft is assumed to operate on a Molniya orbit, with its orbital parameters, inertia matrix and initial conditions detailed in Sect. 2.4.3.5. The nominal mass and inertia of the pursuer are:

$$m_p = 100\,\text{kg}, \ \boldsymbol{J}_p = \begin{bmatrix} 55 & 0.3 & 0.5 \\ 0.3 & 65 & 0.2 \\ 0.5 & 0.2 & 58 \end{bmatrix} \text{kg} \cdot \text{m}^2. \tag{10.71}$$

The initial attitude and angular velocity of the pursuer are

$$\boldsymbol{q}_p(0) = [0.6968, -0.1593, -0.2237, -0.6626]^\top, \ \boldsymbol{\omega}_p(0) = [-0.01, 0.01, 0.02]^\top \text{ rad/s}.$$

The initial relative position and velocity of the pursue with respect to the target are $\boldsymbol{\rho}(0) = [-150, -25, 30]^\top$ m and $\dot{\boldsymbol{\rho}}(0) = [0.1, 0.2, -0.3]^\top$ m/s. The desired relative position is $\boldsymbol{\rho}_d = [-15, 0, 0]^\top$ m (it is not difficult to see that this reference signal does not satisfy the PE condition). The pursuer uses a set of thruster-driven system for 6-DOF pose tracking control (see Sect. 2.4.3.4 for more details for the configuration information). In this configuration, the control force $\boldsymbol{f}_c$ and control torque $\boldsymbol{\tau}_c$ generated by thrusters can be given by $[\boldsymbol{f}_c^\top, \boldsymbol{\tau}_c^\top]^\top = \boldsymbol{D}\boldsymbol{u}_c$, where the thruster configuration matrix $\boldsymbol{D}$ is given in (2.55). Set $d_x = d_y = d_z = 1$ m, and thrust saturation value is $u_{\max} = 10$ N. The control parameters are selected as follows: $k = 0.05$, $c = 3$, $a = 0.1$, $\boldsymbol{K}_c = \text{diag}\{10, 10, 10, 10, 10, 10\}$, $\boldsymbol{\Upsilon} = 0.5\boldsymbol{I}_7$, and $\boldsymbol{\Gamma} = 0.1\boldsymbol{I}_7$. The initial value of adaptive estimation is chosen as $\hat{\boldsymbol{\theta}}(0) = [120, 40, 80, 50, 1, -2, 0]^\top$. The simulation is implemented with a fixed step size of 0.01 s.

10.4.1 Ideal Simulation Campaign

Similar to Sect. 9.4.2, parameter variations, external disturbances, and thruster misalignments are ignored in this subsection. The tracking error responses of the closed-loop system are shown in Figs. 10.1 and 10.2. As can be seen, under the proposed composite learning control method, the pose tracking errors converge to steady states at around 150 s with a smooth transient process (in fact, they are exponentially convergent, which can be seen from the zoom-in view). In addition, although the controller does not consider the unwinding problem, it is clear from Fig. 10.1b that the scalar part q_{e4} of the error quaternion $\boldsymbol{q}_e$ converges to -1 (rather than 1). In fact, if appropriate control parameters are selected such that $\boldsymbol{q}_{ev}$ does not have large overshoot during the transient process, it will usually converge to the equilibrium point corresponding to the shortest rotation path, indicating that the proposed control scheme has a certain anti-unwinding ability. To intuitively illustrate the effectiveness of the proposed control scheme, the 3-D motion trajectory and partial attitude snapshots of the pursuer with respect to the target observed in $\mathcal{T}$ are shown in Fig. 10.3, where both the pursuer and the target are portrayed as the cubes with solar panels. From Fig. 10.3, it can be seen that the pursuer spacecraft can achieve fast and high-precision proximity

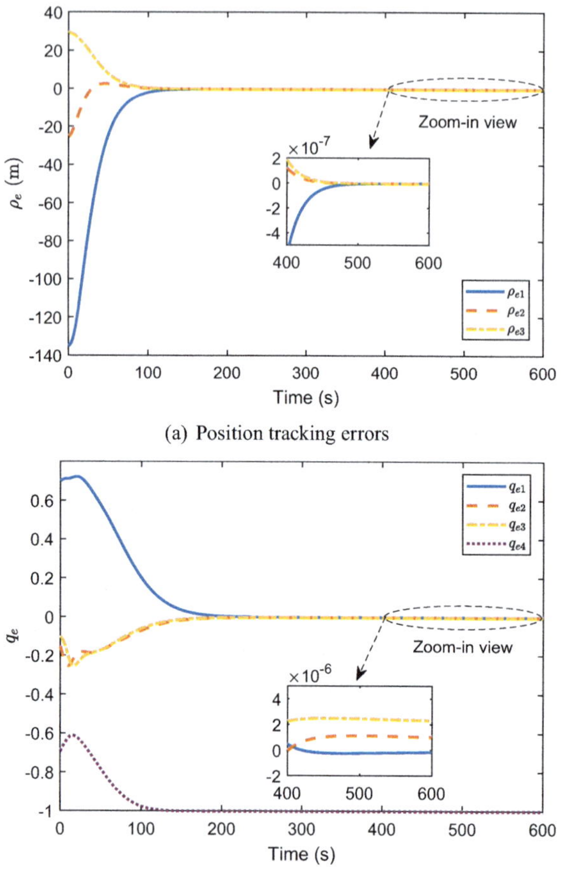

(a) Position tracking errors

(b) Attitude tracking errors

Fig. 10.1 Time responses of the pose tracking errors

operations with a tumbling target. Figure 10.4a depicts the rank of information matrix $\boldsymbol{\Omega}$. It is evident that $\boldsymbol{\Omega}$ becomes full-rank at 0.01 s, which indirectly indicates that the regressor matrix $\boldsymbol{W}_a$ has an IE at the initial stage. However, it is noted that the actual trigger time of the switch is about $t_e = 10$ s, which ensures sufficient accumulation of historical data. According to Theorem 10.2, the parameter estimation error $\tilde{\boldsymbol{\theta}}$ converges exponentially to zero after $t = t_e$, as shown in Fig. 10.4b. To further illustrate the importance of historical information for parameter convergence, the 2D motion trajectory of the parameter estimates (here we take $(\hat{\theta}_1, \hat{\theta}_2)$ as a case study) is plotted in Fig. 10.5, where blue and red arrows denote, respectively, the direction and strength of the parameter updating based on the current data (the direct adaptive part $\boldsymbol{\Upsilon}\boldsymbol{\Phi}^\top \boldsymbol{s}$ of (10.57)) and based on the historical data (the indirect adaptive part $-\boldsymbol{\Upsilon}\boldsymbol{\Gamma}\boldsymbol{\Theta}$ of (10.57)). Through observing the arrows, we find that after the initial time (specifically, from the moment t_e), the indirect adaptive law based on historical data starts to work, which together with the direct adaptive law based on current data

Fig. 10.2 Time responses of the linear/angular velocity tracking errors

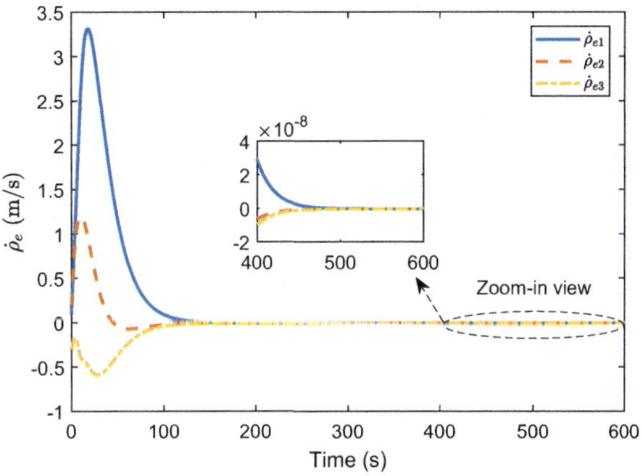

(a) Velocity tracking errors

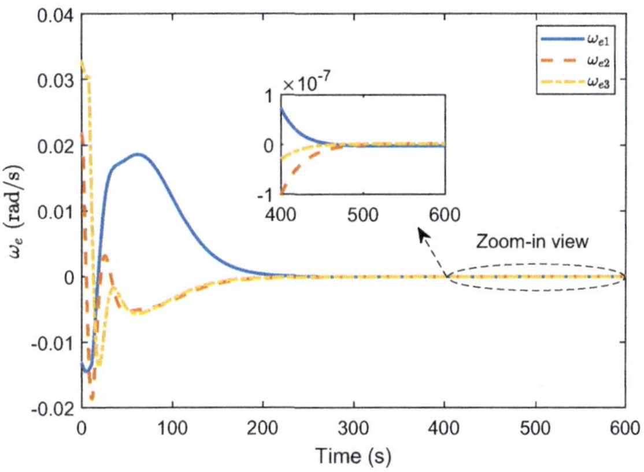

(b) Angular velocity tracking errors

forms two linearly independent directions. As such, the overall adaptive law can derive the parameter estimates to their true values. This reflects the effect of stored historical data on parameter convergence. Since pulse modulation is not considered, only continuous control input is given in Fig. 10.6.

To verify the decoupling property of the proposed composite learning law (10.57) (as stated in Remark 10.4), comparison simulations are conducted by changing the values of adaptive gains Γ_1 and Γ_2. For Case 1, we take $\Gamma_1 = 0.1$ s, 0.05 s, 0.2 s, and keep $\Gamma_i = 0.1, i = 2, 3, ..., 7$ unchanged. The comparison results are shown in Fig. 10.7a. For Case 2, similar to Case 1, we take $\Gamma_2 = 0.1$ s, 0.05 s, 0.2 s and keep $\Gamma_i = 0.1, i = 1, 3, ..., 7$ unchanged. The comparison results are shown in Fig. 10.7b. It can be seen from these two figures that by increasing the gain Γ_1 (resp. Γ_2), we can improve the convergence rate of the parameter estimation error $\tilde{\theta}_1$ (resp. $\tilde{\theta}_2$). The change of Γ_1 (resp. Γ_2) only affects the transient performance of $\tilde{\theta}_1$ (resp. $\tilde{\theta}_2$), and does not affect the convergence property of the other parameter estimation errors. This is

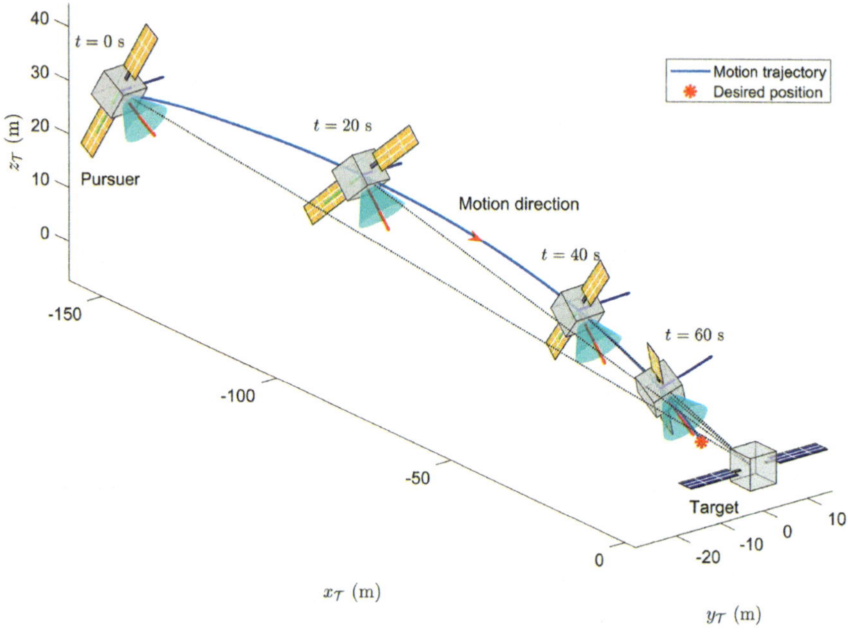

Fig. 10.3 3-D relative motion trajectory of the pursuer with respect to the target observed in $\mathcal{T}$

consistent with the conclusion stated in Remark 10.4. In fact, the above properties are essential for all Γ_i and $\tilde{\theta}_i$, $i = 1, 2, ..., 7$. However, for the sake of brevity, the verification of other parameter estimation errors is not provided here. In general, the DREM-based composite learning law shows the unique decoupling property of parameter estimation. Moreover, the convergence rate of parameter estimates does not depend on the signal excitation strength, so that the selection of adaptive gain is simpler and clearer.

For performance comparison, apart from the composite learning control method proposed in this chapter (denoted as DREM-CLAC), the BGF-based composite adaptive control method presented in Sect. 10.3.2 (denoted as BGF-CAC), and the I&I adaptive control method introduced in [24] (denoted as I&I-AC) are also simulated. The I&I-AC controller is described as follows:

$$\boldsymbol{u}_c = -(\boldsymbol{A}\boldsymbol{D})^{-1}\{\boldsymbol{Y}(\hat{\boldsymbol{\theta}} + \boldsymbol{\beta}) - \gamma \boldsymbol{Y}_f \boldsymbol{Y}_f^{\top}[-(c\mathbf{I}_{6\times 6} - \boldsymbol{K}_c)\boldsymbol{s}_f + \boldsymbol{s}]\}, \tag{10.72}$$

where the adaptive estimation parameter $\hat{\boldsymbol{\theta}}$ and the auxiliary term $\boldsymbol{\beta}$ are given by

$$\dot{\hat{\boldsymbol{\theta}}} = \gamma \boldsymbol{Y}_f^{\top}(c\mathbf{I}_{6\times 6} + \boldsymbol{K}_c)\boldsymbol{s}_f - \gamma \boldsymbol{Y}^{\top}\boldsymbol{s}_f, \tag{10.73}$$

$$\boldsymbol{\beta} = \gamma \boldsymbol{Y}_f^{\top}\boldsymbol{s}_f, \tag{10.74}$$

Fig. 10.4 Time responses of the parameter estimates

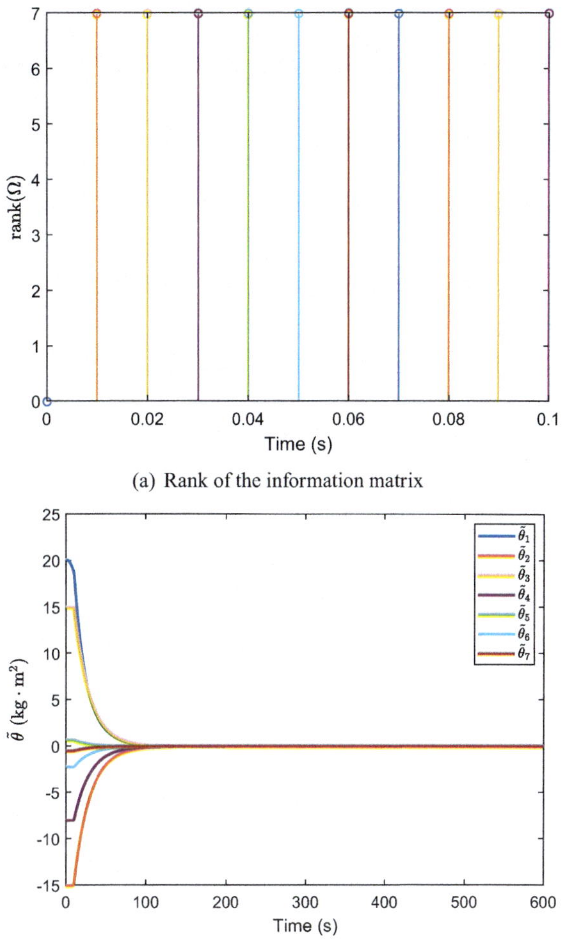

(a) Rank of the information matrix

(b) Parameter estimation errors

where $\gamma > 0$ is the adaptive gain, and $\boldsymbol{Y} \in \mathbb{R}^{6\times 7}$ is the regressor matrix satisfying

$$\boldsymbol{Y}\boldsymbol{\theta} = -\boldsymbol{C}\boldsymbol{s} + \boldsymbol{H} + \dot{\boldsymbol{M}}\dot{\boldsymbol{s}}_f + \dot{\boldsymbol{M}}\boldsymbol{K}_c\boldsymbol{s}_f + \boldsymbol{M}\boldsymbol{K}_c\boldsymbol{s}, \tag{10.75}$$

where $\boldsymbol{K}_c > 0$ is the gain matrix of desired dynamics $\dot{\boldsymbol{s}}_f = -\boldsymbol{K}_c\boldsymbol{s}_f$, and $\boldsymbol{s}_f$ is given by the filter (10.26) with initial conditions $s_i(0)/(c - K_{ci})$, $i = 1, 2, ..., 6$. In addition, $\boldsymbol{Y}_f \in \mathbb{R}^{6\times 7}$ is the filtered regressor matrix, which is derived by the filter $\dot{\boldsymbol{Y}}_f = -c\boldsymbol{Y}_f + \boldsymbol{Y}$, where $c > 0$ is the filter time constant. Notably, c and $\boldsymbol{K}_c$ should be selected such that $c \neq K_{ci}$. The design and analysis of the I&I-AC scheme are similar to the procedure provided in Sect. 8.3.2.

To make a fair comparison, the shared parameters of the other two controllers are set the same with the DREM-CLAC, and the remaining parameters are as follows.

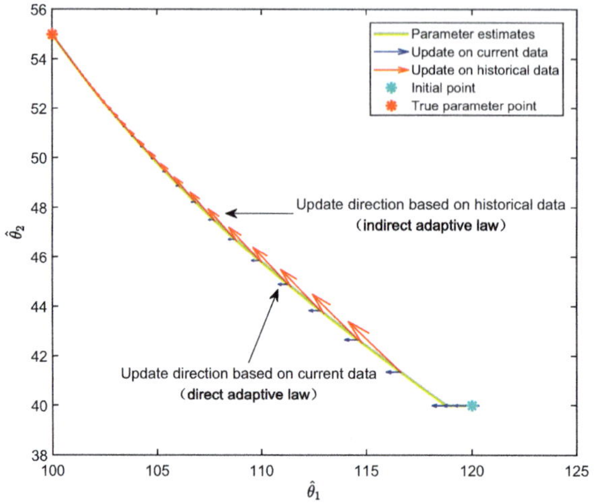

Fig. 10.5 2-D motion trajectory of the parameter estimate pair $(\hat{\theta}_1, \hat{\theta}_2)$

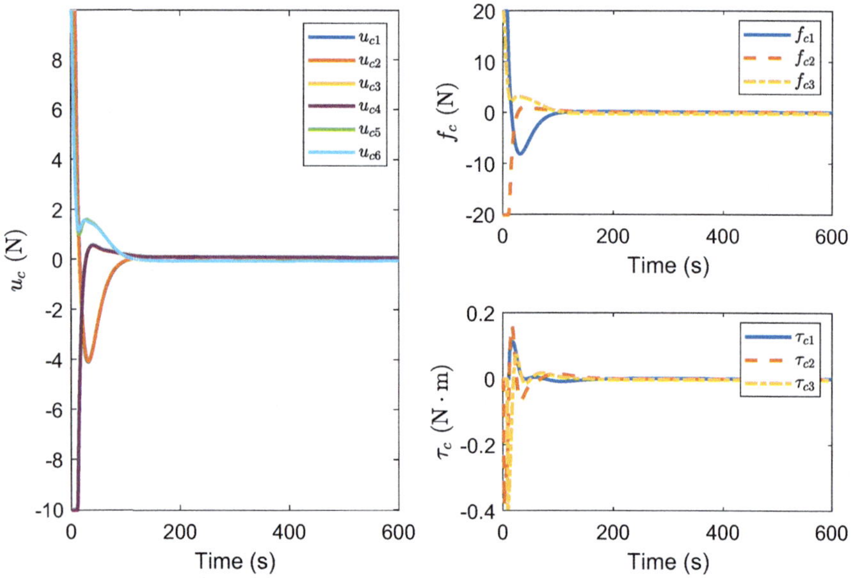

Fig. 10.6 Time responses of the control inputs and the driving forces and torques

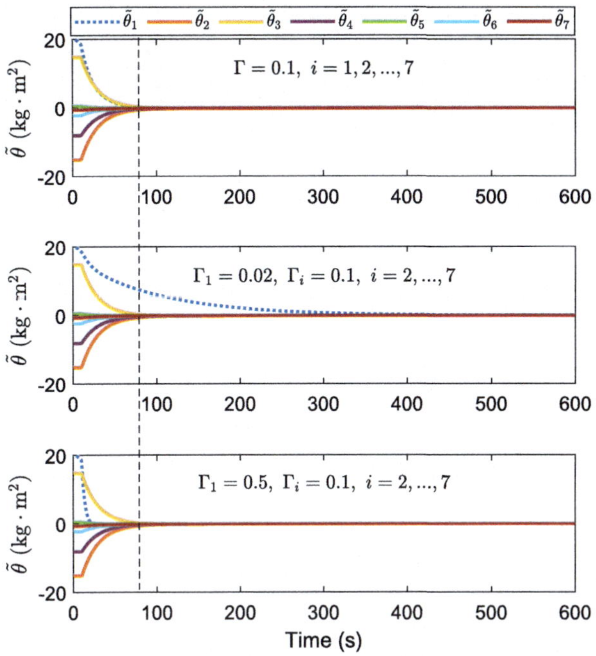

(a) Comparison results under different Γ_1

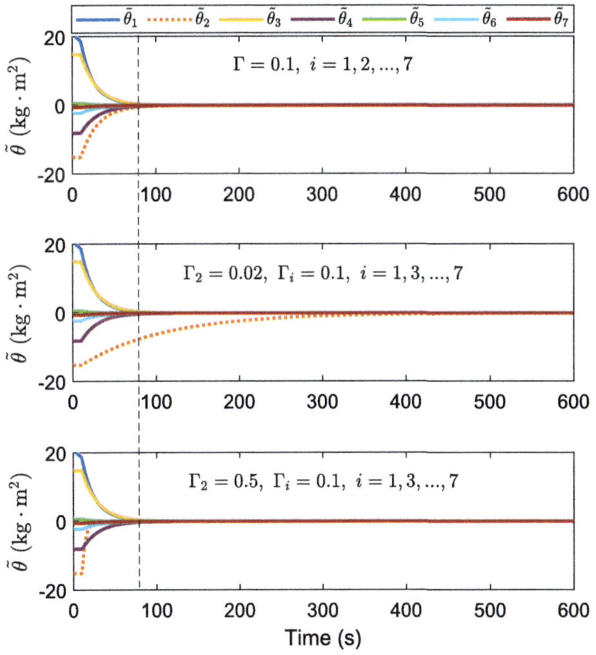

(b) Comparison results under different Γ_2

Fig. 10.7 Time responses of parameter estimation errors under different Γ_1 and Γ_2

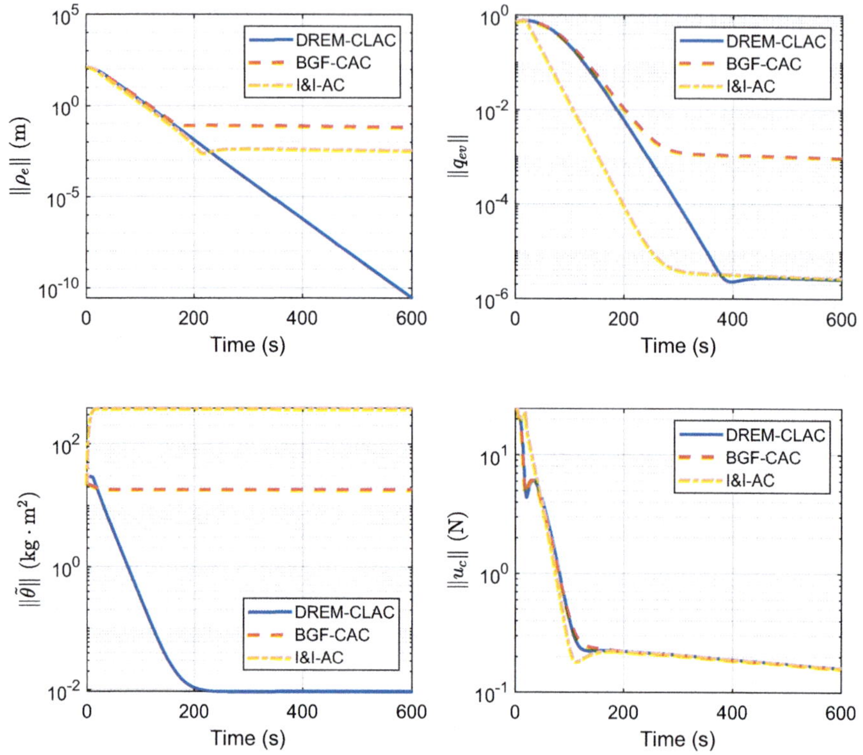

Fig. 10.8 Comparison results of the closed-loop responses

For BGF-CAC, $\lambda_0 = 0.1$, $k_0 = 0.5$, and $\gamma = 0.1$. In order to clearly show the performance of different control methods, all the closed-loop responses are depicted on the semilogarithmic scales, as shown in Fig. 10.8. It is clear that, when compared with the BGF-CAC and I&I-AC methods, the proposed DREM-CLAC scheme not only achieves better transient and steady-state tracking performance, but also ensures exponential convergence of parameter estimation errors. In general, the proposed DREM-CLAC method can synchronously improve parameter convergence and tracking performance if the regressor matrix $\boldsymbol{W}_a$ satisfies a strictly weak IE condition, so that the pursuer spacecraft can better accomplish the proximity operations. To compare the energy consumption of these three methods, energy consumption index is introduced as Energy $= \int_0^t \|\boldsymbol{u}_c(\tau)\|^2 d\tau$. As can be seen from Fig. 10.9, the I&I-AC consumes more energy than the DREM-CLAC and BGF-CLC.

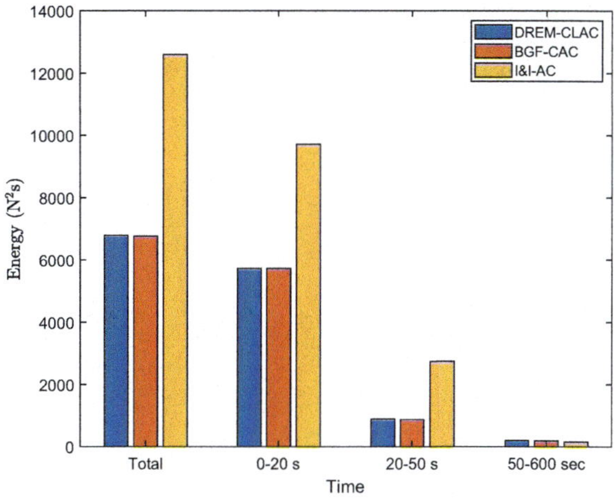

Fig. 10.9 Comparison results of energy consumption

10.4.2 Practical Simulation Campaign

To further verify practical applicability of the proposed control method, a simulation scenario suitable for practical engineering applications is considered in this subsection. Thruster misalignments, pulse modulation, external disturbances, and mass and inertia variations (refer to Sects. 8.4.2 and 9.4.3 for the detailed description) are taken into consideration. To improve robustness and applicability of the proposed composite learning control scheme, the control law (10.23) is modified as:

$$\boldsymbol{u}_c = -(\boldsymbol{A}\boldsymbol{D})^{-1}(\boldsymbol{K}_c\boldsymbol{s} + \boldsymbol{\Phi}\hat{\boldsymbol{\theta}} + d_m \tanh(\boldsymbol{s}/\iota)),$$

where $d_m > 0$ is a rough estimation for the upper bound of uncertainties such as external disturbances, parameter variations, and $\iota > 0$ represents the boundary layer thickness. Note that the above modification does not affect the closed-loop system stability. The control parameters of the proposed controller are the same as Sect. 10.4.1. The design parameters of PWPF are set as: $K_m = 1.5$, $T_m = 0.8$, $\delta_{on} = 0.45$, and $\delta_{off} = 0.15$. The closed-loop responses are shown in Figs. 10.10, 10.11, 10.12 and 10.13.

From Fig. 10.10, it can be seen that the control performance (especially the steady-state performance) of the proposed method degrades to some extent under the practical case, and the pose tracking error can no longer to converge to zero. But nonetheless, the transient tracking performance can still meet the requirements of practical tasks. Note that the transient responses of the linear/angular velocity tracking errors have an obvious chattering, as shown in Fig. 10.11. This is actually caused by the thruster on-off property. As can be seen from Fig. 10.12, the parameter estimation error does not converge to zero. This is mainly because the information matrix for adaptive law proposed in this chapter stops updating at t_e, so that the stored historical information only contains the parameter information within the interval $[0, t_e]$. Con-

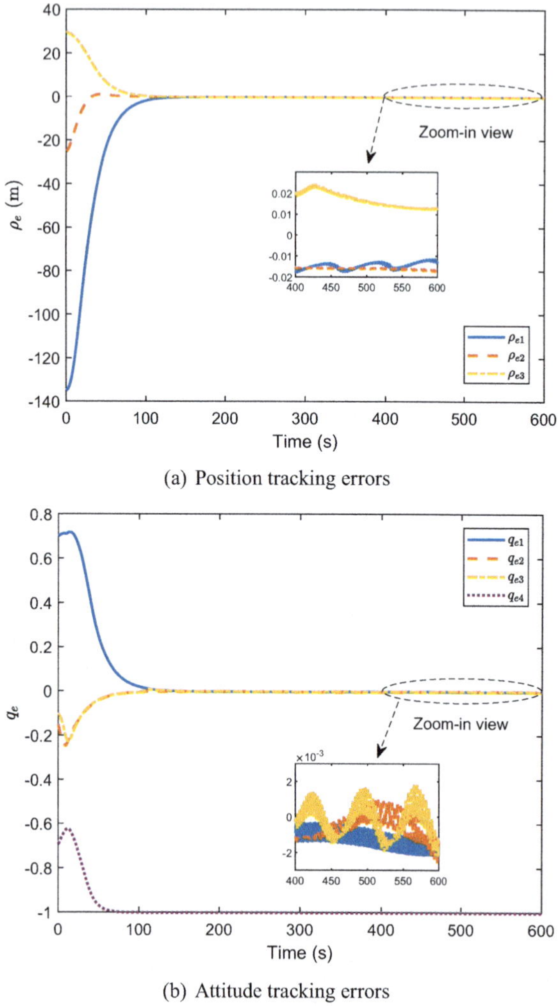

Fig. 10.10 Time responses of the pose tracking errors under the practical case

sequently, if the parameter changes after t_e, the adaptive law cannot learn the true values according to the stored historical information. On the other hand, practical factors such as thruster misalignment, pulse modulation, and external disturbances will also cause that the parameter estimation errors can not converge to zero. In conclusion, the composite learning law proposed in this chapter has some limitations. On the one hand, it cannot accurately identify the time-varying mass and inertia parameters. On the other hand, it has poor robustness to disturbances and misalignments. The pulse control signals of thrusters are given in Fig. 10.13.

Fig. 10.11 Time responses of the linear/angular velocity tracking errors under the practical case

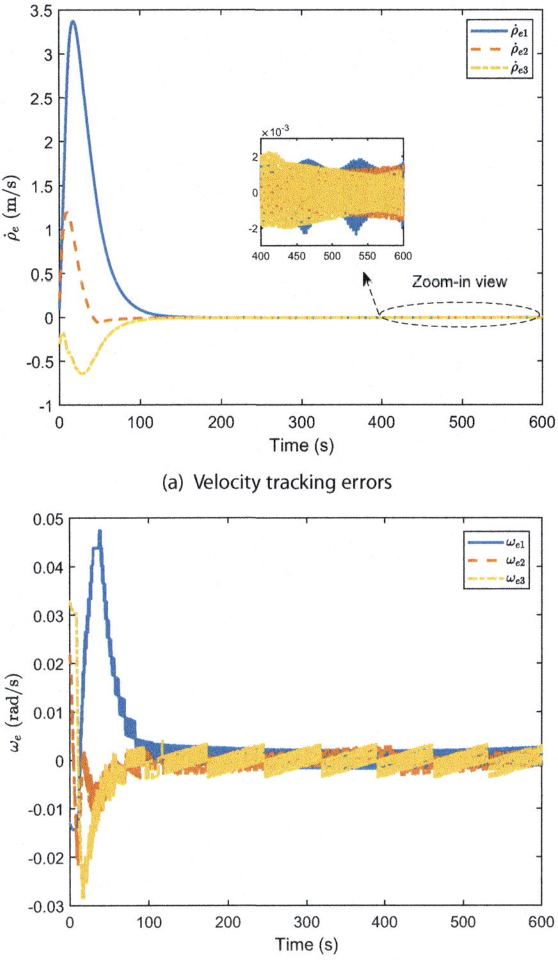

(a) Velocity tracking errors

(b) Angular velocity tracking errors

10.5 Summary

This chapter proposes a composite learning pose tracking control, which can synchronously enhance parameter convergence and tracking performance under a strictly weak IE condition. Based on the CL and DREM techniques, a novel learning control method is proposed. Firstly, a CE-based adaptive control law is given, and the filtered system dynamics is established to avoid the use of unmeasured state derivatives when constructing parameter estimation errors. Then, a traditional composite adaptive law is derived, and on this basis, the composite learning law is further designed. Lyapunov stability analysis shows that if the regressor matrix satisfies the IE condition, the proposed composite learning control scheme can ensure both the tracking errors

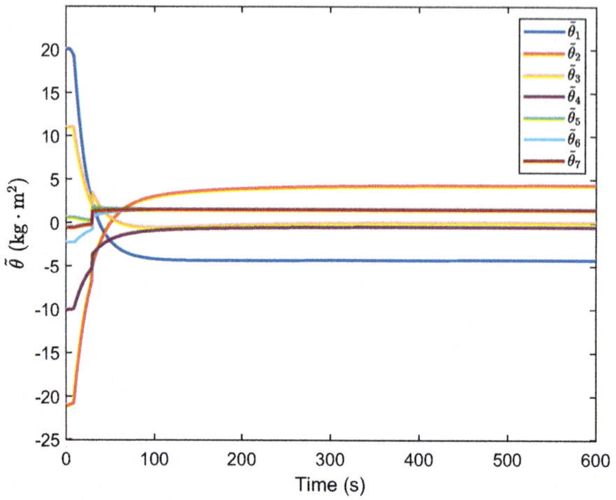

Fig. 10.12 Time responses of parameter estimation errors under the practical case

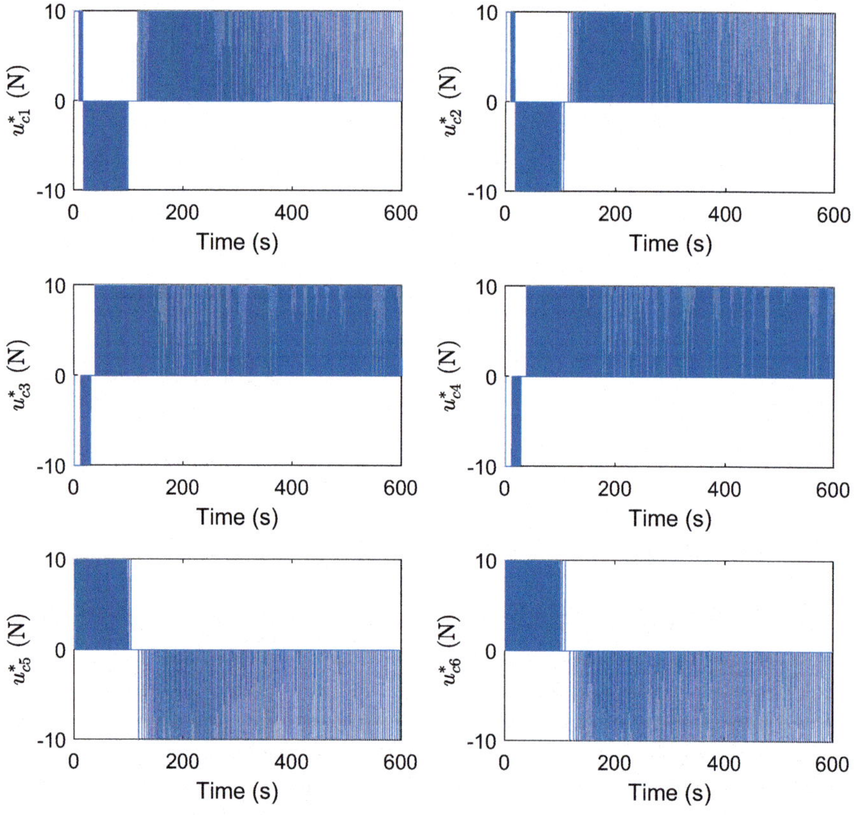

Fig. 10.13 Time responses of thruster outputs

and parameter estimation errors converge to zero by making use of the stored historical information. Moreover, benefiting from the DREM procedure and some special designs, the parameter estimation error dynamics is independent of each other, and parameter convergence rate does not depend on the signal excitation strength, which make the gain selection simpler and clearer. Finally, simulation results verify the effectiveness of the problem control scheme.

References

1. Zhang F, Duan GR (2014) Robust adaptive integrated translation and rotation finite-time control of a rigid spacecraft with actuator misalignment and unknown mass property. International Journal of Systems Science 45(5): 1007–1034
2. Sun L (2016) Passivity-based adaptive finite-time trajectory tracking control for spacecraft proximity operations. Journal of Spacecraft and Rockets 53(1): 46–56
3. Xia K, Huo W (2016) Robust adaptive backstepping neural networks control for spacecraft rendezvous and docking with uncertainties. Nonlinear Dynamics 84(3): 1683–1695
4. Sun L, Huo W, Jiao Z (2016) Adaptive backstepping control of spacecraft rendezvous and proximity operations with input saturation and full-state constraint. IEEE Transactions on Industrial Electronics 64(1): 480–492
5. Sun L, Huo W (2015) Robust adaptive relative position tracking and attitude synchronization for spacecraft rendezvous. Aerospace Science and Technology 41: 28–35
6. Filipe N, Tsiotras P (2015) Adaptive position and attitude-tracking controller for satellite proximity operations using dual quaternions. Journal of Guidance, Control, and Dynamics 38(4): 566–577
7. Dong H, Hu Q, Friswell MI, Ma G (2016) Dual-quaternion-based fault-tolerant control for spacecraft tracking with finite-time convergence. IEEE Transactions on Control Systems Technology 25(4): 1231–1242
8. Hu Q, Chen W, Zhang Y (2019) Concurrent proximity control of servicing spacecraft with an uncontrolled target. IEEE/ASME Transactions on Mechatronics 24(6): 2815–2826
9. Huang Y, Jia Y (2018a) Adaptive fixed-time six-dof tracking control for noncooperative spacecraft fly-around mission. IEEE Transactions on Control Systems Technology (99): 1–9
10. Huang Y, Jia Y (2018b) Robust adaptive fixed-time tracking control of 6-DOF spacecraft fly-around mission for noncooperative target. International Journal of Robust and Nonlinear Control 28(6): 2598–2618
11. Singla P, Subbarao K, Junkins JL (2006) Adaptive output feedback control for spacecraft rendezvous and docking under measurement uncertainty. Journal of Guidance, Control, and Dynamics 29(4): 892–902
12. Slotine JJE, Li W (1989) Composite adaptive control of robot manipulators. Automatica 25(4): 509–519
13. Boyd S, Sastry SS (1986) Necessary and sufficient conditions for parameter convergence in adaptive control. Automatica 22(6): 629–639
14. Tao G (2003) Adaptive control design and analysis. John Wiley & Sons, Hoboken, NJ, USA
15. Chowdhary G, Johnson E (2010) Concurrent learning for convergence in adaptive control without persistency of excitation. In: Proceedings of 49th IEEE Conference on Decision and Control (CDC), Atlanta, GA, USA, pp 3674–3679
16. Valverde A, Tsiotras P (2018) Spacecraft trajectory tracking with identification of mass properties using dual quaternions. In: Proceedings of the AIAA Guidance, Navigation, and Control Conference, Kissimmee, FL, United states, p 1576
17. Mühlegg M, Chowdhary G, Johnson E (2012) Concurrent learning adaptive control of linear systems with noisy measurements. In: Proceedings of the AIAA Guidance, Navigation, and Control Conference, AIAA, Minneapolis, MN, USA, p 4669

18. Parikh A, Kamalapurkar R, Dixon WE (2019) Integral concurrent learning: Adaptive control with parameter convergence using finite excitation. International Journal of Adaptive Control and Signal Processing 33(12): 1775–1787
19. Cho N, Shin HS, Kim Y, Tsourdos A (2017) Composite model reference adaptive control with parameter convergence under finite excitation. IEEE Transactions on Automatic Control 63(3): 811–818
20. Pan Y, Yu H (2018) Composite learning robot control with guaranteed parameter convergence. Automatica 89: 398–406
21. Sastry S, Bodson M (2011) Adaptive control: stability, convergence and robustness. Dover Publications, Mineola, NY, USA
22. Aranovskiy S, Bobtsov A, Ortega R, Pyrkin A (2016) Performance enhancement of parameter estimators via dynamic regressor extension and mixing. IEEE Transactions on Automatic Control 62(7): 3546–3550
23. Ortega R, Aranovskiy S, Pyrkin AA, Astolfi A, Bobtsov AA (2020) New results on parameter estimation via dynamic regressor extension and mixing: Continuous and discrete-time cases. IEEE Transactions on Automatic Control 66(5): 2265–2272
24. Seo D, Akella MR (2008) High-performance spacecraft adaptive attitude-tracking control through attracting-manifold design. Journal of Guidance, Control, and Dynamics 31(4): 884–891
25. Korotina M, Aranovskiy S, Ushirobira R, Vedyakov A (2020) On parameter tuning and convergence properties of the DREM procedure. In: Proceedings of European Control Conference, Saint Petersburg, Russia, pp 53–58
26. Chowdhary G, Johnson E (2008) Theory and flight test validation of long term learning adaptive flight controller. In: Proceedings of the AIAA Guidance, Navigation and Control Conference and Exhibit, AIAA, Honolulu, Hawaii, p 6781
27. Chowdhary G, Yucelen T, Mühlegg M, Johnson EN (2013) Concurrent learning adaptive control of linear systems with exponentially convergent bounds. International Journal of Adaptive Control and Signal Processing 27(4): 280–301

Chapter 11
Reinforcement Learning-Based Pose Control of Spacecraft Under Motion Constraints

11.1 Introduction

Spacecraft Autonomous RPOs are an enabling technology for a broad range of space missions, such as on-orbit servicing, satellite inspection, sample retrieval, active debris removal, and asteroid exploration [1–7]. As the primary requirement of these missions, spacecraft flying safety must be guaranteed during RPOs. This requires the spacecraft to obey multiple complex motion constraints, where the most important two types of constraints are referred as approaching corridor constraint and sensor field-of-view (FOV) constraint in literature. More specifically, a certain safe approaching corridor is introduced to restrict the translational motion trajectory of the pursuer, such that the pursuer does not collide with any components of target spacecraft [8]; while the sensor field-of-view constraint arises from the requirement of autonomous rendezvous and capture sensor system (ARCSS) onboard the puruser [9], that is, the detectable zone of ARCSS must always cover the target to provide real-time relative state information for the pursuer's AOCS.

Due to the practical significance of RPOs, guidance and control methods for spacecraft RPOs with the strong constraint-handling capability have aroused extensive attention from both aerospace industry and academia. In this respect, various guidance and control methods have been reported in the literature, such as artificial potential function (APF) methods [10–15], inverse dynamics in the virtual domain (IDVD) [16], optimization-based method [17, 18], model predictive control (MPC) [16, 19, 20], etc. Generally speaking, these methods can be categorized into two main types: APF-based methods and optimization-based methods. The APF-based methods usually establish virtual high-potential areas for obstacle zones that produce repulsive forces to avoid collision with any obstacles. In [10], a dual-quaternion-based APF was designed to solve the six-degree-of-freedom (6-DOF) pose control problem for spacecraft constrained maneuvers, and the local minimum problem inherent in the APF-based methods was addressed through a judicious selection of control parameters satisfying a mild condition. Zappulla et al. [14] proposed an adaptive APF approach to achieve spacecraft collision-free spacecraft RPOs, and carried out

Q. Hu et al., *Intelligent Autonomous Control of Spacecraft with Multiple Constraints*,
https://doi.org/10.1007/978-981-99-0681-9_11

a hardware-in-loop (HIL) experiments to valid the algorithm practical effectiveness. Huang et al. [11] designed a finite-time control law with full-state constraints by incorporating the tan-type barrier Lyapunov function. It should be pointed out that although the APF-based methods is shown to have strong constraint-handing capability, most existing results usually lack of optimization abilities. They cannot make the balance between the control performance and control cost, resulting in potentially high control costs that are unacceptable for on-orbit RPOs. In this context, constrained optimal control (COC) is a promising alternative solution. A second order cone programming (SOCP) based method was presented in [17] for the spacecraft rendezvous and docking with approaching corridor constraints. However, this method is open-loop and cannot deal with real-time feedback. Although an MPC-based COC approach was proposed in [20] with the capability of feedback control, the receding-horizon characteristic of MPC makes the COC problem solving in real-time become a computationally burdensome task, especially for onboard micro processor which only have very limited computing resources. Thus, it is of significant importance to design a new constrained optimal control scheme, which can efficiently achieve 6-DOF optimal pose tracking for spacecraft RPOs, while strictly obeying the underlying motion constraints.

Theoretically speaking, the optimal control of nonlinear systems usually requires to solve the Hamilton-Jacobi-Bellman (HJB) equation. This is a challenging task and even an accurate numerical solution is hard to be obtained [21]. Besides, the highly nonlinear and coupled model of the 6-DOF dynamics of spacecraft also significantly increase difficulties for this nontrivial task. The reinforcement learning (RL) technique is a new and promising tool to address this challenging problem. RL-based control, which is commonly referred as approximate/adaptive dynamic programming (ADP) [22–24] in the literature, is a powerful data-driven method to solve the optimal control problems of nonlinear systems. The basic idea of RL/ADP-based control is to employ special approximators (such as neural networks) to approximate the cost function and optimal control strategy, and measurement data is implemented in the training process of these approximators. There are many pioneering theoretical works that have emerged based on the ADP framework for the optimal control of various systems [25–27]. However, the constraint-handling capability of these notable results are still immature [28]. They cannot be straightforwardly extended to solve the COC problems for spacecraft RPOs.

Motivated by the above facts, a novel RL-based controller with constraint-handling capabilities is developed in this chapter for spacecraft autonomous RPOs. Two kinds of motion constraints are considered during the whole control process, including the approaching corridor constraint and the sensor field-of-view constraint. Compared with the traditional 6-DOF modeling methods of spacecraft motion [2, 11, 29, 30], dual quaternion can accurately represent the spacecraft 6-DOF motion dynamics while considering the coupling between the rotational and transitional motions. With this in mind, the dual quaternion formalism is employed in this chapter to describe the 6-DOF relative motion of the pursuer with respect to the target. Then, a special dual quaternion based reward function is designed, which not only represents a trade-off between control performance and control cost but also can encode the constraint information into the controller. Besides, by making full use of the underlying properties of dual quaternion to the 6-DOF coupling motion description, an RL-based online learning algorithm is proposed to approximate the optimal control policy, which not only improves the closed-loop performance, but also ensure the satisfaction of underlying motion constraints. Lyapunov stability analysis shows the stability of the closed-loop system. To the best knowledge of the authors, this is the first time an RL-based controller is proposed under the dual-quaternion formalism and applied to the 6-DOF constrained optimal control design for spacecraft constrained RPOs. The advantages of the proposed RL-based pose control scheme are three-fold:

- The proposed method is capable of simultaneously dealing with motion constraints, control performance, and computational efficiency;
- Dual quaternions are used to accurately establish the 6-DOF spacecraft relative motion dynamics in a compact form, such that the algorithm design is more compact. Moreover, the feasibility and applicability of the model-based method can be significantly improved;
- The proposed method is able to rapidly endow a traditional and easy-to-implement controller with optimization and constraint-handling capabilities by online tuning the network weights.

We also show that an easy-to-implement controller under dual-quaternion formulation can be employed as the initial control policy to trigger the learning process, and its boundedness is proved by a special Lyapunov strictification method.

The rest of this chapter is organized as follows. In Sect. 11.2, the concept and operations of the dual quaternion are introduced, and the RPOs control problem is formulated based on the dual-quaternion model. Subsequently, Sect. 11.3 gives the design of the reward function and the development of the RL-based control scheme and the initial control policy. Numerical simulations and analysis for illustrating the superiority of the proposed method are presented in Sect. 11.4. Finally, this chapter is concluded in Sect. 11.5 by some concluding remarks.

11.2 Problem Formulation

In this chapter, dual quaternions are used to describe the pose of the pursuer with respect to (w.r.t.) the target. The reader is referred to Sect. 2.4.4 for the algebraic rules and properties of dual quaternions as well as the dual-quaternion-based spacecraft relative motion dynamics.

11.2.1 Motion Constraints

During the PROs, the pursuer spacecraft should comply with both the approaching corridor and FOV constraints. In this subsection, the aforementioned constraints are discussed in detail.

The FOV constraint is caused by the limited FOV of the optical instruments onboard the pursuer. To ensure the target can always be captured by the pursuer during the RPOs, the angle of line-of-sight (LOS) should be restricted [31]. In general, the FOV constraint can be defined as a cone around the LOS in the pursuer's body frame $\mathcal{P}$, as shown in Fig. 11.1. In Fig. 11.1, the unit vector $\boldsymbol{c}_s$ denotes the boresight of the vision sensor system in the frame $\mathcal{P}$, and α_s represents the maximum allowable LOS angle. To satisfy the FOV constraint, the angle between $\boldsymbol{c}_s$ and $-\boldsymbol{r}_{pt}$ should never be greater than α, which can be described as

$$\frac{-(\boldsymbol{r}_{pt}^{p})^\top}{\|\boldsymbol{r}_{pt}^{p}\|}\boldsymbol{c}_s \geq \cos\alpha_s. \tag{11.1}$$

Aided by the property of quaternions, one has:

$$(\boldsymbol{r}_{pt}^{p})^\top \boldsymbol{c}_s = (\boldsymbol{r}_{pt}^{p} \otimes \boldsymbol{q}_{pt})^\top (\boldsymbol{c}_s \otimes \boldsymbol{q}_{pt}), \tag{11.2}$$

then (11.1) can be further reformulated as:

$$c_1 = -\frac{\hat{\boldsymbol{q}}_{pt} \circ (\hat{\boldsymbol{\Theta}}_s \hat{\boldsymbol{q}}_{pt})}{\|2\varepsilon \circ \hat{\boldsymbol{q}}_{pt}^{s}\|} - \cos\alpha_s \geq 0, \tag{11.3}$$

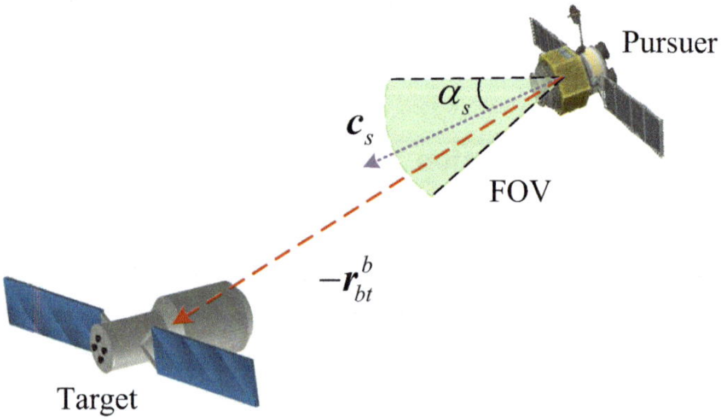

Fig. 11.1 Illustration of field-of-view constraint

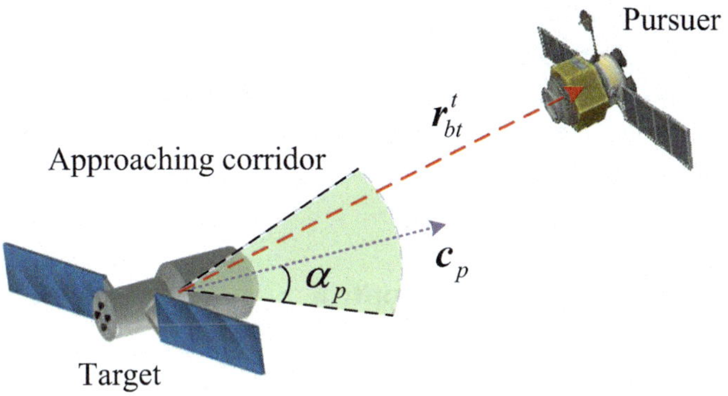

Fig. 11.2 Illustration of approaching corridor constraint

where $\hat{\boldsymbol{\Theta}}_s = \boldsymbol{\Theta}_s^\top \frac{d}{d\varepsilon} + \varepsilon \boldsymbol{\Theta}_s$ with

$$\boldsymbol{\Theta}_s = \begin{bmatrix} -\boldsymbol{S}(\boldsymbol{c}_s) & \boldsymbol{c}_s \\ -\boldsymbol{c}_s^\top & \boldsymbol{0} \end{bmatrix}.$$

Thus, when $c_1 \geq 0$, the FOV constraint can be guaranteed.

In the actual missions, the pursuer should keep in a preassigned zone to avoid obscuring the boresight of observing the target's docking port, as well as collisions with any components of the target. To achieve this, the pursuer should approach the docking port from a certain direction. The approaching corridor constraint is defined as a cone around the central axis (denoted as $\boldsymbol{c}_p$) of docking port that lies in the frame $\mathcal{T}$, as shown in Fig. 11.2. The half-angle of cone is represented by α_p. To satisfies this constraint, one should ensure the following:

$$\frac{(\boldsymbol{r}^t_{pt})^\top}{\|\boldsymbol{r}^t_{pt}\|} \boldsymbol{c}_p \geq \cos \alpha_p. \tag{11.4}$$

Then, (11.4) can be guaranteed by the following inequation:

$$c_2 = \frac{\hat{\boldsymbol{q}}_{pt} \circ (\hat{\boldsymbol{\Theta}}_p \hat{\boldsymbol{q}}_{pt})}{\|2\varepsilon \circ \hat{\boldsymbol{q}}^s_{pt}\|} - \cos \alpha_p \geq 0, \tag{11.5}$$

where $\hat{\boldsymbol{\Theta}}_p = \boldsymbol{\Theta}_p^\top \frac{d}{d\varepsilon} + \varepsilon \boldsymbol{\Theta}_p$ with

$$\boldsymbol{\Theta}_p = \begin{bmatrix} \boldsymbol{S}(\boldsymbol{c}_p) & \boldsymbol{c}_p \\ -\boldsymbol{c}_p^\top & \boldsymbol{0} \end{bmatrix}.$$

11.2.2 Control Objective

The control objective is to develop an online learning control scheme for spacecraft RPOs to achieve the control law evolution for performance optimization, while ensure the satisfaction of both approaching corridor and FOV constraints.

11.3 Learning-Based Pose Control

11.3.1 Reward Function Design

Before proceeding, we first discuss the reward function. The reward function is the feedback of the environment while agents are implementing the corresponding action. The use of a reward signal to formalize the idea of a goal is one of the most distinctive features of reinforcement learning [22]. The basic idea of the reward function design is that giving a high reward (present as a small value herein) to desired states and a low reward to the undesired states (a large value). Further, according to the analysis in Sect. 11.2, the reward function associated with the undesired state is designed as follows:

$$\Upsilon_s = -\beta_1(\hat{\boldsymbol{q}}_{pt} - \hat{\boldsymbol{q}}_I) \circ (\hat{\boldsymbol{Q}}_q(\hat{\boldsymbol{q}}_{pt} - \hat{\boldsymbol{q}}_I)) \log\left(\frac{c_1}{1 - \cos\alpha_s}\right), \tag{11.6}$$

$$\Upsilon_p = -\beta_2(\hat{\boldsymbol{q}}_{pt} - \hat{\boldsymbol{q}}_I) \circ (\hat{\boldsymbol{Q}}_q(\hat{\boldsymbol{q}}_{pt} - \hat{\boldsymbol{q}}_I)) \log\left(\frac{c_2}{1 - \cos\alpha_p}\right), \tag{11.7}$$

where $\hat{\boldsymbol{q}}_I = \boldsymbol{q}_I + \epsilon\boldsymbol{0}_{4\times1}, \boldsymbol{q}_I = [0, 0, 0, 1]^\top$ is identity quaternion, β_1, β_2 are the scale factors interpreted as the "level" of reward, and $\hat{\boldsymbol{Q}}_q = \boldsymbol{Q}_q\frac{d}{d\varepsilon} + \varepsilon\boldsymbol{Q}_q$ is the dual weight matrix. It is noted that, when the highest reward is obtained, the target is at the center of the FOV of vision sensor onboard the pursuer. Contrarily, the reward will rapid decline when the target close to the edge of the sensor FOV. Similarly, the center of the approaching corridor corresponds to the high reward and the edge corresponds to the low reward.

The desired states are set as the target's states, the relevant reward function defined by the form of error dual quaternion and dual angular velocity given by:

$$\Upsilon_{ds} = (\hat{\boldsymbol{q}}_{pt} - \hat{\boldsymbol{q}}_I) \circ (\hat{\boldsymbol{Q}}_q(\hat{\boldsymbol{q}}_{pt} - \hat{\boldsymbol{q}}_I)) + (\hat{\boldsymbol{\omega}}^b_{pt}) \circ (\hat{\boldsymbol{Q}}_\omega\hat{\boldsymbol{\omega}}^b_{pt}), \tag{11.8}$$

where $\hat{\boldsymbol{Q}}_\omega = \boldsymbol{Q}_\omega\frac{d}{d\varepsilon} + \varepsilon\boldsymbol{Q}_\omega$ is a dual weight matrix. The balance between the reward of dual quaternion and dual angular velocity can be adjusted by tuning the dual weight matrix $\hat{\boldsymbol{Q}}_q$, $\hat{\boldsymbol{Q}}_\omega$. Evidently, according to (11.8) the distance from target's states relate to the level of this reward.

Remark 11.1 It is noteworthy that, although the design ideas of reward functions is given a high "penalty" in the prohibited area, it is distinct from the APF-based method (e.g., [10, 14]), the control signal not only related to the current "penalty", but also to the throughout the whole process's "penalty", which will be reflected in the next part. Furthermore, the factors $(-1+\cos\alpha_s)$ and $(-1+\cos\alpha_p)$ are introduced in (11.6) and (11.7), respectively, for adjusting the logarithm operation maps into $[0,+\infty)$. Thus, there is no penalty at the most desired state (that is $\Upsilon_{s\backslash p}=0$).

Summing up the above analysis, the reward functions are constructed by (11.9), considering both desired and undesired states during the RPOs by mapping the states into the corresponding value.

$$\Upsilon = \underbrace{\Upsilon_{ds}}_{\text{desired states}} + \underbrace{\Upsilon_p + \Upsilon_s}_{\text{undesired states}}. \tag{11.9}$$

11.3.2 Optimal Control Solution Analysis

After designing the reward functions, the optimal control solution analysis will be discussed in this part. To formulate the optimal control problem, the dual-quaternion-based relative motion dynamics described by (2.80) and (2.81) are rewritten as the following compact form:

$$\dot{\hat{\boldsymbol{x}}} = \hat{\boldsymbol{F}} + \boldsymbol{G}\hat{\boldsymbol{u}}, \tag{11.10}$$

where $\hat{\boldsymbol{x}} = [\hat{\boldsymbol{e}}^\top, (\hat{\boldsymbol{J}}_p\hat{\boldsymbol{\omega}}_{pt})^\top]^\top$ is motion state with $\hat{\boldsymbol{e}} = \text{vec}(\hat{\boldsymbol{q}}^s_{pt} - \hat{\boldsymbol{q}}^s_I)$ and

$$\hat{\boldsymbol{F}} = \begin{bmatrix} \frac{1}{2}\text{vec}(\hat{\boldsymbol{q}}_{pt} \otimes \hat{\boldsymbol{\omega}}^p_{pt})^s \\ -\hat{\boldsymbol{\omega}}^p_{pt} \times (\hat{\boldsymbol{J}}_p\hat{\boldsymbol{\omega}}^p_{pt}) \end{bmatrix}, \boldsymbol{G} = \begin{bmatrix} \boldsymbol{0}_{3\times 3} \\ \boldsymbol{I}_3 \end{bmatrix}.$$

In space missions, the control cost is also a considerable factor due to the high cost of the energy. The control cost and state error should be both considered in policy design. Therefore, the cost-to-go function of the optimal control $V(\hat{\boldsymbol{x}})$ is defined as the integral of the non-negative reward function $r(\hat{\boldsymbol{x}}, \hat{\boldsymbol{u}}) = \Upsilon(\hat{\boldsymbol{x}}) + \hat{\boldsymbol{u}} \circ \hat{\boldsymbol{u}}$, as will be discussed in the following part.

$$V(\hat{\boldsymbol{x}}) = \int_t^\infty r(\hat{\boldsymbol{x}}(\iota), \hat{\boldsymbol{u}}(\iota))d\iota. \tag{11.11}$$

The optimal control policy is $\hat{\boldsymbol{u}}^*$ (if exists), thus the corresponding cost function is denoted by $V^*(\hat{\boldsymbol{x}})$. Then, $\hat{\boldsymbol{u}}^*$ satisfies

$$H\left(\hat{\boldsymbol{x}}, \hat{\boldsymbol{u}}^*, \hat{\nabla}_{\hat{\boldsymbol{x}}}V^*\right) = 0, \tag{11.12}$$

where the Hamilton equation is defined by

$$H\left(\hat{\boldsymbol{x}}, \hat{\boldsymbol{u}}^*, \hat{\nabla}_{\hat{x}} V^*\right) = \hat{\nabla}_{\hat{x}} V^* \circ (\hat{\boldsymbol{F}} + \boldsymbol{G}\hat{\boldsymbol{u}}) + r(\hat{\boldsymbol{x}}, \hat{\boldsymbol{u}}^*). \tag{11.13}$$

Taking the partial differential of (11.13), we can get the closed-form of $\hat{\boldsymbol{u}}^*$ as follows:

$$\hat{\boldsymbol{u}}^* = -\frac{1}{2}\boldsymbol{G}^\top \hat{\nabla}_{\hat{x}} V^*. \tag{11.14}$$

Substituting (11.14) into (11.13) leads to the following HJB equation:

$$\hat{\nabla}_{\hat{x}} V^* \circ \hat{\boldsymbol{F}} + \Upsilon - \frac{1}{4}(\boldsymbol{G}^\top \hat{\nabla}_{\hat{x}} V^*) \circ (\boldsymbol{G}^\top \hat{\nabla}_{\hat{x}} V^*) = 0. \tag{11.15}$$

One caveat here is that the high nonlinearity of the system model (11.10) increases the intractability of analytically solving the HJB equation (11.15). In the following subsection, an RL-based online controller will be designed to approximate the optimal solution $\boldsymbol{u}^*$ in the following part.

11.3.3 Online Learning Control Algorithm

As discussed in Sect. 11.3.3, a high nonlinearity of the cost function (11.11) makes the HJB equation (11.15) hard to solve. Approximation emerges as a way to deal with such a problem. According to the Weierstrass Approximation theorem [25, 32], a NN that contains a sufficient set of basis functions can be employed to approximate the optimal cost function (11.11), given by

$$V^*(\hat{\boldsymbol{x}}) = \boldsymbol{w}^\top \boldsymbol{\sigma}(\hat{\boldsymbol{x}}) + \epsilon(\hat{\boldsymbol{x}}), \tag{11.16}$$

for $\hat{\boldsymbol{x}} \in \mathcal{X}$, where $\mathcal{X} \subset \hat{\mathbb{R}}^6$ is a compact set. The basis function vector, denoted by $\boldsymbol{\sigma}(\hat{\boldsymbol{x}}) = \left[\sigma_1(\hat{\boldsymbol{x}}), \sigma_2(\hat{\boldsymbol{x}}), \ldots, \sigma_p(\hat{\boldsymbol{x}})\right]^\top \in \mathbb{R}^p$, satisfies

$$\sigma_i(\hat{\boldsymbol{0}}_{6\times 1}) = 0,\ \dot{\sigma}_i(\hat{\boldsymbol{0}}_{6\times 1}) = 0,\ i = 1, 2, \ldots, p.$$

The weight vector of basis function $\boldsymbol{w}$ is a unknown constant vector, and $\epsilon(\hat{\boldsymbol{x}}) \in \mathbb{R}$ is the reconstruction error. Then (11.14) can be equivalently rewritten as

$$\hat{\boldsymbol{u}}^* = -\frac{1}{2}\boldsymbol{G}^\top (\hat{\nabla}_{\hat{x}} \boldsymbol{\sigma}(\hat{\boldsymbol{x}}) \omega + \hat{\nabla}_{\hat{x}} \epsilon(\hat{\boldsymbol{x}})). \tag{11.17}$$

Based on the RL technique, the function of actor-critic is to online approximate the weigh vector $\boldsymbol{w} \in \mathbb{R}^p$. Then, a weight estimation vector $\boldsymbol{w}_{est}$ was employed to construct the estimates of the cost function and control policy:

$$V(\hat{\boldsymbol{x}}, \boldsymbol{w}_{est}) = \boldsymbol{w}_{est}^\top \boldsymbol{\sigma}(\hat{\boldsymbol{x}}), \tag{11.18}$$

$$\hat{\boldsymbol{u}} = -\frac{1}{2}\boldsymbol{G}^\top \hat{\nabla}_{\hat{x}} \boldsymbol{\sigma} \boldsymbol{w}_{est}. \tag{11.19}$$

Subsequently, we further consider the following Bellman error:

$$\delta_b = \hat{\nabla}_{\hat{x}} V \circ (\hat{\boldsymbol{F}} + \boldsymbol{G}\hat{\boldsymbol{u}}) + r(\hat{\boldsymbol{x}}, \hat{\boldsymbol{u}}). \tag{11.20}$$

Recalling (11.15), the Bellman error can be rewritten as

$$\begin{aligned} \delta_b &= \delta_b - H\left(\hat{\boldsymbol{x}}, \hat{\boldsymbol{u}}^*, \hat{\nabla}_{\hat{x}} V^*\right) \\ &= \tilde{\boldsymbol{w}}^\top \boldsymbol{\vartheta} + \epsilon_\delta, \end{aligned} \tag{11.21}$$

where $\boldsymbol{\vartheta} = \hat{\nabla}_{\hat{x}} \boldsymbol{\sigma} \circ (\hat{\boldsymbol{F}} + \boldsymbol{G}\hat{\boldsymbol{u}})$ is defined for expressing simplicity, $\tilde{\boldsymbol{w}} = \boldsymbol{w}_{est} - \boldsymbol{w}$ is the weight error, and ϵ_δ denotes the induced reconstruction error.

It is noted that (11.21) contains the information of $\tilde{\boldsymbol{w}}$, it has been commonly employed to design the learning law of the estimated weight $\boldsymbol{w}_{est}$. Specially, not only the real-time information of δ_b but also the past measurements are utilized. Before proceeding further, we make the following assumptions.

Assumption 11.1 For $x \in \mathcal{X}$, there exist positive constants b_σ, b_{∇_σ} and b_{ϵ_δ}, such that, $\|\sigma\| \leq b_\sigma$, $\|\nabla_x \sigma\| \leq b_{\nabla\sigma}$, and $\epsilon_\delta \leq b_{\epsilon_\delta}$.

Assumption 11.2 For an auxiliary variable defined by $\boldsymbol{\eta} = \boldsymbol{\vartheta}/(\boldsymbol{\vartheta}^\top \boldsymbol{\vartheta} + 1)$, it satisfies a strictly weak IE condition (see Definition 3.1 or [33] for the detailed definition of IE condition), i.e., there exist t_{k1}, t_{k2}, γ_w with $0 \leq t_{k1} \leq t_{k2} \leq t$ and γ_w such that

$$\int_{t_{k1}}^{t_{k2}} \boldsymbol{\eta}(\tau)\boldsymbol{\eta}^\top(\tau) d\tau \geq \gamma_w \boldsymbol{I}_p.$$

Remark 11.2 Assumption 11.1 is a standard assumption, while Assumption 11.2 is much weaker than the restrictive PE condition commonly used in the online RL-based controllers in [34].

Afterward, we introduced an auxiliary variable $\boldsymbol{\Psi}$ to utilize the online data designed as follows:

$$\boldsymbol{\Psi}(t, t_{k2}, t_{k1}) = \boldsymbol{\psi}_1(t_{k2}, t_{k1})\hat{\boldsymbol{w}}_{est} + \boldsymbol{\psi}_2(t_{k2}, t_{k1}), \tag{11.22}$$

with

$$\dot{\psi}_1(t, t_{k1}) = -\kappa\psi_1(t, t_{k1}) + \varphi_1(t), \tag{11.23}$$

$$\dot{\psi}_2(t, t_{k1}) = -\kappa\psi_2(t, t_{k1}) + \varphi_2(t), \tag{11.24}$$

where $\psi_1(t_{k1}) = \mathbf{0}_{p\times p}, \psi_2(t_{k1}) = \mathbf{0}_{p\times 1}, \varphi_1 = \boldsymbol{\eta}\boldsymbol{\eta}^\top, \varphi_3 = \boldsymbol{\eta}/(\boldsymbol{\vartheta}^\top\boldsymbol{\vartheta} + 1), \varphi_2 = r\varphi_3$, and κ is a positive constant. According to (11.22)–(11.24), one has

$$\begin{aligned}\boldsymbol{\Psi}(t, t_{k2}, t_{w1}) &= \int_{t_{k1}}^{t_{k2}} \mathrm{e}^{\kappa(\tau - t_{k2})}(\varphi_1(\tau)\boldsymbol{w}_{est} + \varphi_2(\tau))d\tau \\ &= \boldsymbol{Y}(t_{k2}, t_{k1})\tilde{\boldsymbol{w}} + \epsilon_\Psi,\end{aligned} \tag{11.25}$$

where $\boldsymbol{Y}(t_{k2}, t_{k1}) = \int_{t_{k1}}^{t_{k2}} \mathrm{e}^{\kappa(\tau - t_{k2})}\varphi_1 d\tau$ is an information matrix, which "stores" the information of $\boldsymbol{\eta}$ throughout the time interval $[t_{k1}, t_{k2}]$ and the residual error vector is denoted by $\epsilon_\Psi = \int_{t_{k1}}^{t_{k2}} \mathrm{e}^{\kappa(\tau - t_{k2})}\epsilon_\delta\boldsymbol{\vartheta}/(\boldsymbol{\vartheta}^\top\boldsymbol{\vartheta} + 1)^2 d\tau$. Furthermore, under Assumption 11.2, one has $\boldsymbol{Y}(t_{k2}, t_{k1}) \geq \mathrm{e}^{-\kappa(t_{k2} - t_{k1})}\gamma_w\boldsymbol{I}_m = \gamma_\Psi\boldsymbol{I}_m$.

Introducing the above auxiliary variable (11.25) into the learning law of $\boldsymbol{w}_{est}$ is significantly beneficial to improve learning efficiency. Nevertheless, considering $\boldsymbol{Y}(t_{k2}, t_{k1})$ is positive-define only one has sufficient online data is collected, it is necessary to ensure the boundedness of state first. With regard to this, the following theorem is given as a solution.

Theorem 11.1 *Consider the system defined in (11.10), and the policy defined by (11.19). With Assumption 11.2, design the learning law for* $\boldsymbol{w}_{est}$ *as*

$$\dot{\boldsymbol{w}}_{est} = -\gamma_1\delta_b\varphi_3 - \gamma_2\boldsymbol{\Psi}(t, t_{k2}, t_{k1}), \tag{11.26}$$

where $\gamma_1, \gamma_2 > 0$ *are constants. Then, the weight estimation error* $\tilde{\boldsymbol{w}}$ *as well as the states* $\hat{\boldsymbol{q}}_{pt} - \hat{\boldsymbol{q}}_I$ *and* $\hat{\boldsymbol{\omega}}_{pt}$ *are ultimately bounded, if the condition (11.29) holds.*

Proof Consider the following storage function:

$$L = V^* + \frac{a_1}{2}\tilde{\boldsymbol{w}}^\top\tilde{\boldsymbol{w}}, \tag{11.27}$$

where $a_1 > 0$ is a design constant. Taking the time derivative of (11.27) along (11.11) and (11.26) yields:

$$\begin{aligned}\dot{L} &= \hat{\nabla}_{\hat{\boldsymbol{x}}} V \circ (\hat{\boldsymbol{F}} + \boldsymbol{G}\hat{\boldsymbol{u}}) + a_1\tilde{\boldsymbol{w}}^\top\dot{\tilde{\boldsymbol{w}}} \\ &= -r - \frac{1}{2}\boldsymbol{w}^\top\boldsymbol{\Gamma}\tilde{\boldsymbol{w}} - \frac{1}{4}\boldsymbol{w}^\top\boldsymbol{\Gamma}\boldsymbol{w} + a_1\tilde{\boldsymbol{w}}^\top\dot{\tilde{\boldsymbol{w}}} + \epsilon_1 \\ &\leq -r + \frac{1}{2}\tilde{\boldsymbol{w}}^\top\boldsymbol{\Gamma}\tilde{\boldsymbol{w}} + a_1\tilde{\boldsymbol{w}}^\top\dot{\tilde{\boldsymbol{w}}} + \epsilon_2 \\ &\leq -r - \tilde{\boldsymbol{w}}^\top\boldsymbol{M}\tilde{\boldsymbol{w}} + \epsilon_3,\end{aligned} \tag{11.28}$$

wherein

$$\epsilon_1 = -0.5(\boldsymbol{G}^\top \hat{\nabla}_{\hat{x}} \epsilon) \circ (\boldsymbol{G}^\top \hat{\nabla}_{\hat{x}} \boldsymbol{\sigma} \tilde{\boldsymbol{w}}) + 0.25(\boldsymbol{G}^\top \hat{\nabla}_{\hat{x}} \epsilon) \circ (\boldsymbol{G}^\top \hat{\nabla}_{\hat{x}} \epsilon),$$
$$\epsilon_2 = 0.5(\boldsymbol{G}^\top \hat{\nabla}_{\hat{x}} \epsilon) \circ (\boldsymbol{G}^\top \hat{\nabla}_{\hat{x}} \epsilon),$$
$$\epsilon_3 = \epsilon_2 + 0.5a_1\gamma_2\gamma_\Phi\epsilon_\delta^2 + 0.5a_1\gamma_1\epsilon_\delta^2/(\boldsymbol{\vartheta}^\top\boldsymbol{\vartheta} + 1).$$

In addition, $\boldsymbol{\Gamma}=(\boldsymbol{G}^\top \hat{\nabla}_{\hat{x}} \boldsymbol{\sigma}) \circ (\boldsymbol{G}^\top \hat{\nabla}_{\hat{x}} \boldsymbol{\sigma})$, and $\boldsymbol{M}=0.5(-\boldsymbol{\Gamma} + a_1\gamma_1\boldsymbol{\eta}\boldsymbol{\eta}^\top + a_1\gamma_2\gamma_\Psi \boldsymbol{I}_m)$. Recalling Assumption 11.1, it can be deduced that $\|\boldsymbol{\eta}\| < \frac{1}{2}$ and $\|\boldsymbol{\Gamma}\| < b_\Gamma$, where b_Γ is the a positive constant. Thus, by adjusting γ_1, γ_2 and a_1 to satisfy

$$a_1 > \frac{b_\Gamma}{\gamma_1 + 4\gamma_2\gamma_\Psi}, \tag{11.29}$$

one has $\boldsymbol{M} > 0$. Then, it can be concluded from (11.28) that $\tilde{\boldsymbol{w}}$ as well as $\hat{\boldsymbol{q}}_{pt} - \hat{\boldsymbol{q}}_I$ and $\hat{\omega}_{pt}$ are ultimately bounded, thus completing the proof. □

Remark 11.3 Note that, constant a_1 is a coefficient of $\tilde{\boldsymbol{w}}^\top \tilde{\boldsymbol{w}}$ which is employed just for convergence analysis purpose. Therefore, we do not need to set a value to it. As long as there is an a_1 that satisfies the condition (11.29), a function then can be constructed to guarantee the convergence of the entire system. Hence, in practical applications, parameters γ_1 and γ_2 can be chosen by an empirical way according to the actual situation.

11.3.4 Initial Control Policy

An initial control policy is required to ensure the system states to a compact set $\mathcal{X}$, and it must be represented by the basis functions $\boldsymbol{\sigma}(\hat{\boldsymbol{x}})$. For the spacecraft RPOs problem considered herein, the initial policy designed in (11.30) has the capability of meeting the requirement.

$$\hat{\boldsymbol{u}}_{init} = \hat{k}_p \odot \hat{\boldsymbol{e}} - \hat{k}_d \odot (\hat{\omega}_{pt}^p)^s. \tag{11.30}$$

The coefficients of PD controller are positive constants denoted by $\hat{k}_p = k_{pr} + \varepsilon k_{pd}$, $\hat{k}_d = k_{dr} + \varepsilon k_{dd}$. This PD-like controller can guarantee the asymptotic convergence of system states (though it lacks optimizing and constraint handling abilities). What's more, it can be reconstructed by the following subset:

$$\hat{\nabla}_{\hat{x}} \boldsymbol{\sigma}_{pd}(\hat{\boldsymbol{x}}) = \begin{bmatrix} \operatorname{diag}\left\{[(\boldsymbol{v}_{pt}^p)^\top, (\boldsymbol{\omega}_{pt}^p)^\top]\right\}, & \operatorname{diag}\left\{[\boldsymbol{e}_r^\top, \boldsymbol{e}_d^\top]\right\} \\ \boldsymbol{0}_{6\times 6}, & \operatorname{diag}\left\{[(\boldsymbol{v}_{pt}^b)^\top, (\boldsymbol{\omega}_{pt}^p)^\top]\right\} \end{bmatrix}. \tag{11.31}$$

Thus, the corresponding weights are set to be

$$\boldsymbol{w}_{pd} = [k_{pr}\mathbf{1}_{1\times 3}, k_{pd}\mathbf{1}_{1\times 3}, k_{dr}\mathbf{1}_{1\times 3}, k_{dd}\mathbf{1}_{1\times 3}]^\top,$$

where $\boldsymbol{e}_r$ and $\boldsymbol{e}_d$ represent the real and dual parts of $\hat{\boldsymbol{e}}$, respectively. Herein, the "vector" presentation (like the formulation in [10]) of dual number is used to equivalently construct the basis function.

Remark 11.4 The initial controller is designed based on the dual-quaternion framework. It is distinct to the PD-like control scheme proposed in [10, 35], in which the dual-quaternion error term is denoted by $\mathrm{vec}(\hat{\boldsymbol{q}}_{pt}^{*} \otimes (\hat{\boldsymbol{q}}_{pt} - \hat{\boldsymbol{q}}_{I})^{s})$. Thus, this term is not suit for the initial policy in this framework (this point will be mentioned in Remark 11.6). To deal with this problem, we redesign the PD-like initial controller and give the proof by employing a special Lyapunov strictification method.

Theorem 11.2 *Consider the system described by (2.80) and (2.81). The initial policy designed in (11.30) can guarantee that* $\lim_{t\to\infty} \hat{\boldsymbol{q}}_{pt}(t) = \hat{\boldsymbol{q}}_{I}$ *and* $\lim_{t\to\infty} \hat{\boldsymbol{\omega}}_{pt}^{p}(t) = \hat{\boldsymbol{0}}_{3\times 1}$.

Proof Before proving the above theorem, some algebraic properties of dual quaternions are listed below:

$$\hat{\boldsymbol{a}}^{s} \circ (\hat{\boldsymbol{a}} \times \hat{\boldsymbol{b}}) = \hat{\boldsymbol{0}}_{3}, \quad \hat{\boldsymbol{a}}, \hat{\boldsymbol{b}} \in \hat{\mathbb{R}}^{3}, \tag{11.32}$$

$$\hat{\boldsymbol{q}}_{1} \circ \left(\hat{\boldsymbol{q}}_{2} \otimes \hat{\boldsymbol{q}}_{3}\right) = \hat{\boldsymbol{q}}_{3}^{s} \circ \left[\hat{\boldsymbol{q}}_{2}^{*} \otimes \left(\hat{\boldsymbol{q}}_{1}^{s}\right)\right], \quad \hat{\boldsymbol{q}}_{1}, \hat{\boldsymbol{q}}_{2}, \hat{\boldsymbol{q}}_{3} \in \hat{\mathbb{Q}}. \tag{11.33}$$

The detailed proofs of these properties are given in [36]. To analyze the stability of the closed-loop system, we consider the following Lyapunov-like function candidate:

$$V_{I} = \varrho \hat{k}_{p}^{s} \odot (\hat{\boldsymbol{q}}_{pt} - \hat{\boldsymbol{q}}_{I}) \circ (\hat{\boldsymbol{q}}_{pt} - \hat{\boldsymbol{q}}_{I}) + \frac{\varrho}{2} (\hat{\boldsymbol{\omega}}_{pt}^{p})^{s} \circ (\hat{\boldsymbol{J}}_{p} \hat{\boldsymbol{\omega}}_{pt}^{p}) + N_{I}, \tag{11.34}$$

where $N_{I} = [2\varepsilon\, \mathrm{vec}(\hat{\boldsymbol{q}}_{pt}^{*} \otimes (\hat{\boldsymbol{q}}_{pt} - \hat{\boldsymbol{q}}_{I})^{s})] \circ [\varepsilon(\hat{\boldsymbol{J}}_{p} \hat{\boldsymbol{\omega}}_{pt}^{p})]$ is a cross term defined just for prove purpose. By applying the Binet–Cauchy identity of cross product along with Cauchy–Schwarz inequality, one has:

$$\begin{aligned} V_{I} \geq & \, 2\varrho k_{pd}(1 - q_{0}) + \frac{1}{2} \varrho k_{dd} (\boldsymbol{\omega}_{pt}^{p})^{\top} \boldsymbol{J}_{p} \boldsymbol{\omega}_{pt}^{p} \\ & + (\varrho \frac{k_{pr}}{4} - \frac{m_{p}}{2}) \|\boldsymbol{r}_{pt}^{p}\|^{2} + \frac{m_{p}}{2} (\varrho k_{dr} - 1) \|\boldsymbol{v}_{pt}^{p}\|^{2}. \end{aligned} \tag{11.35}$$

Set $\varrho > \max\left\{2m_{p}/k_{pr},\ 1\right\}$. Then, (11.35) guarantees $V_{I} \geq 0$, and $V_{I} = 0$ only when $\hat{\boldsymbol{q}}_{pt} = \hat{\boldsymbol{q}}_{I}$ and $\hat{\boldsymbol{\omega}}_{pt}^{p} = \hat{\boldsymbol{0}}_{3\times 1}$. As such, V_{I} can be regarded as a valid Lyapunov-like function candidate. By applying the properties (11.32) and (11.33) and substituting (2.80) and (2.81) into (11.30), the time derivative of V_{I} becomes

$$\begin{aligned}\dot{V}_I =& \varrho(\hat{\omega}^p_{pt})^s \circ [-\hat{k}^s_d(\hat{\omega}^b_{pt})^s + ((1-q_0)\boldsymbol{I}_3 + \boldsymbol{q}^\times)\boldsymbol{r}^p_{pt} + \varepsilon\boldsymbol{0}_{3\times1}] \\ &- k_{pr}q_0\|\boldsymbol{r}^p_{pt}\|^2 - k_{dr}(\boldsymbol{r}^p_{pt})^\top\boldsymbol{v}^p_{pt} + m_p\|\boldsymbol{v}^p_{pt}\|^2 \\ =& -(\varrho k_{dr} - 1)m_p\|\boldsymbol{v}^p_{pt}\|^2 - \varrho k_{dd}\|\boldsymbol{\omega}^p_{pt}\|^2 - k_{pr}q_0\|\boldsymbol{r}^p_{pt}\|^2 \\ &+ (\boldsymbol{r}^p_{pt})^\top[(\varrho - \varrho q_0 - k_{dr})\boldsymbol{I}_3 + \varrho\boldsymbol{q}^\times_v]\boldsymbol{v}^b_{pt} \\ \leq& -\varrho k_{dd}\|\boldsymbol{\omega}^p_{pt}\|^2 - (\varrho m_p k_{dr} - m - \frac{\mu_M}{2})\|\boldsymbol{v}^p_{pt}\|^2 \\ &- (k_{pr}q_0 - \frac{\mu_M}{2})\|\boldsymbol{r}^p_{pt}\|^2,\end{aligned} \tag{11.36}$$

with $\mu_M = \|(\varrho - \varrho q_0 - k_{dr})\boldsymbol{I}_3 + \varrho\boldsymbol{q}^\times_v\|$. Thus, by adjusting ϱ, $\hat{k}_p$ and $\hat{k}_d$ to satisfy $k_{pr}q_0 - \frac{\mu_M}{2} \geq 0$ and $\varrho m_p k_{dr} - m_p - \frac{\mu_M}{2} \geq 0$, one has $\dot{V}_I \leq 0$, and $V_I = 0$ only when $\hat{\boldsymbol{q}}_{pt} = \hat{\boldsymbol{q}}_I$, $\hat{\omega}^p_{pt} = \hat{\boldsymbol{0}}_{3\times1}$. According to the Barbalat's lemma [37], it can be concluded that $\lim_{t\to\infty}\hat{\boldsymbol{q}}_{pt}(t) = \hat{\boldsymbol{q}}_I$ and $\lim_{t\to\infty}\hat{\omega}^p_{pt}(t) = \hat{\boldsymbol{0}}_{3\times1}$. □

Remark 11.5 The initial policy is given as a PD-like controller for its simplicity and effectiveness, moreover, it's easy to be reconstructed by a set of simple basis functions. But the initial policy is not limited to the PD-like controller, and the basis is not limited to the simple polynomial-type basis functions. As long as an appropriate set of basis function is designed for reconstructing a given controller, it can evolve into the (sub)optimal controller during the control processing.

Remark 11.6 Reconstructing the initial control policy is tricky work, the basis functions are chosen according to the initial controller design, because it needs states are bounded at the initial stage. So the elements of the basis function should contain the element of the initial controller (such as the terms of PD). The initial controller (11.30) allows us to select the basis functions more convenient. We can employ the terms of the initial controller as a part of basis functions. Then, some other basis functions can be appropriately added to improve the performance of learning. Note that, it is not recommended to use the terms independent of $\hat{\omega}^b_{pt}$ as basis functions here, which will vanish after multiply by $\boldsymbol{G}$.

According to the above analyses, the pose control method proposed in this chapter can be intuitively summarized by a diagram shown in Fig. 11.3.

11.4 Numerical Simulations

In this section, numerical simulations are carried out to illustrate the effectiveness and superiority of the proposed method. The control objective is to drive the pursuer spacecraft to the desired position and attitude relative to the target spacecraft. In the following simulations, the mass and inertia parameters of the pursuer are $m_p = 15$ kg and $\boldsymbol{J}_p = \text{diag}[20.8, 21.1, 32.6]\,\text{kg}\cdot\text{m}^2$, respectively. The structure of the NN is given in (11.31).

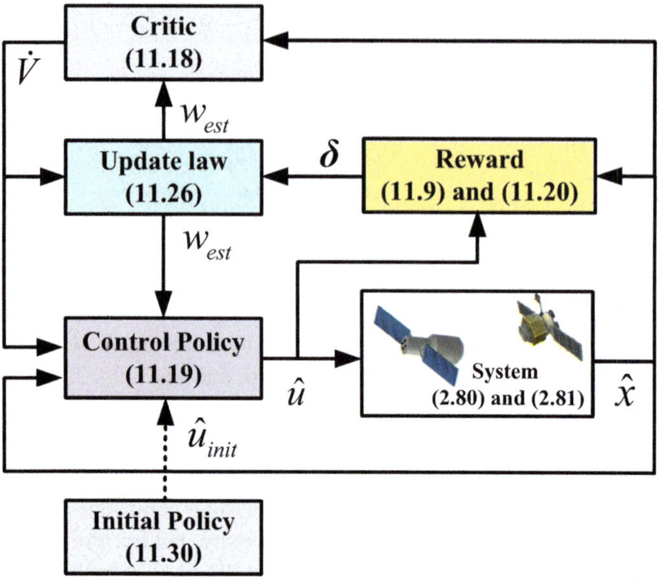

Fig. 11.3 Structure of the system

11.4.1 Point to Point Maneuvers Without Constraints

In this subsection, we consider a point-to-point scenario, in which the pursuer is required to approach a desired pose. The initial relative position and attitude are $\boldsymbol{r}^t_{pt/0} = [500, -185, 163]^\top$m and $\boldsymbol{q}_{pt/0} = [0.3426, -0.2764, 0.1918, 0.8772]^\top$. The desired relative position and attitude are $\boldsymbol{r}^t_{pt/des} = [0, 0, 0]^\top$m and $\boldsymbol{q}_{pt/des} = [0, 0, 0, 1]^\top$, which render $\hat{\boldsymbol{q}}_{pt/des} = [0, 0, 0, 1]^\top + \varepsilon[0, 0, 0, 0]^\top$. Furthermore, the initial relative dual angular velocity of the pursuer w.r.t. the target is assumed to $\hat{\boldsymbol{\omega}}_{pt/0} = [0, 0, 0]^\top + \varepsilon[0, 0, 0]^\top$, and the target is considered to be stationary. The cost function is chosen as: $r(\hat{\boldsymbol{q}}_{pt}, \hat{\boldsymbol{\omega}}_{pt}, \hat{\boldsymbol{u}}_b) = (\hat{\boldsymbol{q}}_{pt} - \hat{\boldsymbol{q}}_I) \circ (\hat{\boldsymbol{Q}}_q(\hat{\boldsymbol{q}}_{pt} - \hat{\boldsymbol{q}}_I)) + (\hat{\boldsymbol{\omega}}^p_{pt}) \circ (\hat{\boldsymbol{Q}}_\omega \hat{\boldsymbol{\omega}}^p_{pt}) + \hat{\boldsymbol{u}} \circ \hat{\boldsymbol{u}}$, where $\hat{\boldsymbol{Q}}_q = \boldsymbol{I}_4 \frac{d}{d\varepsilon} + \varepsilon 2\boldsymbol{I}_4$ and $\hat{\boldsymbol{Q}}_\omega = 5\boldsymbol{I}_3 \frac{d}{d\varepsilon} + \varepsilon 10\boldsymbol{I}_3$.

The control parameters for the initial PD controller are set as: $\hat{k}_p = 0.1 + \varepsilon 0.1$, and $\hat{k}_d = 5 + \varepsilon 5$. For comparison purpose, the PD-like controller is also simulated. To provide a fair comparison and demonstrate the learning results, the initial control scheme is the same as the PD-like controller. The dynamic responses of the proposed controller and initial controller are shown in Figs. 11.4, 11.5 and 11.6 and Figs. 11.7, 11.8 and 11.9, respectively. From these figures, it is clear that both controllers successfully achieve the given control objectives. However, under the proposed controller, the relative position and attitude converge faster than the initial controller; moreover, the performance cost of the proposed method is improved by online learning, which will be witnessed by the control cost comparison in Fig. 11.10. The learning process of the proposed controller is analyzed in Fig. 11.11. It can been seen that the weight

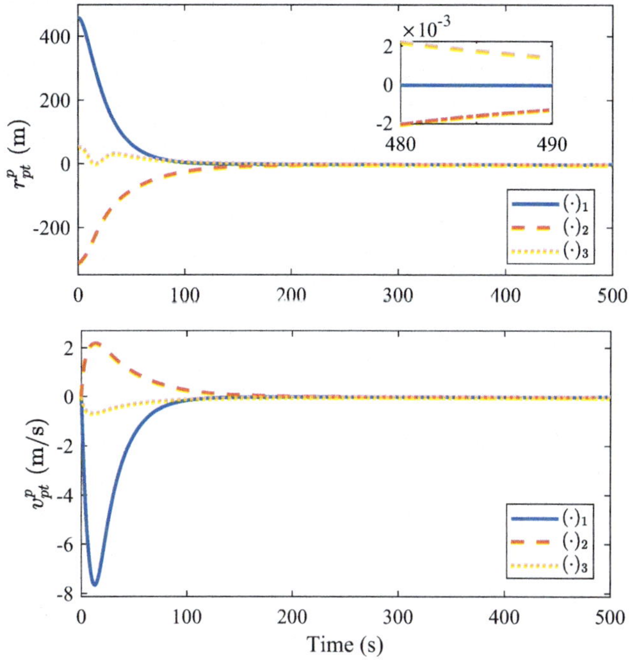

Fig. 11.4 Time responses of the relative position vector under the proposed controller

estimation vector $\hat{\boldsymbol{w}}_{est}$ changes quickly during the initial stage and stabilizes after about 60 s, whilst the Bellman error δ_b also tends to 0 at about 60 s, indicating that the proposed controller tends to be an optimal controller.

11.4.2 Docking to the Target with Constraints

In this subsection, we assume that the target is stationary w.r.t. the inertial frame $\mathcal{I}$, and the pursuer spacecraft is required to approach the desired docking position with a specific attitude, while complying with both the approaching corridor and FOV constraints. In this case, to trigger the constraints, the initial states of pursuer are set as $\boldsymbol{r}^t_{p/0} = [500, -140, 250]^\top$m, $\boldsymbol{q}_{p/0} = [-0.1544, 0.0234, -0.4071, 0.8999]^\top$, $\boldsymbol{v}^p_{p/0} = [-0.5, -2.0, -1.0]^\top$m/s, and $\boldsymbol{\omega}^p_{p/0} = [0, 0, 0]^\top$rad/s. The states of target are set as $\boldsymbol{r}^t_t = [0, 0, 0]^\top$m, $\boldsymbol{q}_t = [0, 0, 0, 1]^\top$, $\boldsymbol{v}^t_t = [0, 0, 0]^\top$m/s, and $\boldsymbol{\omega}^t_t = [0, 0, 0]^\top$ rad/s. In addition, the motion constraints are specified by $c_s = [-1, 0, 0]^\top$, $c_p = [1, 0, 0]^\top$, and $\alpha_p = \alpha_s = \pi/6$ rad. The parameters of the initial controller are chosen as $\hat{k}_p = 0.2 + \varepsilon 1$, and $\hat{k}_d = 5 + \varepsilon 2$.

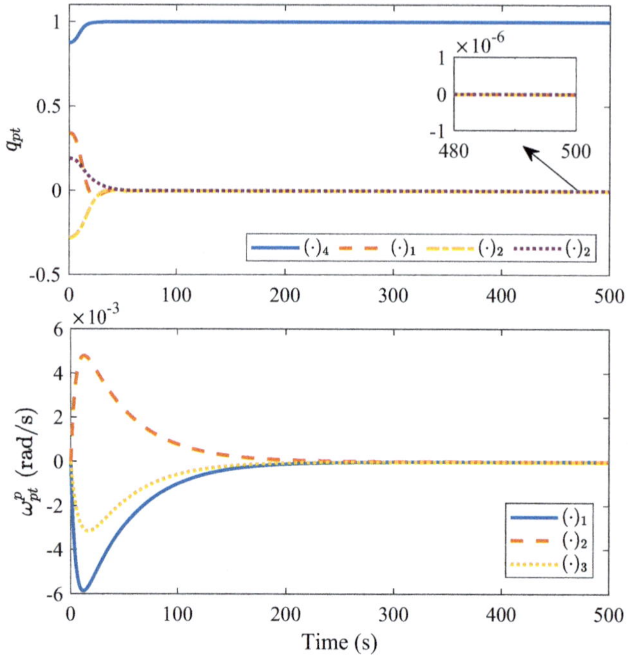

Fig. 11.5 Time responses of the relative attitude vector under the proposed controller

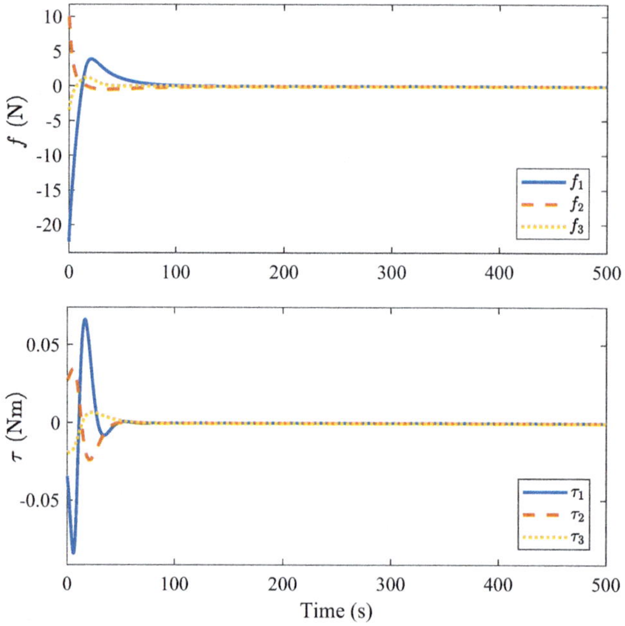

Fig. 11.6 Control inputs of the proposed controller

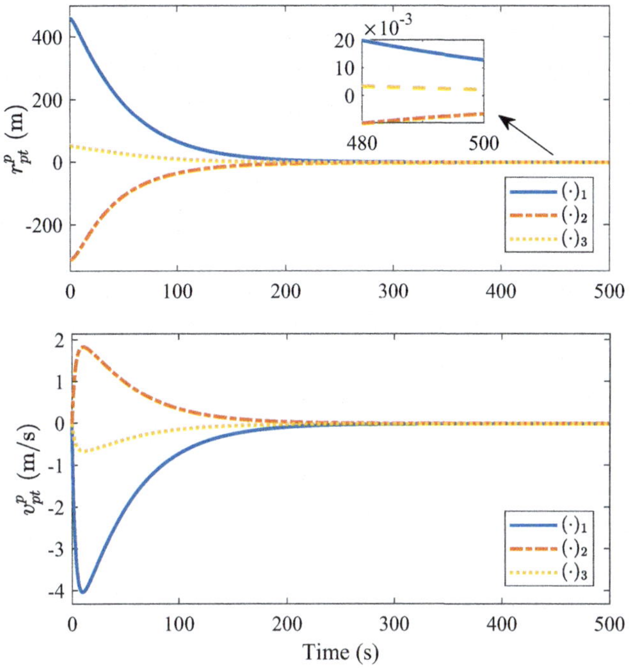

Fig. 11.7 Time responses of the relative position vector under the initial controller

The simulation results of relative states, control inputs, and learning process under the proposed method are shown in Fig. 11.12. It is shown that the proposed controller enables the pursuer to arrive at the desired position with a specific attitude, while complying with the spatial motion constraints. To intuitively show the approaching process, the 3-D motion trajectory of the pursuer w.r.t. the target in the target's body frame $\mathcal{T}$ is depicted in Fig. 11.13, where the instantaneous position and attitude snapshots of the pursuer at different time instances are provided by the craft model, and the LOS cone and approaching corridor are also drawn.

To further show the superiority of the proposed method (denoted here as ADPC), we also simulate the proposed method without considering constraints (denoted as ADPF), the APF-based method in [10] (denoted as APF), and the initial control method (dentoed as PD) for comparison purpose. It should be emphasized that Dong et al. [10], to some degree, addresses a very similar problem as in this chapter, and also use the APF to deal with spatial motion constraints under the dual quaternion mechanism. However, the work of [10] can not account for performance optimization. Thus, from a theoretical viewpoint, our proposed method would display better performance and task completion ability, as will be witnessed in the following simulation results. The comparison results of the four controllers are shown in

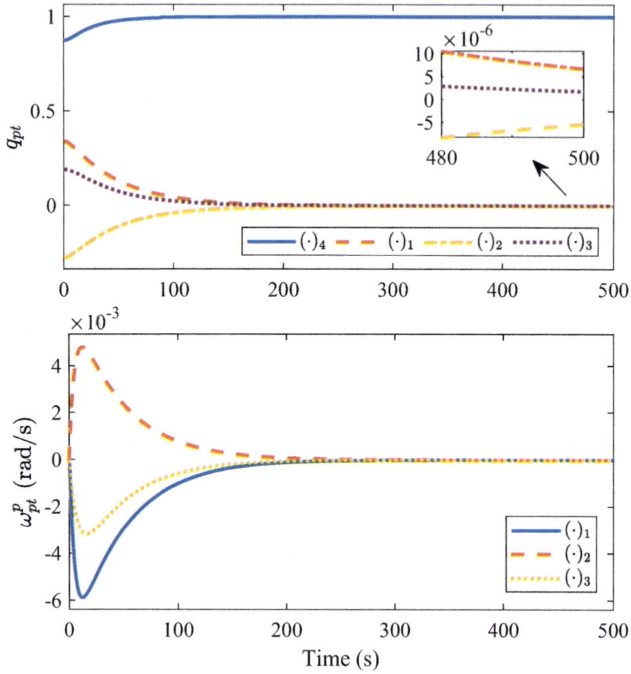

Fig. 11.8 Time responses of the relative attitude vector under the initial controller

Figs. 11.14, 11.15, 11.16 and 11.17. The translational motion trajectories of the pursuer w.r.t. the target during the RPOs are depicted in Figs. 11.14 and 11.15, where the approaching corridor is described by the green cone. The comparison results of (11.3) and (11.5) are shown in Figs. 11.16 and 11.17. From the above figures, it can be seen that, under the PD and ADPF controllers, the pursuer flies out of the approaching corridor and the sensor FOV loses the target. In contrast, both the proposed method and APF method ensure the satisfaction of spatial motion constraints.

The above results show the constraint-handling capabilities of the proposed method and APF method. The following control effort comparison, given in Fig. 11.18, shows the superiority of the proposed method in control cost over the APF method. Define the energy consumption function of force and torque as $H_f = \int \|\boldsymbol{f}^b(t)\|^2 dt$ and $H_\tau = \int \|\boldsymbol{\tau}^b(t)\|^2 dt$, respectively. Figure 11.18 shows that the proposed ADPC consumes less energy than the APF method.

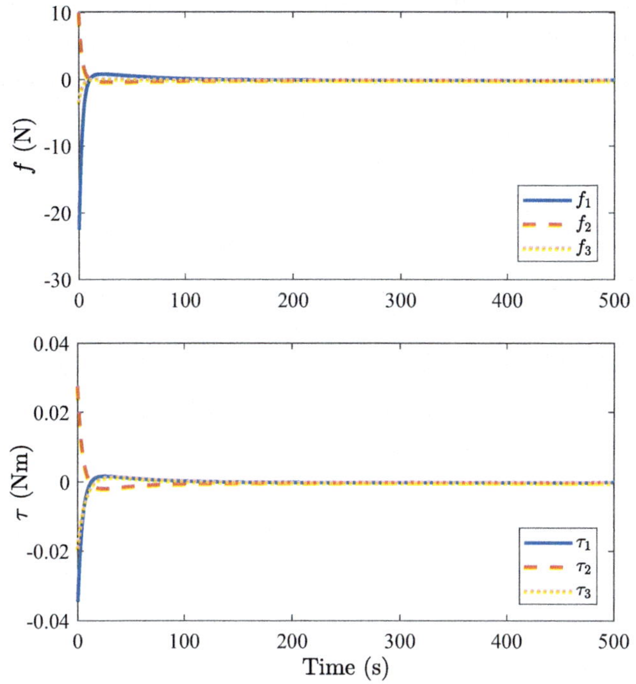

Fig. 11.9 Control inputs of the initial controller

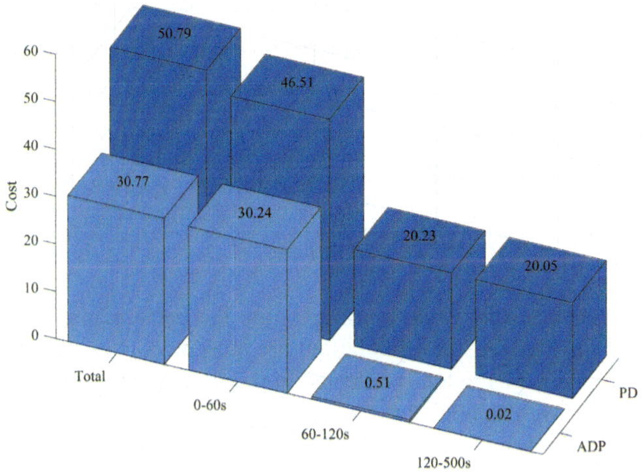

Fig. 11.10 Performance cost comparison between the proposed controller (ADP) and initial PD controller (PD)

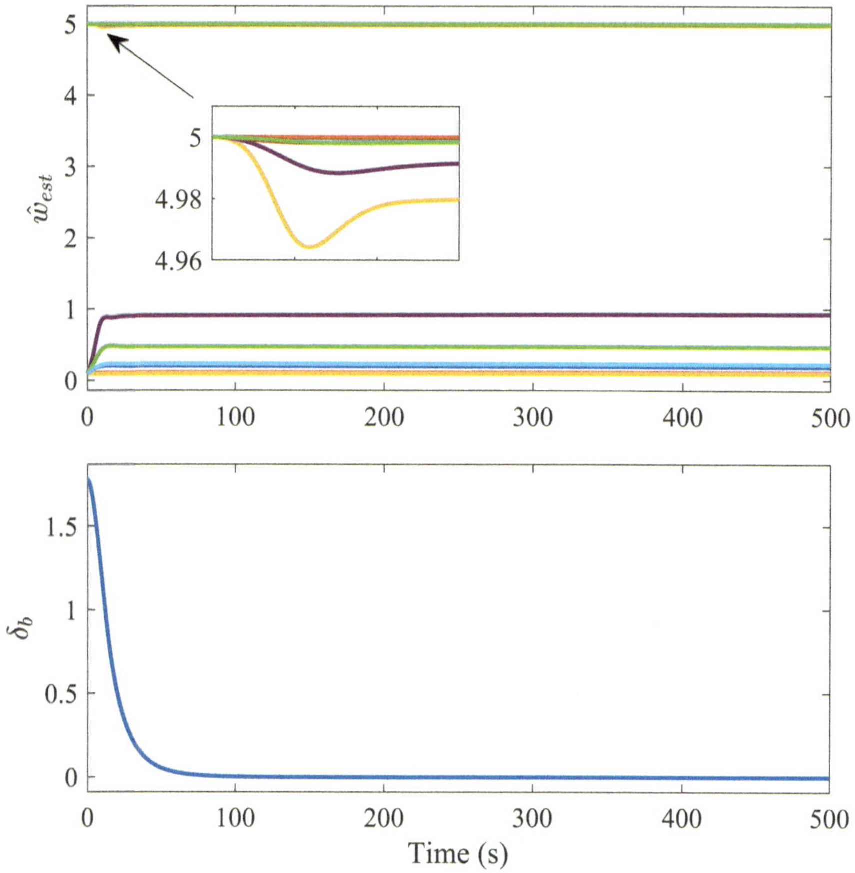

Fig. 11.11 Time responses of $\hat{\boldsymbol{w}}_{est}$ and δ_b

11.4.3 Monte-Carlo Simulations

To further verify the comprehensive insight into the performance of the proposed method, a 500-run Monte-Carlo simulations are presented in this subsection. To this end, the initial states are randomly selected in the ranges listed in Table 11.1. In the simulations, the PWPF modulators [38] are employed for pulse modulation. The prefilter coefficients are chosen as $K_m = 0.8$, and $T_m = 0.1$; the Schmidt Trigger parameters are set as $\delta_{\text{on}} = 0.45$, and $\delta_{\text{off}} = 0.15$; and the thrust magnitude is limited by $u_{max} = 20\text{N}$.

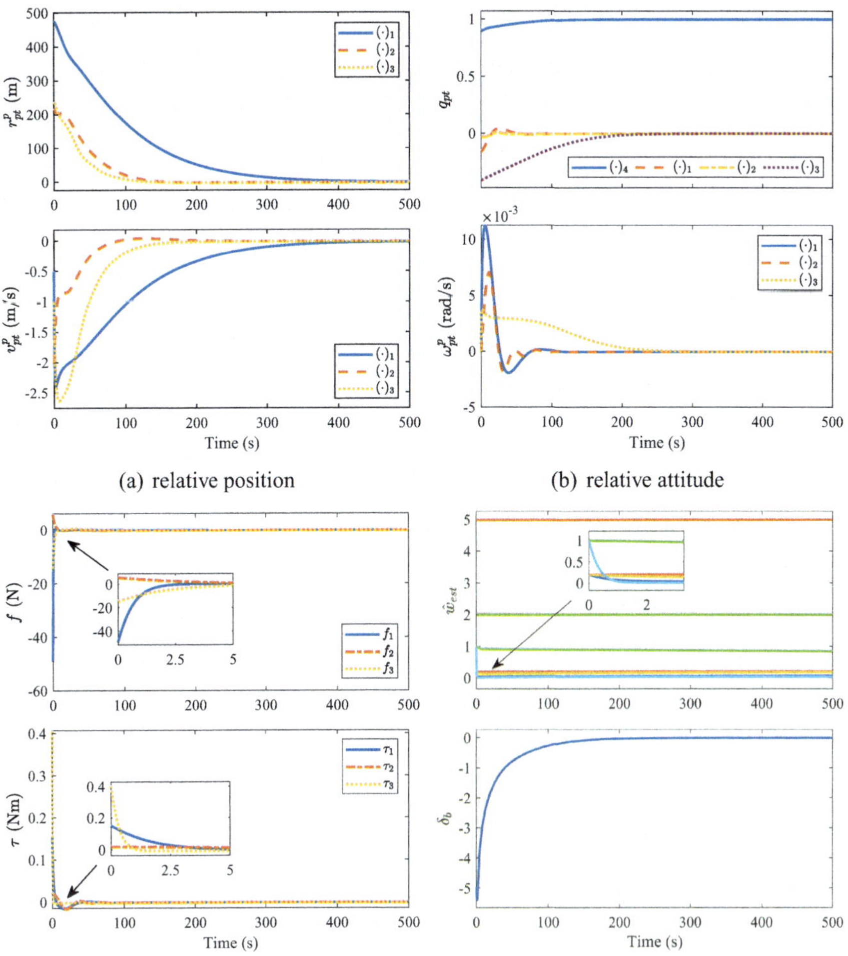

(a) relative position

(b) relative attitude

(c) control inputs

(d) $\hat{\boldsymbol{w}}_{est}$ and $\boldsymbol{\delta_b}$

Fig. 11.12 Dynamic responses of the proposed controller

The results of the Monte-Carlo simulations are summarized in Figs. 11.19 and 11.20. Figure 11.19 presents the overall of the 500-run simulations, from which we can intuitively see that all the trajectories are converged. It can been seen from the subfigure of Fig. 11.19 that the terminal error $\|r^p_{pt}(500)\|$ of each single run is lower than $10^{-0.5}$m, which is admissible in practical applications. The distributions of the maximum FOV angle $\arccos(-(\boldsymbol{r}^p_{pt}/\|\boldsymbol{r}^p_{pt}\|)^\top \boldsymbol{c}_s)$ and approaching corridor angle $\arccos((\boldsymbol{r}^t_{pt}/\|\boldsymbol{r}^t_{pt}\|)^\top \boldsymbol{c}_p)$ (drawn on the corresponding $Y-Z$ planes) of every single run are given in Fig. 11.20. As can be seen, the proposed method can accomplish RPOs under different initial conditions without any constraint violations. It is well

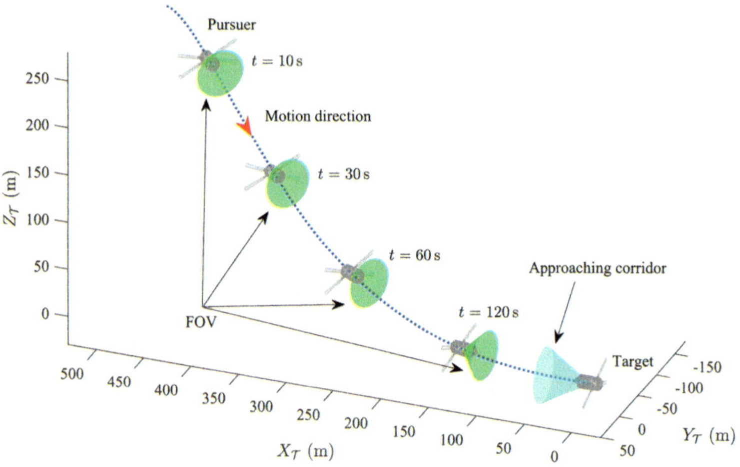

Fig. 11.13 3-D motion trajectory in $\mathcal{T}$ under the proposed controller

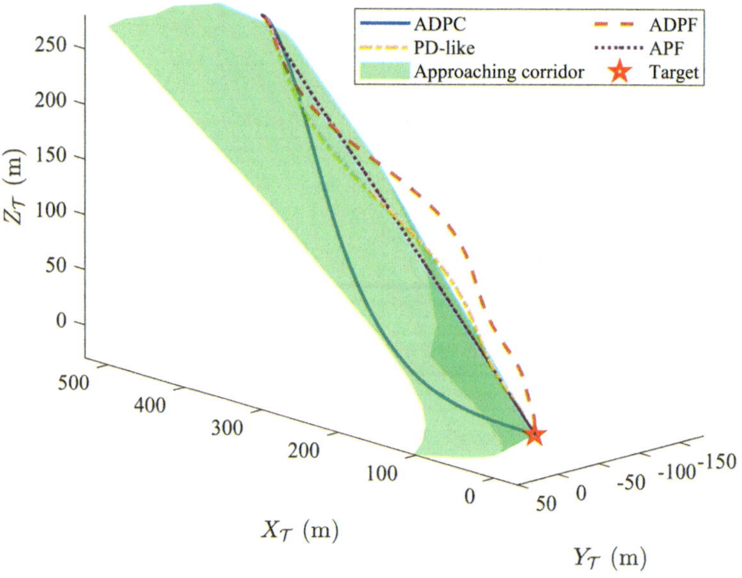

Fig. 11.14 3-D illustration of approaching corridor constraint

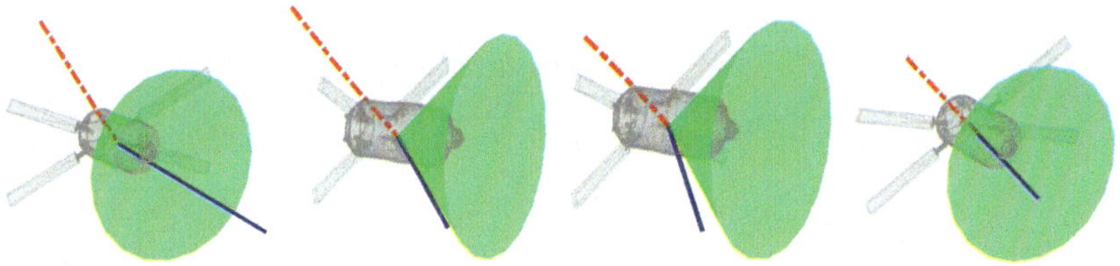

Fig. 11.15 3-D illustration of FOV constraint at 16 s

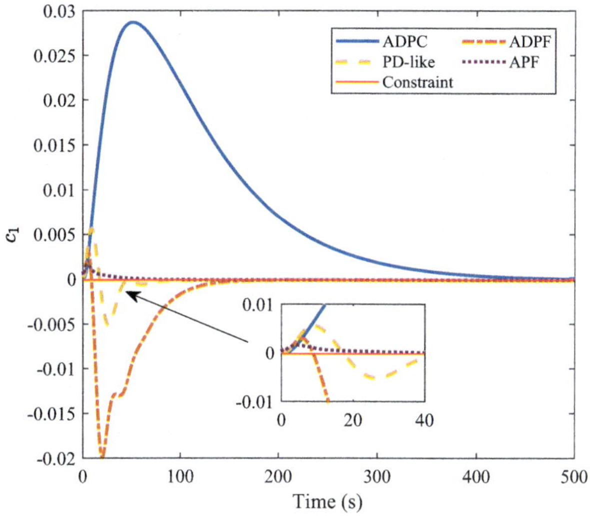

Fig. 11.16 Comparison results of c_1

known that the thruster's on-off nature inevitably leads to control performance degradation. But nonetheless, the Monte-Carlo simulation results are acceptable.

In summary, the above simulation results demonstrate the effectiveness of the proposed method. Compared with the traditional methods, the proposed RL-based pose control scheme not only achieves the fast high-precision convergence of state errors, but also has a constraint-handling capability.

11.5 Summary

An RL-based pose control scheme was developed for the spacecraft RPOs. In the dual-quaternion algebraic framework, a specially designed barrier function was embedded in the reward function to cope with the nontrivial spatial motion constraints. Subsequently, an RL-based control scheme was presented to achieve online

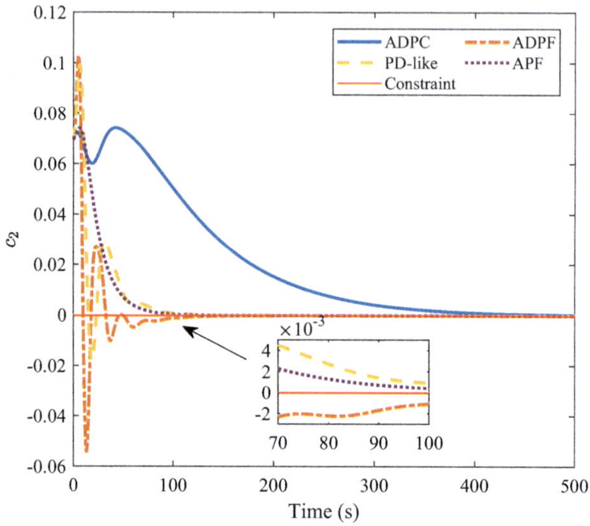

Fig. 11.17 Comparison results of c_2

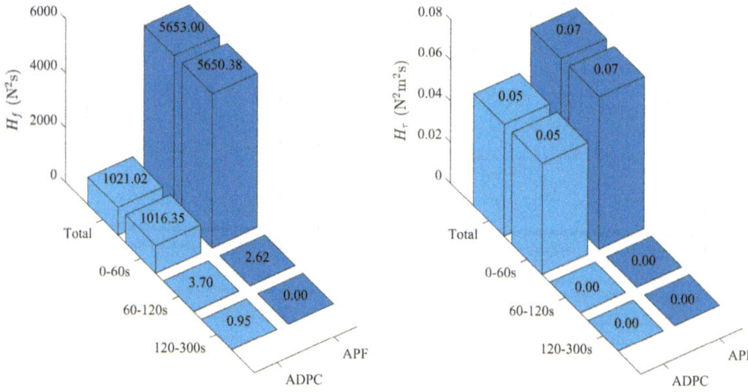

Fig. 11.18 Control effort comparison with APF method

Table 11.1 Ranges of initial states

Parameter	Values/Ranges
r_{mc}, m	$(0, 300)$
θ_{mc}, rad	$(-\pi, \pi)$
$r^b_{pt}(0)$, m	$[500, r_{mc}\cos\theta_{mc}, r_{mc}\sin\theta_{mc}]^\top$
$v^b_{pt}(0)$, m/s	$(-1, 1) \times (-1, 1) \times (-1, 1)$
$\omega^b_{pt}(0)$, rad/s	$(-0.01, 0.01) \times (-0.01, 0.01) \times (-0.01, 0.01)$
$q_{pt}(0)$	Euler angle (y-x-z): $(-\pi/6, \pi/6) \times (-\pi/6, \pi/6) \times (-\pi/6, \pi/6)$

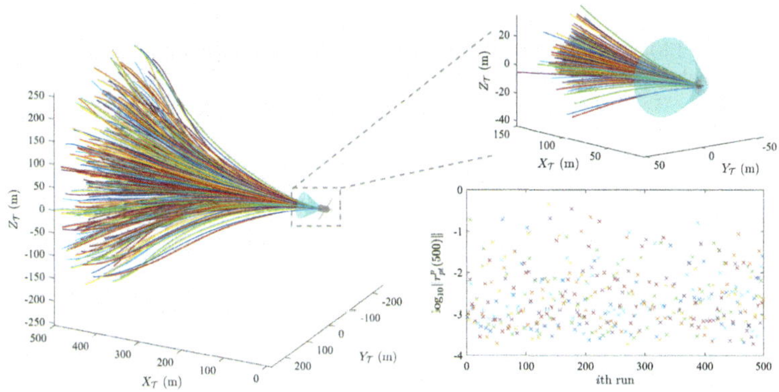

Fig. 11.19 Trajectories and state error of Monte-Carlo simulations

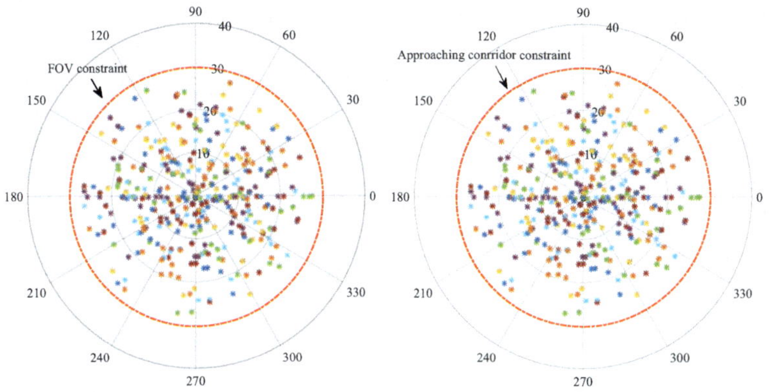

(a) maximum field-of-view angle (b) maximum approaching corridor angle

Fig. 11.20 Monte-Carlo simulations

approximation of the optimal control policy subject to the 6-DOF nonlinear dynamics and motion constraints. The ultimate boundedness of state errors and the network weight estimation errors were shown by Lyapunov stability analysis. The effectiveness and superiority of the proposed method were carefully evaluated through a set of comprehensive numerical simulations.

References

1. Fehse W (2003) Automated rendezvous and docking of spacecraft, vol 16. Cambridge university press
2. Sun L, Huo W (2015) 6-dof integrated adaptive backstepping control for spacecraft proximity operations. IEEE Transactions on Aerospace and Electronic Systems 51(3): 2433–2443

3. Quadrelli MB, Wood LJ, Riedel JE, McHenry MC, Aung M, Cangahuala LA, Volpe RA, Beauchamp PM, Cutts JA (2015) Guidance, navigation, and control technology assessment for future planetary science missions. Journal of Guidance, Control, and Dynamics 38(7): 1165–1186
4. Pirat C, Richard-Noca M, Paccolat C, Belloni F, Wiesendanger R, Courtney D, Walker R, Gass V (2017) Mission design and gnc for in-orbit demonstration of active debris removal technologies with cubesats. Acta Astronautica 130: 114–127
5. Hinkel H, Zipay JJ, Strube M, Cryan S (2016) Technology development of automated rendezvous and docking/capture sensors and docking mechanism for the asteroid redirect crewed mission. In: IEEE Aerospace Conference Proceedings, Big Sky, MT, United states, pp 1–8
6. Li WJ, Cheng DY, Liu XG, Wang YB, Shi WH, Tang ZX, Gao F, Zeng FM, Chai HY, Luo WB, et al. (2019) On-orbit service (oos) of spacecraft: A review of engineering developments. Progress in Aerospace Sciences
7. Vukovich G, Gui H (2017) Robust adaptive tracking of rigid-body motion with applications to asteroid proximity operations. IEEE Transactions on Aerospace and Electronic Systems 53(1): 419–430
8. Breger LS, How JP (2008) Safe trajectories for autonomous rendezvous of spacecraft. Journal of Guidance, Control, and Dynamics 31(5): 1478–1489
9. Weismuller T, Leinz M (2006) Gnc technology demonstrated by the orbital express autonomous rendezvous and capture sensor system. In: 29th annual AAS guidance and control conference, American Astronautical Society, pp 06–016
10. Dong H, Hu Q, Akella MR (2017) Dual-quaternion-based spacecraft autonomous rendezvous and docking under six-degree-of-freedom motion constraints. Journal of Guidance, Control, and Dynamics 41(5): 1150–1162
11. Huang Y, Jia Y (2019) Adaptive finite-time 6-dof tracking control for spacecraft fly around with input saturation and state constraints. IEEE Transactions on Aerospace and Electronic Systems 55(6): 3259–3272
12. Shao X, Hu Q, Shi Y, Yi B (2022) Data-driven immersion and invariance adaptive attitude control for rigid bodies with double-level state constraints. IEEE Transactions on Control Systems Technology 30(2): 779–794
13. Shao X, Hu Q (2021) Immersion and invariance adaptive pose control for spacecraft proximity operations under kinematic and dynamic constraints. IEEE Transactions on Aerospace and Electronic Systems 57(4): 2183–2200
14. Zappulla R, Park H, Virgili-Llop J, Romano M (2018) Real-time autonomous spacecraft proximity maneuvers and docking using an adaptive artificial potential field approach. IEEE Transactions on Control Systems Technology (99): 1–8
15. Bevilacqua R, Lehmann T, Romano M (2011) Development and experimentation of lqr/apf guidance and control for autonomous proximity maneuvers of multiple spacecraft. Acta Astronautica 68(7-8): 1260–1275
16. Virgili-Llop J, Zagaris C, Park H, Zappulla R, Romano M (2018) Experimental evaluation of model predictive control and inverse dynamics control for spacecraft proximity and docking maneuvers. CEAS Space Journal 10(1): 37–49
17. Lu P, Liu X (2013) Autonomous trajectory planning for rendezvous and proximity operations by conic optimization. Journal of Guidance, Control, and Dynamics 36(2): 375–389
18. Xin M, Pan H (2012) Indirect robust control of spacecraft via optimal control solution. IEEE Transactions on Aerospace and Electronic Systems 48(2): 1798–1809
19. Jewison C, Erwin RS, Saenz-Otero A (2015) Model predictive control with ellipsoid obstacle constraints for spacecraft rendezvous. IFAC-PapersOnLine 48(9): 257–262
20. Lee U, Mesbahi M (2014) Dual quaternion based spacecraft rendezvous with rotational and translational field of view constraints. In: AIAA/AAS Astrodynamics Specialist Conference, p 4362
21. Bertsekas DP (2015) Dynamic programming and optimal control. Athena scientific Belmont, MA
22. Sutton RS, Barto AG (2018) Reinforcement learning: An introduction. MIT press

23. Jiang Y, Jiang ZP (2017) Robust adaptive dynamic programming. John Wiley & Sons
24. Wang D, He H, Liu D (2017) Adaptive critic nonlinear robust control: A survey. IEEE Transactions on Cybernetics 47(10): 3429–3451
25. Vamvoudakis KG, Lewis FL (2010) Online actor–critic algorithm to solve the continuous-time infinite horizon optimal control problem. Automatica 46(5): 878–888
26. Wei C, Luo J, Dai H, Duan G (2018) Learning-based adaptive attitude control of spacecraft formation with guaranteed prescribed performance. IEEE Transactions on Cybernetics 49(11): 4004–4016
27. Kiumarsi B, Vamvoudakis KG, Modares H, Lewis FL (2018) Optimal and Autonomous Control Using Reinforcement Learning: A Survey. IEEE Transactions on Neural Networks and Learning Systems 29(6): 2042–2062
28. Görges D (2017) Relations between model predictive control and reinforcement learning. IFAC-PapersOnLine 50(1): 4920–4928
29. Hu Q, Shao X, Chen WH (2018) Robust fault-tolerant tracking control for spacecraft proximity operations using time-varying sliding mode. IEEE Transactions on Aerospace and Electronic Systems 54(1): 2–17
30. Wang X, Shi P, Wen C, Zhao Y (2020) Design of Parameter-Self-Tuning Controller Based on Reinforcement Learning for Tracking Noncooperative Targets in Space. IEEE Transactions on Aerospace and Electronic Systems 56(6): 4192–4208, 10.1109/TAES.2020.2988170
31. Cheng Y, Crassidis JL, Markley FL (2006) Attitude estimation for large field-of-view sensors. The Journal of the Astronautical Sciences 54(3-4): 433–448
32. Abu-Khalaf M, Lewis FL (2005) Nearly optimal control laws for nonlinear systems with saturating actuators using a neural network hjb approach. Automatica 41(5): 779–791
33. Chowdhary G, Mühlegg M, Johnson E (2014) Exponential parameter and tracking error convergence guarantees for adaptive controllers without persistency of excitation. International Journal of Control 87(8): 1583–1603
34. Ioannou PA, Sun J (2012) Robust adaptive control. Courier Corporation
35. Filipe N, Tsiotras P (2015) Adaptive position and attitude-tracking controller for satellite proximity operations using dual quaternions. Journal of Guidance, Control, and Dynamics 38(4): 566–577
36. Dong H, Hu Q, Akella MR, Mazenc F (2019) Partial lyapunov strictification: Dual-quaternion-based observer for 6-dof tracking control. IEEE Transactions on Control Systems Technology 27(6): 2453–2469
37. M. Krstic IK, Kokotovic PV (1995) Nonlinear and Adaptive Control Design. Wiley, New York
38. Song G, Buck NV, Agrawal BN (1999) Spacecraft vibration reduction using pulse-width pulse-frequency modulated input shaper. Journal of Guidance, Control, and Dynamics 22(3): 433–440

Conclusion

This chapter concludes the book with a review of our major findings, along with our ideas about the future of the field.

A.1 General Conclusion

With the rapid development of space technology and continuous deepening of human space exploration, some emerging space missions, such as on-orbit servicing, spacecraft formation flying, deep space exploration, etc., have received considerable attention and investment from major space powers in recent years. In these applications, the spacecraft is usually required to autonomously and safely perform precision attitude and orbital motions in a complex space environment, while complying with complex constraints, such as state constraints, physical constraints, performance constraints, etc. This imposes higher autonomy, safety, and precision requirements for spacecraft AOCS. Multi-constraint attitude and orbit control technology is the key to ensure the autonomous, stable and high-precision operation of spacecraft in complex space environments and the success of emerging space missions. The past decades have witnessed remarkable progress in the field spacecraft multi-constraint attitude and orbit control, while most of the existing solutions have a limited constraint-handling capability and, moreover, are susceptible to parameter uncertainties, multi-source disturbances, and actuator faults. In view of this, it is thus necessary to develop safe, robust, and high-performance controllers with strong constraint-handling capability for the spacecraft AOCS. However, thus far, it still remains an open problem to deal with multiple types of constraints in the presence of parameter uncertainties, multi-source disturbances, and actuator faults. In recent years, with the vigorous development of the new generation of AI technology and onboard micro processors, using the AI technology to endow the spacecraft AOCS with higher computational efficiency as well as stronger complex problem solving and constraint-handling capabilities has

Q. Hu et al., *Intelligent Autonomous Control of Spacecraft with Multiple Constraints*,
https://doi.org/10.1007/978-981-99-0681-9

attracted considerable attention from both the aerospace industry and academia. The AI technology provides a promising way to enhance the autonomy and intelligence of spacecraft AOCS.

This book gathers a collection of the authors' recent research results that can reflect the up-to-date theoretical and technological advances in the field of intelligent autonomous control of spacecraft under multiple constraints. Several adaptive (can be viewed as a low-level intelligent technique) and learning-based attitude, orbit, or pose control schemes are presented to enable the spacecraft to autonomously and safely accomplish the given missions, while complying with some underlying constraints, despite the presence of parameter uncertainties, multi-source disturbances, and even actuator faults. According to the intelligent control methods proposed in Chaps. 3–11, the main results and innovations of this book are summarized as follows.

In Chap. 3, a data-driven adaptive control scheme is proposed for attitude reorientation of uncertain spacecraft under forbidden pointing and angular velocity constraints. An I&I adaptive control framework is developed to remove the so-called realizability condition, and is further extended a data-driven counterpart to achieve on-line inertia identification in the absence of PE. The salient features of the proposed data-driven controller are three-fold: (i) it achieves asymptotic convergence of gradient-related terms and angular velocity, while ensuring the satisfaction of both attitude and angular velocity constraints; (ii) it preserves all the key properties of the I&I adaptive control methodology, thus exhibiting better performance than the CE-based adaptive controllers; (iii) it guarantees exponential parameter convergence under a strictly weak IE condition. Moreover, the parameter convergence rates across all entries are independent of the excitation level and can be tuned separately.

In Chap. 4, two kinds of FTC methods are proposed spacecraft attitude reorientation under attitude and angular velocity constraints. Firstly, an adaptive robust FTC scheme with saturated virtual control is proposed in the framework of backstepping, wherein both attitude and angular velocity constraints are addressed using judiciously constructed (i)BLFs. In particular, an uniform strong controllability assumption is established, and its sufficient conditions are provided along with a feasibility analysis based on the Monte Carlo simulation method. Secondly, a learning-based approximate optimal FTC scheme is proposed, wherein a cost function that accounts for constraints and faults is constructed. In the RL framework, a single-critic NN is developed to online exactly approximate the cost function under a weak IE condition. The proposed learning-based FTC method enables the spacecraft to achieve energy-optimal attitude reorientation, while complying with both attitude and angular velocity constraints, regardless of actuator faults.

In Chap. 5, a NN-based fault diagnosis scheme is proposed for SGCMGs of spacecraft operating in periodic orbit. Firstly, an NDO is designed to online learn the environmental disturbance torques, and the learning results are used to decouple environmental disturbances and actuator faults, which can effectively improve the fault diagnosis accuracy. Considering that when multiple actuators simultaneously occur faults, the fault isolation and estimation cannot be achieved using only spacecraft attitude information, an information-fusion-based diagnosis scheme is proposed, where a group of adaptive estimators are designed to achieve fault isolation and preliminary fault estimation by fusing the spacecraft attitude and gimbal position information. Finally, based on the disturbance and fault estimation results, an adaptive sliding mode controller is derived to achieve the active fault-tolerant control purpose.

In Chap. 6, an RL-based dynamic control allocation scheme is proposed for attitude stabilization of spacecraft equipped with CMGs, with the aim of saving energy and avoiding the configuration singularity problem of CMGs. A null-space based control allocation scheme is proposed to decouple the outer loop control from the inner loop control allocation, such that no control error is introduced into the control allocation problem. Then, by using the control linearization assumption, the CMGs and attitude dynamics are modeled as an augmented system, the control allocation is transformed into a dynamic problem, for which the cost function is designed and transformed into the Bellman equation. Subsequently, an integral RL algorithm based on Off Policy strategy is designed to estimate the parameters, instead of obtaining the analytical solution of the PDE. A salient feature of the proposed control allocation scheme is that it does not require system model and the adjustability of disturbances.

In Chap. 7, an adaptive optimal tracking control scheme is proposed for the leader-follower spacecraft formation flying system. By using the ADP technique, a continuous near optimal tracking control scheme was designed, where a critic-only NN is established to approximate the optimal cost function, so as to solve the HJB equation. By combining the parameter projection rule and gradient descent algorithm, a semi-global adaptive update law is designed to adaptively update the critic network. Subsequently, an event-triggered mechanism without Zeno-free behavior is established such that the near optimal tracking controller is implemented only when specific events occur, thus significantly reducing the execution frequency of the control commands and avoiding unnecessary resources expenditure. Furthermore, by defining an input-based triggering error, the Lipschitz continuity assumption of the controller that commonly used in the existing literature is removed, making the control scheme easier to be implemented.

In Chap. 8, an adaptive prescribed performance pose tracking control scheme is proposed for spacecraft RPOs with a tumbling target, under spatial motion constraints and mass/inertia uncertainties. By introducing a class of ATPFs, the desired performance specifications and spatial motion constraints are tactfully transformed into the performance boundaries of the pose tracking errors. Then, a non-CE adaptive prescribed performance controller is derived using the BLF in conjunction with the error transformation technique. The designed controller ensures that the transformed errors remain within the specified ranges in the presence of mass and inertia uncertainties, indicating the satisfaction of both motion and performance constraints. Hence, the proposed control method can enable the pursuer to accomplish the RPOs in a preset time with overshoot and accuracy tolerance less predefined levels, whilst ensuring that both path and FOV constraints are satisfied. Moreover, the underlying singularity in attitude extraction of the LOS frame can be strictly avoided, by properly choosing the performance boundaries for the position tracking errors.

On the basis of Chap. 8, Chap. 9 further considers relative linear/angular velocity constraints for spacecraft RPOs with a tumbling target. Two APFs free of local minima problem are tactfully constructed to deal with both spatial kinematic and dynamic constraints. Then, an APFs-based adaptive pose control scheme is proposed, where the I&I adaptive design philosophy is adopted to circumvent the realizability condition that is required for most existing adaptive control approaches, but nonetheless may not hold under dynamic constraints under dynamic constraints, and meanwhile, to establish an attracting manifold, whereby the deterministic case of closed-loop performance can be asymptotically recovered. Moreover, the dynamic scaling technique is introduced to overcome the integrability obstacle that arises in the traditional I&I adaptive control design. The proposed control method ensures that pursuer successfully accomplishes the RPOs, while complying with both kinematic and dynamic constraints, despite the presence of mass and inertia uncertainties.

In Chap. 10, a composite learning pose tracking control scheme that can synchronously enhance parameter convergence and tracking performance is proposed for spacecraft RPOs with a tumbling target. First, a CE-based adaptive control law is designed, and the filtered system dynamics is established to construct the parameter estimation errors using only easily-obtained signals. Then, a composite learning law is designed based on the CL and DREM techniques, which can greatly relax the requirement of parameter convergence on the excitation condition and exactly identify the mass and inertia parameter online by making full use of the stored historical information. It is shown that the proposed composite learning control method can ensure that both the tracking errors and parameter estimation errors exponentially converge to zero, if the regressor matrix satisfies a strictly weak IE condition. Moreover, benefiting from the DREM procedure and some special designs, the parameter estimation error dynamics are independent of each other, and parameter convergence rate does not depend on the excitation strength, making the gain selection simpler.

In Chap. 11, an RL-based pose control method is developed for the spacecraft RPOs using the dual quaternion representation formalism. A specially designed barrier function was incorporated into the reward function to deal with spatial motion constraints. Then, an RL-based pose controller is designed in the dual-quaternion algebraic framework to achieve online approximation of the optimal control policy subject to 6-DOF nonlinear dynamics and motion constraints. It is shown that the state errors and network weight estimation errors are ultimately bounded. In addition, a dual-quaternion-based PD-like poase controller is introduced as the initial control policy to trigger the online learning process, which is proved table by a special Lyapunov strictification method. Compared with the adaptive pose controllers presented in Chaps. 8 and 9, the proposed RL-based pose control scheme not only achieves spacecraft RPOs under spatial motion constraints, but also balances the control performance and control cost.

A.2 Future Work

Intelligent autonomous control technology has significant academic value and application prospect in ensuring safe autonomous flying of spacecraft under complex constraints. In this respect, this book reports several advanced adaptive and learning control schemes for the spacecraft AOCS, which have significant advantages over the existing works in the literature. Although recent research on multi-constraint intelligent autonomous control of spacecraft has reached an impressive state, yet there are still many open questions that require further studies, and new application areas are emerging. Future research could focus on the following aspects:

(1) Spacecraft motion planning with multi-objective and multi-constraint. The traditional motion planning methods usually pre-plan a feasible trajectory according to the initial states and the desired states of the spacecraft before the mission is executed. However, complex and uncertain space environments and changes in mission goals caused by emergencies will make it difficult to execute pre-planning normally. Thus, it is necessary to study rapid autonomous re-planning methods for spacecraft to quickly response to pre-planning failure, such that the mission objectives is achieved as much as possible under the premise of ensuring the spacecraft safety. In addition, as space missions become more and more complex, the spacecraft motion planning problem gradually develops from the traditional single-objective planning to multi-objective multi-constraint planning. Performance indicators such as execution time and energy consumption need to be considered, and the constraints are more complex. Apart from the common static hard constraints, there may also exist dynamic soft constraints (e.g., dynamic orientation constraints, and time duration limit for entering forbidden zones). It is a problem worth exploring in the future to study multi-objective multi-constraint motion planning technology in complex and uncertain environments.

(2) Flexible task reconfiguration and multi-constraint control technologies under actuation capability degradation. Severe actuator faults (e.g., total failure and stuck) will lead to actuation capability degradation of the spacecraft AOCS, and even cause the system to become under-actuated, which seriously affect the successful execution of space missions and the spacecraft safety in complex environments. Most of the existing works assume that the spacecraft AOCS system remains over-actuated or full-actuated under actuator faults, and focus on ensuring the stability and steady-state performance of the closed-loop system. However, there are very few studies on the quantitative analysis of the remaining actuation capacity of the AOCS system under faulty conditions. From an engineering viewpoint, the actuation capability degradation may cause the given mission to be impossible to complete. Thus, online autonomous task reconfiguration (e.g., mission degradation) is required to ensure that the spacecraft can accomplish the mission by virtue of the remaining actuation capacity. On the other hand, actuation capability degradation, especially when the AOCS becomes under-actuated, will significantly reduce the feasible space of the spacecraft position and attitude maneuvering paths, which brings great challenges to the multi-constraint motion planning and control design of spacecraft. Consequently, it is urgent to carry out research on the key technologies of flexible task reconfiguration and multi-constraint control under actuation degradation, and make breakthroughs in the quantitative evaluation of control system capabilities, flexible task reconfiguration, and feasible space and controllability analyses of the under-actuated AOCS.

(3) Multi-constraint motion planning and control based on "digital twin + AI". With the significant improvement in computing power of the onboard microprocessor, it will be an inevitable trend to apply AI technology represented by machine learning for the design of spacecraft AOCS. However, most of the existing AI methods are exploration-driven, that is, they need to accumulate a large amount of "action-feedback" data for learning and training, so as to form an optimal strategy. However, for space missions, it is unrealistic to obtain rich empirical data by conducting a large amount of "trial and error", due to fuel and safety concerns. The emergence of digital twin technology provides a promising solution to this problem. By digitally mirroring the complex environment and spacecraft AOCS, the digital twin technique can conduct the virtual mapping of constraint scenarios, actuators, sensors, etc., and then achieve environment simulation, constraint evolution, fault prediction and diagnosis, etc. Therefore, the use of digital twin technology can allow intelligent algorithms to accumulate nearly the same exploration experience as the real environment, and update the twin model in real-time based on the sensing data. It will be an important direction for future development to realize the online autonomous spacecraft motion planning and control, via the "digital twin + AI" model.

(4) Game-based attitude takeover control of spacecraft using cellular satellites. With the development of microsatellite technology, the use of multiple cellular satellites attached to the failed spacecraft to implement takeover control provides a new development direction for emerging space missions, like on-orbit servicing. After the cellular satellites attach to the target spacecraft, the formed combined spacecraft has redundant actuators and high reliability. To control cellular satellites in a coordinate way and minimize the control cost, a cooperative game process is characterized for each cellular satellite, with the aim of achieving the overall optimal goal, while minimizing the cost of singular cellular satellite. Traditional control methods are difficult to achieve such a goal. A cooperative control strategy based on differential games provides a solution to this problem. Considering that the mass property of the combined body are unknown, and that the numerical optimization problem transformed from multi-body game control is relatively complex, it is difficult to solve it using the existing model-based numerical solving tools. The RL technology provides an effective way to solve the multi-agent game problem. The RL-based cooperative game control technique is the basis for cellular satellite takeover missions, and is also a new direction for the development of spacecraft on-orbit servicing in the future.

(5) Distributed coordinated planning and control for spacecraft formation flight under multiple constraints. In recent years, with the rapid development of onboard computing, communication, sensing and some other technologies, it has become a new trend to use multiple small satellites to form a large-scale spacecraft formation system, which can accomplish emerging space exploration and imaging tasks that can hardly be accomplished by the traditional single spacecraft system. Compared with a single spacecraft system, the formation flying system has the advantages of short research and development cycle, good applicability, high fault tolerance, and strong space task protection and survivability capabilities under space offensive and defensive confrontation; moreover, it also plays a vital role in the fields of scientific research, national economy and national defense construction. During the on-orbit

flying of the spacecraft formation system, it is necessary to maintain communication and avoid collisions, which brings complex constraints to the design of coordinated planning and control algorithms. At the same time, the space environment is often highly dynamic, and the tasks may also change, which put forward higher requirements for the on-orbit adaptability of the planning and control algorithms. In addition, the formation flight mission may suffer from some emergency situations, such member failures or damages, new spacecraft replenishment, etc., which will bring great challenges to the task coordination, flight safety, and communication topology of the formation system. Most of the existing coordinated planning and control methods for formation systems do not consider complex constraints, and most of them adopt centralized processing methods, which have defects such as poor real-time performance, slow response, poor consistency, etc. Hence, it is urgent to design distributed coordinated planning and control strategies with a strong constraint-handling capability for spacecraft formation flying, and an integrated "task re-planning + configuration reconfiguration" planning and control framework, in order to ensure flight safety and stability of the spacecraft formation system, even under the condition of member failure or damage.

(6) Experimental verification for aerospace engineering. In this book, the proposed intelligent autonomous attitude and orbit control schemes are only validated by numerical simulations, and there is still a certain gap from the practical engineering requirements. It is urgent to establish semi-physical or even full-physical simulation platforms and present the corresponding test and evaluation techniques, in order to carry out ground experiment validation for the practical effectiveness of the developed algorithms. This is vital to promote the engineering applications of the proposed control methods.

The manufacturer's authorised representative in the EU is Springer Nature Customer Service Centre GmbH, Europaplatz 3, 69115 Heidelberg, Germany. If you have any concerns regarding our products, please contact ProductSafety@springernature.com

Printed and bound by CPI Group (UK) Ltd, Croydon, CR0 4YY
01/10/2026
02225870-0008